Preface

As the world enters the AI era, auditors face new challenges and opportunities that require them to use the right tools and technologies to stay ahead. With over 20 years of experience in audit and business data analytic, I have gained a deep understanding of the importance of transitioning from traditional audit operations to smart auditing that involves proactive warning or prediction using AI. Smart auditing enables businesses to obtain valuable insights and identify potential risks that traditional auditing methods may not detect. By using AI to analyze large amounts of structured and un-structured data, auditors can provide more accurate and insightful audits, as well as assist with compliance and risk management.

JCAATs is a new audit software for smart auditing. It utilizes AI language Python and can run on Windows and MAC operating systems. It offers traditional computer-aided audit tools (CAATs) data analysis functions along with AI functions such as text mining, machine learning, and data crawling, resulting in smarter audit analysis. The software allows for the analysis of large amounts of data and an open data architecture that enables interfacing with various databases, cloud data sources, and different file types, making data collection and integration more convenient and faster. Additionally, the multiple language and visual user interface makes generating Python audit programs simple and easy, even for auditors not familiar with Python language. Integrating with open-source Python program resources enables more scalability and openness for audit programs, eliminating the limitations of only a few software programs.

This textbook explains the use of technology such as big data analysis, text mining, and machine learning in auditing through practical cases. Readers will gain an understanding of data analysis and smart audit and their advancements. JCAATs, which includes data fusion technology and an OPEN DATA connector, helps auditors quickly obtain heterogeneous data for audit operations, enhancing effectiveness and efficiency. The textbook includes exercise data for practicing with JCAATs to fully experience intelligent auditing practices. It's suitable for professionals like accountants, auditors, legal and compliance personnel, risk management, and information security, as well as managers at all levels, college teachers, and students with data analysis needs.

Sherry Huang ICCP, CEAP, CFAP, CIA, CCSA
Jacksoft Ltd., Taipei, Taiwan
2023/03/15

Readers Guide

For Lecturers

To cultivate AI programming skills in business schools and other institutions, the International Computer Auditing Education Association (ICAEA) recommends starting with No Code. This approach allows students to develop audit applications to solve real-world problems without needing to write any code, giving them practical experience and enhancing their professional value. Once proficient in No Code, students can progress to Read Code training, where they gain an understanding of coding logic. Finally, Write Code training enables them to create their own AI applications. This approach prepares students for a future working environment centered on data analysis and smart auditing, where they can collaborate with AI auditing robots. As the AI era presents new challenges to the education of business professionals, colleges must adopt innovative and effective approaches to prepare their students for the workforce of the future.

This book is designed to be a comprehensive guide for learners to develop their skills in big data analysis, text mining, and machine learning using the JCAATs AI audit software. The book is structured to provide a progressive learning experience, with chapters covering different concepts and applications, as well as exercises and practice questions. The book also includes simulated independent audit case exercises to help learners apply what they have learned to real-world business environments and develop their problem-solving abilities. The operating manuals for each audit instruction enable teachers to deliver No Code courses that are more practical and aligned with real-world applications. Overall, this book aims to provide a powerful knowledge system for smart auditing and equip learners with the knowledge and skills needed for the modern workplace.

This book provides an educational version of the JCAATs - AI audit software for trial use, with a multi-language interface. This allows students to install the software on their personal computers for operational practice and learning. Since the JCAATs software is based on Python, it is easier to incorporate external resources into advanced Read Code and Write Code teaching, providing students with the necessary tools to develop their skills in big data analysis, text mining, and machine learning for the future auditing.

For students

As technology continues to advance, computer auditing plays an increasingly important role in ensuring the reliability and accuracy of financial statements and other key business information. Computer audit software enables auditors to analyze large amounts of data quickly and effectively, allowing them to identify potential areas of risk or fraud. It is essential for those entering the field of computer auditing to have a strong understanding of the latest computer audit software and to be able to use these tools effectively in practice.

This book offers an educational version of the software for trial use, which features a multi-language interface. Students can install the software on their personal computers for practice and learning, providing them with more opportunities for self-study. The book includes simulated independent audit case exercises, which go beyond theoretical knowledge transfer and enable students to integrate course content with practical applications more effectively. This approach makes learning more engaging and interesting.

Trail education software and exercise dataset

By scanning the QR code located in the image below, readers can access the teaching resource portal for this book. Once they enter the unique registration code, they can browse a variety of learning materials, download exercise datasets, and even apply for a trial use of the educational software version.

Login to download the test data and trail software

CAATs Professional Certificate

The **International Certified CAATs Practitioner (ICCP)** designation is a personal, professional certification to signify that you possess CAATs foundation knowledge and skills. Individuals qualifying for this designation must satisfy and substantiate the extensive skills and knowledge requirements established by the International Computer Auditing Education Association (ICAEA). It is the most common and fundamental certification for CAATs. The software currently used in the certification exams includes ACL, IDEA, JCAATs, etc.

- **ICCP Exam Method**

This closed-book, 120 minutes-long certification exams is divided into two components: a Knowledge Inventory and two short computer auditing case studies. The case study is CAATs free independent. You can use any CAATs, such as ACL, IDEA, JCAATs, Picalo, etc., to attend the exam.

1. **Multiple choice questions, worth 60 points in total -**

   Test the knowledge of the use of CAATs computer-aided audit tools and the basic concepts of JCAATs audit instructions.

2. **Example operation, 40 points in total -**

   The on-board test is conducted in the computer classroom, one person has one computer. Before the test, each computer will be pre-installed with JCAATs software, test data files and test questions. The scoring method is based on the answers you fill in and the planning process.

Table of Contents

Python Based Computer-Assisted Audit Techniques (CAATs)

Data Analysis and Smart Audit

Chapter 1. Introduction to Data Analysis and Smart Audit

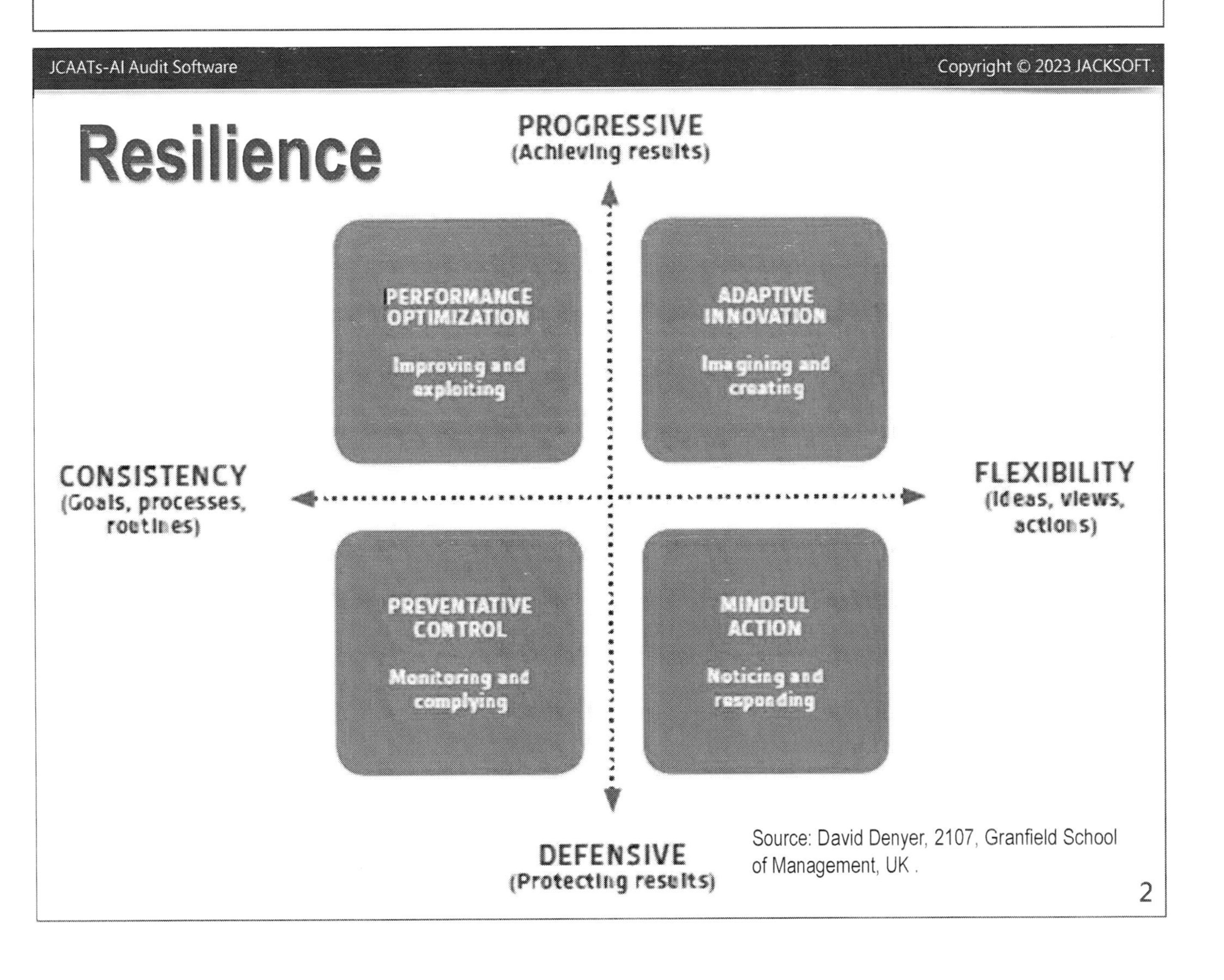

Where is Auditors' Future Job?

Source: Published by the Oxford Martin Programme on Technology and Employment, 2013

Bring on the personal trainers

Probability that computerisation will lead to job losses within the next two decades, 2013 (1=certain)

Job	Probability
Recreational therapists	0.003
Dentists	0.004
Athletic trainers	0.007
Clergy	0.008
Chemical engineers	0.02
Editors	0.06
Firefighters	0.17
Actors	0.37
Health technologists	0.40
Economists	0.43
Commercial pilots	0.55
Machinists	0.65
[illegible]	[illegible]
Retail salespersons	0.92
Accountants and auditors	0.94
Telemarketers	0.99

Accountants and auditors 0.94

Source: "The Future of Employment: How Susceptible are Jobs to Computerisation?" by C.Frey and M.Osborne (2013)

Source : 2014 The Economist

Accounting Education Shift

AACSB (The Association to Advance Collegiate Schools of Business) :
to create the next generation of great leaders. AACSB provides internationally recognized, specialized accreditation for business and accounting programs at the bachelor's, master's, and doctoral level.

Standard A7 is a new standard for Accounting Program: (Information Technology Skills And Knowledge for Accounting Graduates):
data sharing, data analytics, data mining, data reporting and storage within and across organizations

Standard A7 White Paper

AACSB International Accounting Accreditation Standard A7: Information Technology Skills and Knowledge for Accounting Graduates: An Interpretation

An AACSB White Paper issued by:

AACSB International Committee on Accreditation Policy
AACSB International Accounting Accreditation Committee

Source: AACSB

AICPA CPA Exam changes 2021

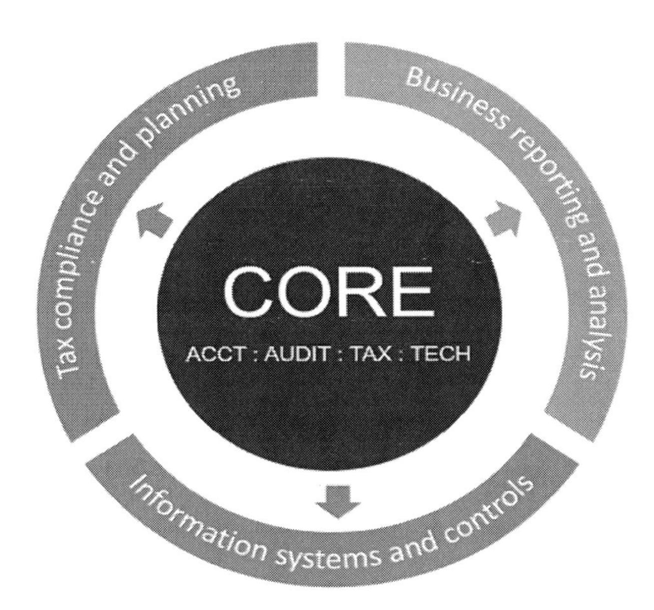

Data Analytics +
Digital Mindset

= CPA

Source: AICPA, 2019

JBOT in Action

You are either the one creating automation ... or you are the one being automated.

APAC CIOoutlook RegTech

JACKSOFT
RegTech Bots in Action

The fintech ecosystem in the Asia Pacific is witnessing significant growth in recent years. This wave of new technologies is also changing the behavior of financial regulators, specifically around money laundering evaluations by Asia Pacific Group on Money Laundering (APG),which is having a profound impact on how financial transactions are conducted worldwide. As a leading RegTech and computer auditing company in the Asia Pacific, JACKSOFT specializes in innovative RegTech solutions leveraging AI. With partners scattered across Tokyo and Dubai, the Taiwan-based company has served more than 800 organizations and 1000+ licenses for the RegTech software. The company's client base comprises big international public accounting firms and internationally renowned banks and financial institutions.

JACKSOFT's array of RegTech products comprise JGRC-Governance, Risk and Compliance System series, and JBOT-Intelligent Robot for Compliance series including JBOT for anti-money laundering (AML) or JAML as well as a solution for combating the financing of terrorism (CFT)."Effective AML/CFT regimes are essential to protect the integrity of markets and the global financial framework. For this, clients need to comply with a set of procedures, laws, and regulations through their internal control system," says Sherry Huang, CEO of the company. JAML aims to assist auditors to comply with these regulations and check millions of transactions and customer data, which is a knowledge sensitive and high-risk area. This way, JAML saves time and costs in carrying out complicated tasks.

JAML combines AI, data analytics and open data extraction technology that helps in processing not only English but also complicated Chinese language, such as Traditional Chinese and Simple Chinese, by using text mining and sound mining technology. With more than 200 expert AML/CFT rules within the JAML, the Robot learns from the existing money laundering fraud patterns and generates useful features to detect frauds into components like knowing your customer, blacklists, large currency transaction, and suspected money laundering. What is more, users can verbally ask questions pertaining to AML to JAML, and it will reply through speech, dashboard, or an analysis report.

JACKSOFT has built a new service model with accounting firms for JAML. The innovative business model has successfully been implemented in several cases. One such example is of Crowe Global-Taiwan, an international accounting firm that audited AML/CFT compliance for an international Bank, Taiwan in 2018. The firm forecasted three-month auditor manpower to complete the project. Within three days, JAML was implemented and required only one-month auditor manpower to finish the project. The JAML assistant reduced more than 60 percent auditor manpower with improved audit quality.

> JBOT aims to assist auditors to comply with these regulations and check millions of transactions and customer data, which is a knowledge sensitive and high-risk area

The JAML system implementation is based on a continuous monitoring and auditing approach. The new service model with the CPA firms also makes a win-win situation between the company and accounting firms. JACKSOFT also sees the Regtech as a big potential feature and will continue to invest in the new Regtech such as Self-Assessment System, IT Security/FCPA/GDPR Compliance, and such. The company has been grateful for the assistance and research grants support provided by Ministry of Economic Affairs, Taiwan over several years. This innovation has shown how RegTech can be used in the Asia Pacific financial institution to assistant AML/CFT compliance, even when language and culture different. "This not only has enormous potential for cost savings but is also changing financial regulators' behavior in the entire area industry," adds Huang.

JACKSOFT is helmed by Huang who has worked as a Chief internal auditor and Chief accountant for public companies over the years. She has also significantly contributed toward the growth of the overall financial and auditing industry in the Asia Pacific.

Source: APAC CIOoutlook, Top 10 RegTech Consulting/Services Companies - 2019

Capturing Opportunity

Source: Delen, Dursun & Ram, Sudha. (2018).

Current Status of CAATs from IIA Survey

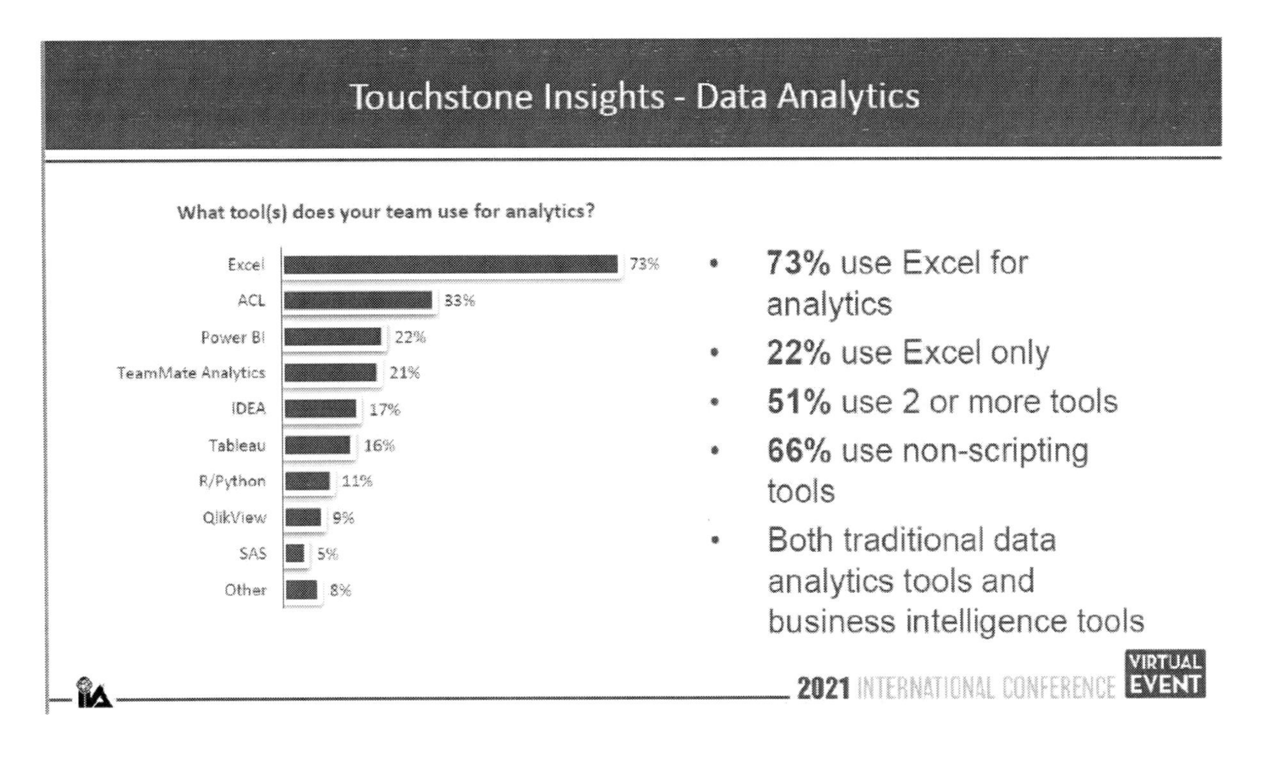

Source: Phil Leifermann, Shagen Ganason. (2021).

JCAATs-AI Audit Software
Copyright © 2023 JACKSOFT.

Audit Data Analytic Activities

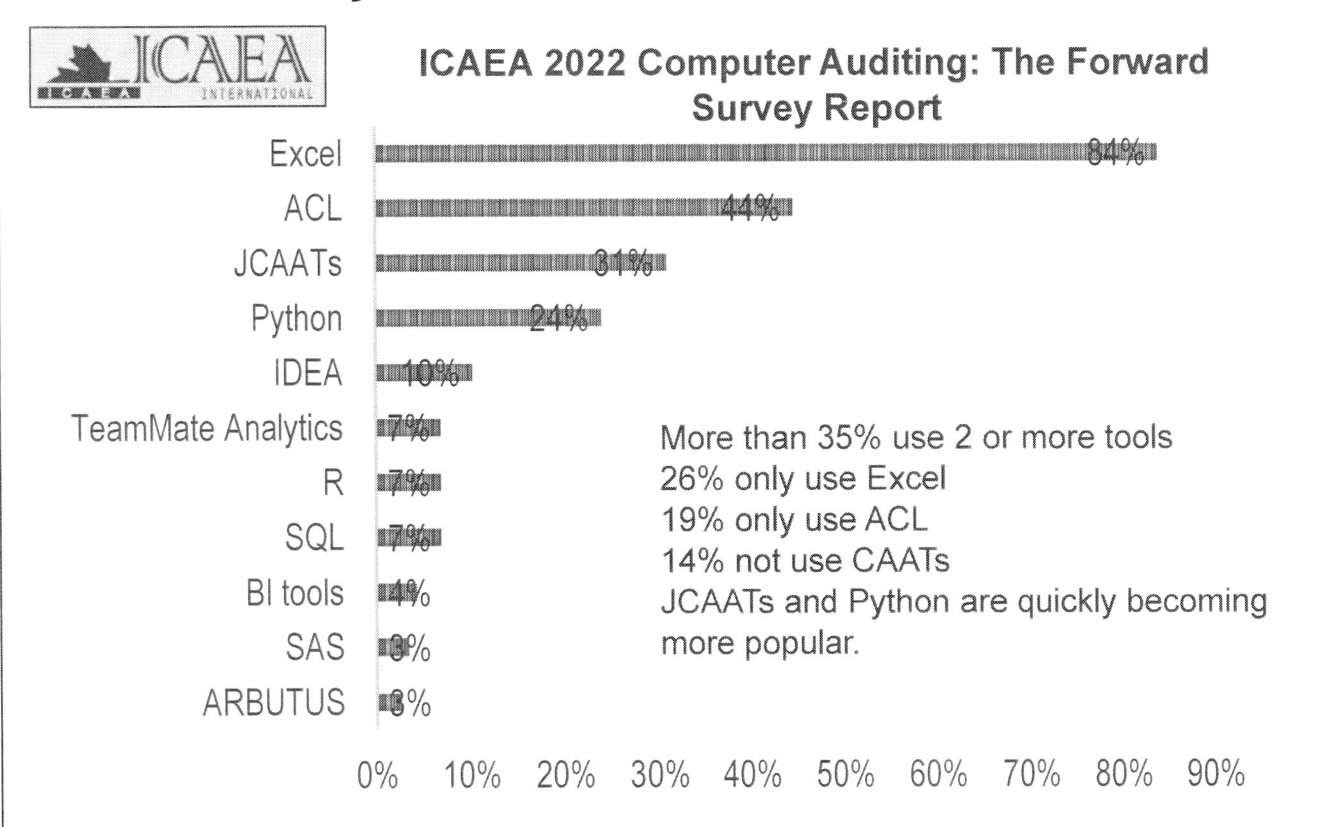

Total 117 participates from 16 countries.

Source: ICAEA, 2023

CAATs - General Audit Software History

ACL - 1987
Vancouver, Canada

Dr. Hart Will CPA
ACL Founder, Canada

IDEA - 1989
Toronto, Canada

Dwight Wainman CPA
IDEA Founder,
Canada

JCAATs - 2017
London, UK

Sherry Huang CIA, CEAP
Jacksoft Ltd.,
Taiwan

Source: Hart W., 1983; Huang S & Huang S.M., 2017

Python

- Python is the most popular programming language for AI, it's one of the hottest languages going around, and it's also easy to learn!
- Python is an open-source programming language that was first released in 1991.
- It's available for a variety of operating systems and can be used for general-purpose programming for both large and small projects.
- Python supports many programming paradigms such as imperative, functional, and procedural.
- Python's simple coding style makes it the preferred language for those beginning to learn how to code.

Source: https://www.python.org/

Python Code Example from AICPA

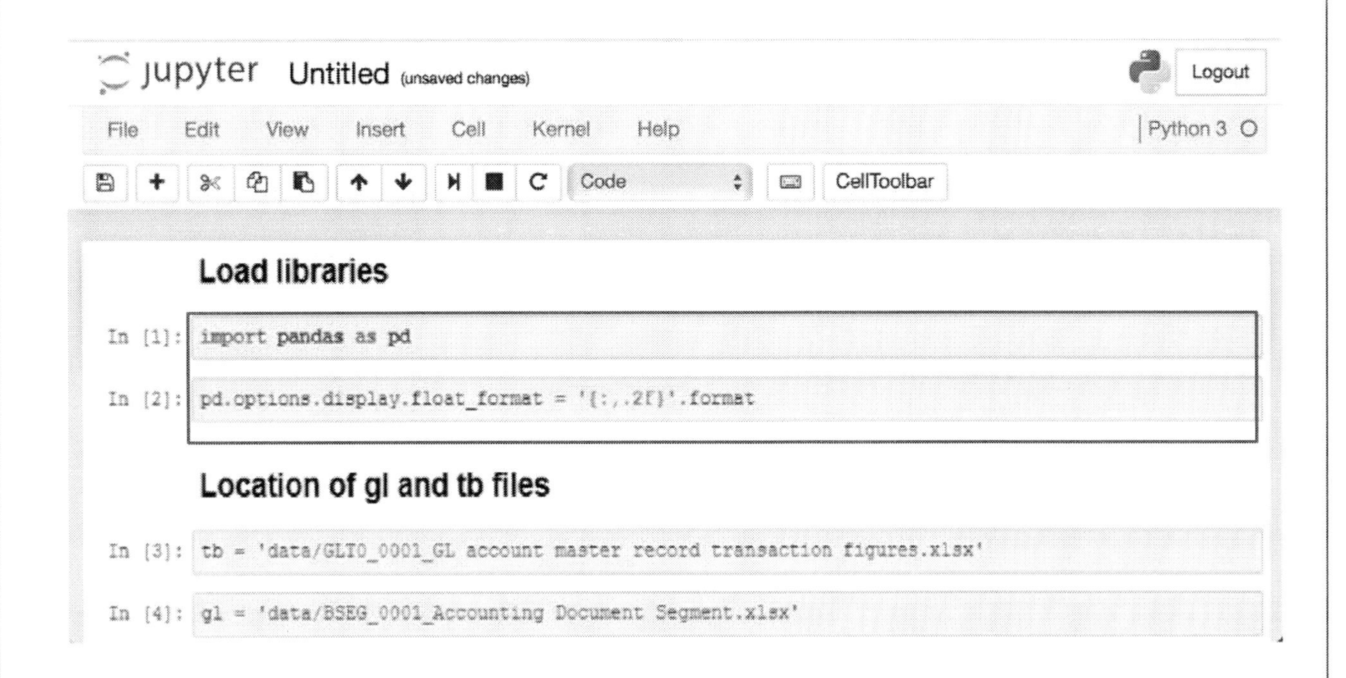

Source: AICPA, Audit Data Standards.

AICPA
Upgrade the Financial Statement Audit with Python

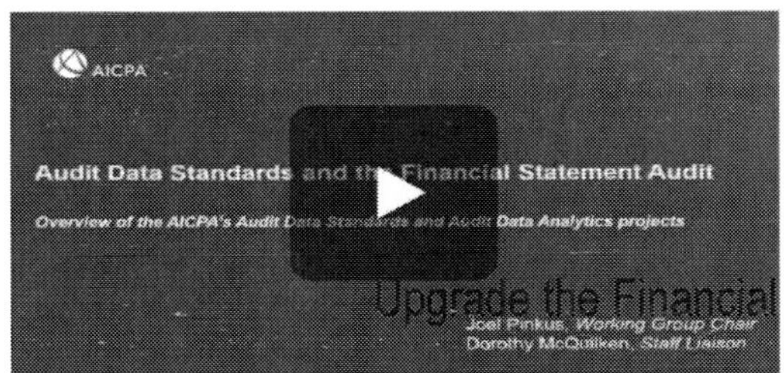

Audit Data Standards and the Financial Statement Audit

This video discusses the Audit Data Standards and some of the other projects and initiatives going on in the area of Audit Data Analytics.

Upgrade the Financial Statement Audit with Audit Data Analytics

This video illustrates how Python, an open source programming language, can be used to apply the AICPA's Audit Data Standard formatting to a data set, and how to develop routines to further analyze the Audit Data Standard standardized data set.

Learn more about Python

Access Python Routines at GitHub

Source: AICPA, Audit Data Standards.

ICAEA Mission

Studying Computer Auditing

I' M INTERESTED IN STUDYING COMPUTER AUDITING. WHAT CAN YOU TELL ME?

Why ICCP® Cert...

ICAEA

YouTube

The purposes of the society are:

- to give the potential computer auditors an overview of the main activities of computer audit and the role of computer auditor;
- to assist the potential computer auditors to become a successful computer auditor;
- to define the competency of a computer auditor;
- to define the learning map of a computer auditor;
- to evaluate the professional quality of a computer auditor.

A Not-Profit-Organization for the computer auditing education ecosystem.

Source: ICAEA, 2023

ICAEA Professional Certificates

ICAEA Code of Ethics and Professional Practice

- **Knowledge Inventory by Selection Questions**
- **Case Studies by Computer Hand-on Practice**

ICAEA Continuing Professional Education (CPE) Requirements

1. **For Practitioner:** A practitioner who is performing computer auditing functions must complete a total of 12 hours of acceptable CPE annual period.
2. **For Professional:** A professional who is performing computer auditing functions must complete a total of 12 hours of acceptable CPE annual period.

Source: ICAEA, 2023

2022 ICCP Competency Framework

Knowledge
- Technology Mindset
- Computer Assisted Audit Techniques (CAATs)
- Advanced Technology: Text Analytic and Machine Learning

Skill
- Using CAATs Skill
- Hands-on Practice Case Study

Attitude
- Legal/Guidance Issues for Computer Auditors
- Ethical Issues for Computer Auditors

Source: ICAEA, 2023

ICCP Exam Review

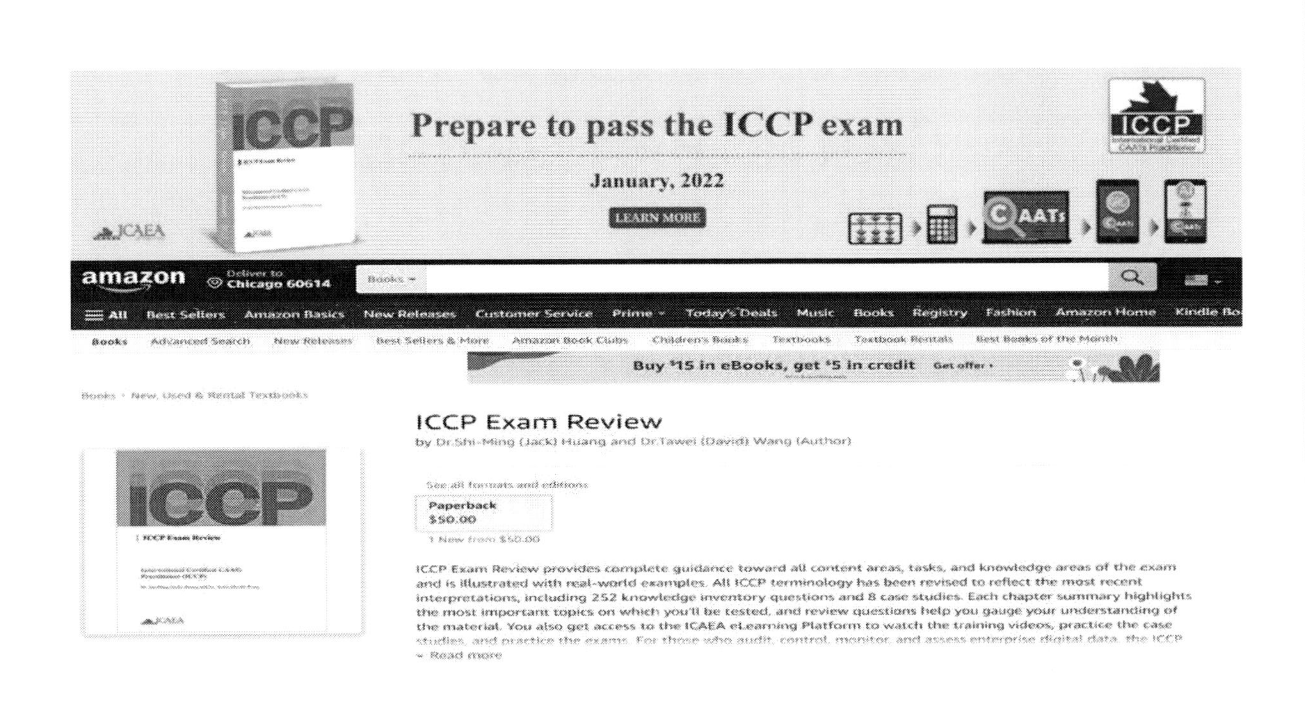

Source: ICAEA, 2023

Prepare to Pass ICCP Exam

PREPARE TO PASS THE ICCP EXAM · YOUR WAY

SELF-STUDY ACL PROGRAM

Comprehsive eLearning program for Case Studies of ICCP Exam Review Solution by using CAATs - ACL.

Learn More >

SELF-STUDY IDEA PROGRAM

Comprehsive eLearning program for Case Studies of ICCP Exam Review Solution by using CAATs - IDEA.

Learn More >

SELF-STUDY JCAATs PROGRAM

Comprehsive eLearning program for Case Studies of ICCP Exam Review Solution by using CAATs - JCAATs.

Learn More >

"A journey of a thousand miles begins with a single step"

Welcome You to Become CAATs Professional

Past MODEL

TO_BE MODEL

Audit Data Analytic Case Contest

INTERNATIONAL COMPUTER AUDITING EDUCA...

Students Learn Data Analytic Skills in New Case Competition
Sixteen students from DePaul, Loyola and Northe .....

INTERNATIONAL COMPUTER AUDITING ED... ...

The workshop was organized by Dr. Jack Huang, Dr. Shaio-Yan Huang, and his students, going to attend the second international data analytics case competition in No ......

Source: ICAEA , 2021

Computer Auditing: The Way Forward

Computer Auditing: The Way Forward

1) Adoption
2) Solutions
3) Audit quality
4) Robotic process automation, artificial intelligence, and emerging issues
5) Teaching materials or instructor training materials

Wang, T. and Huang, S., 2019, Computer Auditing: The Way Forward, International Journal of Computer Auditing , Vol.1, No.1, pp.1- 3.

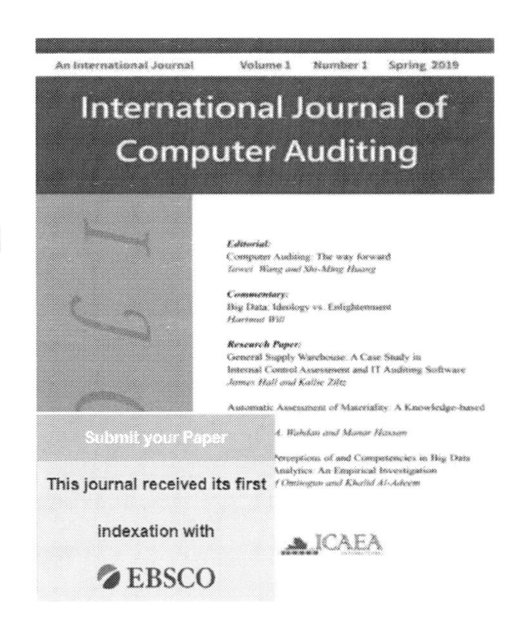

Source: Huang, S.M. and Wang, T., 2021

Continuous Auditing/Monitoring Architecture

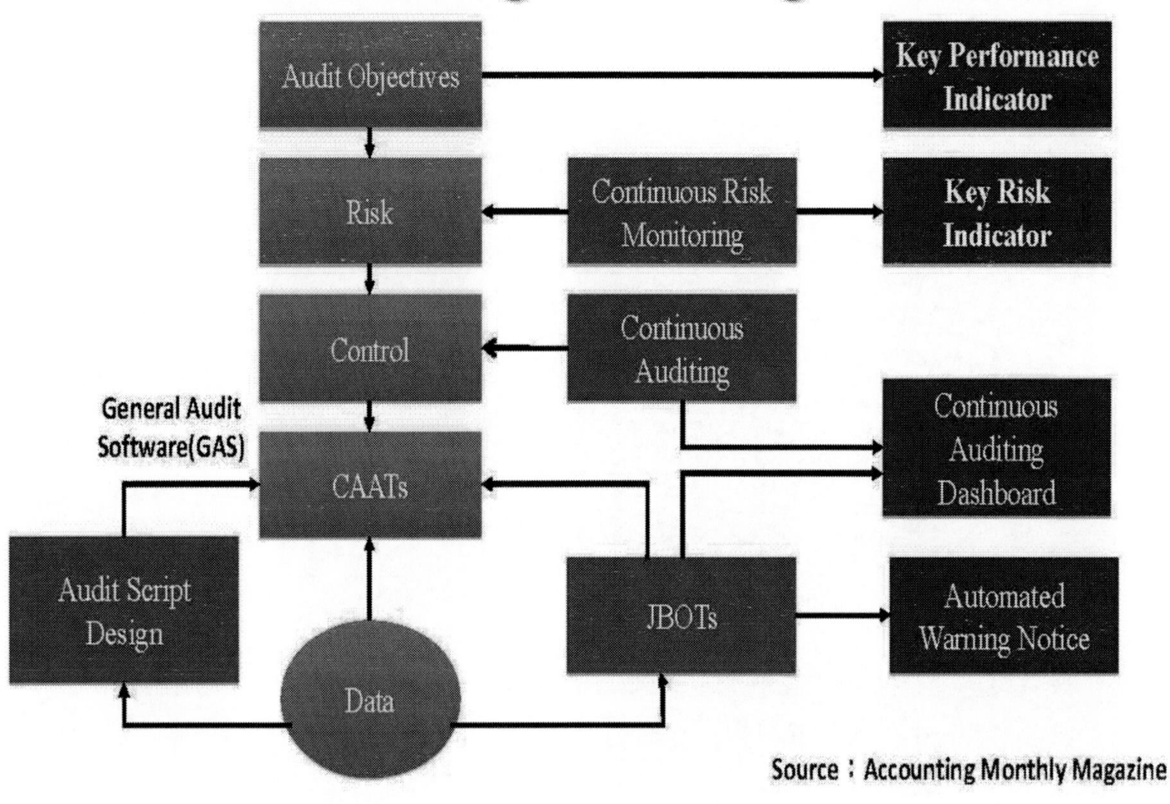

Source：Accounting Monthly Magazine

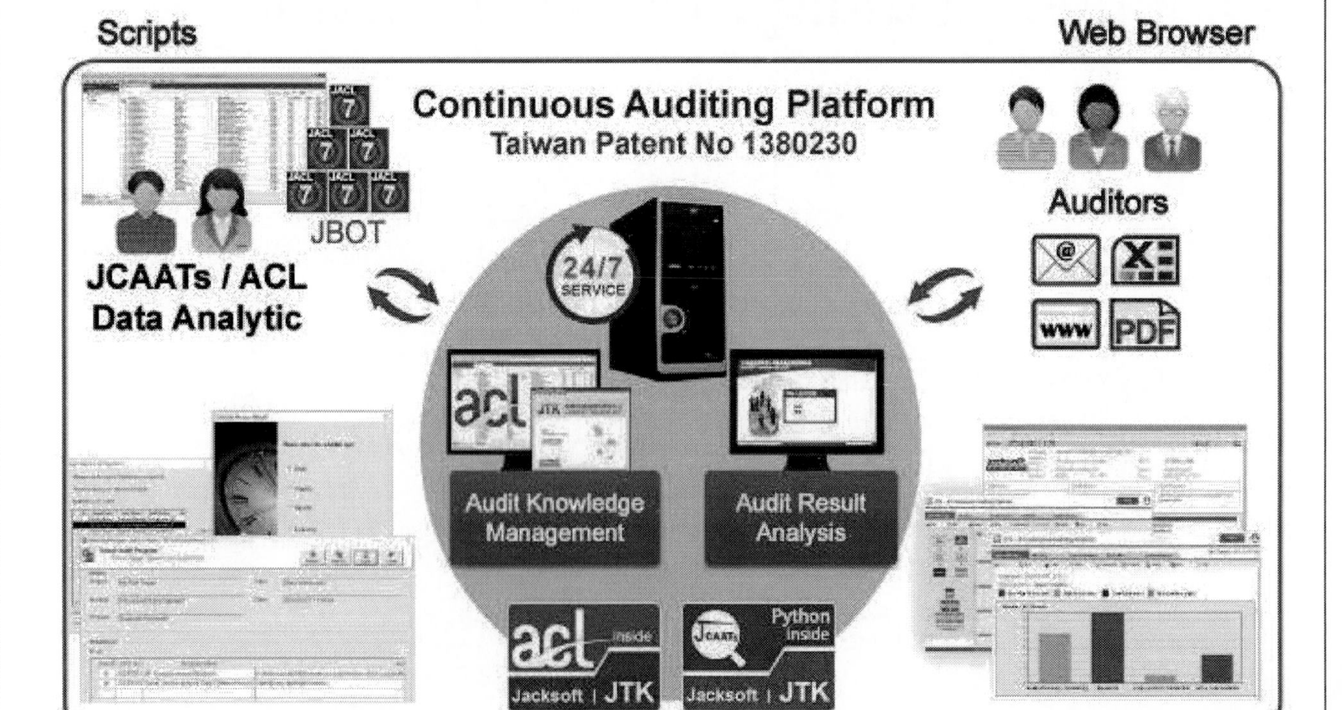

Ten Commandments for Computer Auditing

1. Accept full responsibility for their own computer auditing work.
2. Keep private any confidential information gained in their professional work.
3. Ensure that they are qualified for any project on which they work or propose to work.
4. Ensure an appropriate method is used for any computer auditing project on which they work or propose to work.
5. Ensure that their computer auditing scripts and related modifications meet the highest professional standards possible.
6. Maintain integrity and independence in their professional judgment.
7. Not engage in deceptive financial practices such as bribery, double billing, or other improper financial practices.
8. Participate in lifelong learning regarding the practice of their profession.
9. Assist audit team members in professional development and become more audit efficient.
10. Promote public knowledge of computer auditing.

Source: ICAEA

JCAATs - AI Audit Software

Chapter 1 - Exercise

( ) 1-1 Which of the following is the most common computer language that an IT professional uses to perform queries in a relational database?

a) Java
b) Python
c) SQL
d) C
e) ACL

( ) 1-2 Jane is an audit manager who performs a new audit project in her unfamiliar environment. What should her first step be for the project?

a) Design the audit procedure for every system or function.
b) Make a group of compliance tests and practical tests.
c) Collect background information regarding the new audit project.
d) Relocate audit manpower and economic resources.
e) Take a tour to visit the project environment.

JCAATs-AI Audit Software
Copyright © 2023 JACKSOFT.

Chapter 1 - Exercise

( ) 1-3 When enterprises adopt computer-based information systems, which of the following basic principles remain applicable, from the auditing perspective?

Ⅰ. The design and implementation of confirmation and control tests
Ⅱ. Internal control objectives
Ⅲ. Financial statements declaration
Ⅳ. Considerations of the inherent risks and the controlled risks
Ⅴ. Audit technique

a) I, II, III, and IV
b) I, III, IV, and V
c) II and III
d) II, III, and IV
e) I, II, III, IV, and V

Chapter 1 - Exercise

( ) 1-4 Which of the following functions is not the purpose of Generalized Audit Software (GAS)?

a) Verify calculations and grand totals
b) Execute complicated calculations
c) Select uncommon data, as defined by the auditors
d) Generate reports and machine-readable export files
e) Easily modify data

( ) 1-5 want to add a date field, C_Date, with the format YYYY-MM-DD. Which of the following is a correct JCAATs function to generate the result?

a) date(YYYY-MM-DD)
b) datetime(YYYY-MM-DD)
c) dt.date(YYYY-MM-DD)
d) todate(YYYY-MM-DD)
e) datadate(YYYY-MM-DD)

Chapter 1 - Exercise

( ) 1-6 You have two tables in JCAATs: one containing vendor information and the other containing a government-released company blacklist. How can you determine if your company is listed in the blacklist and therefore at high risk?

a) Use the Merge command to combine the two tables.
b) Use the Extract command to extract relevant information from the two tables.
c) Use the Join command to combine the two tables based on a shared key field.
d) Use the Verify command to check for matches between the two tables.
e) Use the Summarize command to aggregate information from the two tables.

JCAATs - AI Audit Software

Python Based Computer-Assisted Audit Techniques (CAATs)

Data Analysis and Smart Audit

Chapter 2. Overview

Outline:

1. Introduction
2. JCAATs User Interface
3. JCAATs Architecture

JCAATs - AI Audit Software

1. Introduction

AI Audit Software

- JCAATs has developed a cutting-edge audit software for AI using the Python language. This software goes beyond adhering to AICPA Audit Data Standards and includes traditional computer-aided audit tool (CAATs) data analysis functions. Moreover, it offers advanced AI features such as Text Mining, Machine Learning, and Web Crawling. This innovative software elevates audit analysis to a new level of smart audit.
- JCAATs is incredibly powerful yet straightforward to operate, and it can analyze vast amounts of data through an open data structure. It is compatible with numerous databases, cloud sources, various types of files, and the ACL software project, making audit data collection and fusion easier and faster.
- Users who are not familiar with Python can also operate it easily to export Python audit programs through the visualization user interface. Additionally, it can integrate with a wide range of free Python program resources, making the audit program more expandable and open, without being limited to a few software options.

JCAATs AI New Audit

The world's first universal Audit software that can run by Mac and Windows

Multiple Language and Visualization User Interface

Modern Tools for Modern Time

JCAATs- Python Based AI Audit Software

Machine Learning & AI

Outlier Analysis | Cluster Analysis | Learning | Forecast | Trend Analysis

Data Fusion

- Import multiple files at once
- ODBC Database Interface
- OPEN DATA Crawler
- Cloud Service Connector
- SAP ERP

JCAATs

Text Mining

- Fuzzy Join
- Fuzzy Duplicates
- Keyword
- Text Cloud
- Sentiment

Visualization Analysis | Data Verify | Relate Join | Analytical Review | Data Analysis

Big Data Analysis

AI Audit Ecosystem

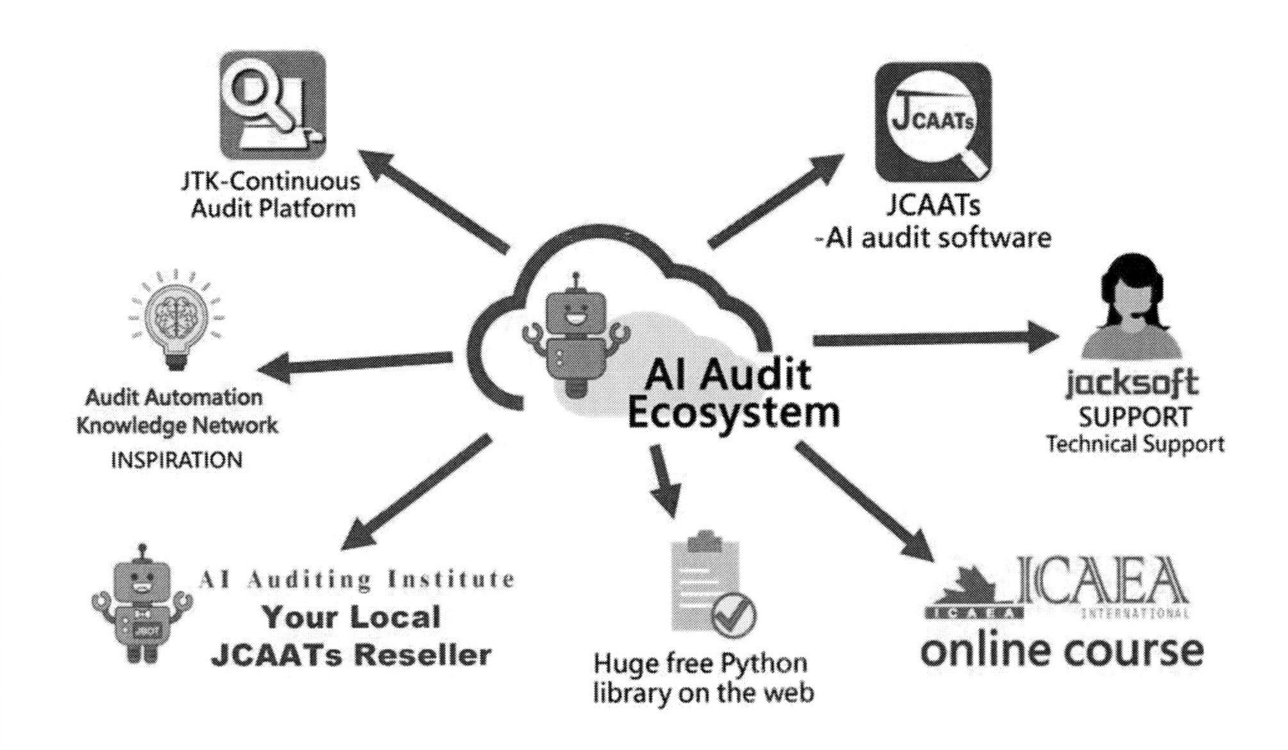

JCAATs - AI Audit Software

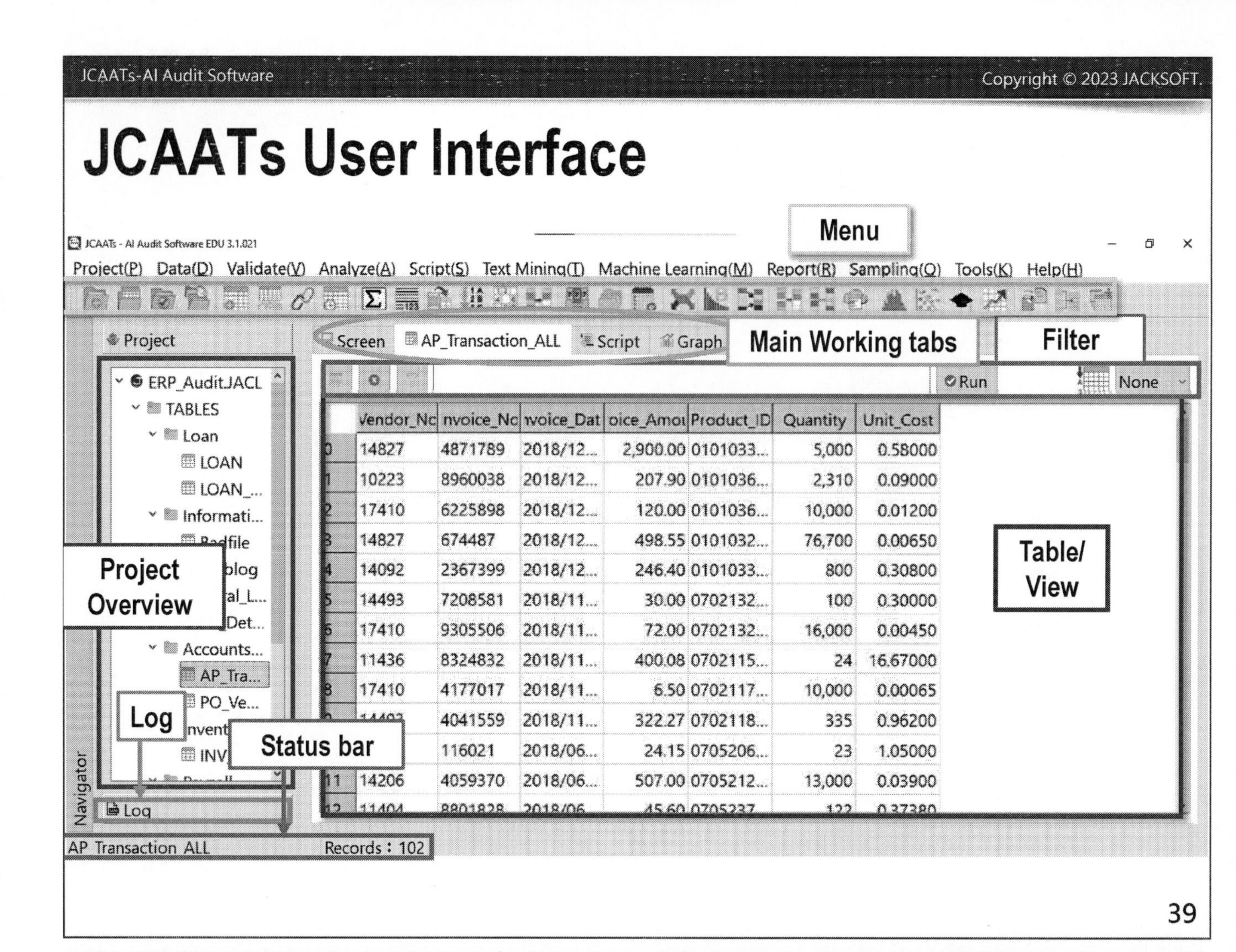

JCAATs Main Interface Elements

- **Main Working tabs**
  - Included main screen, table, program and result diagram
- **Project**
  - Display project related to table and program
- **Log**
  - Display log record
- **Menu**
  - Display JCAATs related to command
- **Status bar**
  - Display main table related to information

Main

- **Main Screen**
  - Display command result
- **Table**
  - Display the properties of the working table
- **Script**
  - Display the properties of working program
- **Graph**
  - Display graphical table and results of analysis

Project Navigator

1. Project

Display the current project's related tables and scripts. By creating folders, it is possible to effectively manage related projects.

2. Log

Record entire log for subsequent reference or review, the operation track can be transferred to a **Script** to improve operation efficiency and achieve the goal of automation.

Status bar

- Display data count
- Display data filtering conditions

JCAATs - AI Audit Software

Menu Bar – List Audit Commands

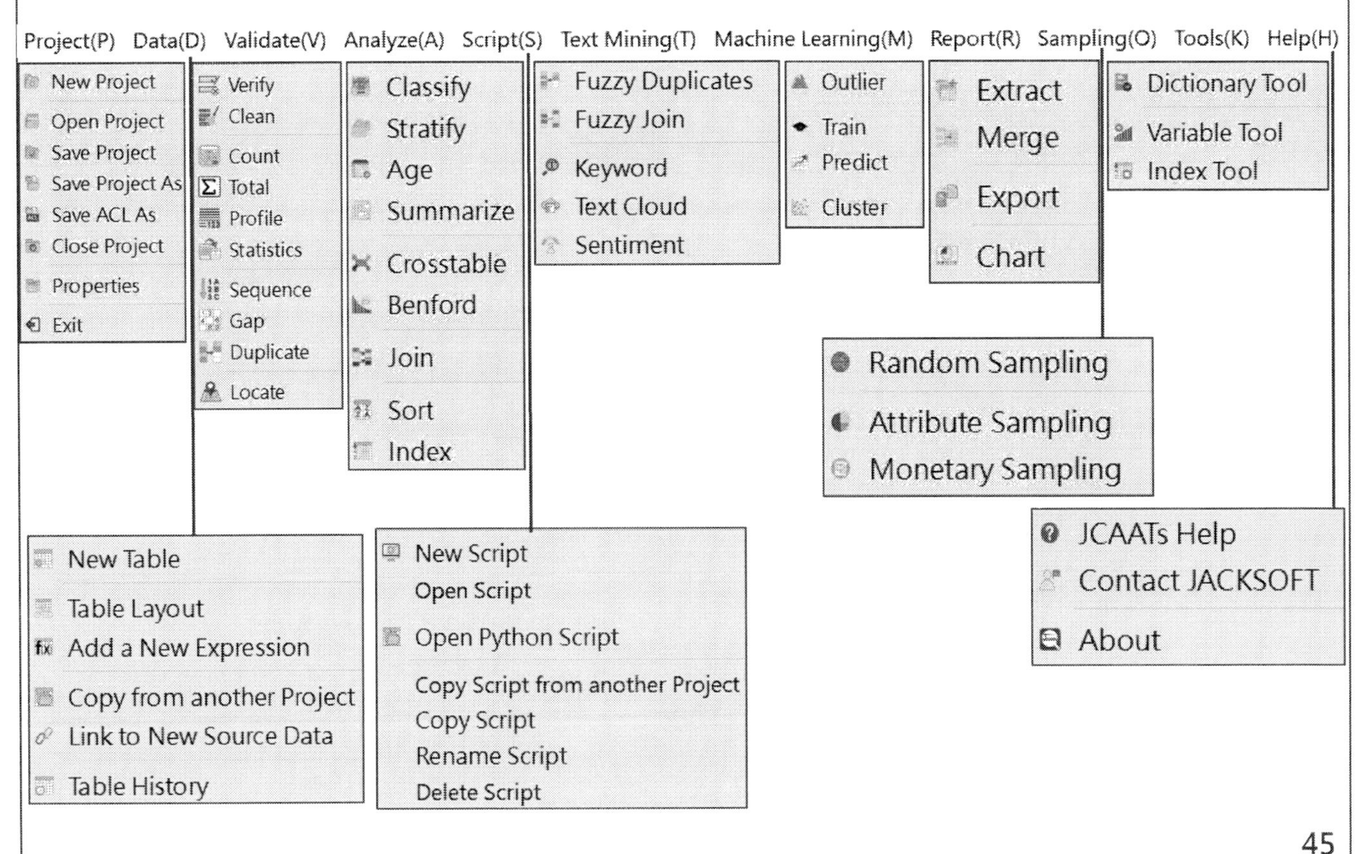

Components of JCAATs Project

- Project：
  Organization data analysis and results of analysis
- Components of Project:
  - Tables
  - Scripts
  - Logs
  - Folders

JCAATs Project Structure

- Project
  - Scripts
  - Tables
  - Folders

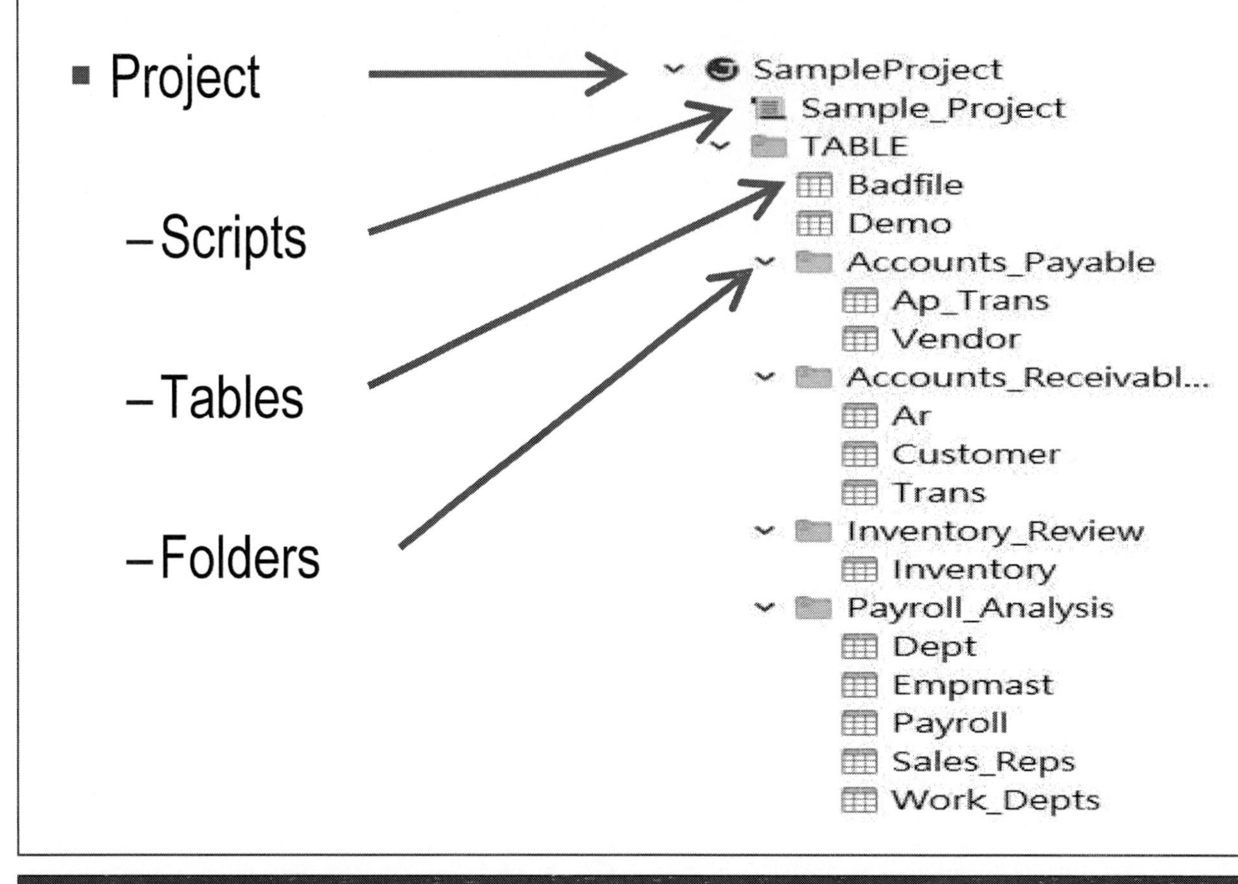

Project Navigator - Log

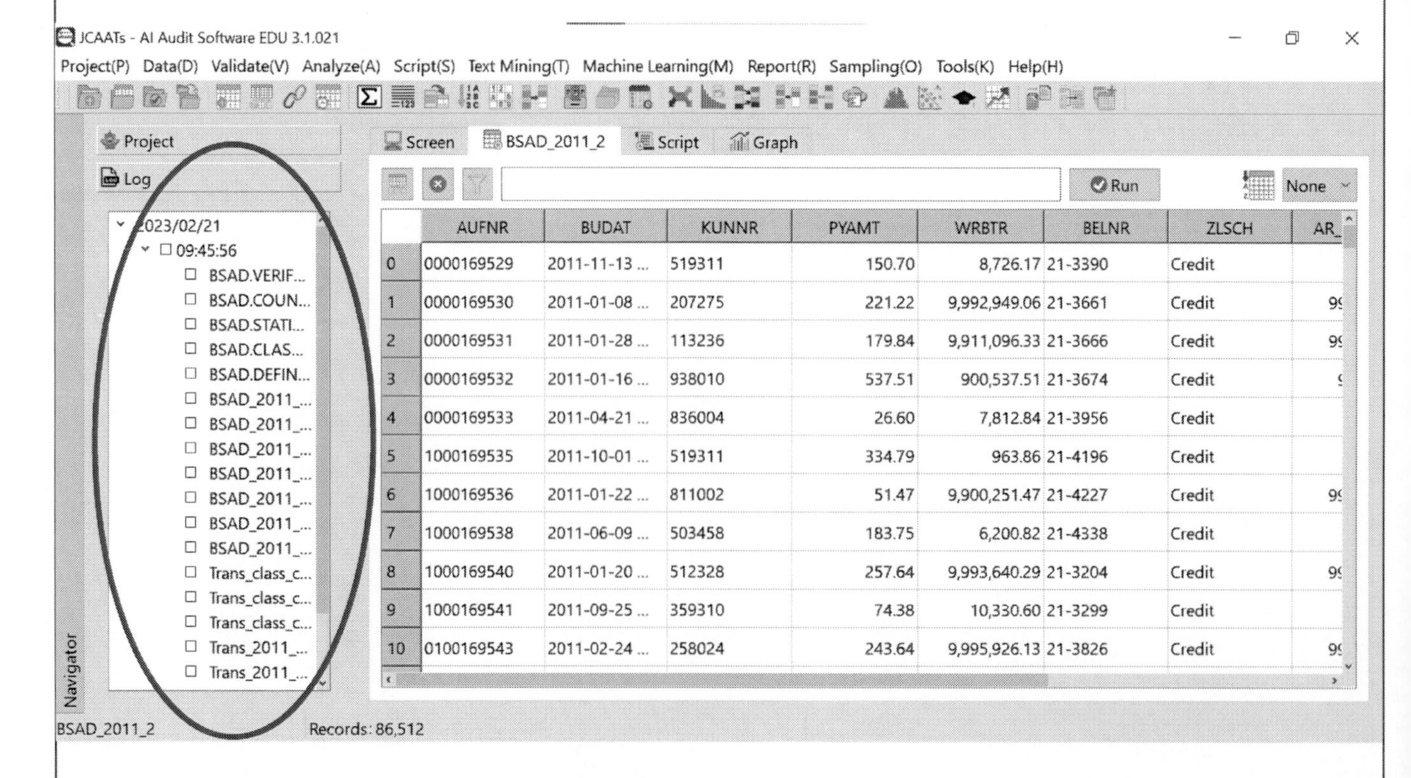

	AUFNR	BUDAT	KUNNR	PYAMT	WRBTR	BELNR	ZLSCH
0	0000169529	2011-11-13 ...	519311	150.70	8,726.17	21-3390	Credit
1	0000169530	2011-01-08 ...	207275	221.22	9,992,949.06	21-3661	Credit
2	0000169531	2011-01-28 ...	113236	179.84	9,911,096.33	21-3666	Credit
3	0000169532	2011-01-16 ...	938010	537.51	900,537.51	21-3674	Credit
4	0000169533	2011-04-21 ...	836004	26.60	7,812.84	21-3956	Credit
5	1000169535	2011-10-01 ...	519311	334.79	963.86	21-4196	Credit
6	1000169536	2011-01-22 ...	811002	51.47	9,900,251.47	21-4227	Credit
7	1000169538	2011-06-09 ...	503458	183.75	6,200.82	21-4338	Credit
8	1000169540	2011-01-20 ...	512328	257.64	9,993,640.29	21-3204	Credit
9	1000169541	2011-09-25 ...	359310	74.38	10,330.60	21-3299	Credit
10	0100169543	2011-02-24 ...	258024	243.64	9,995,926.13	21-3826	Credit

JCAATs Architecture

Input file definitions
The **table layout** describes the data structure allows you to define the fields.

Tables/Views
Display visualization data that can be used in a template for exporting a printed report

Tool: Index & Variable
Can manage index, variable and other settings information used in the project

Scripts
Allows you to automate repetitive queries and complex analysis and let you customize an application.

Log
Log can record all your operation on JCAATs just like **black box**

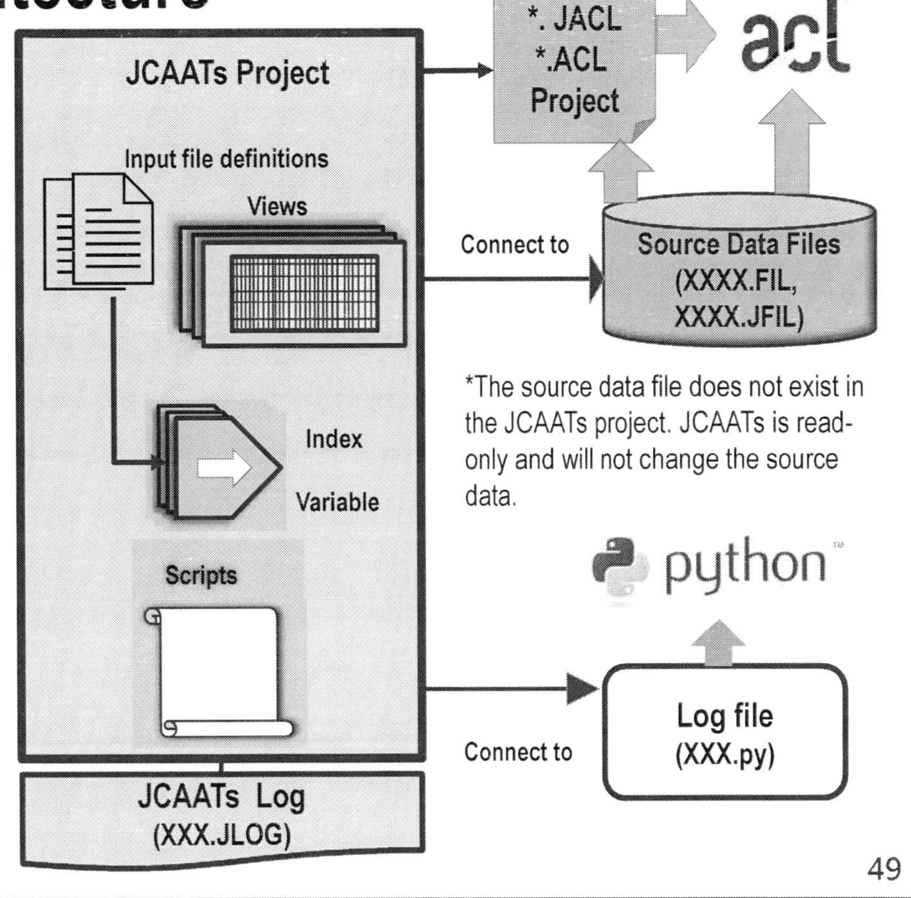

JCAATs File Organization

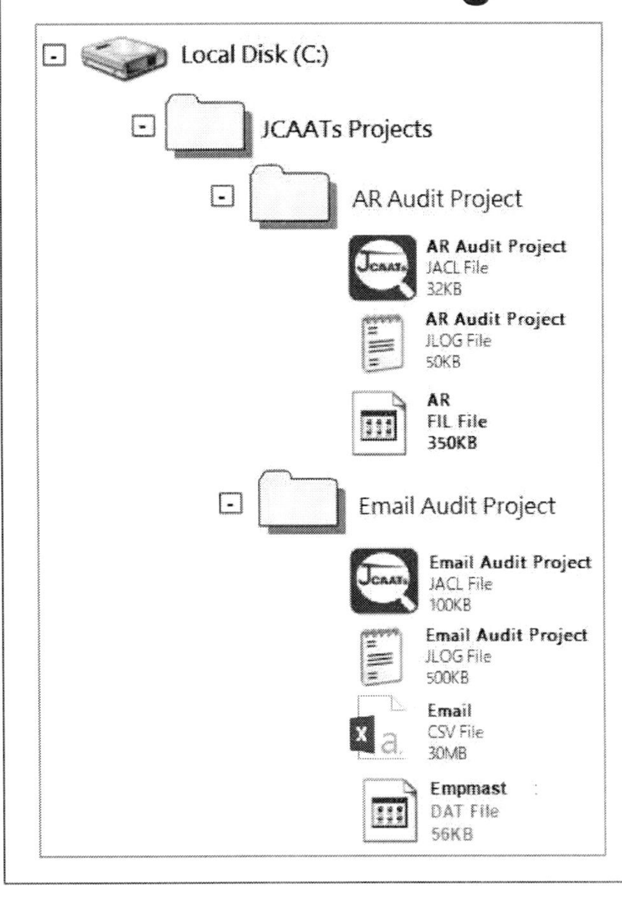

- Organize your hard drive by creating a new Windows folder for each new JCAATs Project
- The location of your JCAATS Project is your Default Working Folder
- All new files created by JCAATS as a byproduct of your analysis will be stored automatically in the Default Working Folder

JCAATs - AI Audit Software

Chapter 2 - Exercise

Chapter 2 - Exercise

( ) 2-1 Which of the following items are components of a JCAATs Project?

Ⅰ. Tables
Ⅱ. Scripts
Ⅲ. Folders
Ⅳ. Workspaces
Ⅴ. Files

a) I, II, III, IV, and V
b) I, II, IV, and V
c) I, II, III, and IV
d) I, II, IV, and V
e) I, II and III

Chapter 2 - Exercise

( ) 2-2 What is the correct basic workflow for conducting data analysis using JCAATs?

Ⅰ. Determine the appropriate sampling method.

Ⅱ. Import data files into JCAATs tables.

Ⅲ. Conduct audit data analytic using techniques such as Data Analysis, Text Mining and Machine Learning, and output the results into a results table.

Ⅳ. Create a script to automate the analysis process.

Ⅴ. Export the analysis results, or import them into a reporting tool.

a) I, II, and III.
b) II, III, and IV.
c) II, III, and V.
d) II, III, IV, and V.
e) I, II, III, IV, and V.

Chapter 2 - Exercise

( ) 2-3 Which command in JCAATs can be used to create a folder within a JCAATs project?

a) The MKDIR command
b) The MD command
c) The os.makedirs command
d) The self.SET_FOLDER command
e) The SET DIRECTORY command

( ) 2-4 Which of the following statements about JCAATs folders is INCORRECT?

a) JCAATs project folders are distinct from Windows folders.
b) Creating a JCAATs project folder does not create a corresponding Windows folder.
c) There is no direct link between JCAATs project folders and Windows folders.
d) It may be challenging to maintain a consistent structure between JCAATs project folders and Windows folders.
e) None of the above.

Chapter 2 - Exercise

( ) 2-5 What is a possible way to harmonize the structure of JCAATs project folders and Windows folders?

a) Manually create and maintain a parallel folder structure.

b) Use a function on the JCAATs menu to set the link between JCAATs project folders and Windows folders.

c) Set the link between JCAATs project folders and Windows folders, and then select Window > Folder.

d) Set the link between JCAATs project folders and Windows folders, and then select Tools > Options.

e) There is no need to have a direct parallel between JCAATs project folders and Windows folders for your audit workflow.

JCAATs-AI Audit Software
Copyright © 2023 JACKSOFT.

Chapter 2 - Exercise

( ) 2-6 Which of the following statements regarding JCAATs variables is correct?

a) There are two types of variables in JCAATs, called system-generated variables and user-defined variables..

b) There are two types of variables in JCAATs: one is called a local variable for the current JCAATs Analytics project, and the other is a global variable for all JCAATs Analytics projects.

c) There are two types of variables in JCAATs: one is called a local variable for the currently open table, and the other is a global variable for all tables within the current JCAATs Analytics project.

d) All types of variables remain in your computer's memory until you delete them or until the Analytics project is closed.

e) A variable supports the null value.

Python Based Computer-Assisted Audit Techniques (CAATs)

Data Analysis and Smart Audit

Chapter 3 . Project

Outline:

1. Smart Audit Project Procedure
2. Audit Commands
3. Project Navigator & Project Property
4. Main Screen & Result Graph

JCAATs - AI Audit Software

1. Smart Audit Project Procedure

Smart Audit Project

- **Smart Audit** is the use of big data and advanced audit analytics tools embedded with artificial intelligence functions. .
- A JCAATs project represents a smart audit project and is the highest level of organization in the process. The main project file stores most of the project information, such as **table formats**, **script names**, **variables**, and **folders**, while data are stored outside the project in native JCAATs data files and scripts are stored outside the project in **Python** language files.
- This architecture ensures that data and scripts are controlled at the operating system level, and information security remains consistent with your IT system. All command operations are recorded in a log file.

Smart Audit Project Procedure

➢ Through JCAATs AI Audit Software, effectively complete Projects, including six phases.

- Planning
- Data Import
- Data Validation
- Data Preparation
- Smart Analysis
- Reporting

➢ With these powerful tools, you can easily share your results with stakeholders and collaborate effectively on the next steps of the audit process.

Smart Audit Project Procedure

Step 1: Planning

1. Planning before started
2. Plan the steps to achieved the project's purpose in order to proceed to the next five stages
3. Clearly describe the purpose to develop a strategy and a time budget
4. Identity data sources, and data quality requirements.

Smart Audit Project Procedure

Step 2: Data Access/Import

1. Design the data required strategy plan
2. List the data sources which Include the data source location (such as IP address, URL, etc), source system (such as database name, ERP name, etc.), and transformation method (such as ODBC, FTP, files, etc.)
3. Perform Data Import command to process of acquiring and integrating data into the project.

Smart Audit Project Step and Procedure

Step 3: Data Validation

1. Test data integrity, otherwise the results may be incorrect or incomplete.
2. JCAATs provides various tools, such as commands and functions, that allow users to ensure data integrity more conveniently.
3. JCAATs **Clean** commands that assists analysts in identifying the missing values more quickly, ensuring that follow-up data analysis is conducted with clean data.

Smart Audit Project Step and Procedure

Step 4: Data Preparation

1. The is where the imported data is transformed into a format that is suitable for analysis or prediction, such as by merging tables, filtering records, or creating new computed fields.

2. JCAATs provides various tools, such as commands and functions for data preparation. All analysis commands can produce table results which assists analysts to prepare their data in an easy and quick way.

3. JCAATs **Extract** command involve data clean mechanism which assists analysts to purify their data in an easy way.

Smart Audit Project Step and Procedure

Step 5: Smart Analysis

1. The Smart Analysis step includes not only traditional data analysis commands but also new text mining and machine learning commands.
2. You can perform the tests necessary to achieve your audit objectives by using various analysis commands.
3. You can perform the text analysis to identity risk or achieve your audit objectives by using various text mining commands.
4. You can perform the prediction analysis to identity risk or achieve your **Pre-Auditing** objectives by using various machine learning commands.

Smart Audit Project Step and Procedure

Step 6: Reporting

1. Generate data tables from the analysis results and make related reports.
2. Multiple data tables can be quickly merged into a new data table through data fusion technology.
3. Table can be exported to various file types such as CSV, TXT, Excel, ODS, JSON, XML and others.
4. The data can be visualized in 2D, 3D, and 4D using the **Chart** command, allowing you to explore and communicate your findings in a more impactful and interactive way.

JCAATs - AI Audit Software

2. Audit Commands

JCAATs - Project

- Each JCAATs project includes
  - File Type*.**JACL**
  - Log Type*.**JLOG**
- **JCAATs can import ACL Project** to conduct audit analysis.
- **JCAATs can be saved as an ACL Project** for use by ACL software
- Project properties display the current folder, scripts, and other related information.

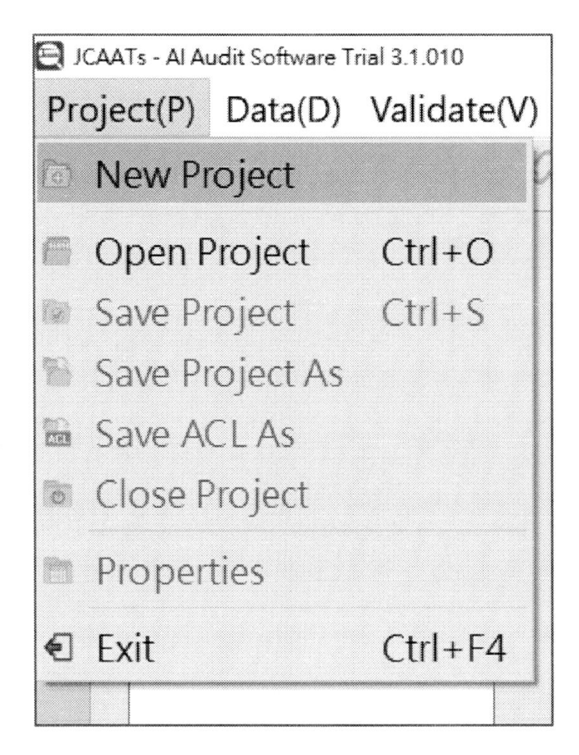

One Project One Folder, Easy and Convenient to manage

New Project

One folder per project, create a new folder in file management first, then input project name to create a new project.

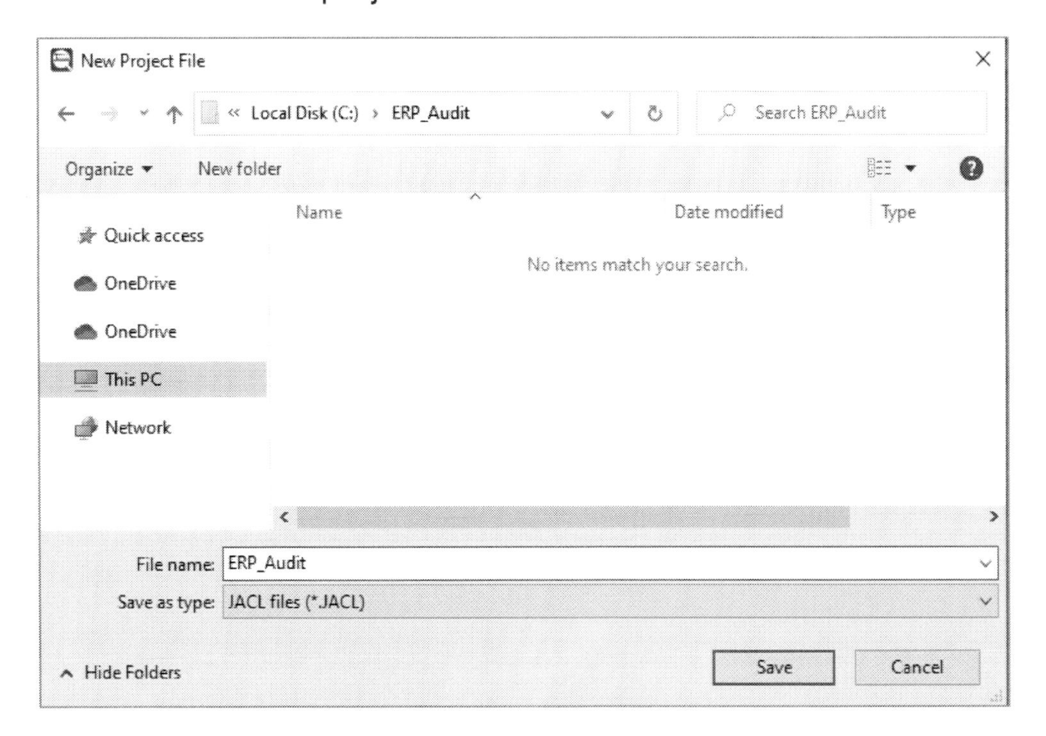

Open Project

- JCAATs can choose to open JCAATs Project(*.JACL)

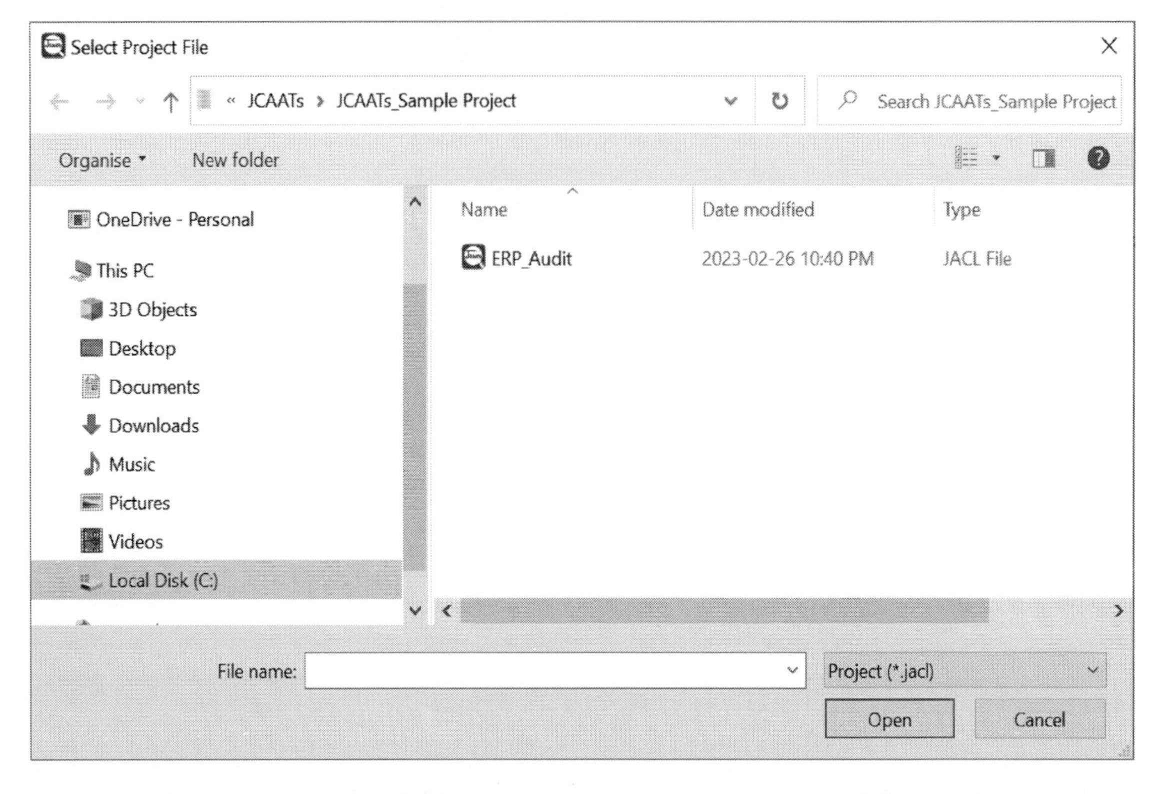

Save Project as & Save ACL as

- " Save Project as “ is to save a new project file with a new name.
- " Save ACL Project as “ is to save the project as an ACL project file. It will only save tables, but not scripts to the ACL project file. It allows tables to be shared between JCAATs and ACL.

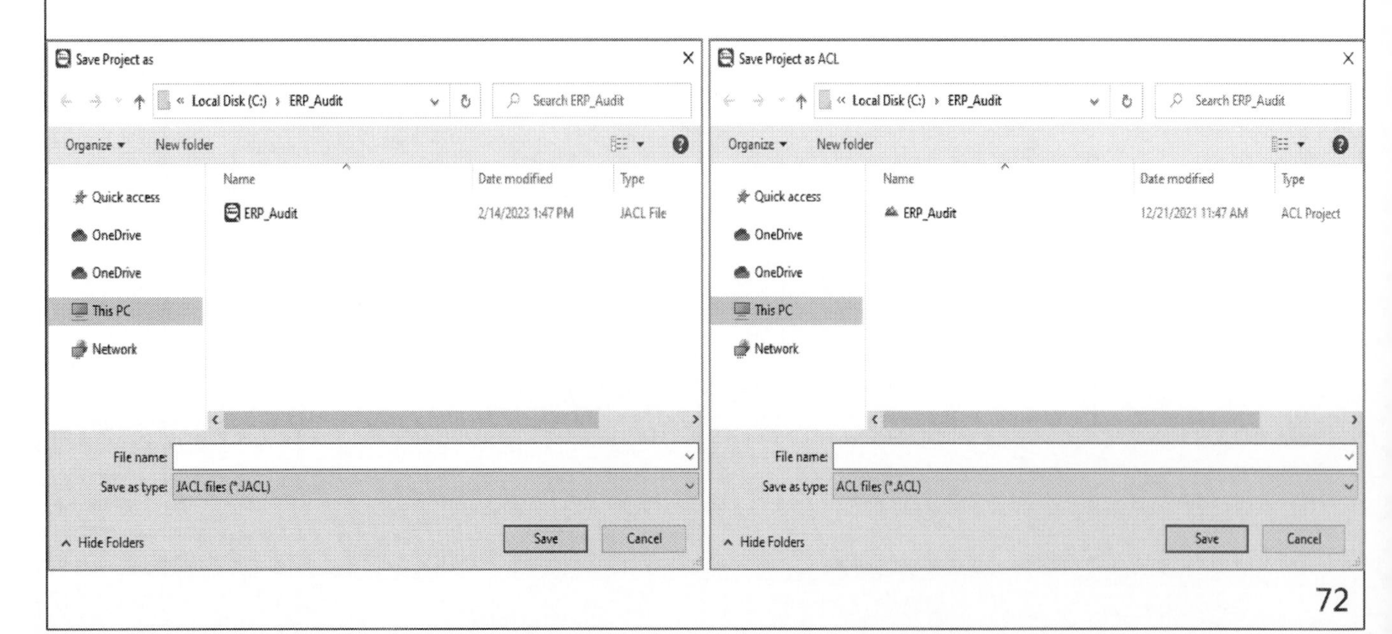

JCAATs - AI Audit Software

3. Project Navigator & Project Property

Project Property

- The Project Property displays the current project properties, including the table list and script list. These properties can be saved as a text file for use in other systems.

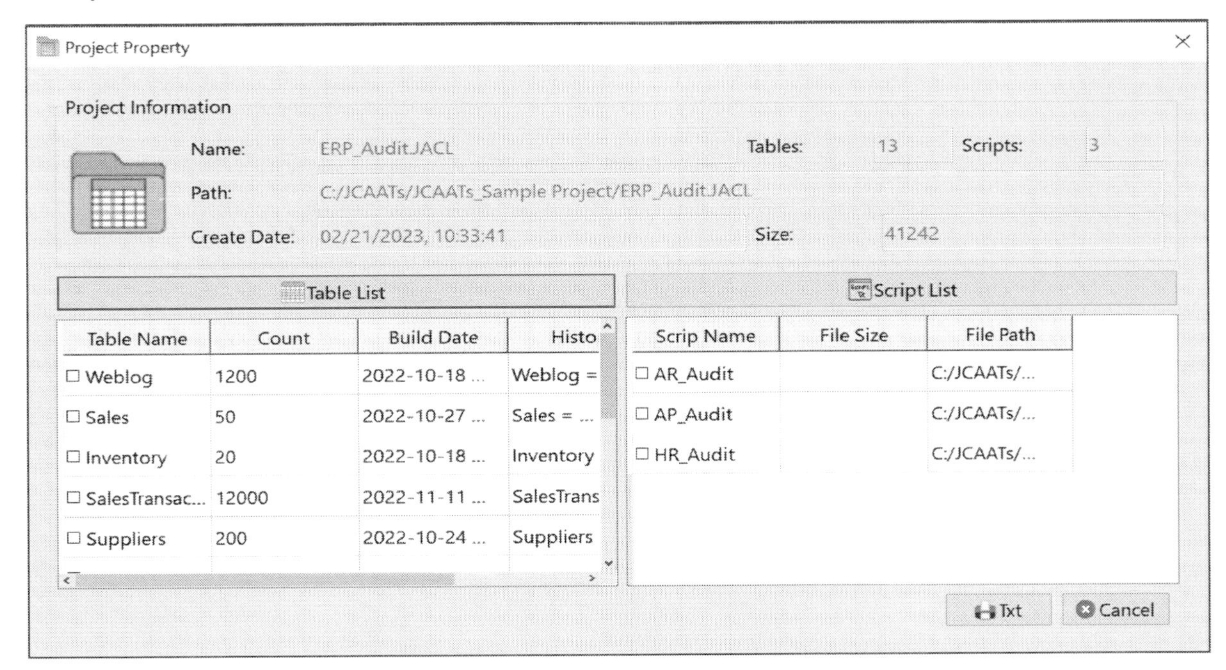

Project Navigator

Two modules including Project and Log. Each module has its own functions which can be accessed by right-clicking the mouse button.

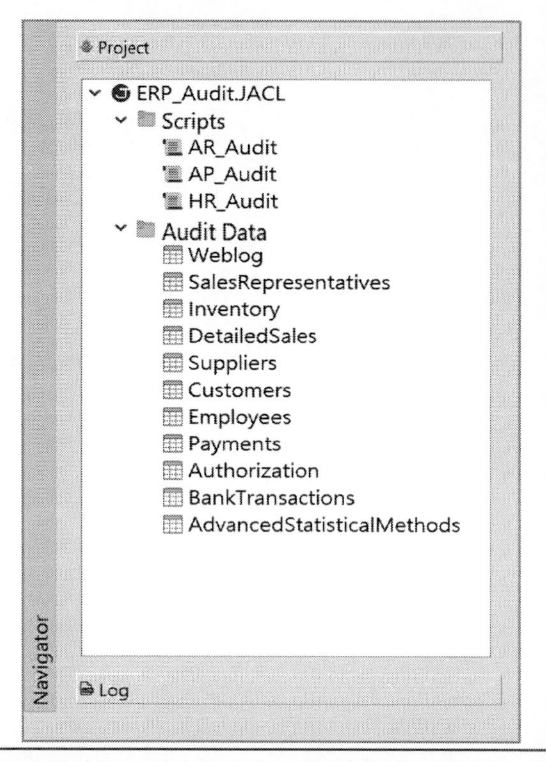

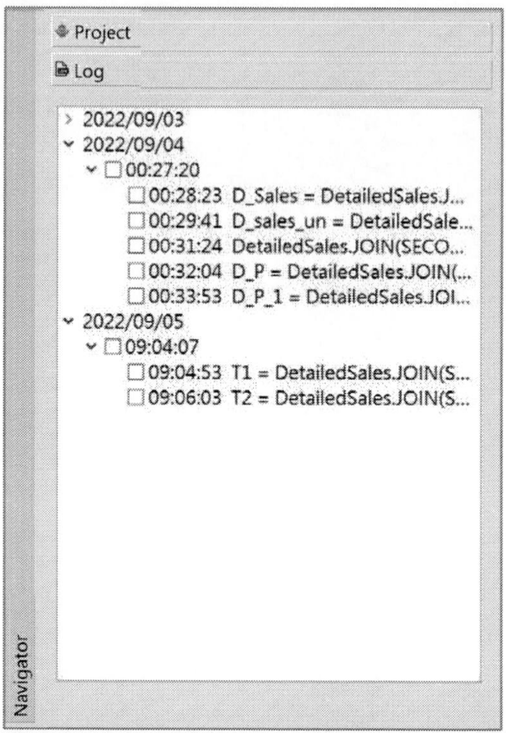

Project Navigator - Project

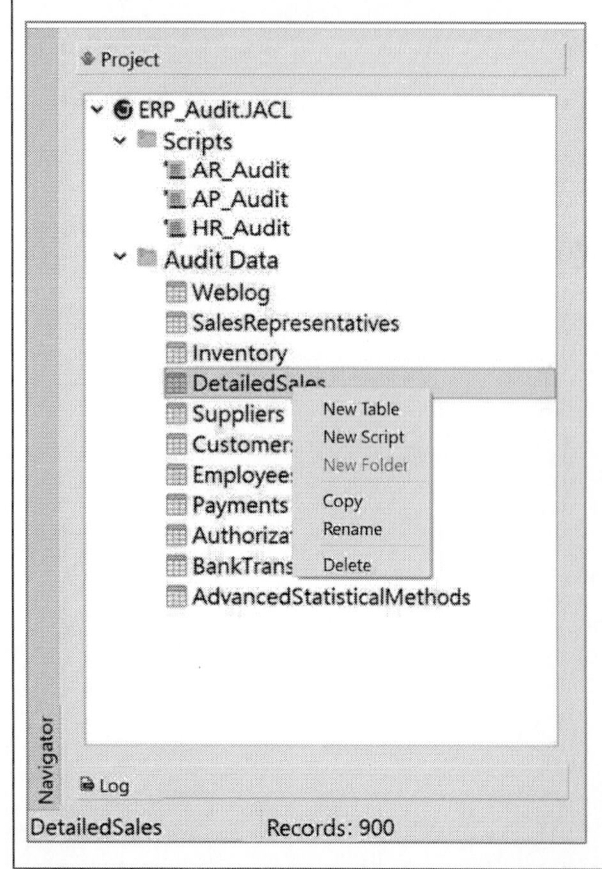

- **New Table:** Display New Table function interface
- **New Script:** Display New Script interface, input name to edit script.
- **New Folder:** Create a new folder by entering a name and move the entire table or script to this folder.
- **Copy:** Choose the table or script that you want to copy and input the new name.
- **Rename:** Revise folder, table and script names.
- **Delete:** Delete tables, scripts, and folders in the project. (Deleting a folder will not delete the tables and scripts within it.) This is a security measure. To delete the actual file, it must be removed from OS.

Project Navigator - Log

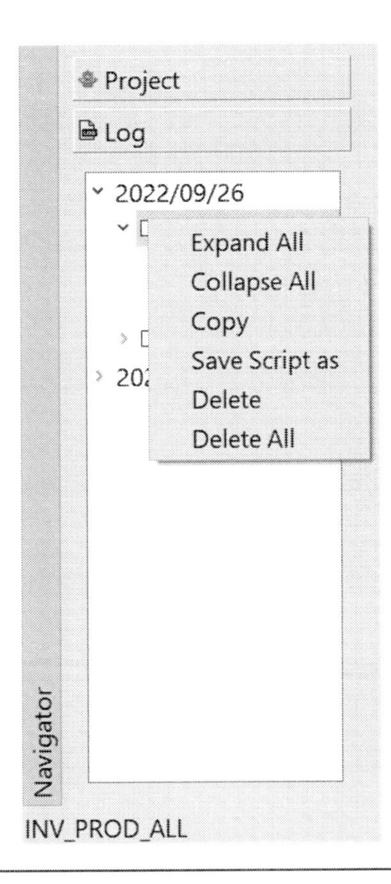

The tree structure includes three categories: Project Data, Project Time, and Command Properties.

- **Expand All:** Expand all the log record details.
- **Collapse All:** Collapse all log record details and only show the date.
- **Copy:** Select the demand log record and copy them to the system clipboard.
- **Save Script as:** Select the demand log record and choose to save it as a script. This script will be displayed under the project.
- **Delete:** Delete the selected log. Deleting a log on the upper layer will also delete the logs below it.

JCAATs - AI Audit Software

4. Main Screen & Result Graph

Main Screen

Project(P) Data(D) Validate(V) Analyze(A) Script(S) Text Mining(T) Machine Learning(M) Report(R) Sampling(O) Tools(K) Help(H)

Project

ERP_AuditJACL
- Weblog
- Sales
- Inventory
- SalesTransactions
- Suppliers
- Customers
- Payments
- Authorization
- BankTransactions
- StatisticalReports
- Employees
- Payroll
- DetailedSales

Log

Navigator

Screen | Table | Script | Graph

Run | Clear | Clean

Welcome to JCAATs 3.1 - a new Computer Assisted Audit Techniques (CAATs) for AI era.

- **Main Screen can input commands and display the implemented results of those commands.**
- After sending the data (similar to Python), display the results of both the top ten and last ten data.
- Results of partial commands can be **drilled down** to obtain detailed information.
- Click on the command to display the command dialog, making it easy to use again.
- The result information will be accumulated, and you can slide up and down to view the results of each command.
- Clear screen: Select “ clear ” if you don’t want to display too much information

Example:
Main Screen after Command Execution

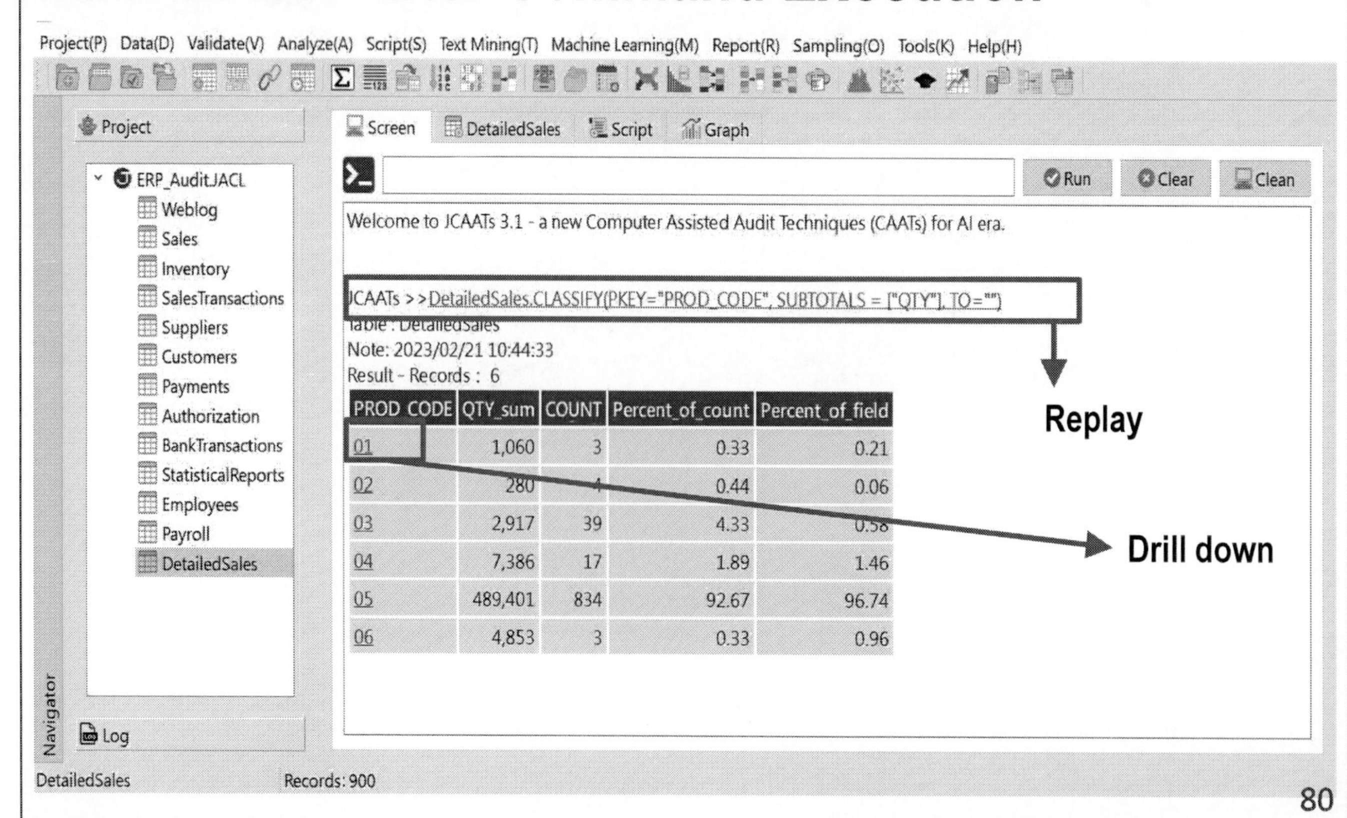

PROD_CODE	QTY_sum	COUNT	Percent_of_count	Percent_of_field
01	1,060	3	0.33	0.21
02	280	4	0.44	0.06
03	2,917	39	4.33	0.58
04	7,386	17	1.89	1.46
05	489,401	834	92.67	96.74
06	4,853	3	0.33	0.96

Example: Main Screen after Click Log

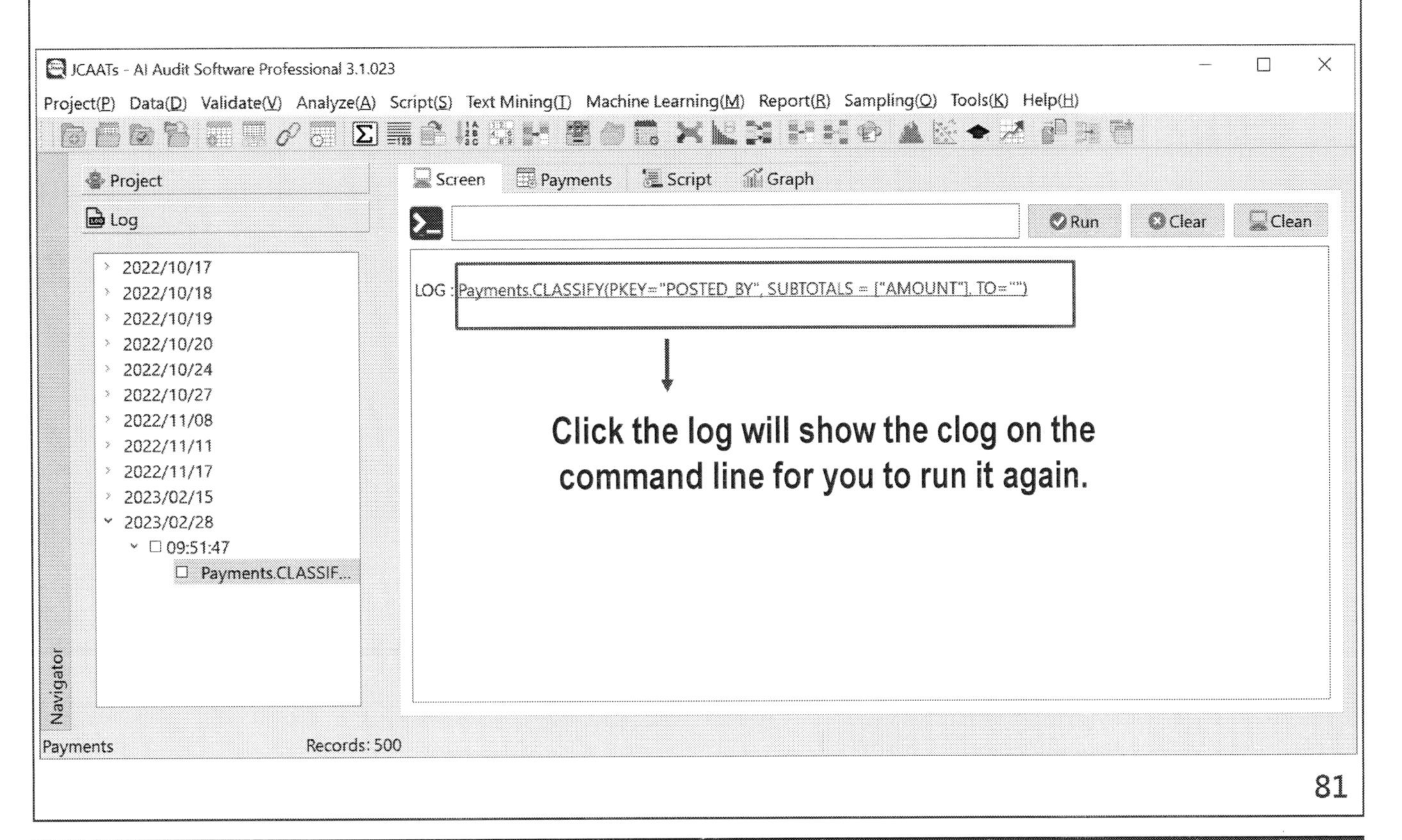

JCAATs-AI Audit Software Copyright © 2023 JACKSOFT.

Result Graph

- **Improve visualization analysis:** The graph for each table is displayed a **plot**, which basically includes **bar chart, line chart, scatter chart**, and other charts. The X-axis represents the index, while the Y-axis represents numeric values. Various commands such as **Classify, Stratify, Age, Summarize, Crosstable, Benford, Text cloud, Cluster** will export exclusive diagrams.

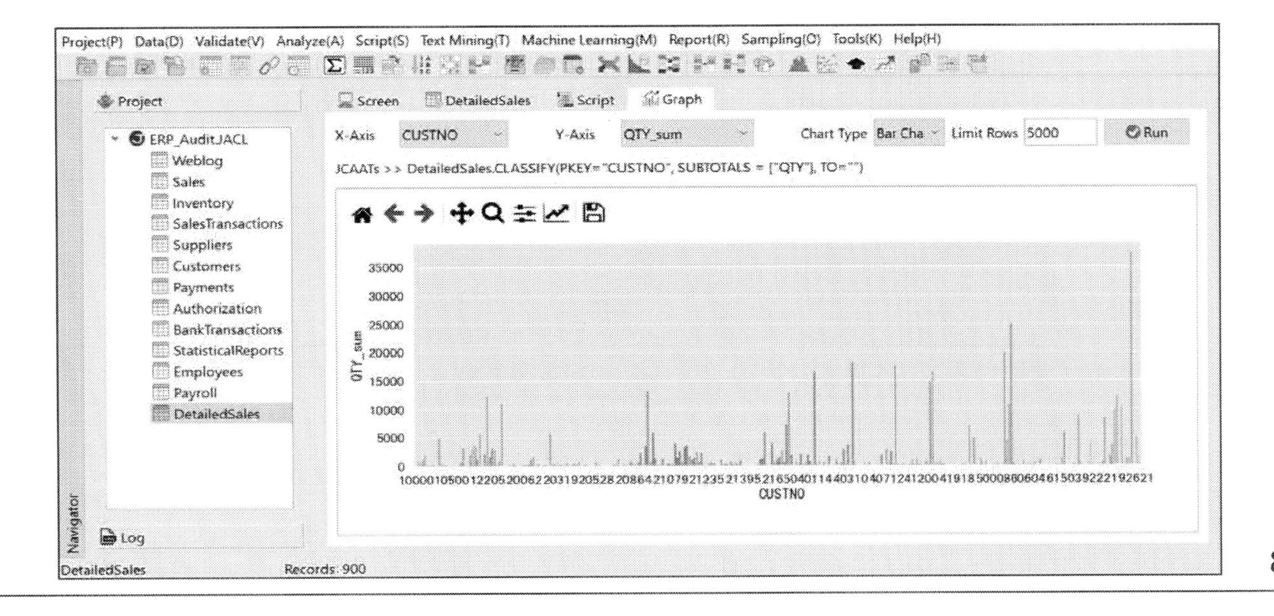

System Commands

- JCAATs also **offer some system commands** to control the project's system environment. As follows are the main commands for the project:
  - **self.SET_FOLDER(Folder)**
    This command sets up the current folder and corresponds to a system variable.
  - **self.SCREEN_CLEAR()**
    This command clears the current command property and is the same as the 'clear screen' button.

Example: Command Line

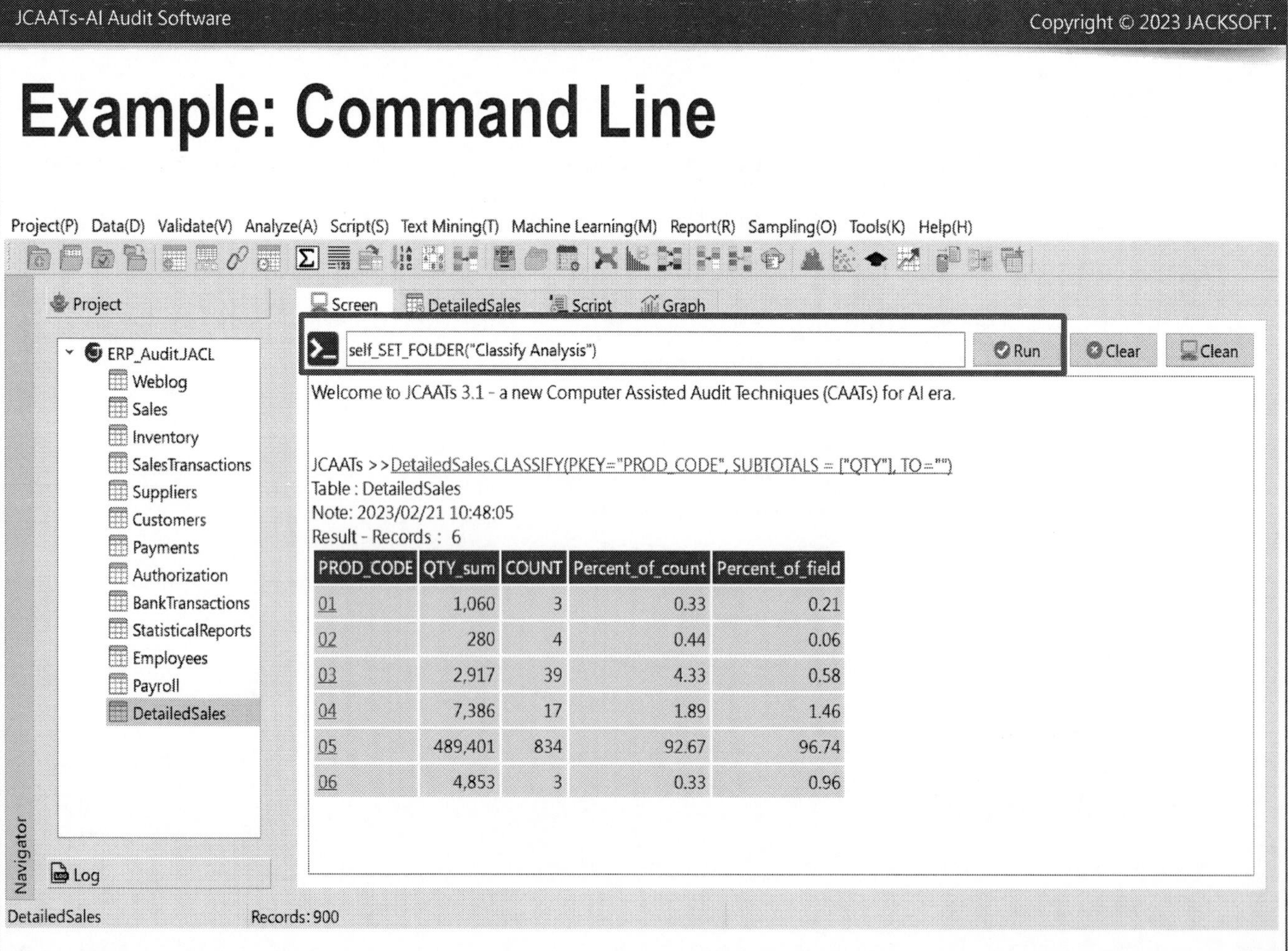

PROD_CODE	QTY_sum	COUNT	Percent_of_count	Percent_of_field
01	1,060	3	0.33	0.21
02	280	4	0.44	0.06
03	2,917	39	4.33	0.58
04	7,386	17	1.89	1.46
05	489,401	834	92.67	96.74
06	4,853	3	0.33	0.96

JCAATs - AI Audit Software

Chapter 3 - Exercise

Chapter 3 - Exercise

( ) 3-1 Which of the following data types is included in JCAATs?

a) TEXT
b) BOOLEAN
c) NUMERIC
d) DATETIME
e) All of the above

( ) 3-2 Which of the following statements regarding the JCAATs data read method is incorrect?

a) The data read method determines whether the resulting JCAATs table reads data from a JCAATs data file (.jfil) or directly from the data source.
b) Data in a .jfil file is static and must be manually updated, whereas data sources that are accessed directly are updated with the most current information each time the JCAATs table is opened.
c) JCAATs data files (.jfil) created from imported data are also treated as read-only by JCAATs. JCAATs cannot alter .jfil files, except for refreshing the file from the data source.
d) .jfil files are completely separate from the data source used to create them. Therefore, deleting a .jfil file does not affect the data source.
e) JCAATs tables read directly from the data source data can add, update, or delete data in a data source.

Chapter 3 - Exercise

( ) 3-3 What is not a factor influencing the sample size for monetary unit sampling in JCAATs?

a) Upper Error Limit
b) Materiality
c) Population
d) Expected Total Errors
e) Confidence level

( ) 3-4 Which of the following is NOT one of the new audit commands in JCAATs?

a) Fuzzy Duplicate
b) Fuzzy Join
c) Text Cloud
d) Neural Network
e) Keyword

JCAATs Learning notes：

Python Based Computer-Assisted Audit Techniques (CAATs)

Data Analysis and Smart Audit

Chapter 4 . Data

Outline:

1. How to obtain data
2. Import Data
3. Example: Import Employees.csv
4. Copy From and Link New Source
5. Table Layout and History

JCAATs - AI Audit Software

1. How to obtain data

How to Obtain Data

- Partner with IT
- Educate yourself and your staff
- Identify available data:
  - Meet with IS and Auditee.
  - Review data dictionary
  - Obtain reports from area being analyzed
  - Meet with data entry personnel
- Determine available data formats
  - ACL Project File (*.acl)
  - Excel, ODS (*.xlsx, *.xls, *.ods)
  - ODBC-compliant data sources
  - Fix Length Flat File
  - PDF files
  - PDF Table files
  - Delimited files
  - XML, JSON
  - OPEN DATA
  - Optional API (Ex. FTP、Email、SAP)

Access Data

- Determine your objectives
- Request the data
  - Data request letter
  - Summary Report
  - Record Layout
  - Control Totals
- Transferring the data
  - Access to production database
  - Access to a copy of the data
  - User access to source data
- **Open data**
  - Access Web data file
  - Access data layout
  - Download source data

Retrieve Data

- **Create a JCAATs project and create a data table format to retrieve data.**
- **JCAATs provides a powerful function to input a single data file or multiple data files simultaneously.**
- **JCAATs has two ways to retrieve data:**
- IMPORT DATA
  - Automatic layout
  - Manual layout
  - External definition (ODBC, API....)
- **Copy From and Link New Source**
  - Another JCAATs Projects
  - ACL Projects

JCAATs - AI Audit Software

2. Import Data

IMPORT Data Steps

JCAATs mainly divided into five steps to import a data file:

① Platform ② File Type ③ Property ④ Field ⑤ Finish

Step 1 : **Platform** - Define the way to import data sources.

Step 2 : **File Type** - Confirm the format of file type.

Step 3 : **Property** - Define data's encoding type, starting line, etc.

Step 4 : **Field** - Define data type, display format, etc.

Step 5 : **Finish** - Confirm file path and table name.

Data Source Platform

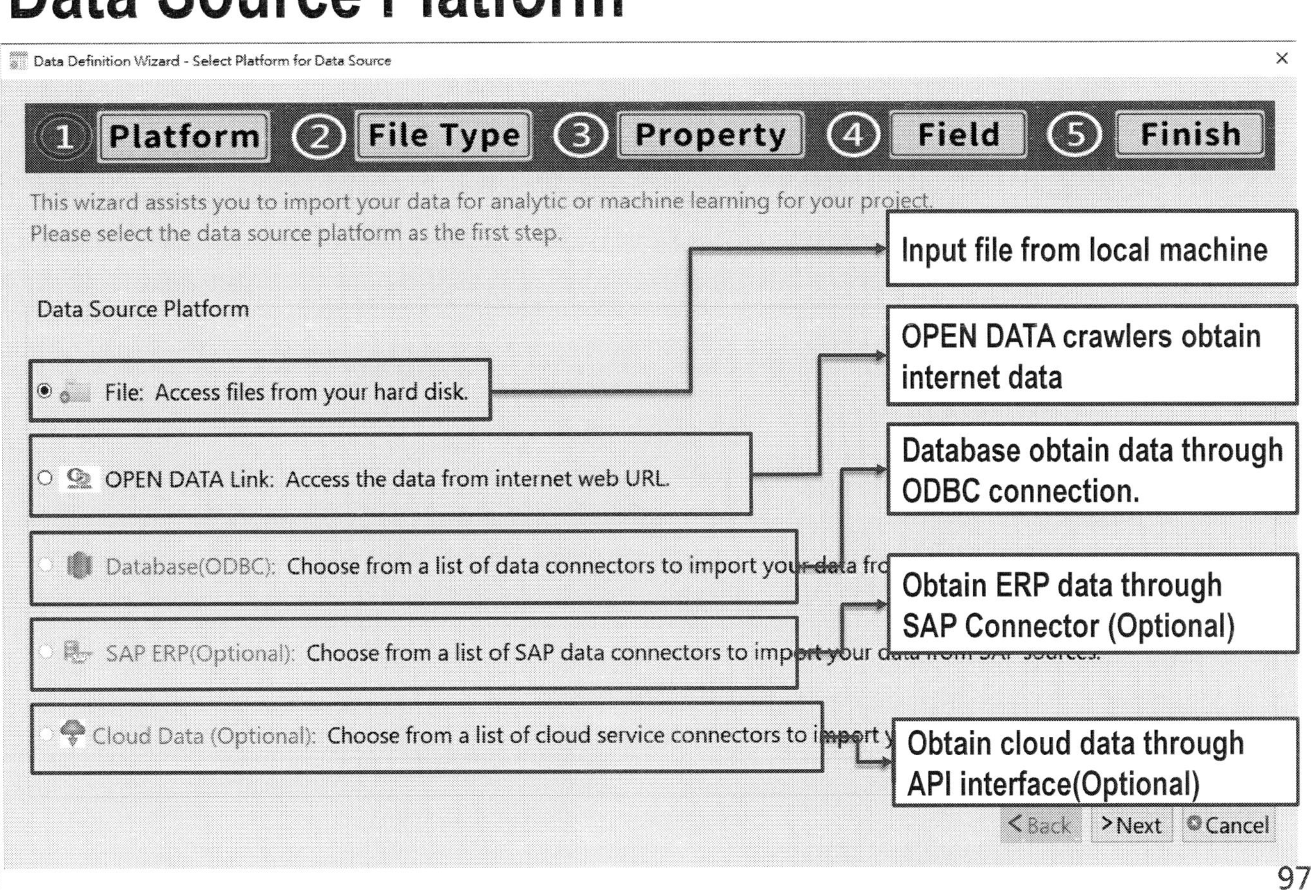

Platform: Import files from Local Machine

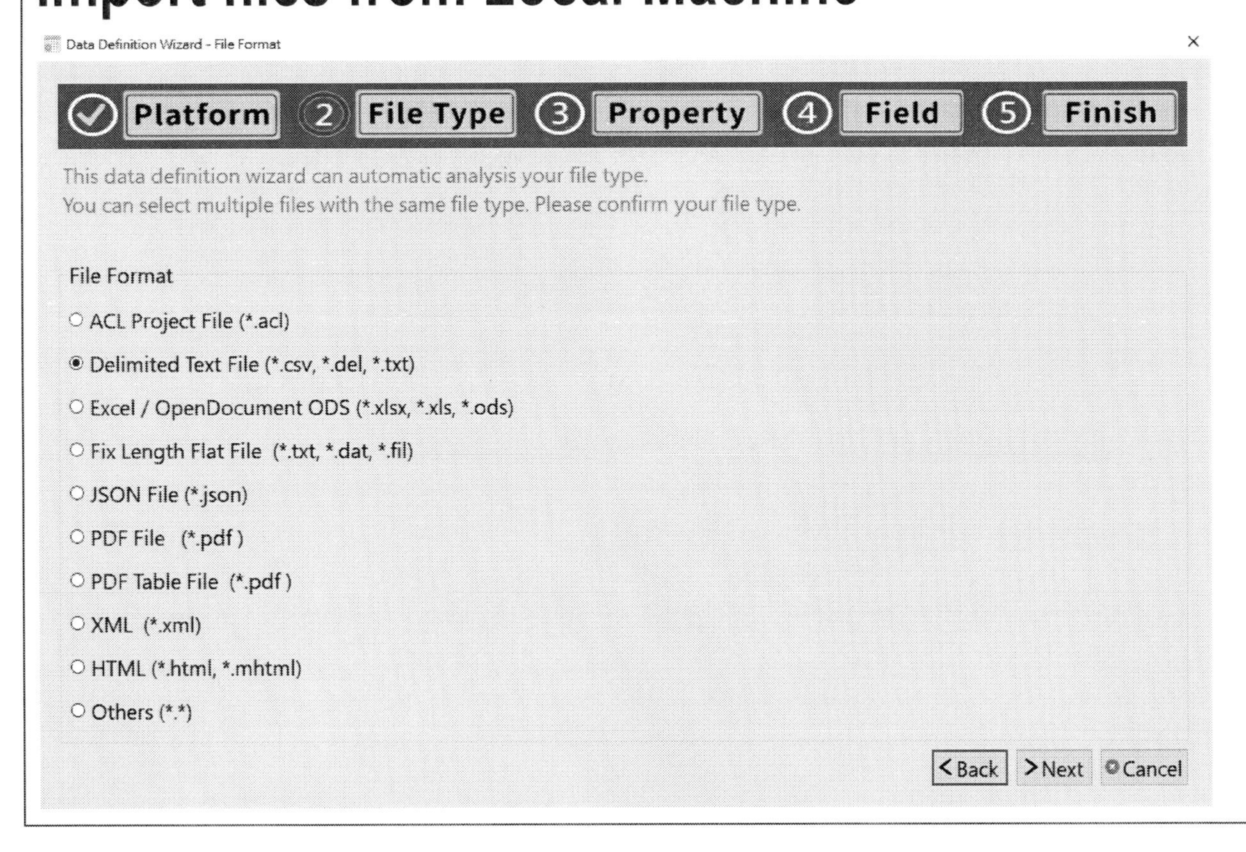

Import Data File Types

1. **ACL:** Select the ACL project file (*.acl) to import tables into the project.
2. **Delimited:** Select a Delimited Text File. (*.csv, *.del, *.txt)
3. **MS Excel/OpenDocument:** Select a spreadsheet file such as Excel or ODS (*.xls, *.xlsx, *.ods).
4. **Flat File:** Select a fixed-length flat file (*.txt, *.dat, *.fil) and define the input data format using the fixed-length flat file.
5. **JavaScript (JSON):** Select a JSON format data file (*.json) and define the input data format using the JSON standard.
6. **PDF:** Select a PDF format data file (*.pdf), and import each line of text into the project as a text format.
7. **PDF-Table:** Select a PDF format data file (*.pdf). Files can be identified automatically by AI and imported into the project in table format.
8. **XML:** Select a XML format data file (*.xml) and define the input data format using XML standard.
9. **HTML:** Select a HTML format data file (*.html, *.mhtml) and define the input data format of a table on the web file.
10. **Other:** Other non-standard formats.

Input multiple data files

- **Import multiple files:**
  - JCAATs provides a function to import multiple files simultaneously, which can greatly save time and improve work efficiency.
  - During the import process, you will need to define each file format.
- **The applicability of reusing Table Format:**
  - The file has the same table structure.
  - Receive a file that has the same table structure regularly.
- **The steps to reuse Table Format:**
  - **Copy** the definition of the same table format with different table name.
  - **Link** a new source data rom another JCAATs project.

ODBC Connect to Database

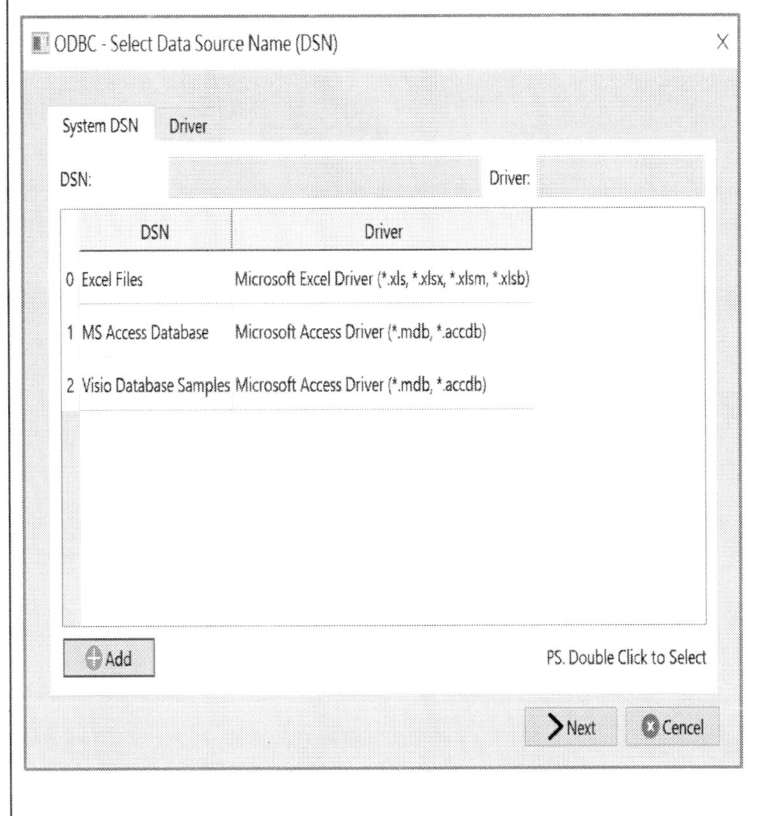

1. **JCAATs uses a 64 bits ODBC system.**
2. Lists the configured ODBC data sets and drivers on the current computer system.
3. Recommends users to use Microsoft ODBC 64-bit to manage platform downloads, drivers, and set the DSN data source.
4. The database data dictionaries that JCAATs currently connect to via ODBC are: Access, MS SQL, MySQL, Oracle, DB2, Teradata, and others. Users can add other database dictionaries by adding modules.

OPEN DATA

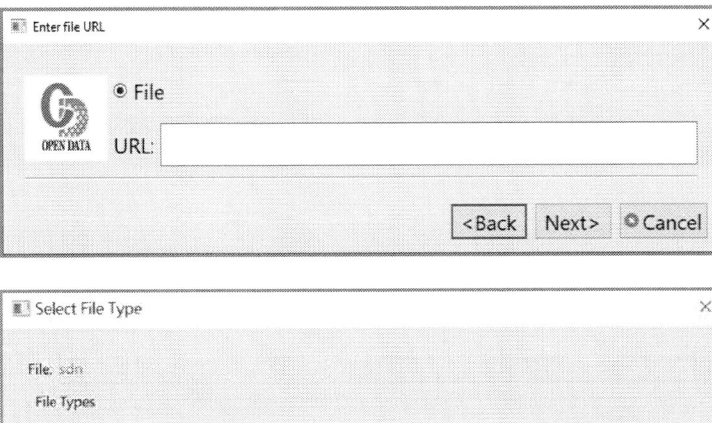

Select File Type

File: sdn

File Types

- Delimited Text File (*.csv, *.del)
- Excel / OpenDocument ODS (*.xlsx, *.xls, *.ods)
- Fix Length Flat File (*.txt, *.dat, *.fil)
- JSON File (*.json)
- PDF File (*.pdf)
- PDF Table File (*.pdf)
- XML (*.xml)
- Others

<Back Next> Cancel

1. JCAATs provides a specified URL data file import function. Users need to confirm file's data type.
2. **JCAATs professional edition provides a web crawler function.** All files of the same type in the URL specified by the user can be imported into JCAATs at once.
3. Please ensure that the internet connection is smooth and you are able to connect to the web address in order to use this function.
4. Since there is generally a large amount of data on the internet, a large network bandwidth and time are required to download the data.

JCAATs - AI Audit Software

Step 1: Platform

- Select "Data" and choose "New Table"
- Choose "File" as the data source platform
- Click "Next"

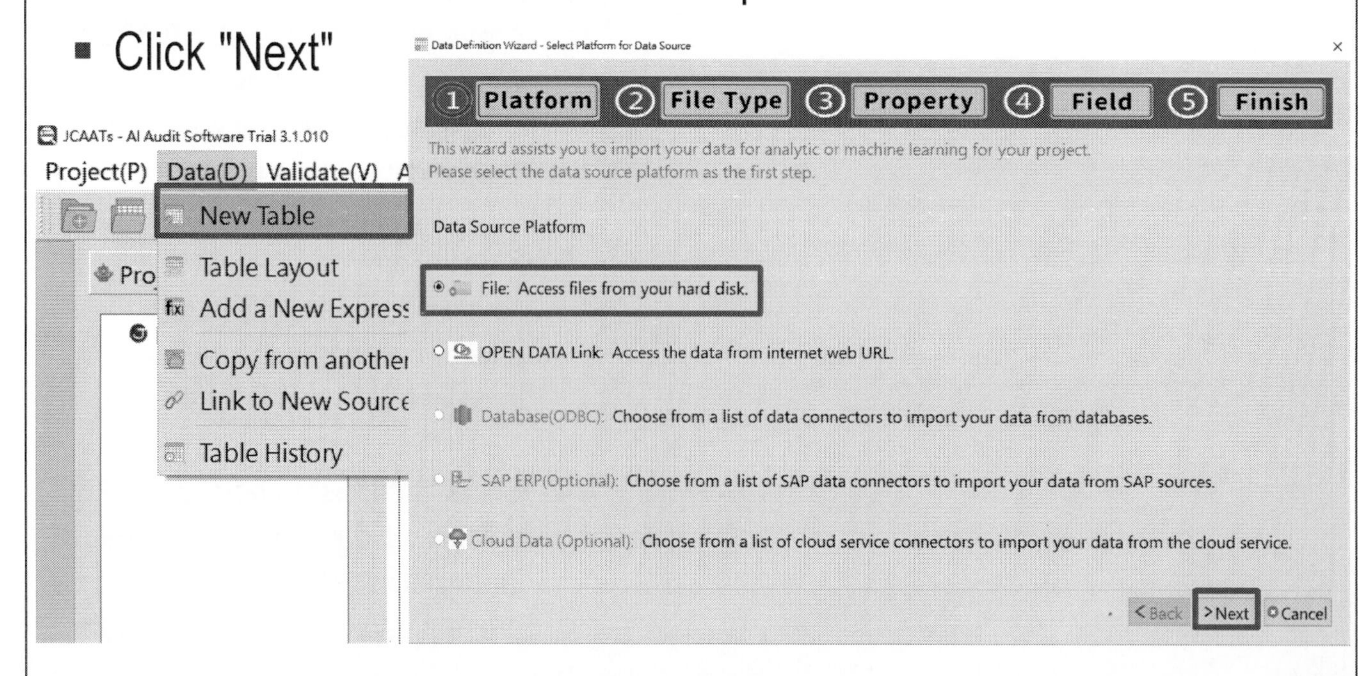

Step 2: File Type - Open Data File

- Open the data file and select "Employees.csv".
- Click "Open"

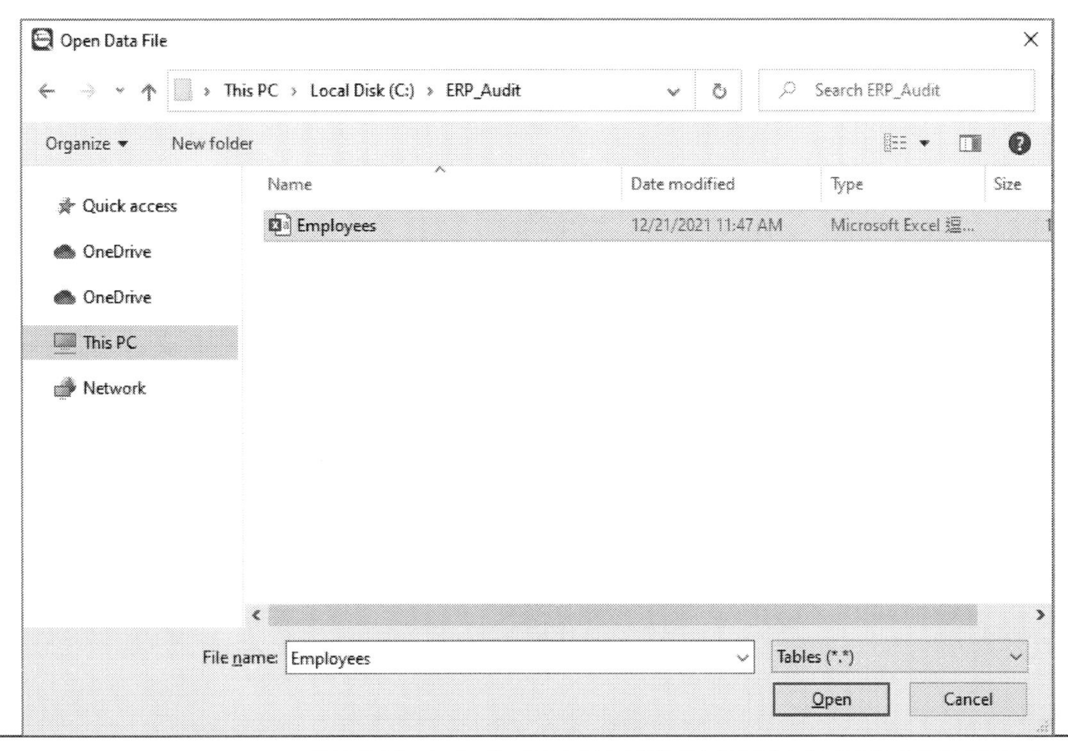

Step 2: File Type - File Format

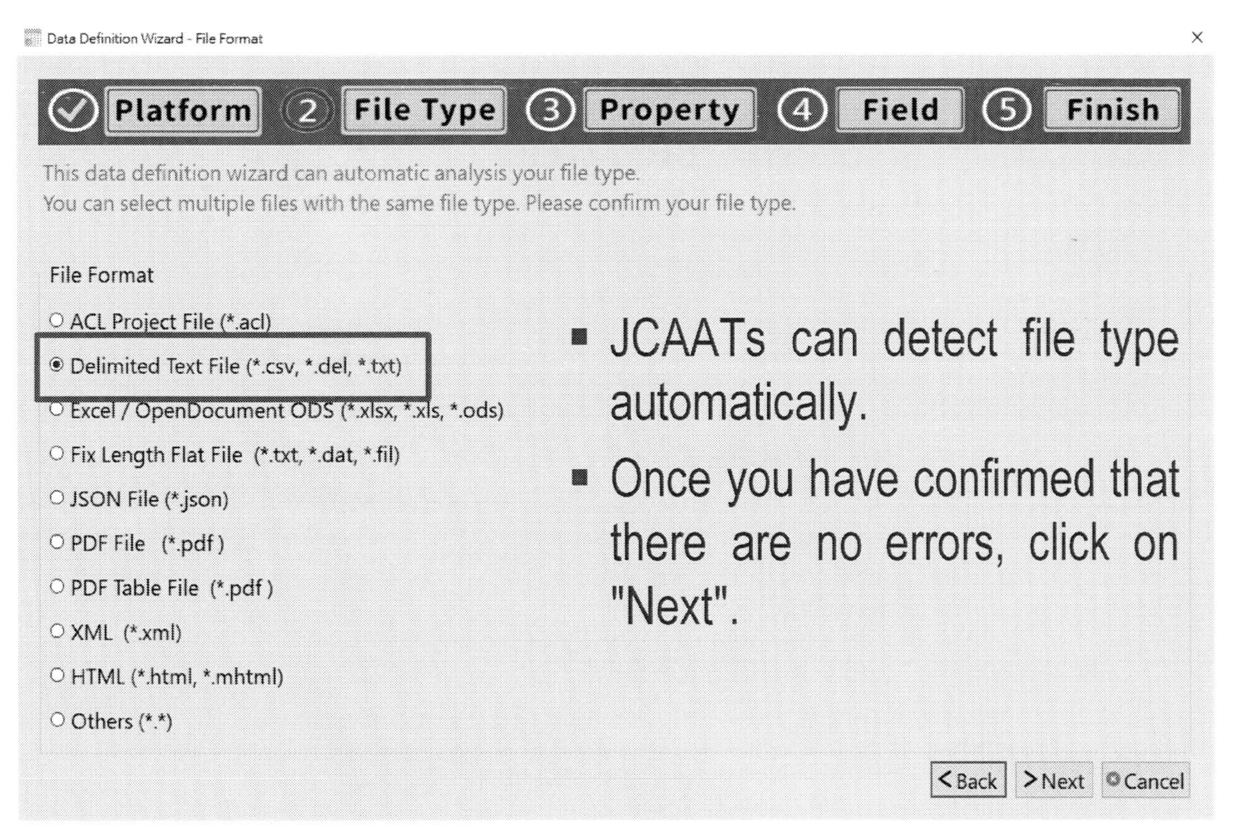

- JCAATs can detect file type automatically.
- Once you have confirmed that there are no errors, click on "Next".

Step 3: Property

- **JCAATs will automatically detect the character encoding method of the file.** Unless an error is confirmed, it is recommended to use the system's result to avoid errors.
- **Confirm the number of rows.**
- **Confirm that the first column is the field name.**
- **Confirm the delimiter.**
- If there are multiple files, you can select the file name and set the data characteristics separately.
- **Click "Next".**

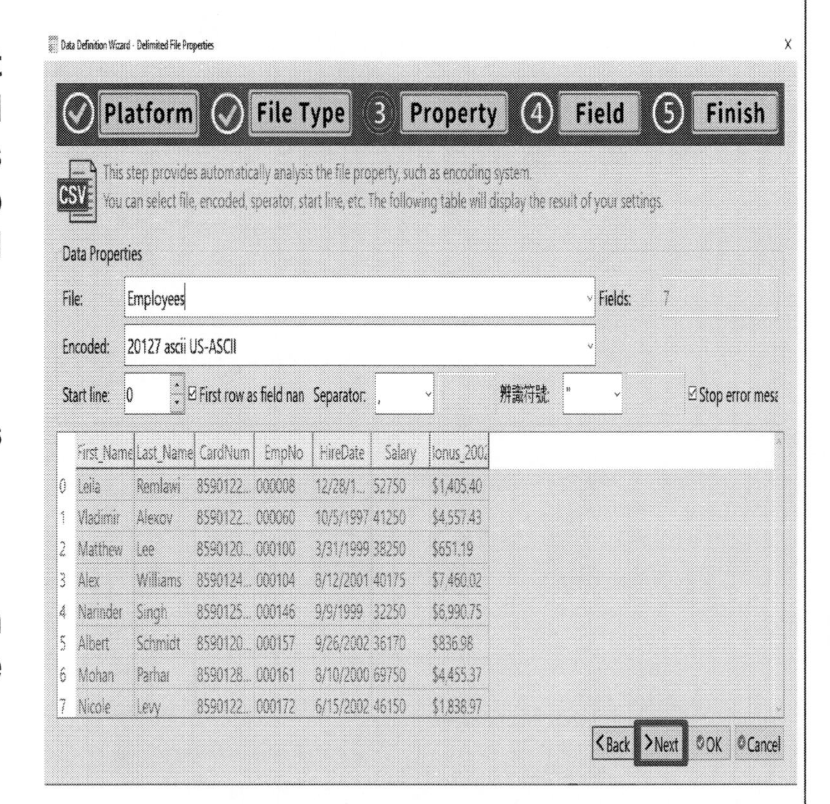

Step 4: Field

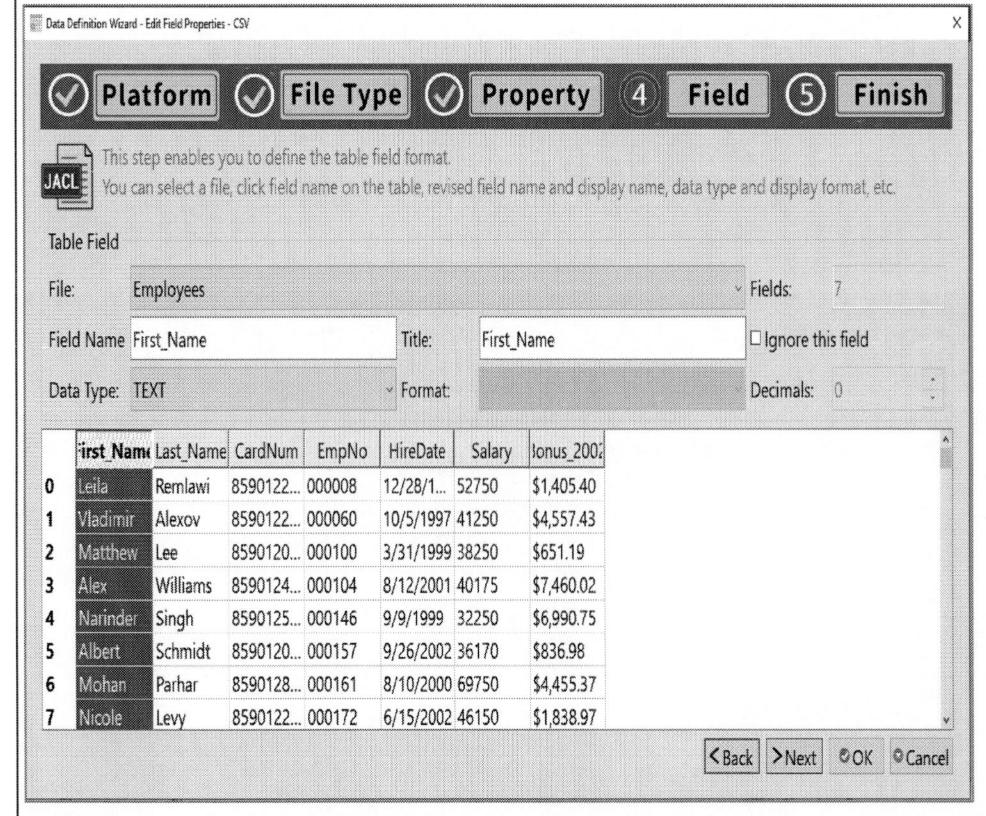

- Set up each field on JCAATs, such as the table name, field, data type, format, and others.
- Click on each field name to access different field settings.
- If fields are not set, JCAATs will set them to text fields.
- After setting up, click "Next".

Step 5: Finish

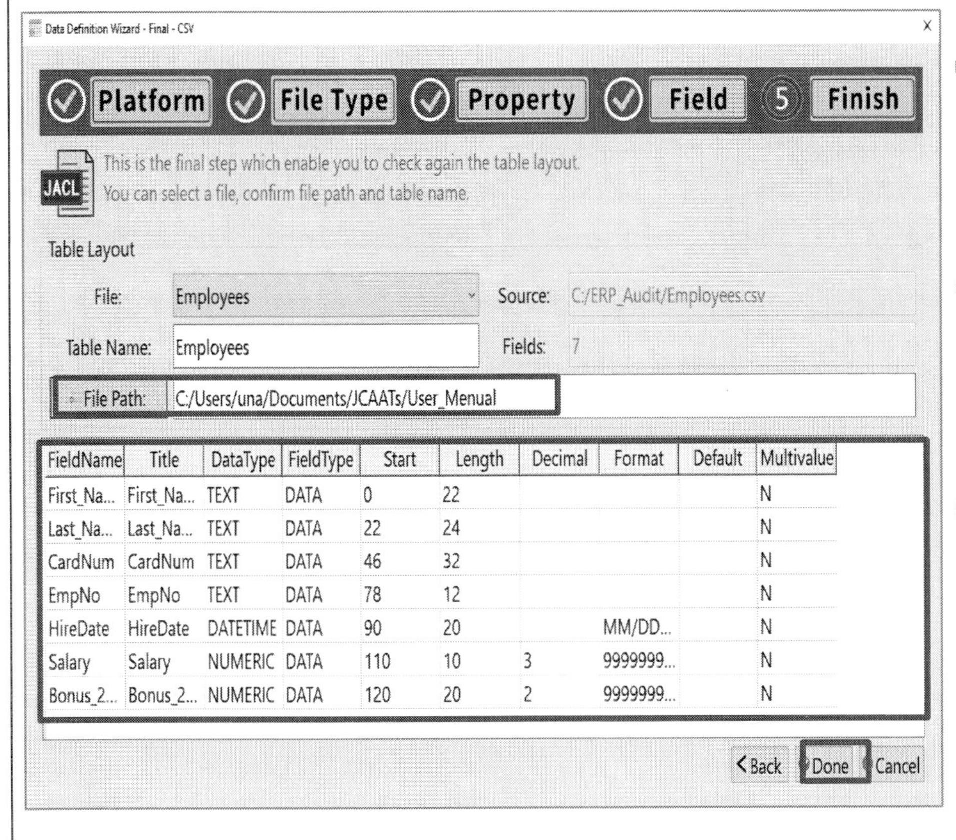

FieldName	Title	DataType	FieldType	Start	Length	Decimal	Format	Default	Multivalue
First_Na...	First_Na...	TEXT	DATA	0	22				N
Last_Na...	Last_Na...	TEXT	DATA	22	24				N
CardNum	CardNum	TEXT	DATA	46	32				N
EmpNo	EmpNo	TEXT	DATA	78	12				N
HireDate	HireDate	DATETIME	DATA	90	20		MM/DD...		N
Salary	Salary	NUMERIC	DATA	110	10	3	9999999...		N
Bonus_2...	Bonus_2...	NUMERIC	DATA	120	20	2	9999999...		N

- Confirm **the name of the data** file on JCAATs.
- The **default data storage path** is the project folder.
- **Confirm the file format and type**. If there are no mistakes, select "Done".

Complete Import Employees.csv

- You can confirm that the table has been successfully imported into JCAATs by checking the display until the waiting process is finished.

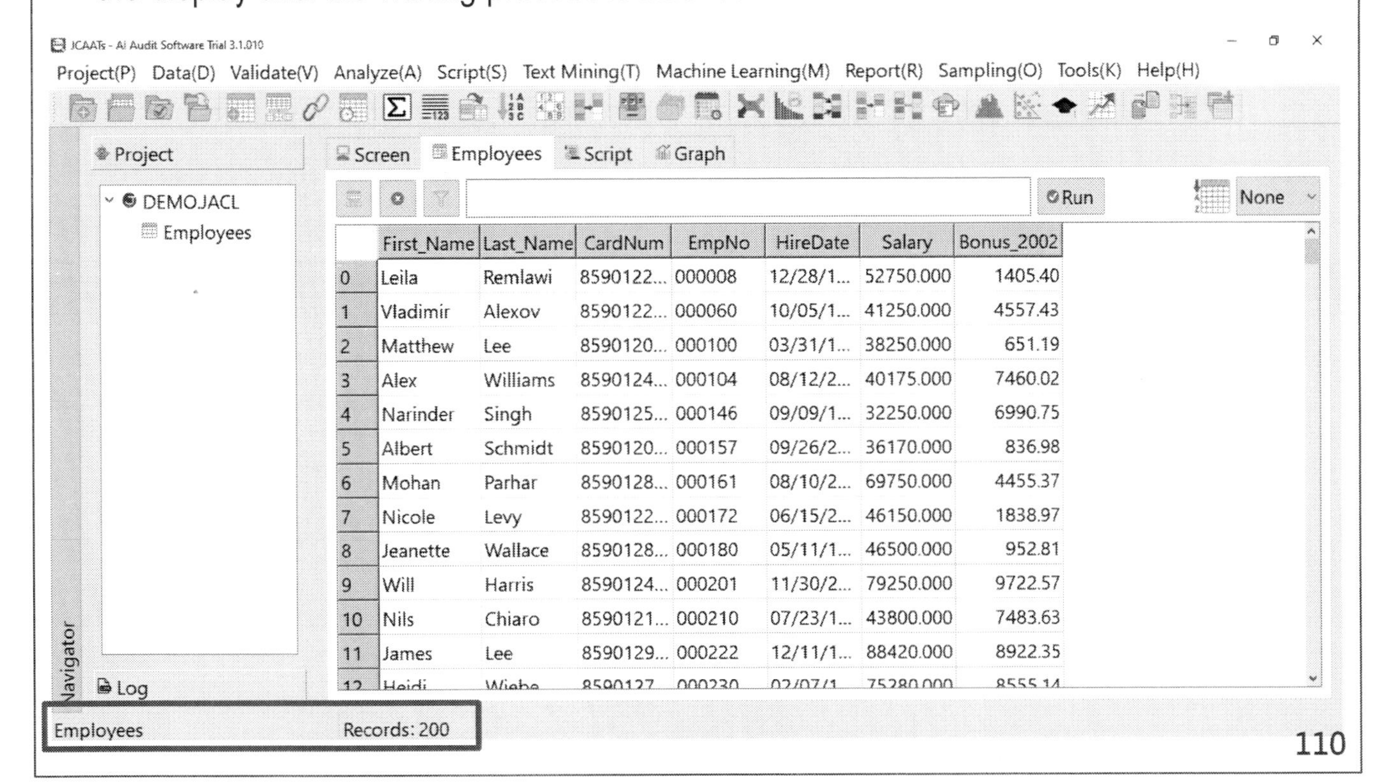

	First_Name	Last_Name	CardNum	EmpNo	HireDate	Salary	Bonus_2002
0	Leila	Remlawi	8590122...	000008	12/28/1...	52750.000	1405.40
1	Vladimir	Alexov	8590122...	000060	10/05/1...	41250.000	4557.43
2	Matthew	Lee	8590120...	000100	03/31/1...	38250.000	651.19
3	Alex	Williams	8590124...	000104	08/12/2...	40175.000	7460.02
4	Narinder	Singh	8590125...	000146	09/09/1...	32250.000	6990.75
5	Albert	Schmidt	8590120...	000157	09/26/2...	36170.000	836.98
6	Mohan	Parhar	8590128...	000161	08/10/2...	69750.000	4455.37
7	Nicole	Levy	8590122...	000172	06/15/2...	46150.000	1838.97
8	Jeanette	Wallace	8590128...	000180	05/11/1...	46500.000	952.81
9	Will	Harris	8590124...	000201	11/30/2...	79250.000	9722.57
10	Nils	Chiaro	8590121...	000210	07/23/1...	43800.000	7483.63
11	James	Lee	8590129...	000222	12/11/1...	88420.000	8922.35

JCAATs - AI Audit Software

4. Copy From and Link New Source

Reuse the existing project tables

- JCAATs provides a function to Copy From Another Project tables which includes JCAATs projects and ACL projects.
- This will can greatly save time and improve work efficiency.
- The data security also can be more controlled.
- **The steps to reuse Table Format:**
  - **Step 1: Copy from Another Project** ◦ **Input** table format from other JCAATs projects or ACL Projects.
  - **Step 2: Link a New Data Source** ◦ The table format to new source data which can be JCAATs data source file (*.JFIL) or ACL data source file (*.fil).

Step 1: Select Project (JCAATs or ACL)

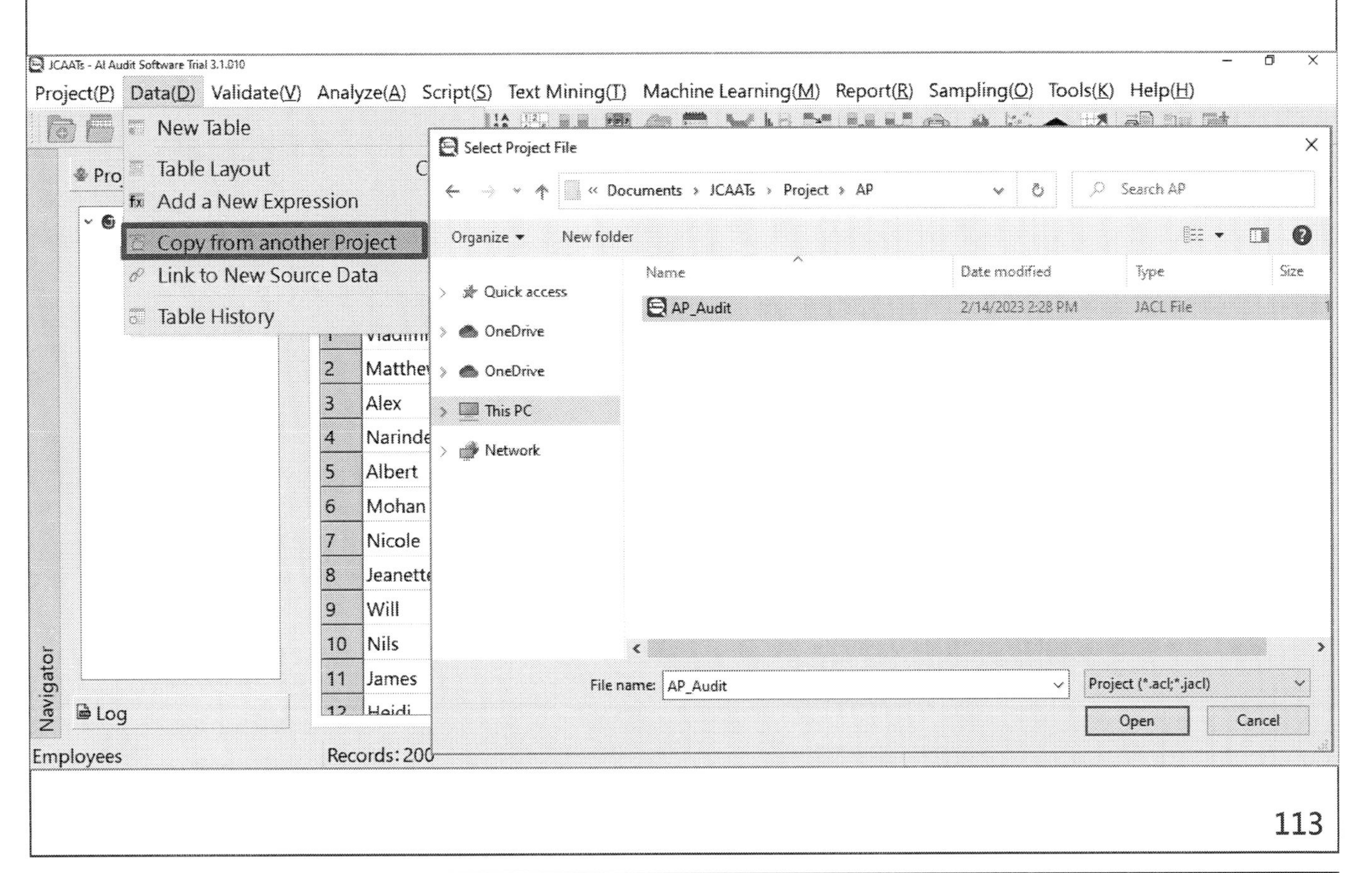

Step 1: Select Tables (multiple selection)

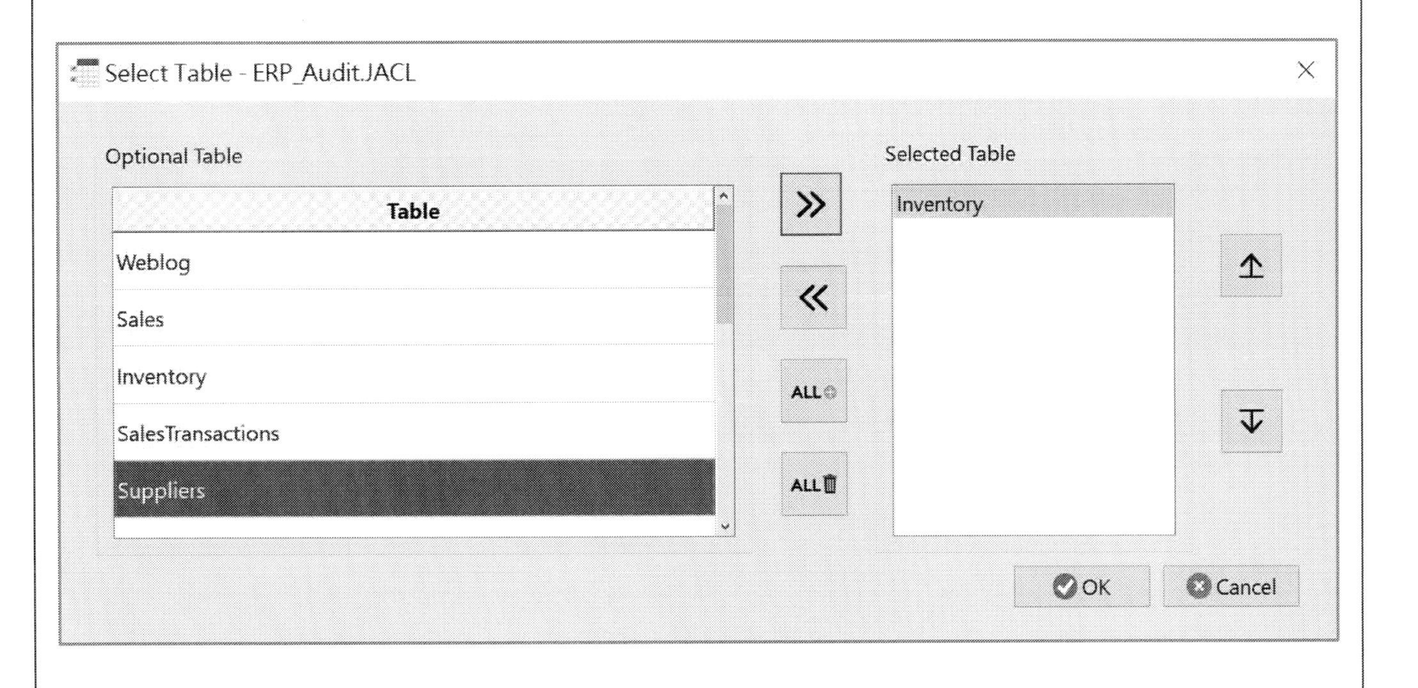

Step 1: Import Table Layout

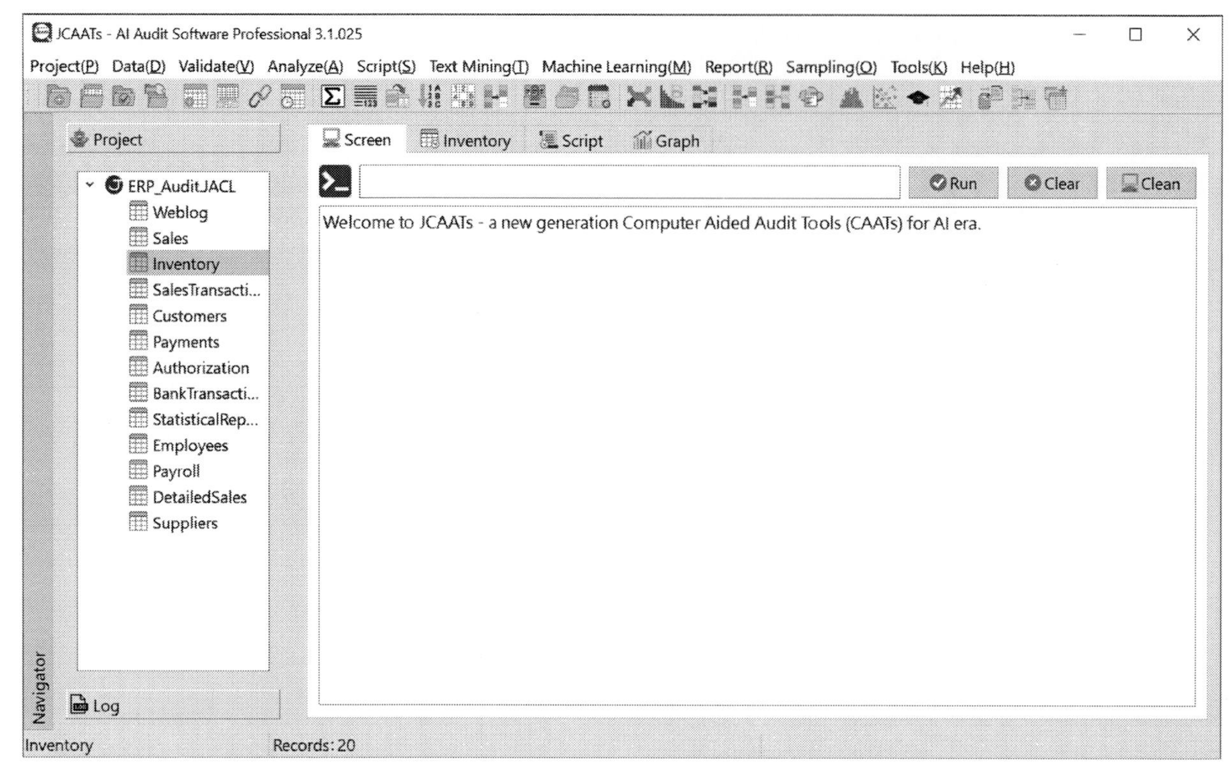

Step 2:
Select a Data Source File(*.JFIL or *.Fil)

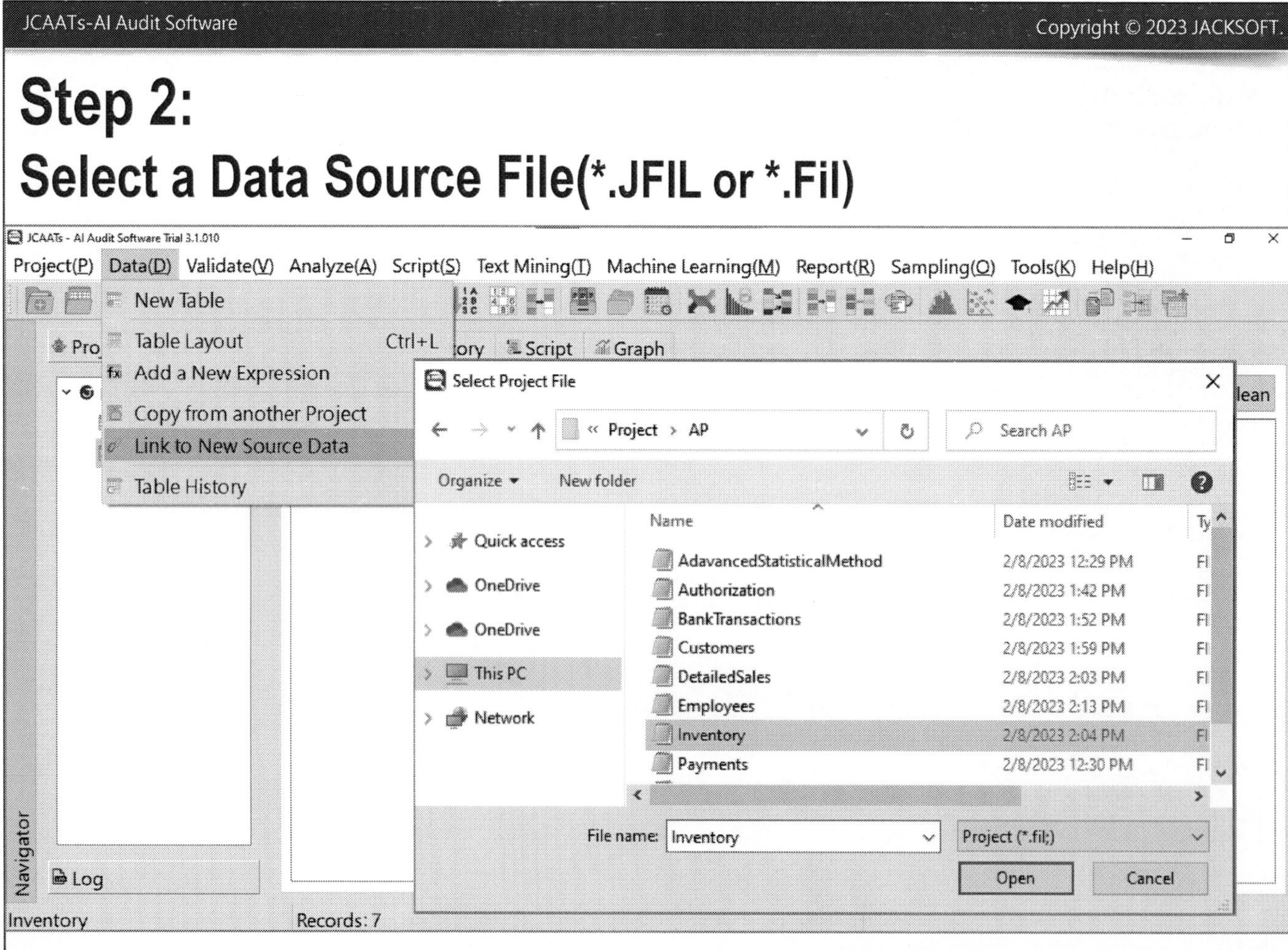

Step 2: The Result

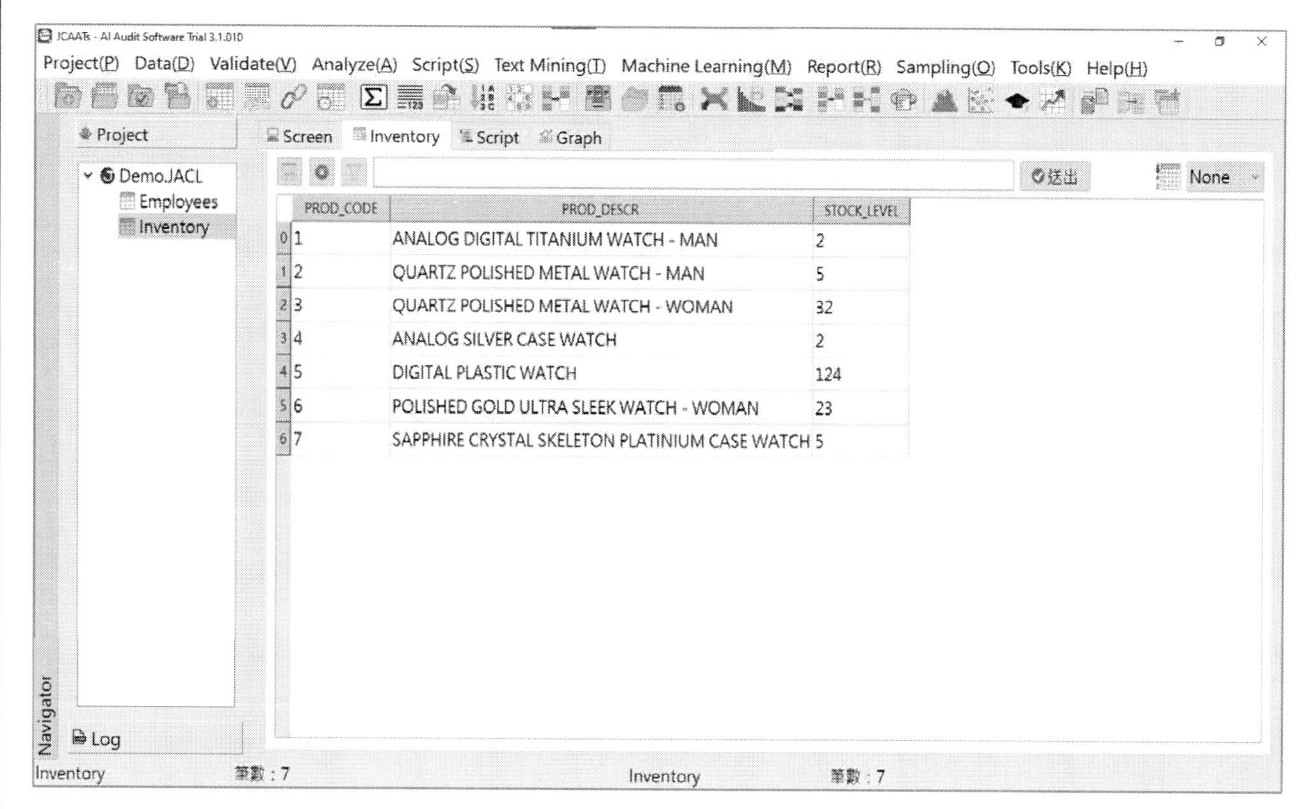

	PROD_CODE	PROD_DESCR	STOCK_LEVEL
0	1	ANALOG DIGITAL TITANIUM WATCH - MAN	2
1	2	QUARTZ POLISHED METAL WATCH - MAN	5
2	3	QUARTZ POLISHED METAL WATCH - WOMAN	32
3	4	ANALOG SILVER CASE WATCH	2
4	5	DIGITAL PLASTIC WATCH	124
5	6	POLISHED GOLD ULTRA SLEEK WATCH - WOMAN	23
6	7	SAPPHIRE CRYSTAL SKELETON PLATINIUM CASE WATCH	5

JCAATs - AI Audit Software

Where is Table Layout Command?

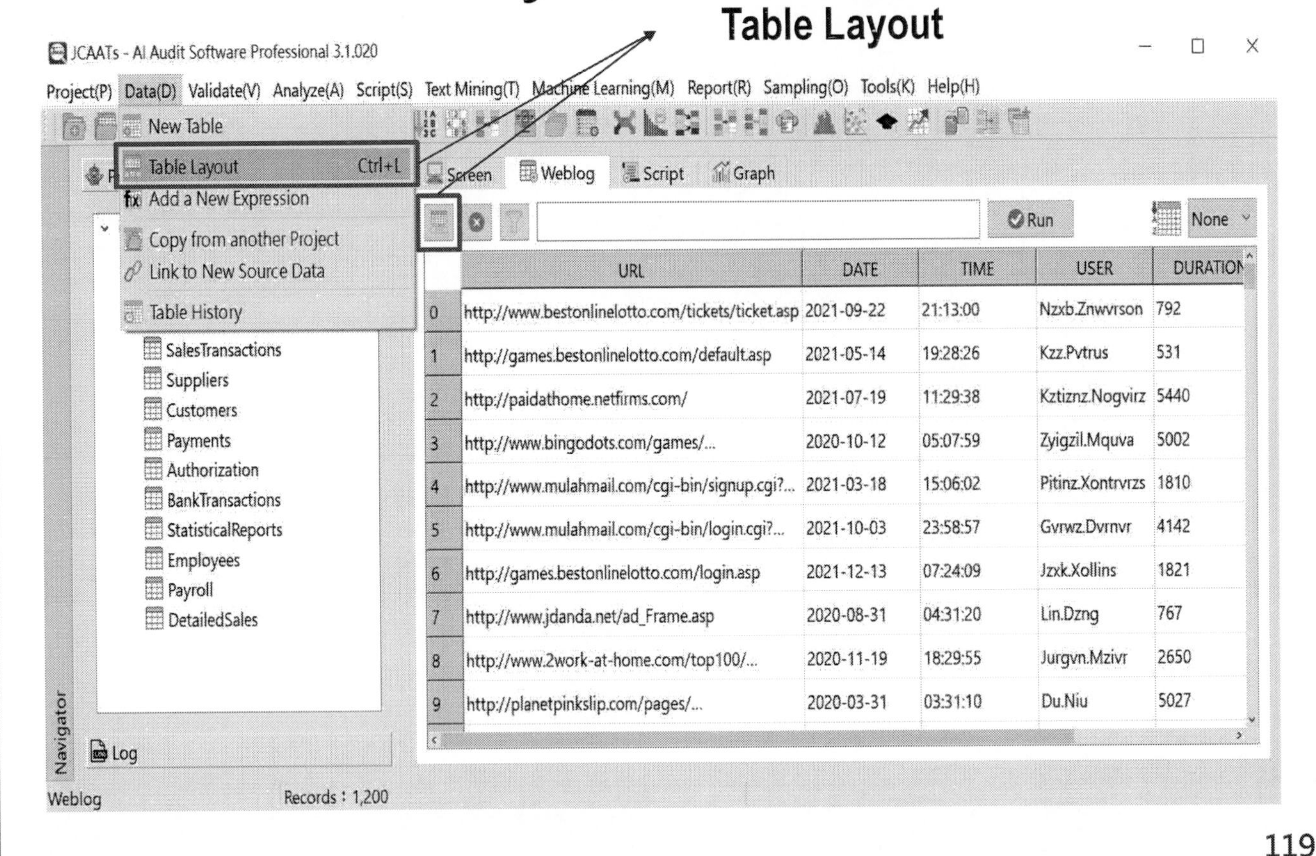

	URL	DATE	TIME	USER	DURATION
0	http://www.bestonlinelotto.com/tickets/ticket.asp	2021-09-22	21:13:00	Nzxb.Znwvrson	792
1	http://games.bestonlinelotto.com/default.asp	2021-05-14	19:28:26	Kzz.Pvtrus	531
2	http://paidathome.netfirms.com/	2021-07-19	11:29:38	Kztiznz.Nogvirz	5440
3	http://www.bingodots.com/games/...	2020-10-12	05:07:59	Zyigzil.Mquva	5002
4	http://www.mulahmail.com/cgi-bin/signup.cgi?...	2021-03-18	15:06:02	Pitinz.Xontrvrzs	1810
5	http://www.mulahmail.com/cgi-bin/login.cgi?...	2021-10-03	23:58:57	Gvrwz.Dvrnvr	4142
6	http://games.bestonlinelotto.com/login.asp	2021-12-13	07:24:09	Jzxk.Xollins	1821
7	http://www.jdanda.net/ad_Frame.asp	2020-08-31	04:31:20	Lin.Dzng	767
8	http://www.2work-at-home.com/top100/...	2020-11-19	18:29:55	Jurgvn.Mzivr	2650
9	http://planetpinkslip.com/pages/...	2020-03-31	03:31:10	Du.Niu	5027

Table Layout

Display the data structure through Table Layout

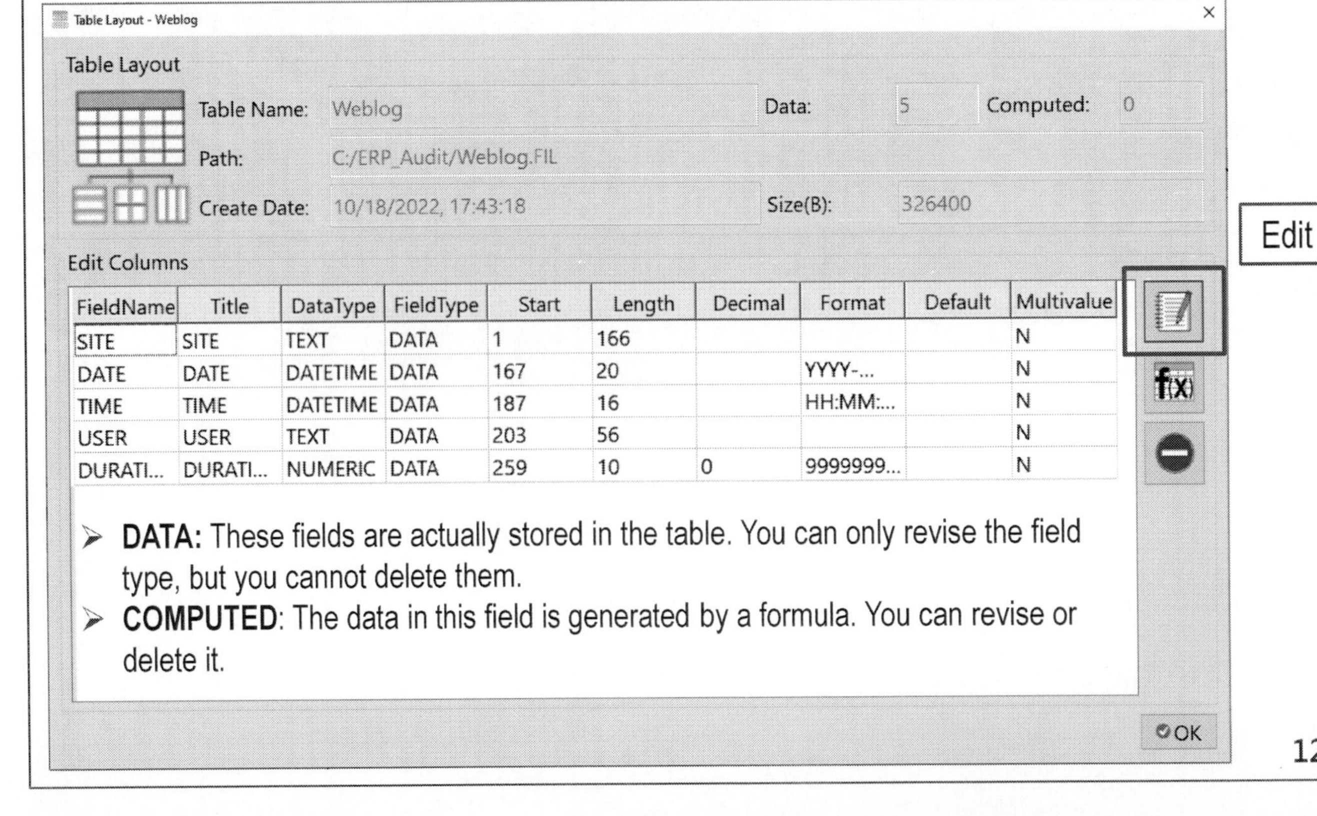

FieldName	Title	DataType	FieldType	Start	Length	Decimal	Format	Default	Multivalue
SITE	SITE	TEXT	DATA	1	166				N
DATE	DATE	DATETIME	DATA	167	20		YYYY-...		N
TIME	TIME	DATETIME	DATA	187	16		HH:MM:...		N
USER	USER	TEXT	DATA	203	56				N
DURATI...	DURATI...	NUMERIC	DATA	259	10	0	9999999...		N

- **DATA:** These fields are actually stored in the table. You can only revise the field type, but you cannot delete them.
- **COMPUTED**: The data in this field is generated by a formula. You can revise or delete it.

Table Layout - Edit

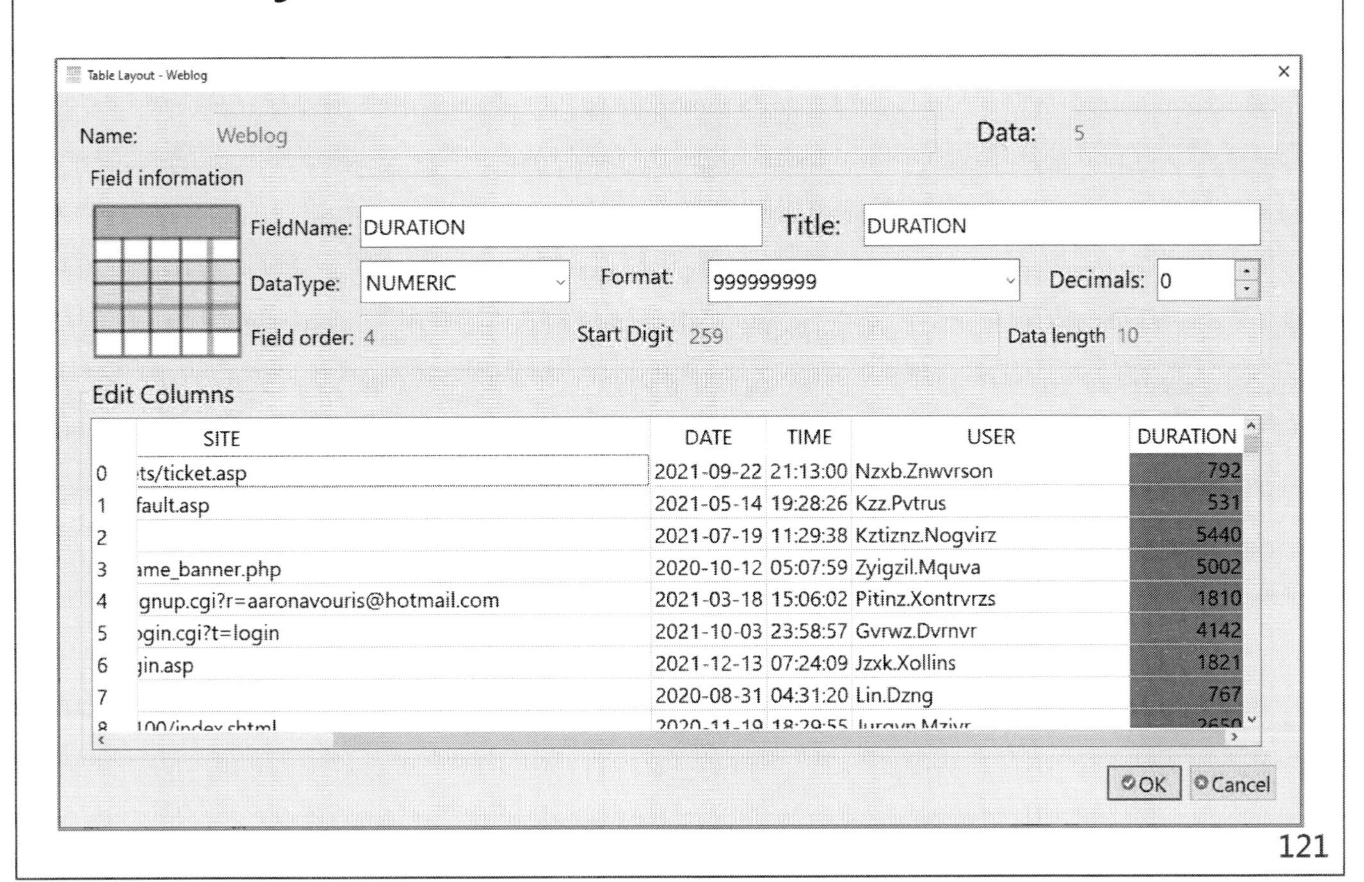

JCAATs-AI Audit Software
Copyright © 2023 JACKSOFT.

Table History

- Display the time at which this dataset was created and the command that was generated for this project.

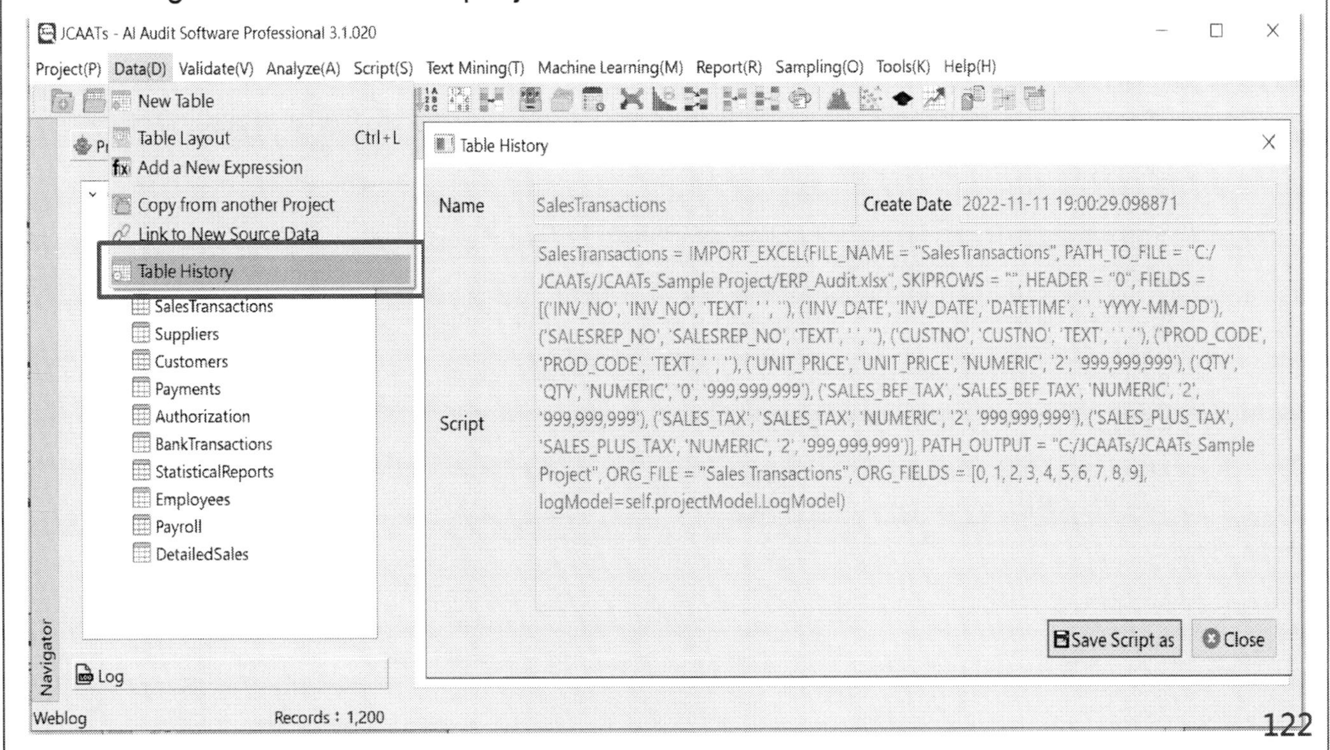

JCAATs - AI Audit Software

Chapter 4 - Exercise

Chapter 4 - Exercise

( ) 4-1 Which of the following procedures is the best way to copy an existing JCAATs table from another JCAATs project into the current JCAATs project?

a) Export the table from the other JCAATs project using the EXPORT command and import it into the current JCAATs project using the IMPORT command.

b) Use the "Link to New Source Data" function in the current JCAATs project to select the table data source file.

c) Copy the table from another JCAATs project and paste it into the current project.

d) Select "Copy from Another Project > Table" in the current JCAATs project to choose another JCAATs project table, and then use "Link to New Source Data" to link to another data source file.

e) Export the table from the other JCAATs project using the "Export to Another Project > Table" option, and then use "Link to New Source Data" in the current JCAATs project to select the table data source file.

Chapter 4 - Exercise

( ) 4-2 When importing the following data source using JCAATs, which file would provide the table layout definition automatically?

a) DB2
b) dBase File
c) Excel File / Access File
d) ODBC Source file
e) All of the above

( ) 4-3 When auditors use ODBC middleware to import audit data, which of the following are they dealing with?

a) The driver
b) The domain names (DNS)
c) The data source name (DSN)
d) The website name (URL)
e) None of the above

Chapter 4 - Exercise

( ) 4-4 Which of the following statements is correct for delimited files?

a) A delimited file is a flat file used to store data, where each line uses spaces to force every field to the same width.
b) A delimited file is a text file used to store data, where each line has fields separated by the delimiter.
c) A delimited file must contain field names in the first row.
d) A delimited file uses commas, tabs, or other characters as delimiters to separate the values.
e) Each line of text is separated by a line feed (LF) character.

( ) 4-5 It is important to verify the data's date population to start your audit project. How is the data range in JCAATs determined?

a) DATE Command
b) PROFILE Command
c) STATISTICS Command
d) AGE Command
e) CHART Command

Chapter 4 - Exercise

( ) 4-6 Which JCAATs command should be used to remove one or more specified substrings from a string?

a) str.len() function
b) str.replace() function
c) str.upper() function
d) str.capitalize() function
e) None of the above

( ) 4-7 What method does JCAATs use for numerical calculations?

a) Two-point operation
b) Fixed point arithmetic
c) Floating-point operation
d) Point operation
e) None of the above

Chapter 4 - Exercise

( ) 4-8 The Profile command provides summary statistics for one or multiple numeric fields. Which of the following is true about the information provided from the command?

a) Total value
b) Absolute value
c) Minimum and Maximum value
d) Mean value
e) All of the above

( ) 4-9 Which of the following methods is not commonly used by auditors to assess the expected total errors of a parent?

a) Analyzing sample experience values based on previous years
b) Conducting a preliminary assessment of the control value for the year
c) Deriving a pre-estimate from a single test sample
d) Setting the expected total errors on a random basis
e) Checking that the expected total errors are lower than the materiality threshold

Chapter 4 - Exercise

( ) 4-10 Which of the following JCAATs commands is unable to generate an analysis chart?

a) Summarize
b) Stratify
c) Classify
d) Age
e) All of the above.

( ) 4-11 What is text mining?

a) The process of turning structured text into unstructured text
b) The process of transforming unstructured text into a structured format to uncover patterns and insights
c) The process of analyzing unstructured text to create new patterns and insights
d) The process of transforming text into a random format
e) All of the above

Chapter 4 - Exercise

( ) 4-12 What is NLP in Artificial Intelligent?

a) The process of analyzing numerical data
b) The process of transforming structured text into unstructured text
c) The process of transforming unstructured text into a structured format to uncover patterns and insights
d) The field of computer science and AI that focuses on interactions between computers and human language
e) Normal Logistic Process

( ) 4-13 What are the Machine Learning commands offered by JCAATs?

a) Train, Predict, Outlier, and Cluster
b) Train, Evaluate, Clean, and Deploy
c) Train, Predict, Compare, and Evaluate
d) Train, Cluster, Evaluate, and Improve
e) None of the Above

Chapter 4 - Exercise

( ) 4-14 What is the purpose of the TEXTCLOUD command in JCAATs?
- a) To detect fraudulent activity in sales transactions
- b) To perform named entity recognition
- c) To create a graphical representation of text data
- d) To analyze sentiment in text data
- e) All of the above

( ) 4-15 What is a text cloud?
- a) A tool used to detect fraudulent activity in sales transactions
- b) A graphical representation of text data where the size of each word corresponds to its frequency or importance
- c) A command in JCAATs for named entity recognition
- d) A machine learning algorithm for sentiment analysis
- e) None of the Above

Chapter 4 - Exercise

( ) 4-16 Which of the following JCAATs commands can generate an XML output file?
- a) Summarize
- b) Report
- c) Export
- d) Classify
- e) Statistics

( ) 4-17 Which of the following statements is correct regarding the JCAATs Export Command?
- a) JCAATs can generate an XBRL report file.
- b) JCAATs can generate a Crystal Report file.
- c) JCAATs can generate an IBM EBCDIC file.
- d) JCAATs can generate a SAS file.
- e) None of the above.

Chapter 4 - Exercise

( ) 4-18 Which of the following JCAATs functions can be used to convert a date field into a text field?

a) dt.days
b) str.replace()
c) str.strip()
d) str.zfill()
e) astype(str)

( ) 4-19 I want to EXPORT a text field, Name, with a length of 20 and no spaces to the front. Which of the following is a correct JCAATs function to perform the result?

a) Name.str.strip(20)
b) Name.pad(20)
c) Name.str.pad(20,"right")
d) Name.str.zfill(20)
e) Name.str.isspace(20)

Chapter 4 - Exercise

( ) 4-20 Which of the following functions can be used to show interval data that a user wants to know?

a) str.isupper()
b) str.contains(pat, na=False)
c) between(min, max)
d) isin([values])
e) str.involve(min, max)

JCAATs Learning notes：

JCAATs-AI Audit Software
Copyright © 2023 JACKSOFT.

JCAATs Learning notes：

Python Based Computer-Assisted Audit Techniques (CAATs)

Data Analysis and Smart Audit

Chapter 5 . Expression

Outline:

1. Elements of Smart Audit
2. Expressions
3. Filters
4. Add a New Expression
5. Example: Add a Computed Fields

1. Elements of Smart Audit

Elements of Smart Audit

- Commands
- Expressions
  --Filters
  --Computed fields
- Functions
- Variables
- Dictionary
- Knowledge model

Elements of Smart Audit

- **Commands**
- Expressions
  --Filters
  --Computed fields
- Functions
- Variables
- Dictionary
- Knowledge model

Commands

- ➢ Predefined routines that can be used for Smart Analysis purposes.
- ➢ Some parameters are optional.
- ➢ Most commands are located under the menu bar or tool bar.
- ➢ Results can be sent to table, screen, or graph.

Elements of Smart Audit

- Commands
- **Expressions**
  --Filters
  --Computed fields
- Functions
- Variables
- Dictionary
- Knowledge model

Expressions

- ➢ Statements can be used to create filters or computed fields.
- ➢ Filters apply logical conditions to data, generating a subset that meets the specified criteria and is labeled as True or False.
- ➢ Computed fields execute expressions that generate values not explicitly stored in the table file.

Elements of Smart Audit

- Commands
- Expressions
  --Filters
  --Computed fields
- **Functions**
- Variables
- Dictionary
- Knowledge model

Functions

- ➢ Predefined routines are built-in functions that can be incorporated into expressions.
- ➢ Approximately 100 functions are available to achieve simple or complex objectives.

Elements of Smart Audit

- Commands
- Expressions
  --Filters
  --Computed fields
- Functions
- **Variables**
- Dictionary
- Knowledge model

Variables

- ➢ A named memory space is used to store data.
- ➢ It can be character, numeric, date, or logical type.

Elements of Smart Audit

- Commands
- Expressions
  --Filters
  --Computed fields
- Functions
- Variables
- **Dictionary**
- Knowledge model

Dictionary

- Dictionaries are commonly used for text mining.
- JCAATs has default dictionaries based on the English language.
- Users can upload their own dictionaries.
- A stop word dictionary may be used for tokenization purpose. Additionally, two types of dictionaries commonly used in sentiment analysis are positive and negative dictionaries.

Elements of Smart Audit

- Commands
- Expressions
  --Filters
  --Computed fields
- Functions
- Variables
- Dictionary
- **Knowledge model**

Knowledge model

- Knowledge models are used for Machine Learning.
- After executing the TRAIN command for supervised learning, JCAATs will generate a knowledge model with a .JKM file extension, which represents the training result.
- The user needs to select a knowledge model in order to perform the PREDICT command.

2. Expressions

Expressions

- An expression is a set of operators and values used to perform various operations on data, which include:
  - Calculations
  - Specify logical conditions
  - Generate values that don't exist in the data
- Expression can consist of the following set of elements:
  - Data fields
  - Operators
  - Constants
  - Functions
  - Variables

Expression - Filter and Calculated field

When constructing an expression, it is important to determine whether the desired outcome should be a filter or a computed field.

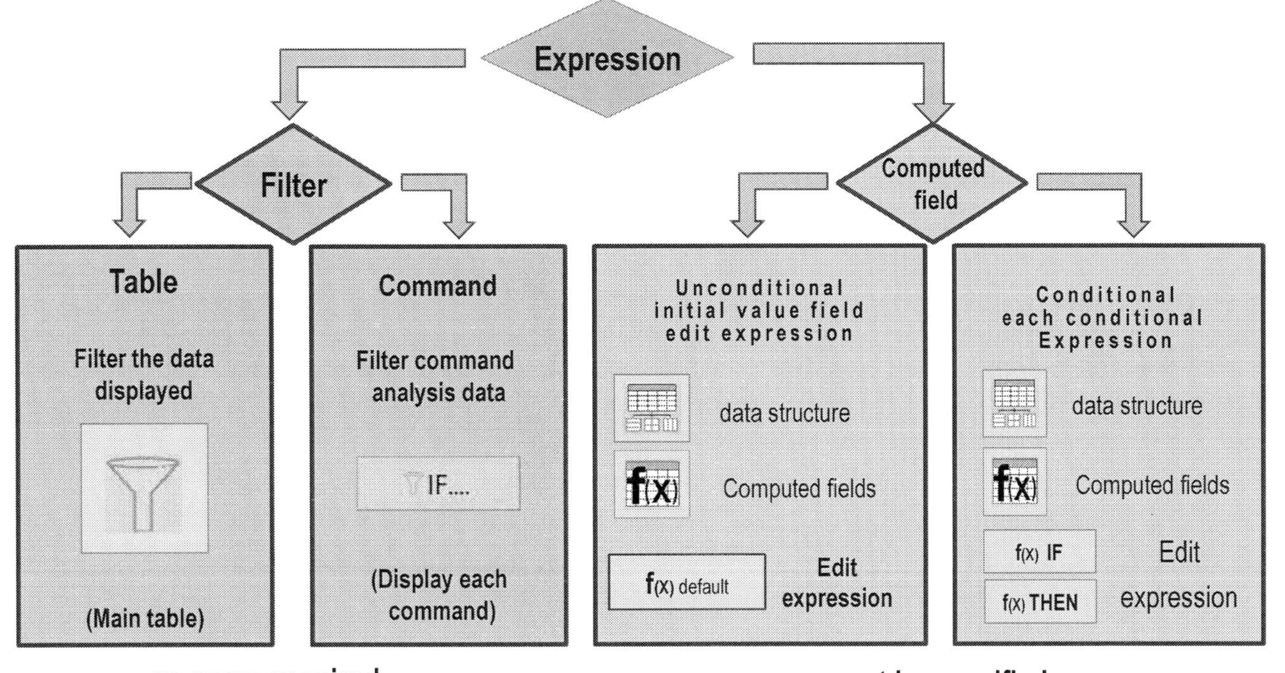

Filters

- A filter is a logical expression that evaluates to true (T) or false (F)
- A filter allows you to select the specific data you want to work with.
- A filter can be thought of as similar to a query.
- Two kinds of filters ：
- **Global filter**: It is defined on the main screen and is applied to the data being analyzed.
- **Local filter**: Iis defined within a command and is applied to the execution of that command.
- Global filters can also be activated using the Quick Filter feature on the table view.

Computed Fields

- The Computed Field is for storing an expression.
- The computed field is not the actual field that is stored in the original dataset.
- When the table is opened, the expression of the computed field is executed and its result is displayed.
- The results displayed in the computed field do not modify or affect the original data.
- The result of a Computed Field's execution can be a value of text, numeric, datetime, or Boolean data type..
- Computed fields serve four major purposes:
  - ✓ Executing math expressions.
  - ✓ Converting data types in the fields.
  - ✓ Performing text replacement.
  - ✓ Creating Boolean test results

ICON on the Main Working Area

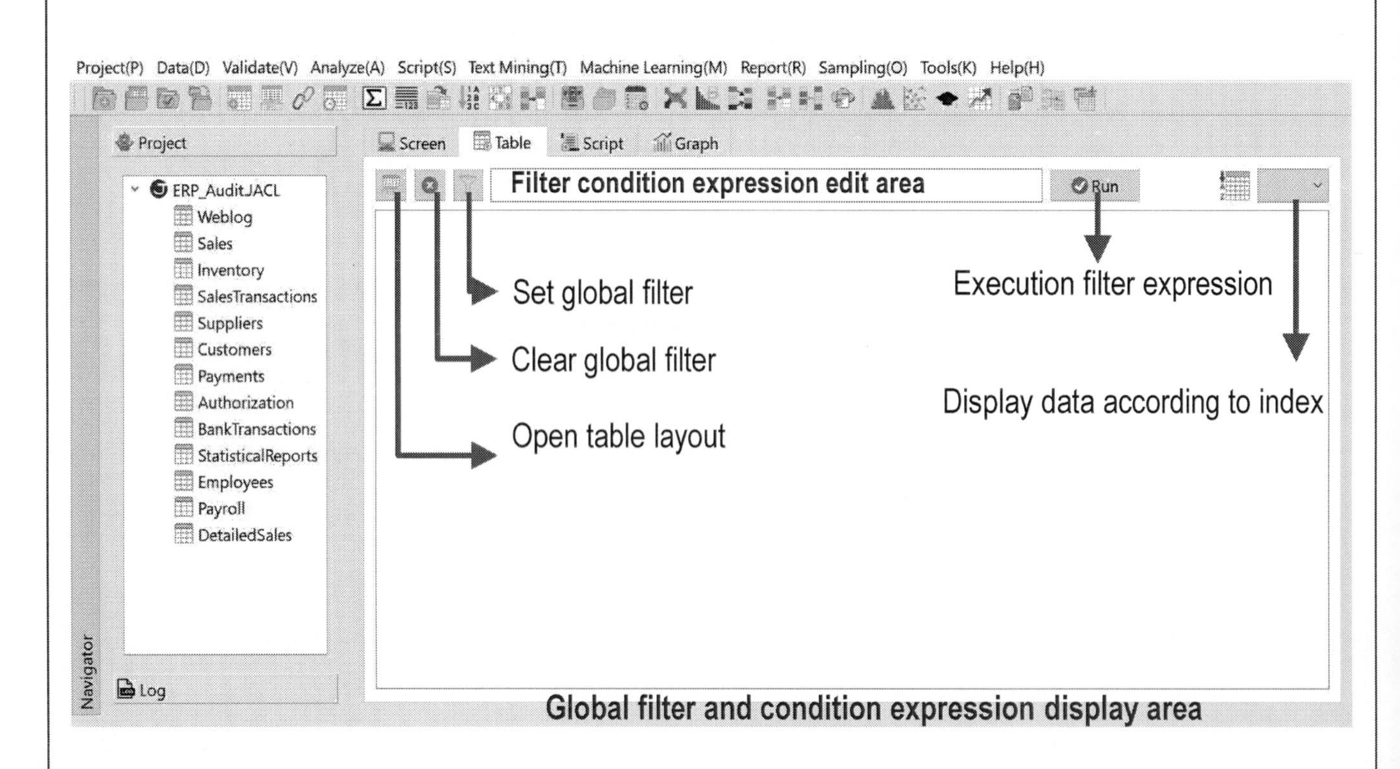

JCAATs - AI Audit Software

3. Filters

Filter Editor interface

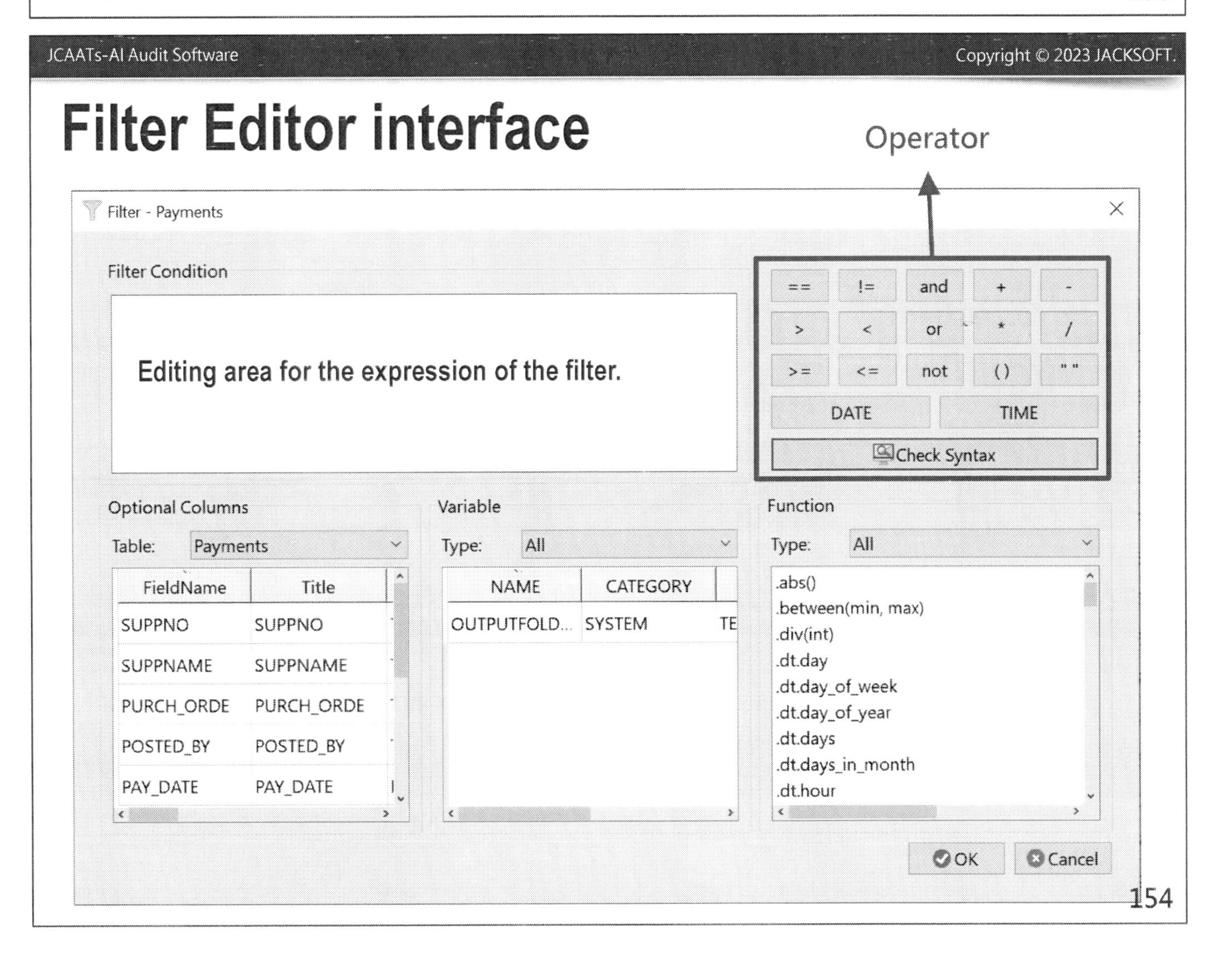

Operators within Filter Editor

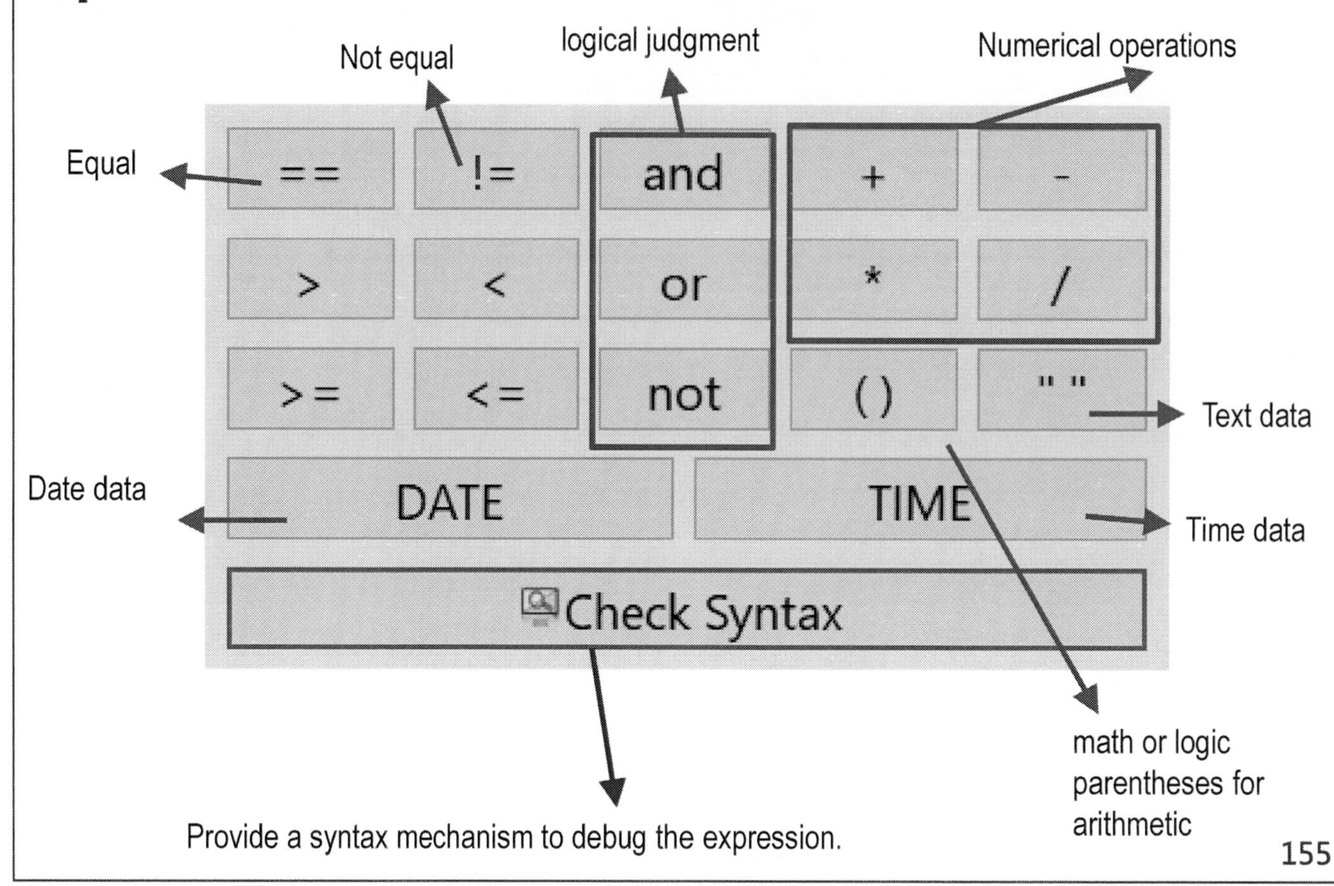

Function Syntax

"Function syntax" refers to the specific way in which a function is written, including the function name, input parameters, and output value.

1. No parameter

This type of syntax is mostly used with date functions. Examples of the syntax include .dt.day_of_week, .dt.days, .dt.days_in_month, and others.
The syntax requires placing the field name at the front without listing the set.

- For example, to get the day of the week of the SDATE field, use the following expression: **SDATE.dt.day_of_week**

2. Null parameter

This type of syntax is mostly used with numeric fields and includes methods such as .abs(), .max(), .isna(), and others. The syntax requires placing the field name at the front.

- For example, to get the absolute value of the PRICE field, use the following expression: **PRICE.abs()**

Function Syntax

3. One text parameter

This type of syntax includes methods such as .str.match(pat), .str.extract(pat), .str.endswith(pat), .str.startswith(pat), and others. The syntax requires placing the field name at the front and entering text in parentheses.

- For example, to check if the CITY field equals "TAIPEI", use the following expression: **CITY.str.match("TAIPEI")**

4. One integer parameter

This type of syntax includes methods such as .div(int), .round(int), .mod(int), and others. The syntax requires placing the field name at the front and entering a numeric value in parentheses.

- For example, to get the remainder of dividing the PRICE field by 2, use the following expression: **PRICE.mod(2)**

Function Syntax

5. Two text parameter

This type of syntax includes methods such as .str.replace(pat, repl). The syntax requires placing the field name at the front , entering text in parentheses, and entering the second string of text after the comma.

- For example, to replace “New York" with “NY" in the CITY field, you would use the following expression: **CITY.str.replace ("New York", "NY")**

6. One parameter and one statement

This type of syntax includes methods such as .str.contains(pat, na=False). The syntax requires placing the field name at the front , entering text in parentheses, and adding the condition as required to evaluate to True or False. Na=False define that Null value is False.

- For example, to check if the ADDRESS field contains "TAIPEI" and not empty values, you would use the following expression: **ADDRESS.str.contains("TAIPEI", na=False)**

Function Syntax

7. Multiple parameters

This type of syntax includes methods such as .str.slice(start=None, stop=None, step=None). The syntax requires placing the field name at the front and entering the corresponding content according to the parameters in the parentheses.

➢ For example, to take the characters of the ADDRESS field from position 0 to 100, and then take every third character, you would use the following expression: ADDRESS.str.slice(start=0, stop=100, step =5).
As a result, "**1**26 J**o**hn S**t**reet " would be changed to "1ot".

Function Syntax

8. Function with fields as parameter

The parameters of a filtering function include pat, int, or val, which can use variables or constants. When using variables, the syntax for it is "var" (short for variable). The functions @find(col, val), @find_multi(col, [val]), and @shift(col, int, up=False) use "col" to represent the field. "val" can be a text, field or variable. The instructions are as follows:

- **@find(DESC, KEYWORD):** The DESC field contains the value KEYWORD.
- **@find(ADDRESS, "TAIPEI"):** The address contains "TAIPEI".
- **@find_multi(DESC, [KEYWORD, "Penalty"]):** The DESC field contains the value KEYWORD or the word "Penalty".
- **@find_multi(ADDRESS, ["TAIPEI", "TAICHUNG"]):** The address contains "TAIPEI" or "TAICHUNG".
- **@shift(Invoice_No, 1, up=False):** Takes the previous value of the Invoice_No field.
- **@shift(Invoice_No, 1, up=True):** Takes the next value of the Invoice_No field.

Function syntax

JCAATs functions are categorized into several groups, including Text, Math, Datetime, Financial, and others.

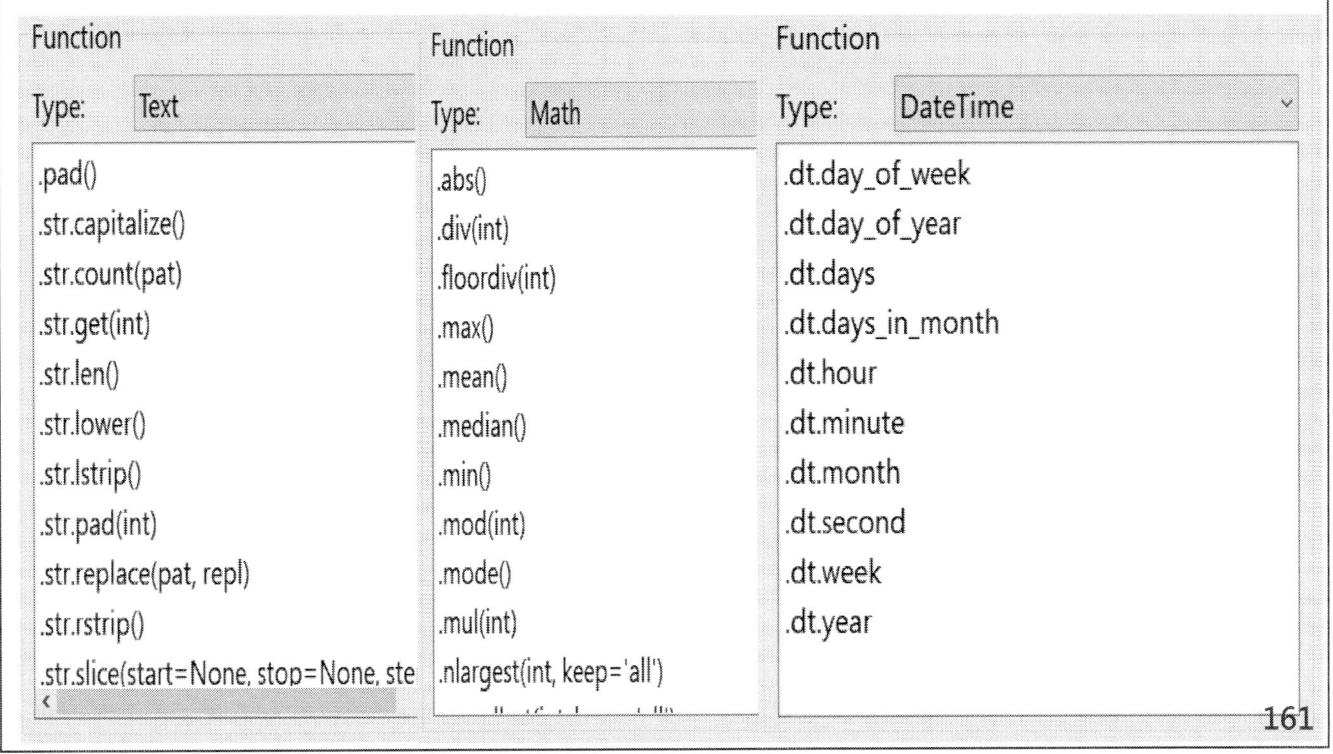

Example: Open a Filter

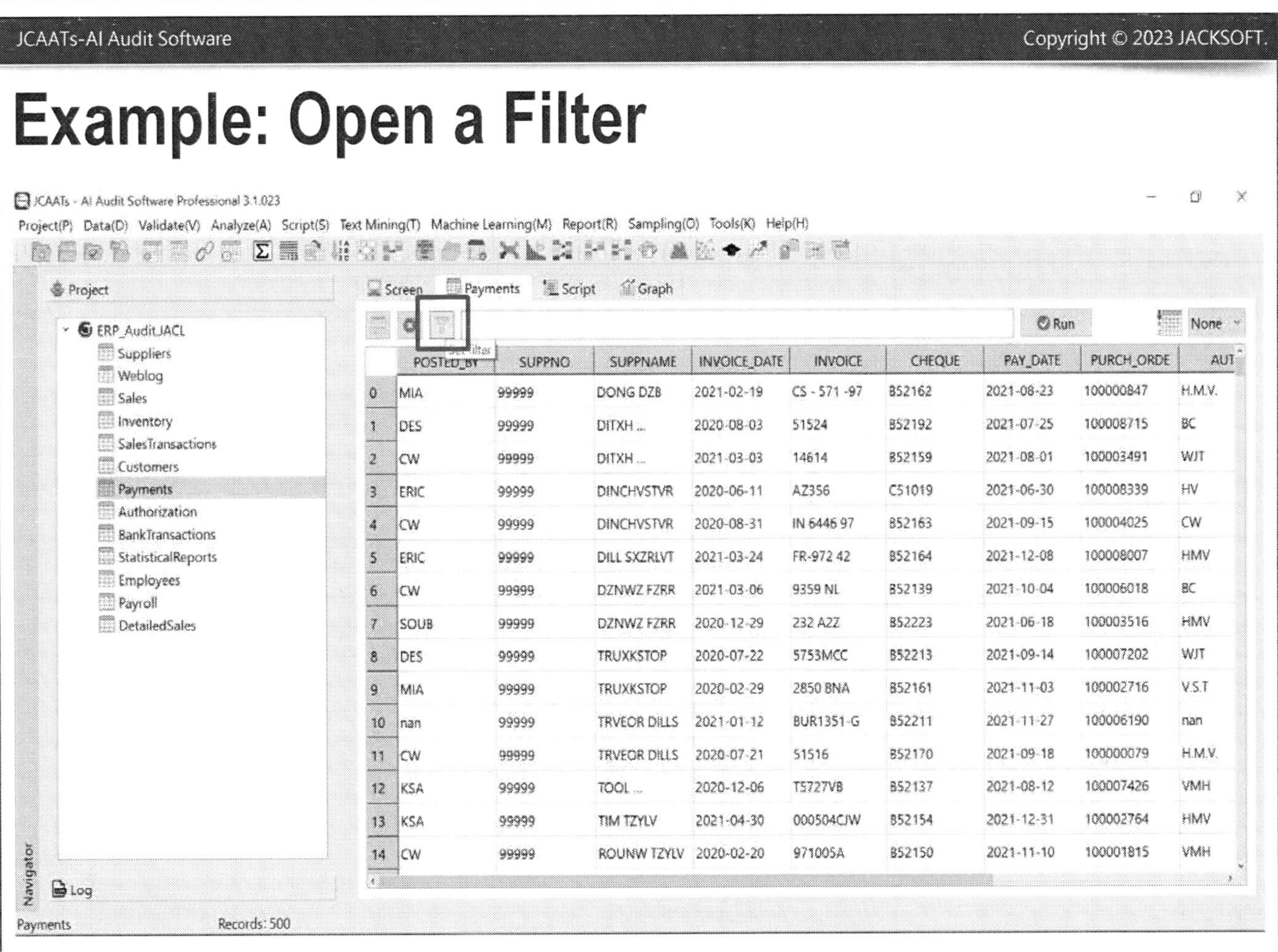

Example: Filter Editor

Open Payments table and set a global filter to display that MEMO field contains "Gold".

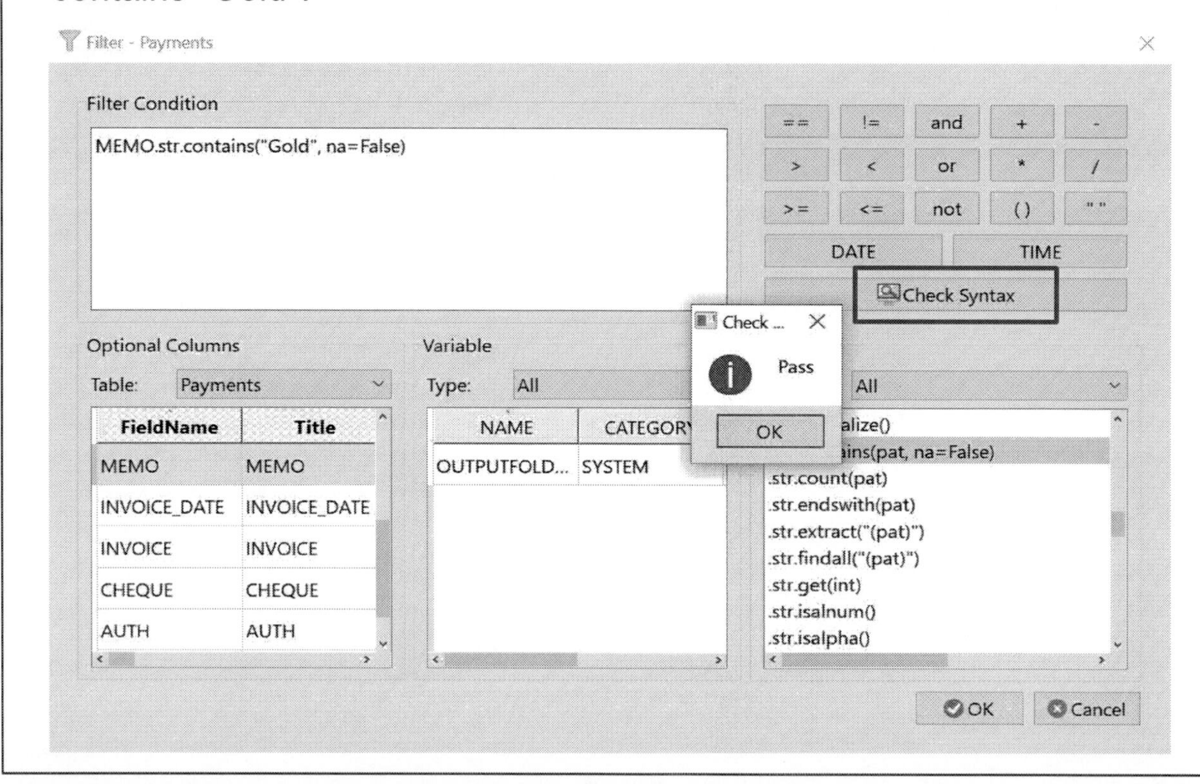

Example: Result View

Press " OK " then display the table result. The status column displays the number of transactions and filter conditions.

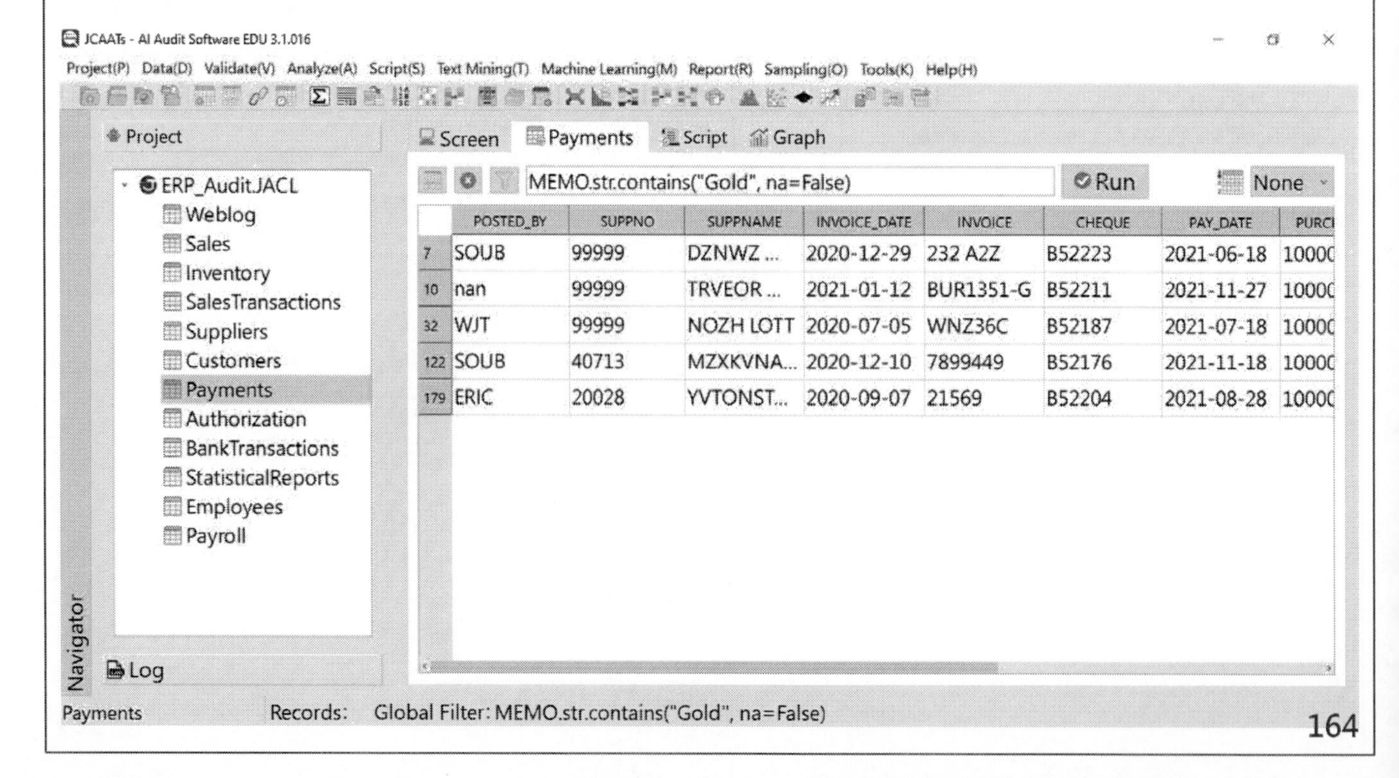

	POSTED_BY	SUPPNO	SUPPNAME	INVOICE_DATE	INVOICE	CHEQUE	PAY_DATE	PURC
7	SOUB	99999	DZNWZ ...	2020-12-29	232 A2Z	B52223	2021-06-18	1000C
10	nan	99999	TRVEOR ...	2021-01-12	BUR1351-G	B52211	2021-11-27	1000C
32	WJT	99999	NOZH LOTT	2020-07-05	WNZ36C	B52187	2021-07-18	1000C
122	SOUB	40713	MZXKVNA...	2020-12-10	7899449	B52176	2021-11-18	1000C
179	ERIC	20028	YVTONST...	2020-09-07	21569	B52204	2021-08-28	1000C

4. Add a New Expression

Open add a new expression

There are three methods to add a new expression in JCAATs. The first method is to select the 'Add a New Expression' function from the Menu. The second method is to select the 'Table Layout' function from the Menu and then select 'Add a Computed Field'. The third method is to click on the 'Table Layout' icon.

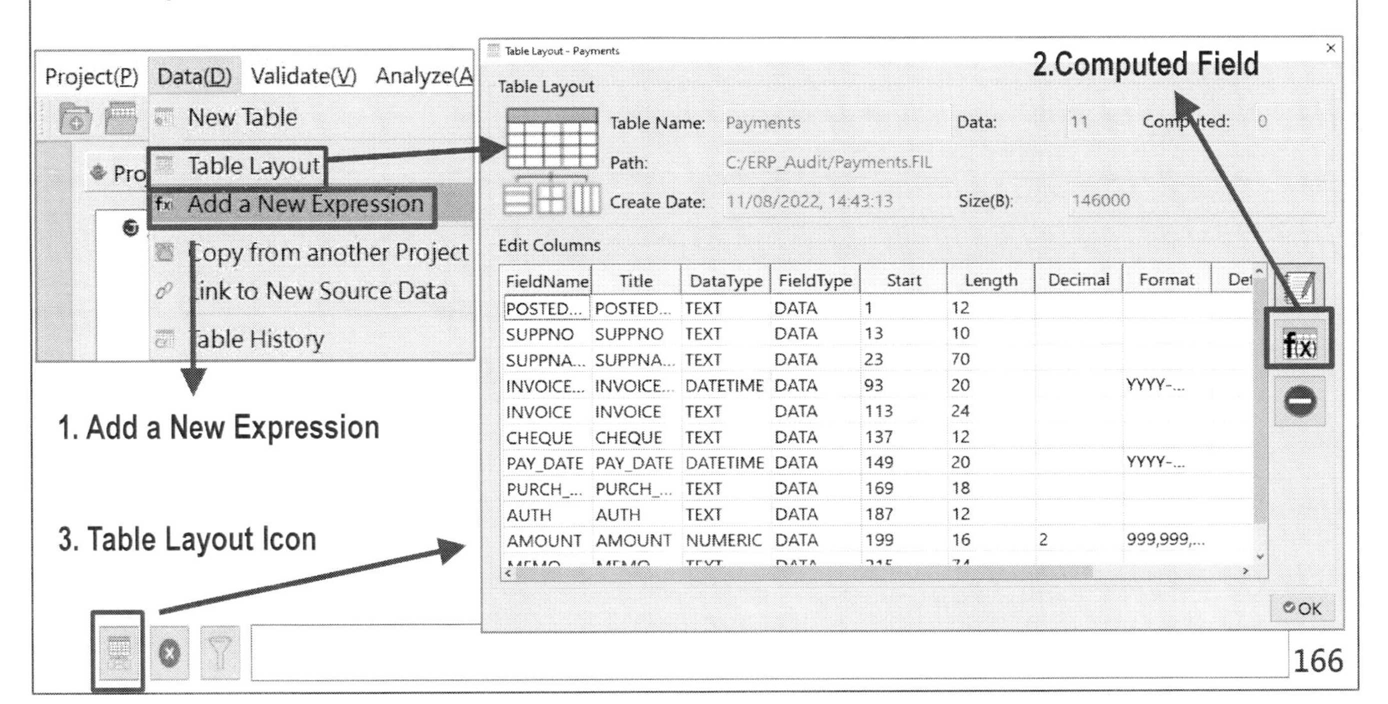

Add a New Expression- Computed Field

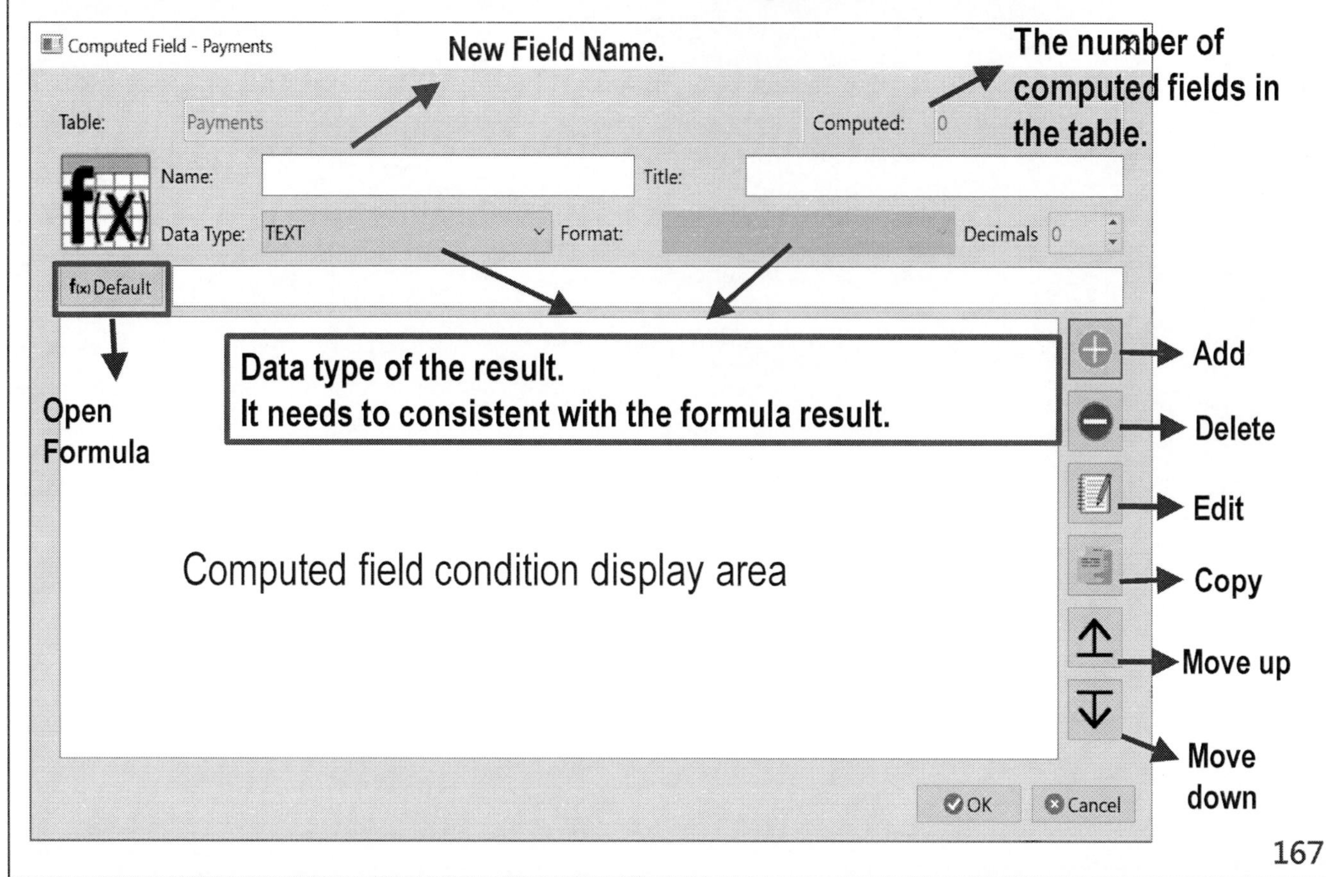

Condition Manager

The condition manager will appear after press the **Add** button or **Edit** button in Computed Field interface.

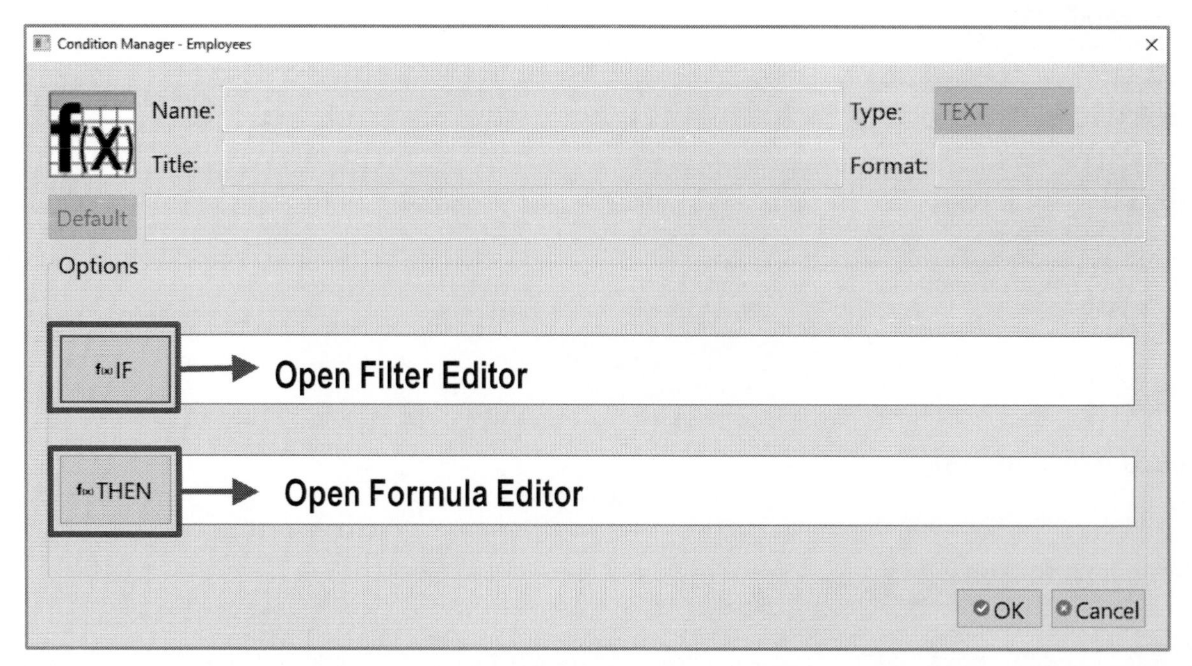

Formula Editor

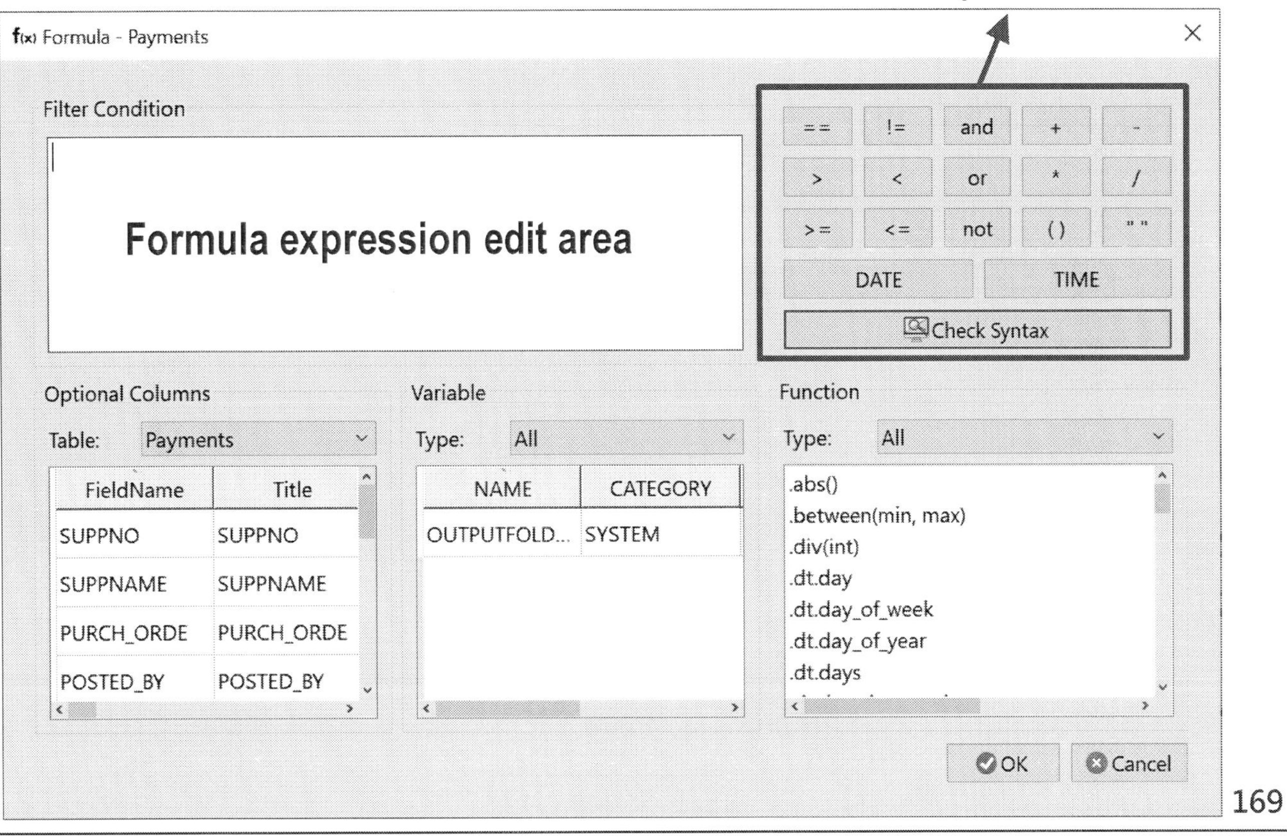

JCAATs - AI Audit Software

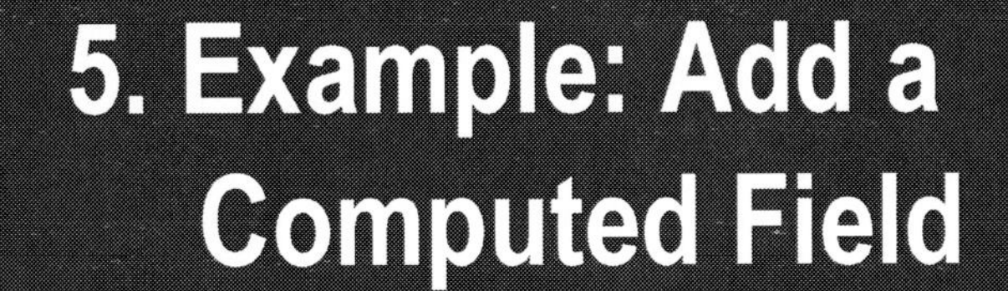

Example: Add a Computed Field

You need add a computed field named pay_level to prepare your audit data.
The expression of the computed field is:
Default value as pay_level = "C ".
IF AMOUNT > 50000 THEN pay_level = "A" ,
IF AMOUNT > 10000 AND AMOUNT <=50000 THEN pay_level = "B"

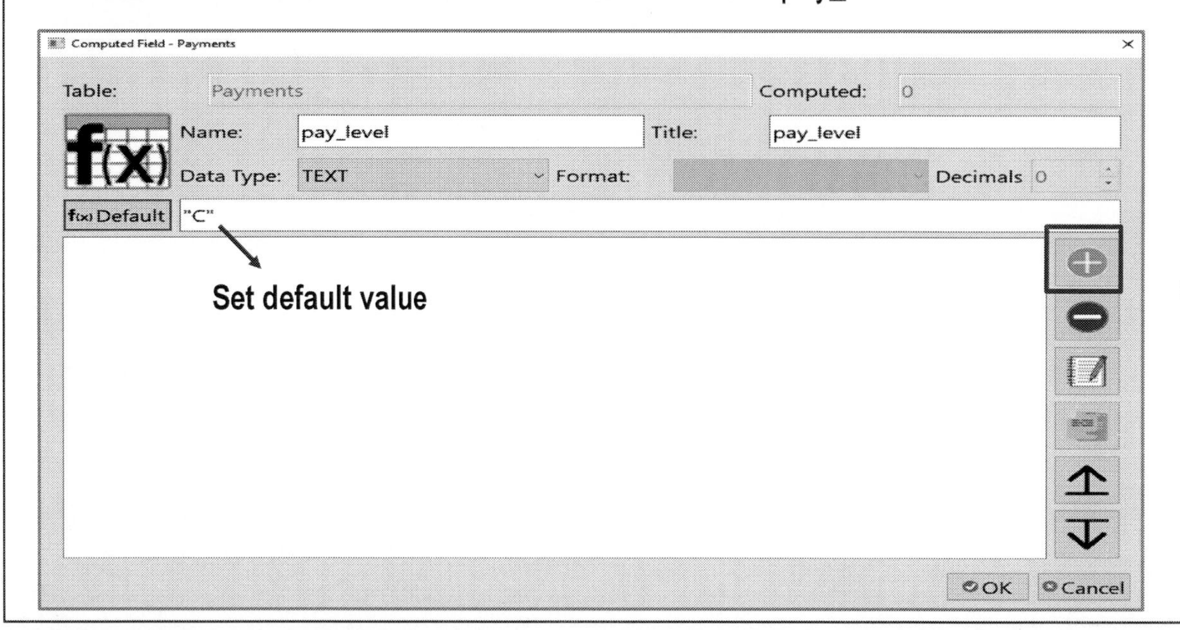

Add a Condition

JCAATs-AI Audit Software Copyright © 2023 JACKSOFT.

Example: Add a Condition - Condition Manager

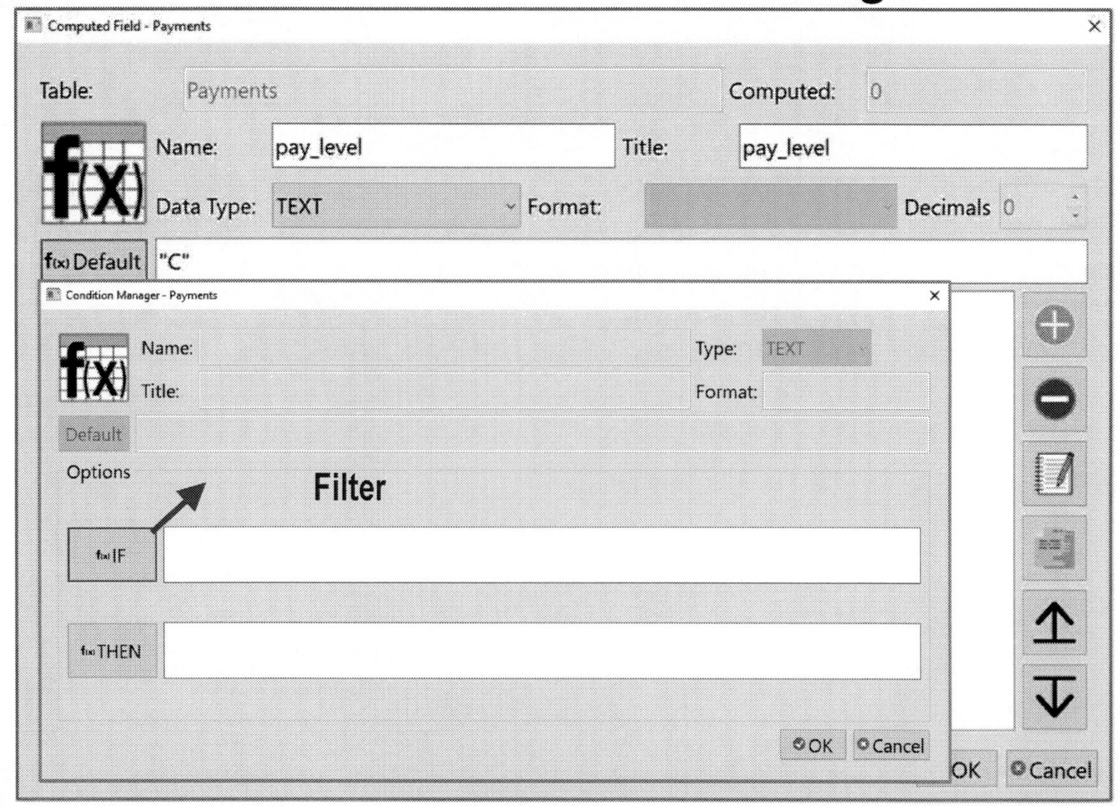

Example: Condition Manager – IF

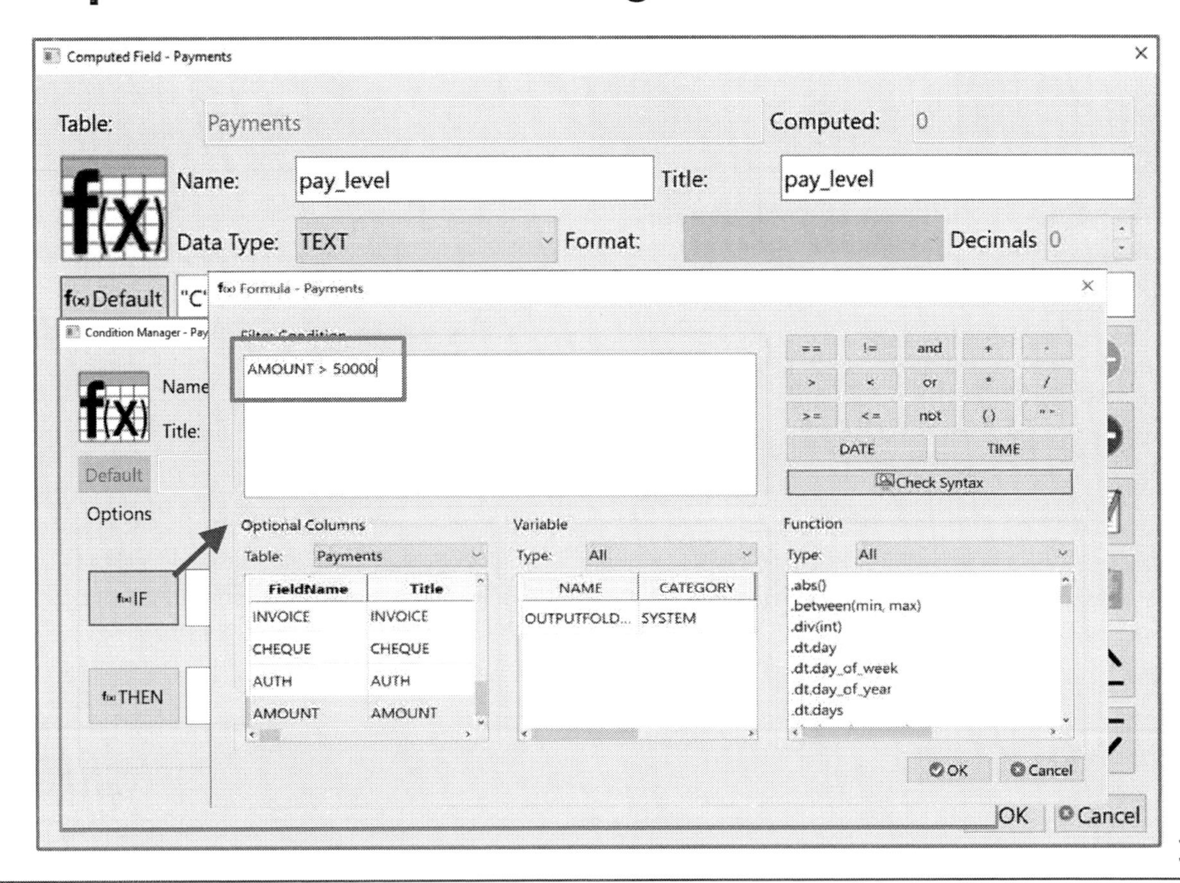

Example: Condition Manager – THEN

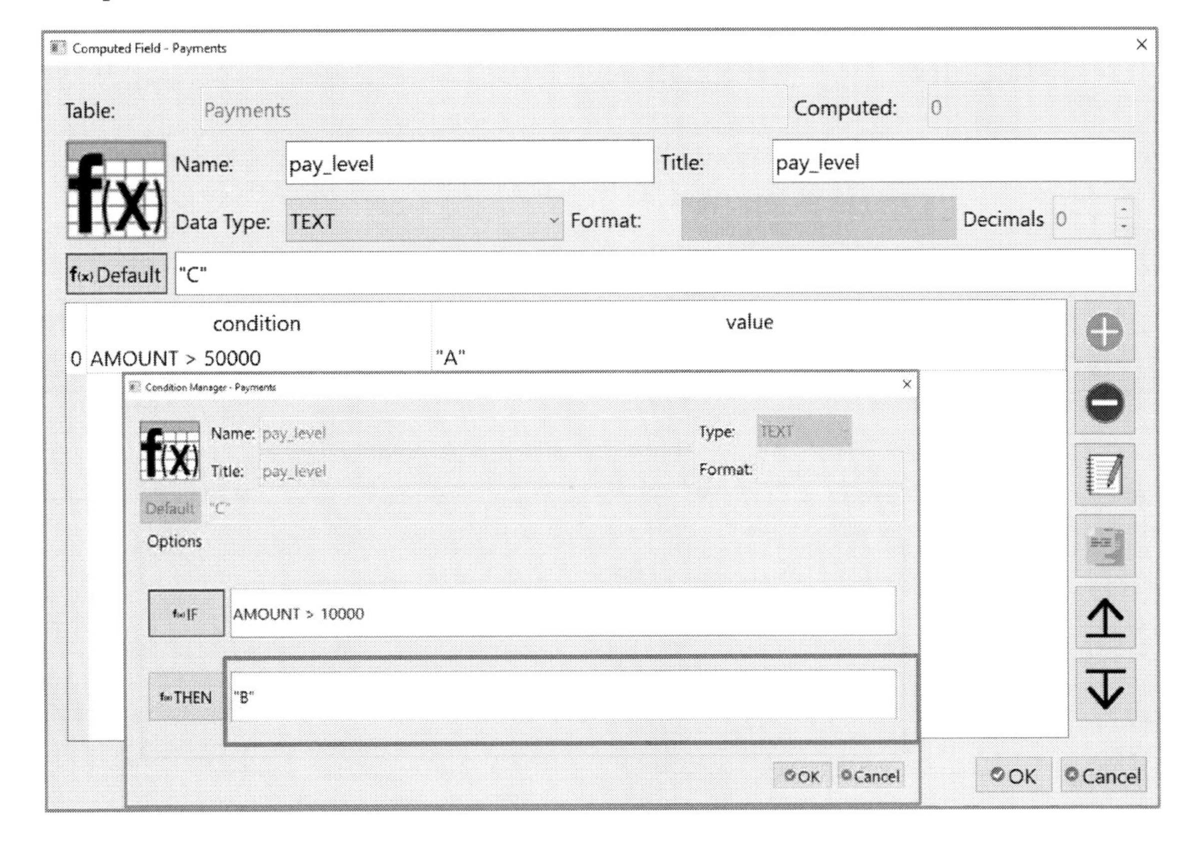

Example: A Computed Field

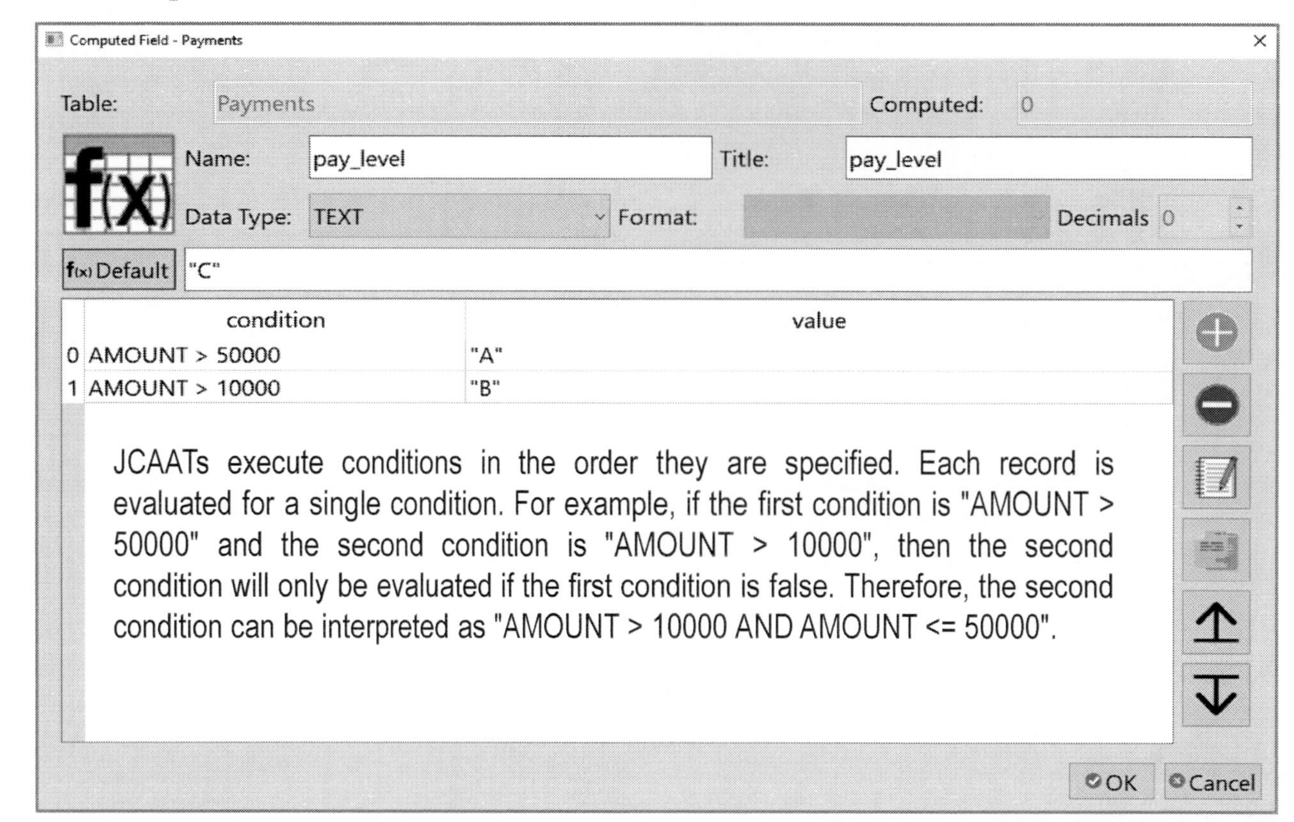

Example:

Table Layout with a new computed field

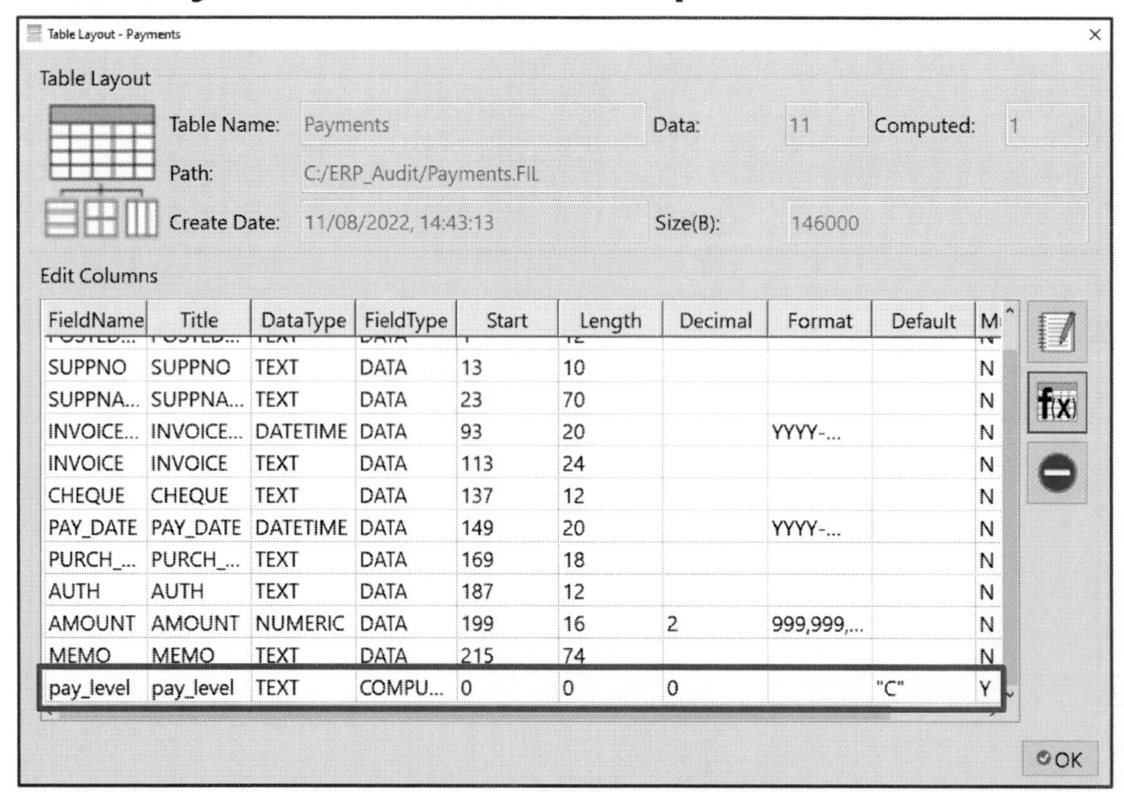

FieldName	Title	DataType	FieldType	Start	Length	Decimal	Format	Default	M
SUPPNO	SUPPNO	TEXT	DATA	13	10				N
SUPPNA...	SUPPNA...	TEXT	DATA	23	70				N
INVOICE...	INVOICE...	DATETIME	DATA	93	20		YYYY-...		N
INVOICE	INVOICE	TEXT	DATA	113	24				N
CHEQUE	CHEQUE	TEXT	DATA	137	12				N
PAY_DATE	PAY_DATE	DATETIME	DATA	149	20		YYYY-...		N
PURCH_...	PURCH_...	TEXT	DATA	169	18				N
AUTH	AUTH	TEXT	DATA	187	12				N
AMOUNT	AMOUNT	NUMERIC	DATA	199	16	2	999,999,...		N
MEMO	MEMO	TEXT	DATA	215	74				N
pay_level	pay_level	TEXT	COMPU...	0	0	0		"C"	Y

Example:

Table with the Computed field

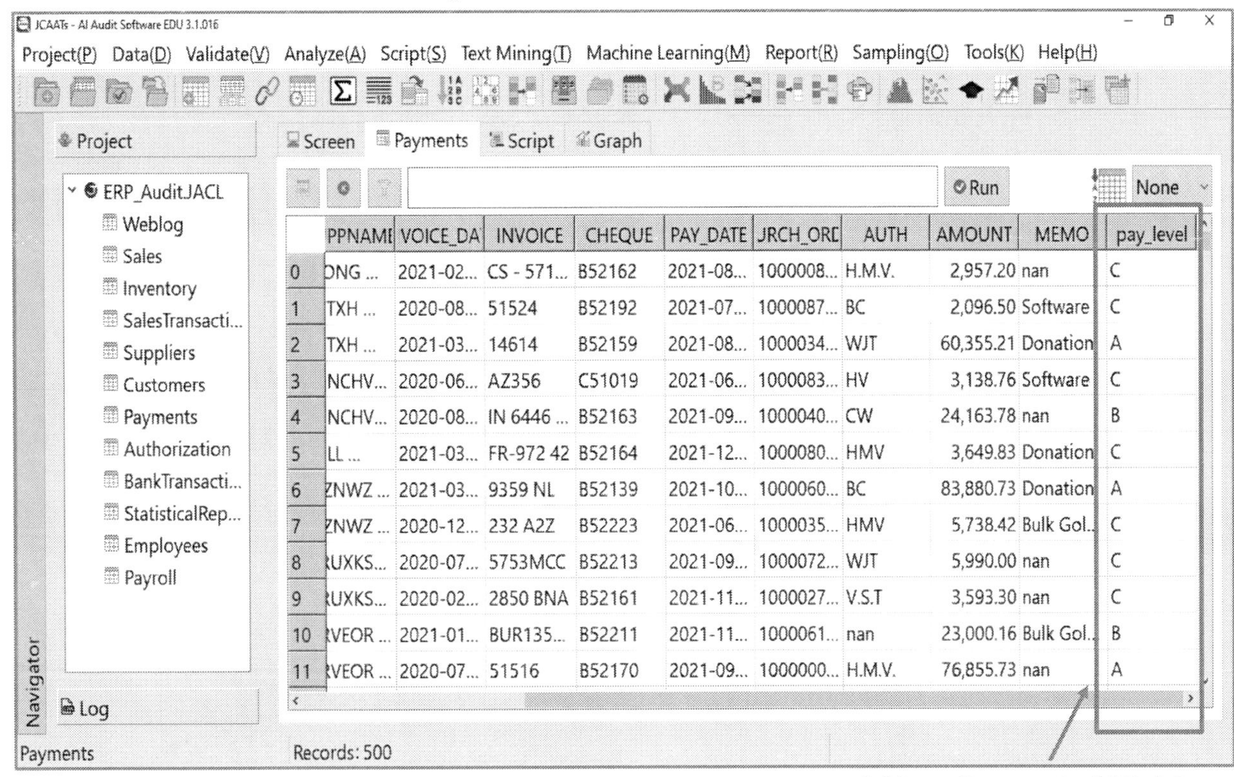

A New Computed Field

JCAATs - AI Audit Software

Chapter 5 - Exercise

Chapter 5 - Exercise

( ) 5-1 How can you get the remainder of a numeric field in JCAATs?

a) Define a field and use the function ".mod()".
b) Classify a field and output a table.
c) Statistic a field and output a table.
d) Define a field and use the function ".divide_by()".
e) None of the above.

( ) 5-2 You have two text fields named "X" and "Y" in JCAATs. You want to combine the contents of these two fields into a new text field. Which expression should you use to create the new field?

a) X+Y
b) X&Y
c) X*Y
d) X==Y
e) X!=Y

Chapter 5 - Exercise

( ) 5-3 You are importing an ASCII data file into your JCAATs. JCCATs is an UNICODE version software. The Invoice_Number field is a numeric field and is padded to the left with zeros in the source file (e.g., 000450), but it displays as 450 in the CAATs. How do you convert the data to display as 000450 in the CAATs?

a) "000"+ Invoice_Number
b) Invoice_Number.string(6)
c) Invoice_Number.zoned(6)
d) Invoice_Number.str.pad(6,"left","0")
e) Invoice_Number.str.sub(1,6)

Chapter 5 - Exercise

( ) 5-4 Inconsistency in data formats can be a significant issue for data quality, such as when social security numbers (SSNs) are stored in various formats. To determine if SSNs in your data are in the standard format of 999999999, which JCAATs function should you use?

a) SSN.str.isalnum() and SSN.str.match("999")
b) SSN.str.isalpha()
c) SSN.str.isnumeric() and SSN.str.len() == 9
d) SSN.str.replace(" -/", "")
e) @find(SSN, "\D")

JCAATs-AI Audit Software
Copyright © 2023 JACKSOFT.

Chapter 5 - Exercise

( ) 5-5 Which of the following operations in JCAATs would produce the same result as multiplying a Field by 2?

a) Field / 2
b) Field // 2
c) Field % 2
d) Field ** 2
e) Field * 2

( ) 5-6 In JCAATs, what is the purpose of a Computed Field?

a) Perform mathematical calculations on existing fields
b) Convert data types or formats of existing fields
c) Substitute words or phrases in existing fields
d) Apply logical operations to existing fields
e) All of the above

JCAATs Learning notes：

JCAATs - AI Audit Software

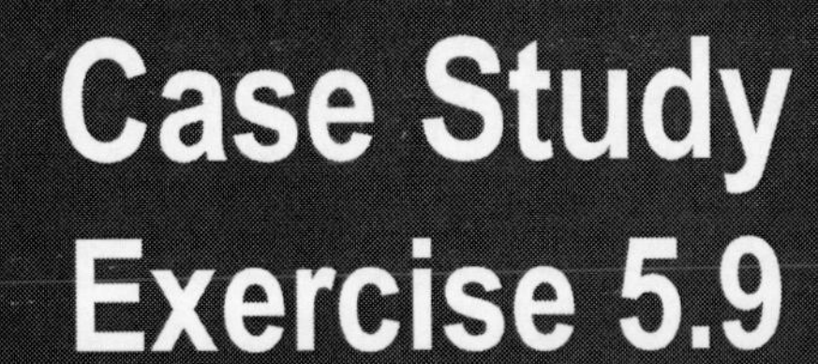

Case Study
Exercise 5.9

Chapter 5 - Exercise Questions

- Exercise 5.9: Create a new project file named 'Proj100', and then define a new table in the project called 'Sales_Record', which is a sales record file for a company. Please define the table according to the following table format (Table Layout):

Field Name	Type	Start	Length	Decimal	Note
Customer ID	TEXT	1	11	0	
Shipping Date	DATE	12	13	0	
Shipping Number	TEXT	25	14	0	
Sales Amount	NUMERIC	39	12	2	
Uncollected AR	NUMERIC	51	14	0	
Shipping Region	TEXT	65	24	0	

Exercise 5.9

- To ensure data quality, please perform the following data integrity validation process, answer the following test questions, and briefly describe the results.
  - Verify the customer data format:
    Test results:
    ○ Validity
    ○ Invalidity, please list the errors detected____________________
  - Control Total validation
    - ➢Number of records in 2009: ____________________
    - ➢Total Revenue in 2009: ____________________
    - ➢Number of records in 2010: ____________________
    - ➢Total Revenue in 2010: ____________________

Exercise 5.9

- To ensure data quality, please perform the following data integrity verification and answer the test results, and briefly describe the analysis method and results.
  - Sequence test: Check whether the shipping numbers are all in sequence order,
    - ○ Yes
    - ○ No, please list the shipping numbers that are not in order:

      ____
  - Duplicate test: Are there any duplicate shipping numbers?
    - ○ No
    - ○ Yes, please list the shipping numbers that are duplicates:

      ____

JCAATs-AI Audit Software
Copyright © 2023 JACKSOFT.

Exercise 5.9

- **Data analysis**
  - Sales income analysis by region:
    - ➢ Which region has the top revenue:

      ____
  - Analysis of accumulated transaction amount and outstanding debt:
    - » List of customers whose accumulated transaction amount is greater than 300,000:

      ____
    - » List of customers whose accumulated outstanding debt exceeds 200,000:

      ____

Exercise 5.9

- Summarize sales amount by shipment date and analyze whether there is a special concentration of revenue on specific dates (such as the end of the month) or months (such as the end of the year) (i.e., when the number of transactions is greater than 10).

- Data Analysis
  - Analysis of Sales Revenue by Region:
    - » If we only analyze the sales amount of Taipei, Keelung, Taichung, Chiayi, and Kaohsiung regions, which region has the highest sales revenue: ______________________________

Exercise 5.9

- Analysis of Receivables by Region:
  - » Based on December 31, 2010, perform an analysis of receivable aging, and calculate the total amount of receivables that are over 180 days, 240 days, 360 days in each of the four regions.

    0~180 days: ______________________________

    181~240 days: ______________________________

    241~360 days: ______________________________

    Over 361 days: ______________________________

Exercise 5.9
Verify the customer data format

- Open "Sales_Record" Table
- Click on "Validate>Verify"

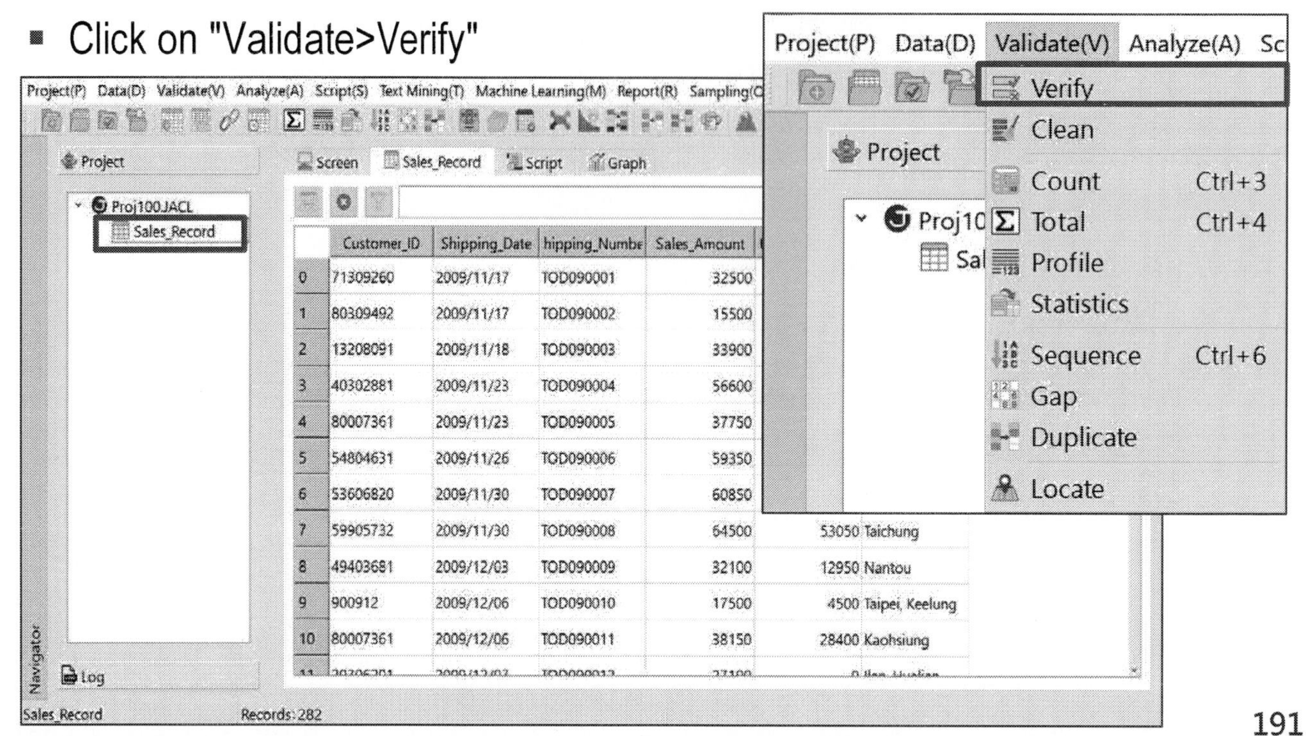

	Customer_ID	Shipping_Date	hipping_Numbe	Sales_Amount
0	71309260	2009/11/17	TOD090001	32500
1	80309492	2009/11/17	TOD090002	15500
2	13208091	2009/11/18	TOD090003	33900
3	40302881	2009/11/23	TOD090004	56600
4	80007361	2009/11/23	TOD090005	37750
5	54804631	2009/11/26	TOD090006	59350
6	53606820	2009/11/30	TOD090007	60850
7	59905732	2009/11/30	TOD090008	64500
8	49403681	2009/12/03	TOD090009	32100
9	900912	2009/12/06	TOD090010	17500
10	80007361	2009/12/06	TOD090011	38150

Exercise 5.9
Verify the customer data format

- Click on "Verify on"
- Click on "Add all"

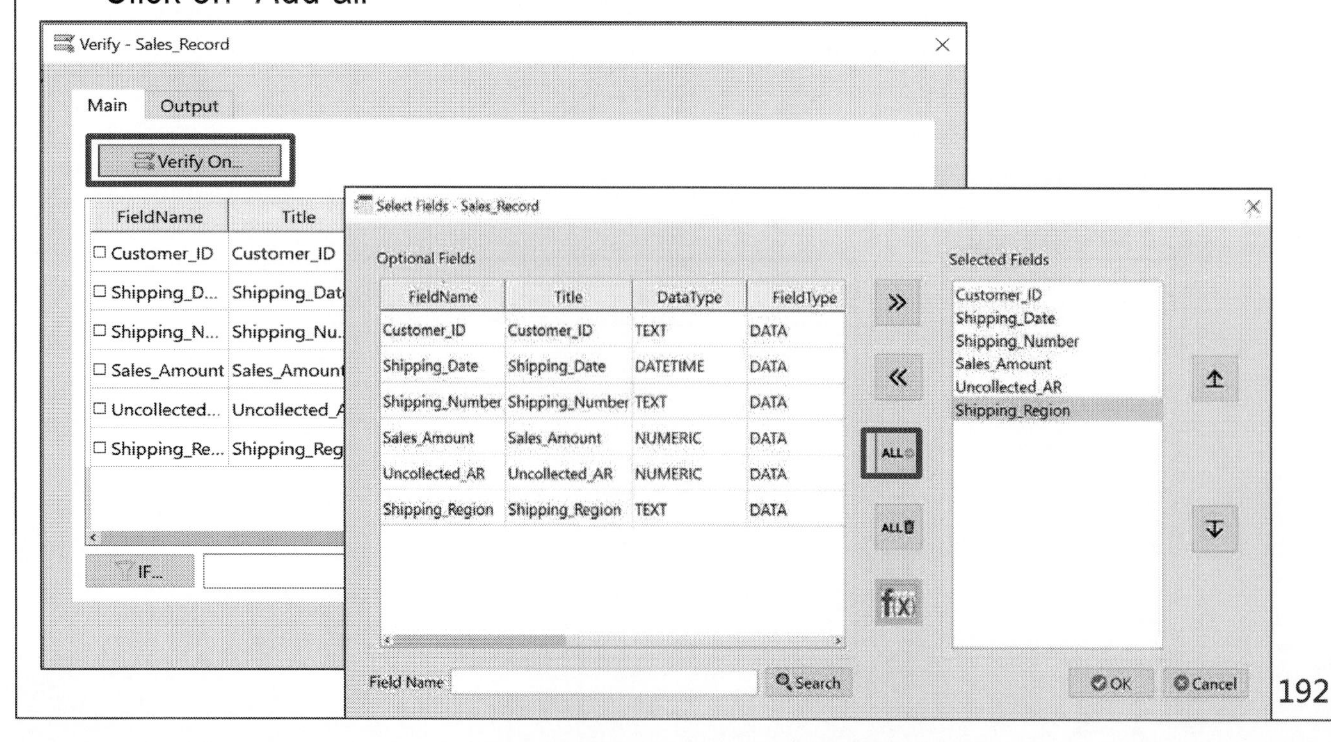

FieldName	Title	DataType	FieldType
Customer_ID	Customer_ID	TEXT	DATA
Shipping_Date	Shipping_Date	DATETIME	DATA
Shipping_Number	Shipping_Number	TEXT	DATA
Sales_Amount	Sales_Amount	NUMERIC	DATA
Uncollected_AR	Uncollected_AR	NUMERIC	DATA
Shipping_Region	Shipping_Region	TEXT	DATA

Exercise 5.9
Verify the customer data format

- Output to Screen

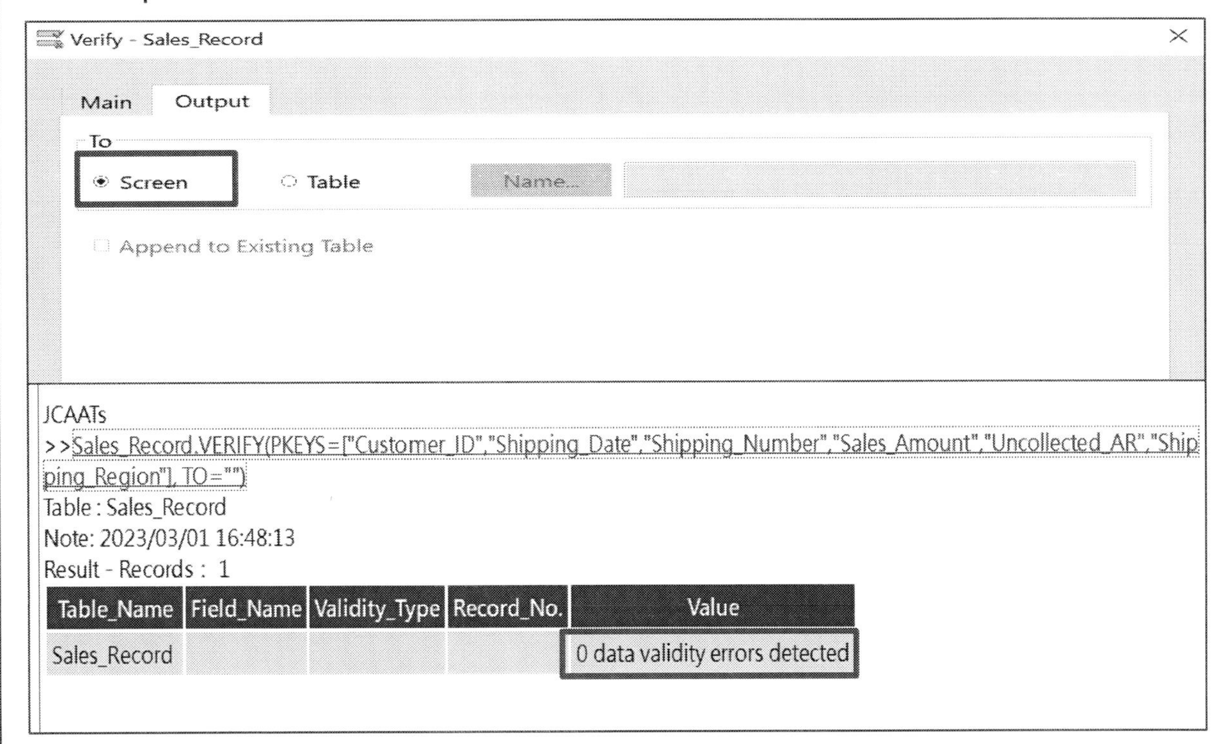

Exercise 5.9
Control Total validation Number of records in 2009

- Open "Sales_Record" Table
- Click on "Validate>Count"

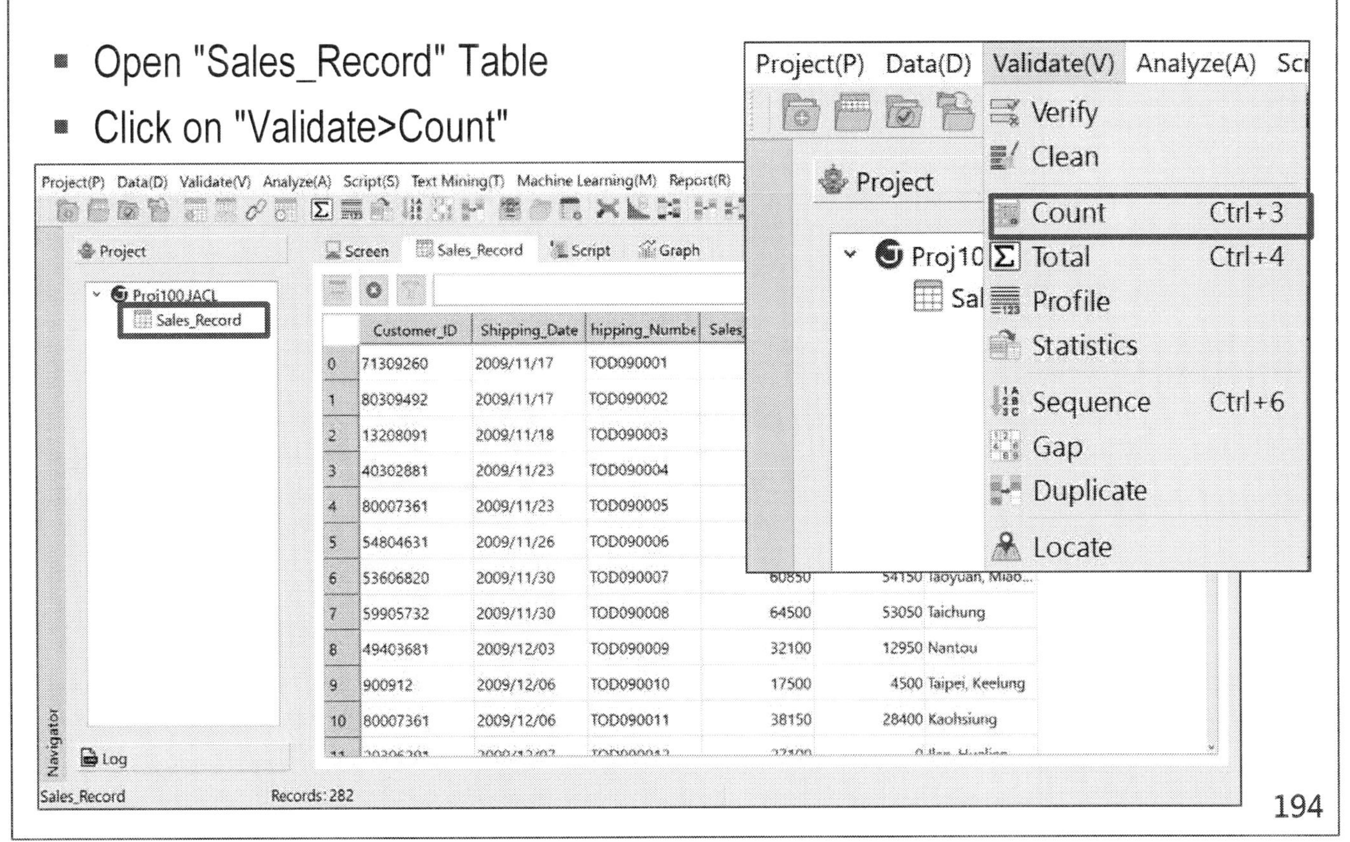

Exercise 5.9
Control Total validation Number of records in 2009

- Add a Filter Condition:
  Shipping_Date.between(date(2009-01-01), date(2009-12-31))

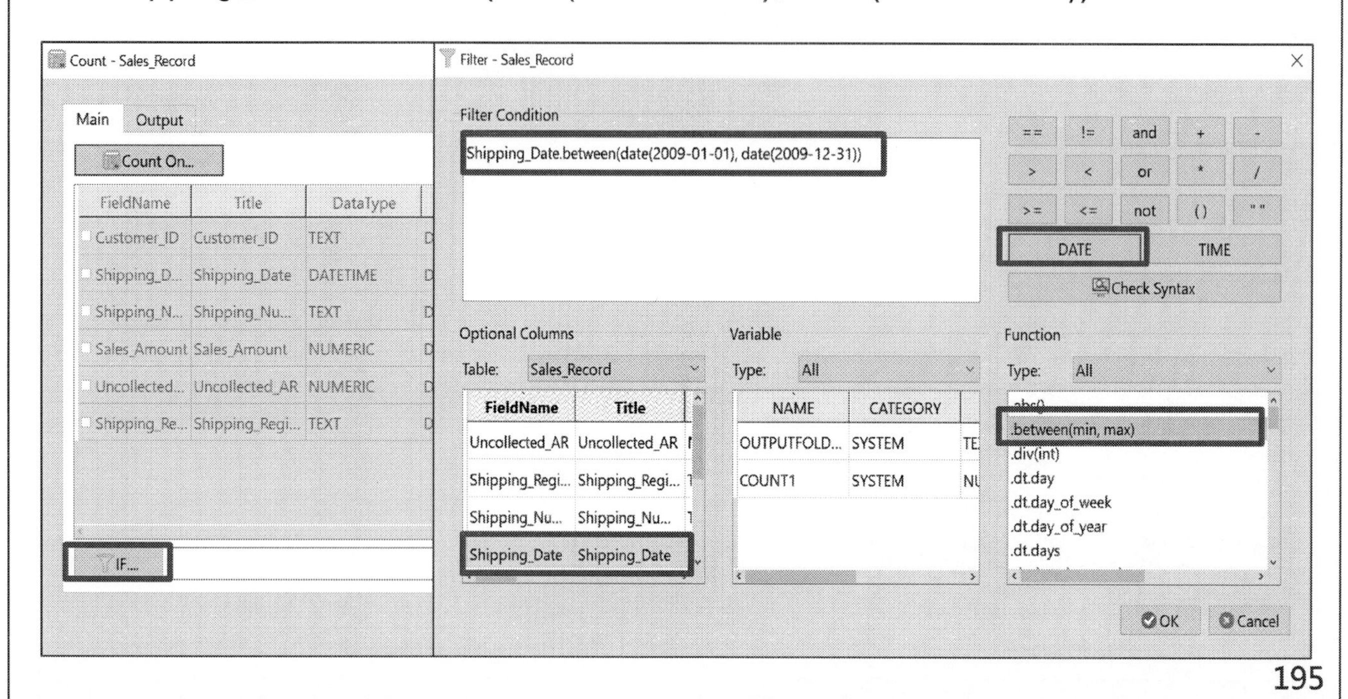

Exercise 5.9
Control Total validation Number of records in 2009

- Output to Screen

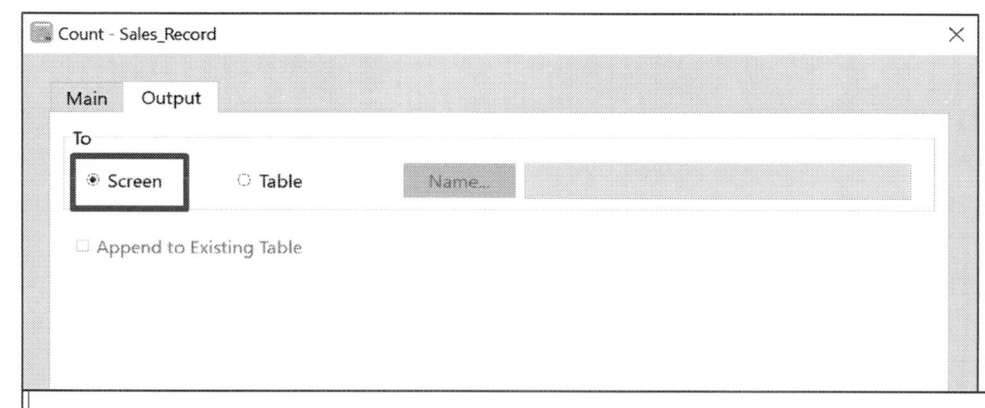

JCAATs >>Sales_Record.COUNT(IF = ["Shipping_Date.between(date(2009-01-01), date(2009-12-31)) "])
Table : Sales_Record
Note: 2023/03/01 17:05:42
Result - Records : 1

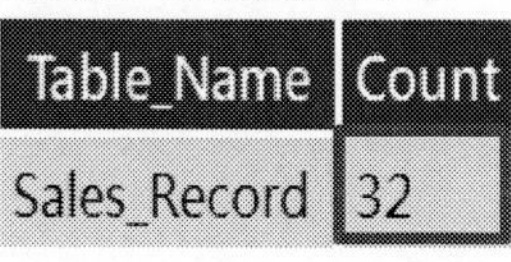

Table_Name	Count
Sales_Record	32

Exercise 5.9
Control Total validation Total Revenue in 2009

- Add a Filter Condition:

  Shipping_Date.between(date(2009-01-01), date(2009-12-31))

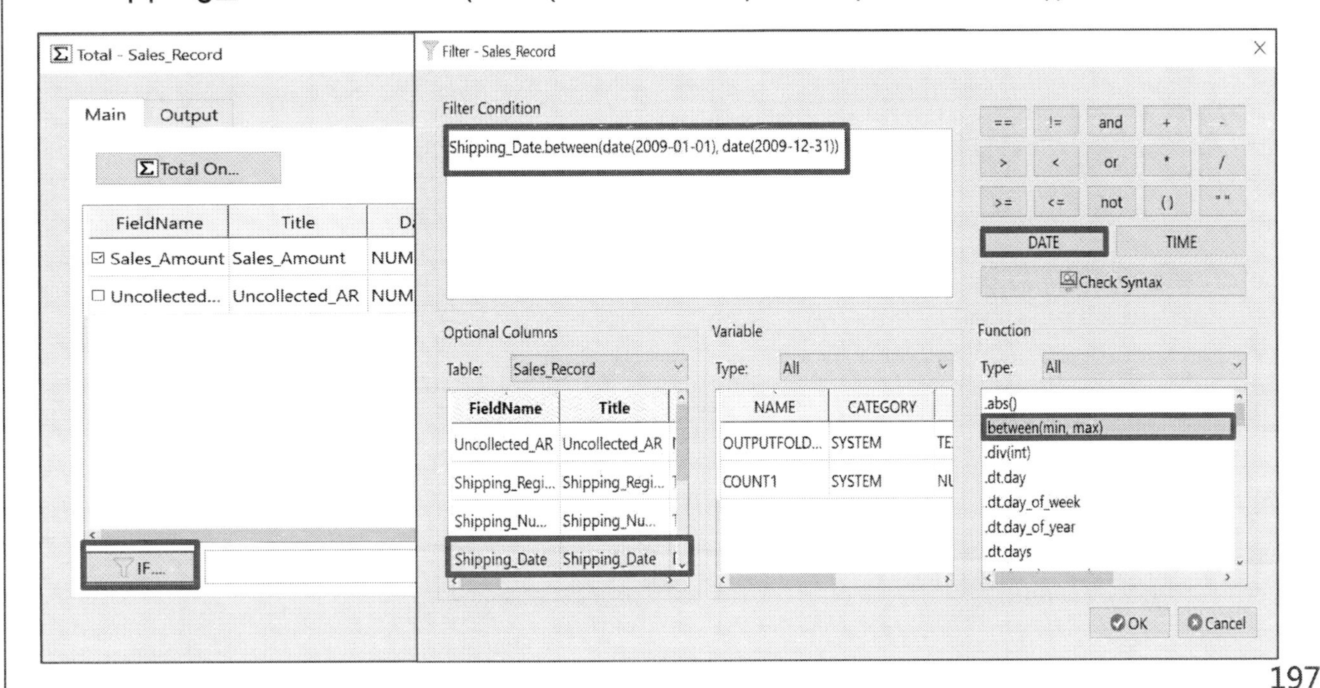

JCAATs-AI Audit Software Copyright © 2023 JACKSOFT.

Exercise 5.9
Control Total validation Total Revenue in 2009

- Output to Screen

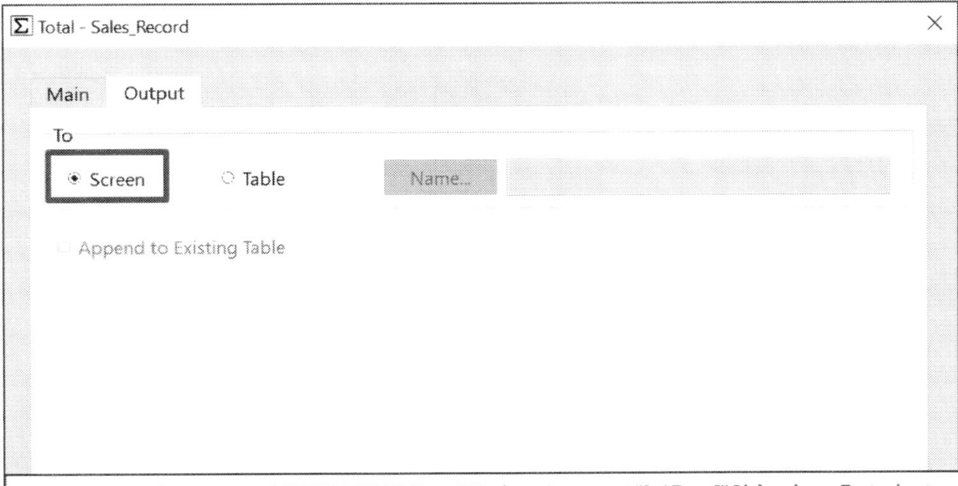

JCAATs >>Sales_Record.TOTAL(PKEYS = ["Sales_Amount"], IF = ["Shipping_Date.between(date(2009-01-01), date(2009-12-31))"])
Table : Sales_Record
Note: 2023/03/01 17:09:45
Result - Records : 1

Table_Name	Field_Name	Total
Sales_Record	Sales_Amount	1,340,450

Exercise 5.9
Control Total validation Number of records in 2010

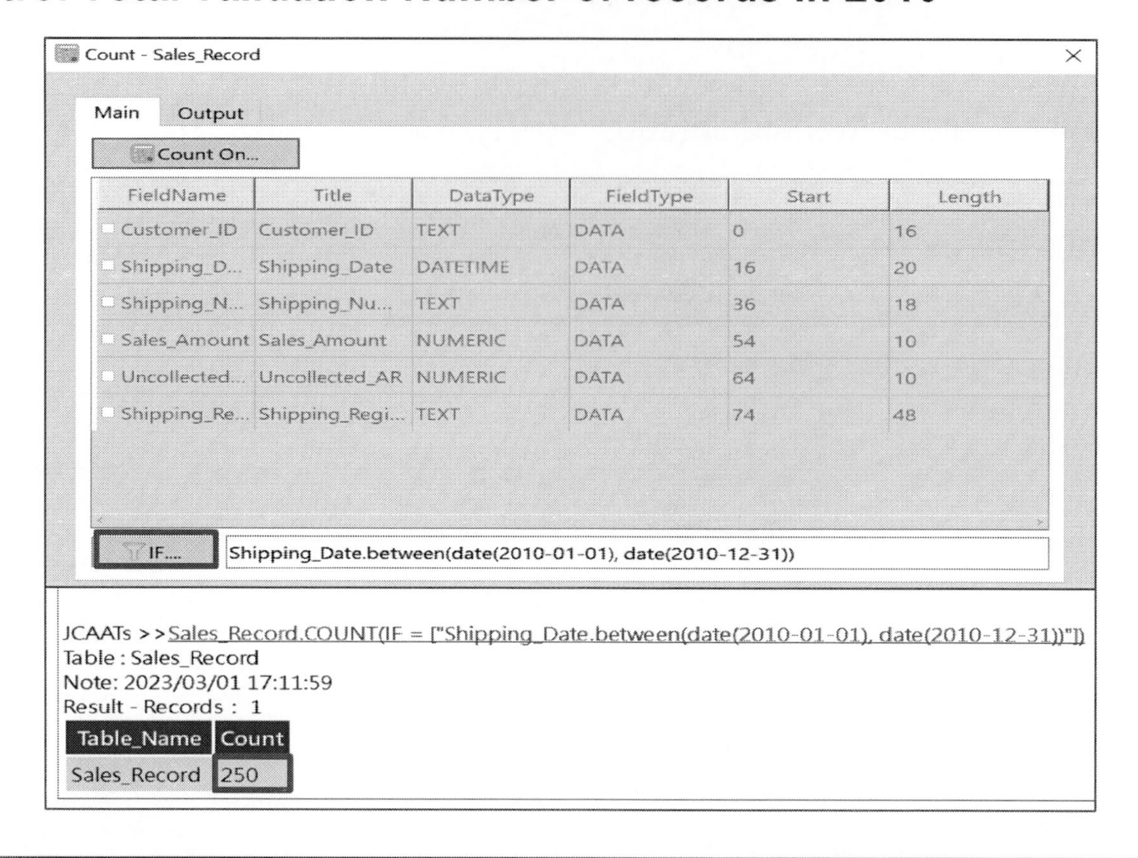

JCAATs-AI Audit Software Copyright © 2023 JACKSOFT.

Exercise 5.9
Control Total validation Total Revenue in 2010

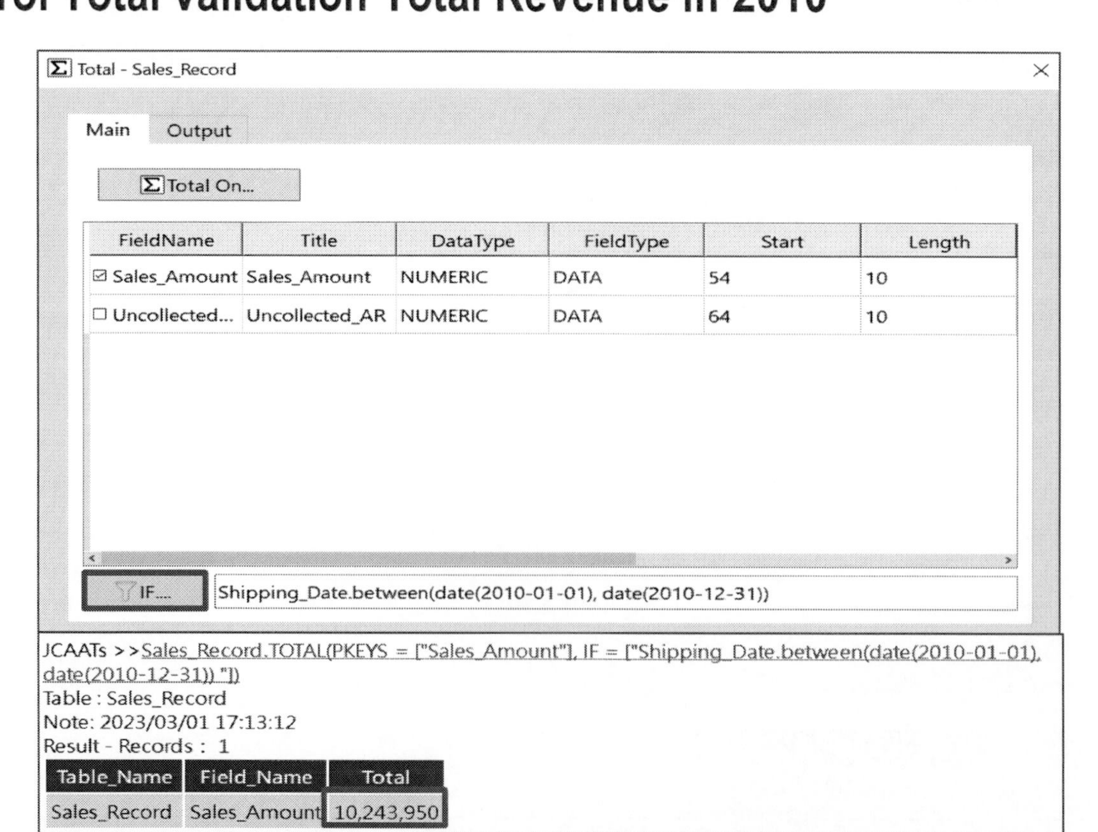

Exercise 5.9
Sequence test: Check whether the shipping numbers are all in sequence order.

- Open "Sales_Record" Table
- Click on "Validate>Sequence"

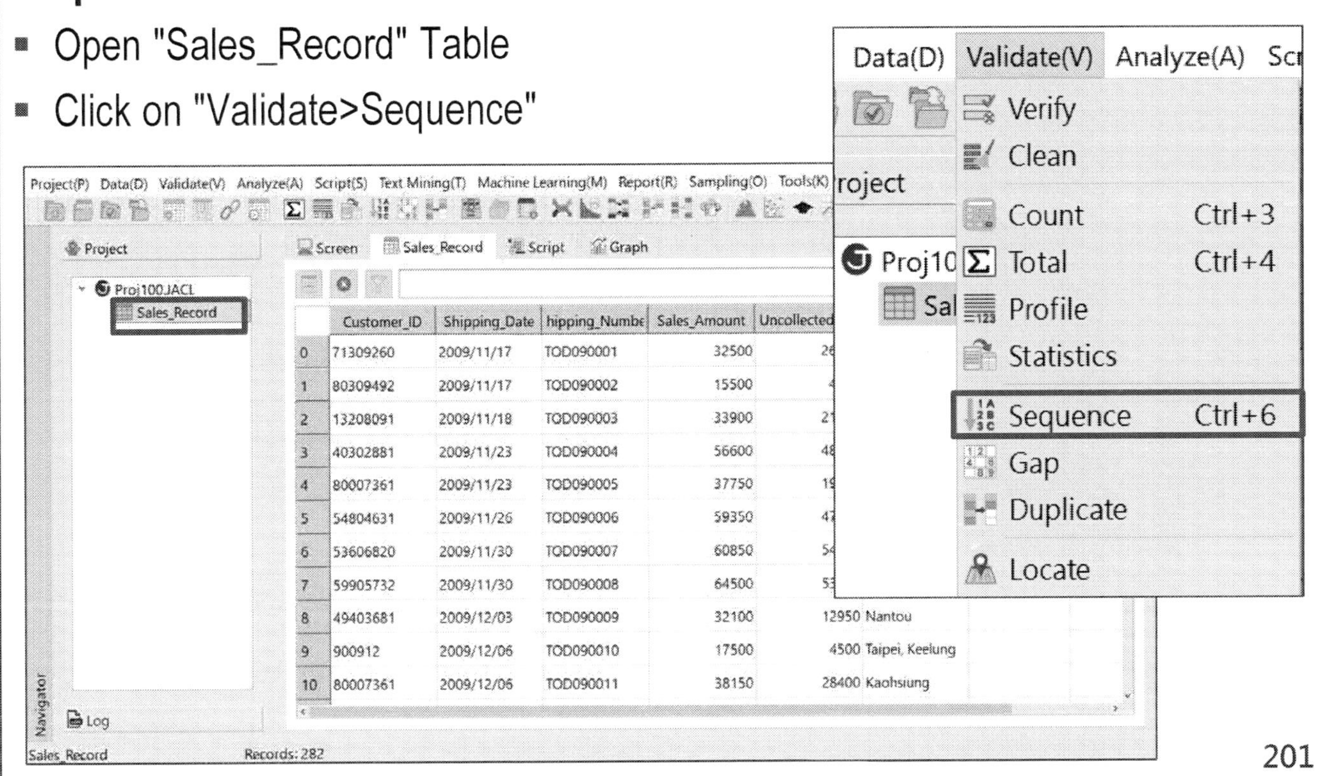

Exercise 5.9
Sequence test: Check whether the shipping numbers are all in sequence order.

- Select the field we need. (Shipping_Number)

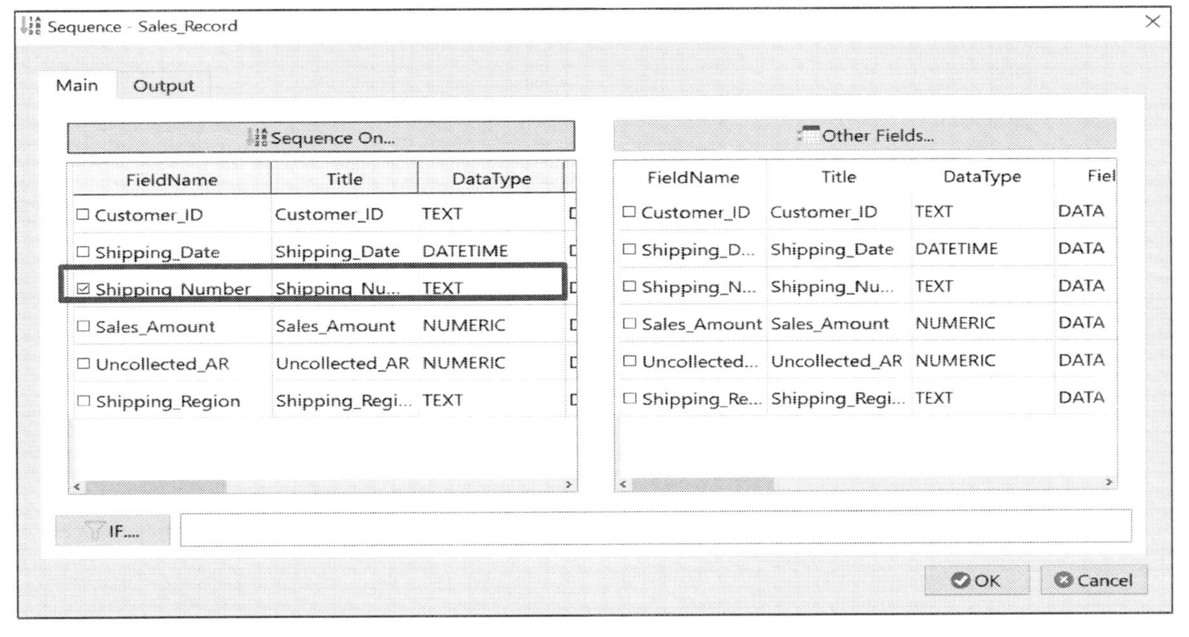

Exercise 5.9
Sequence test: Check whether the shipping numbers are all in sequence order.

- Output to Screen

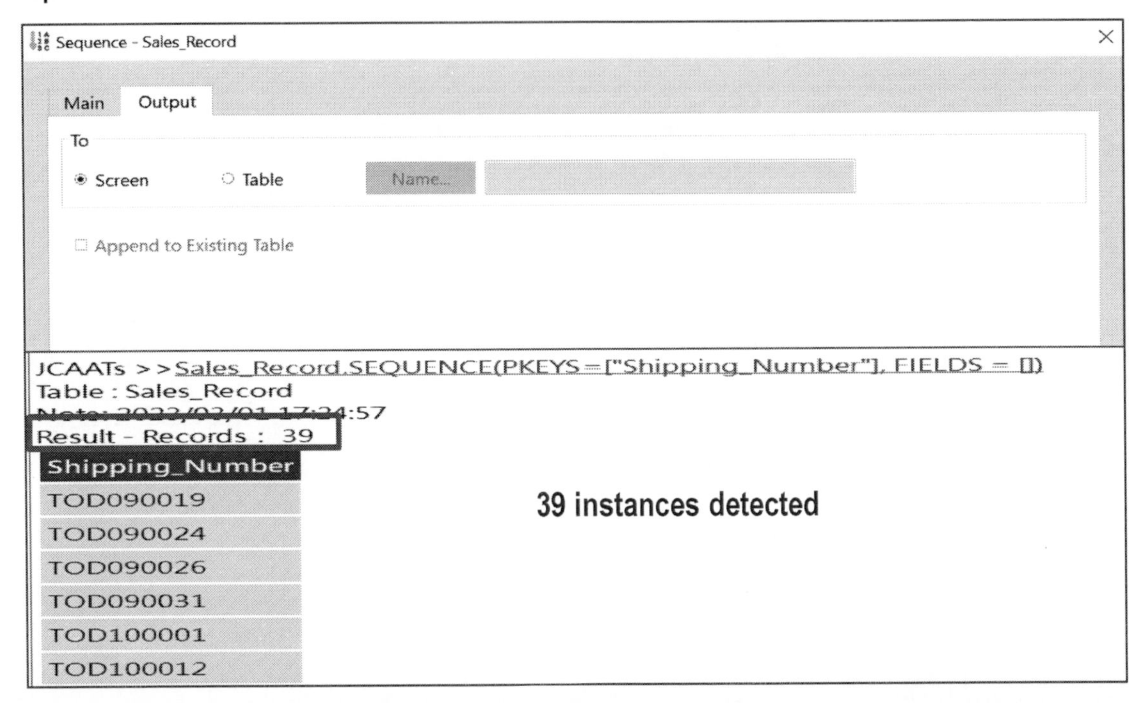

Exercise 5.9
Duplicate test: Are there any duplicate shipping numbers?

- Click on "Validate>Duplicate "
- Select the field we need. (Shipping_Number)

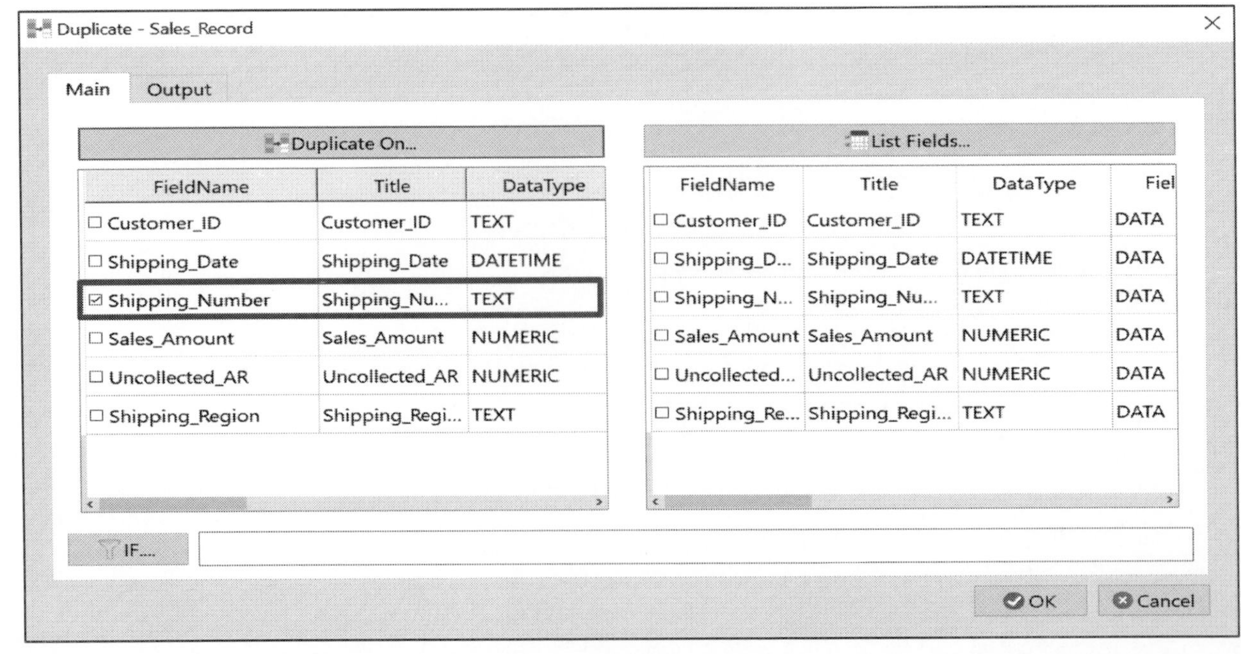

Exercise 5.9
Duplicate test: Are there any duplicate shipping numbers?

- Output to Screen

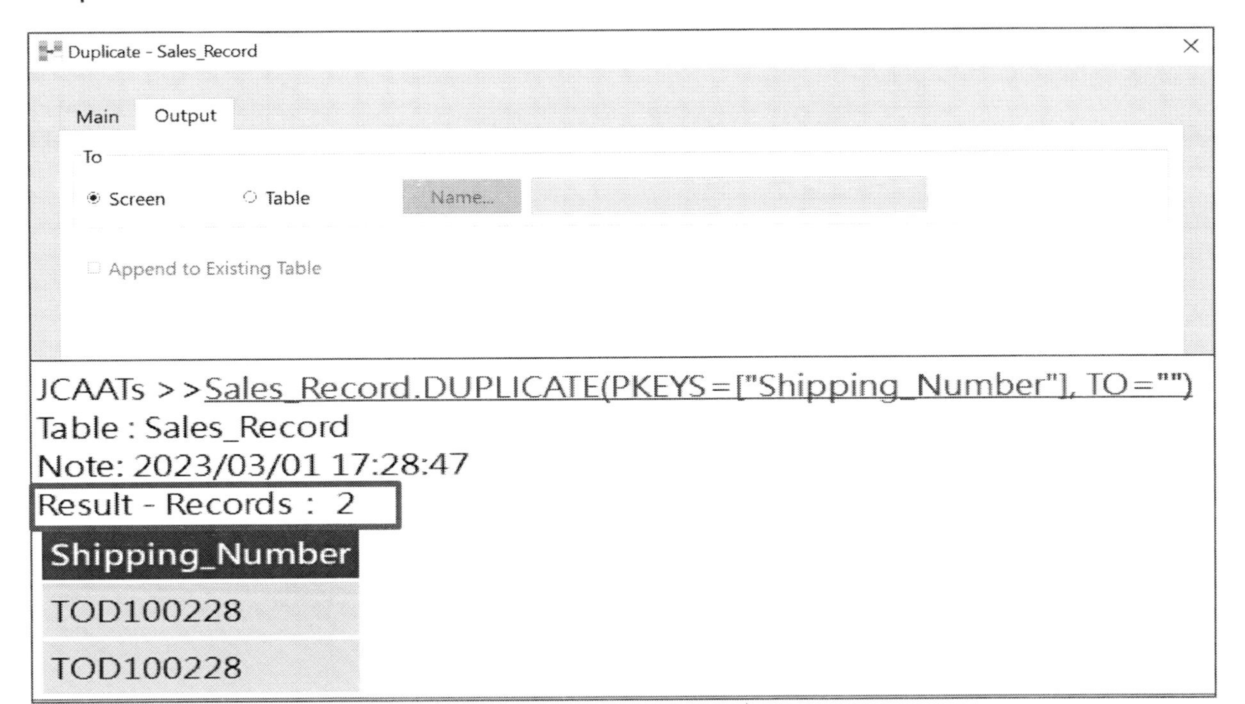

Exercise 5.9
Sales income analysis by region: Which region has the top revenue

- Click on " Analyze>Classify "
- Select the field we need to classify on. (Shipping_Region)
- Subtotal Sales Amount

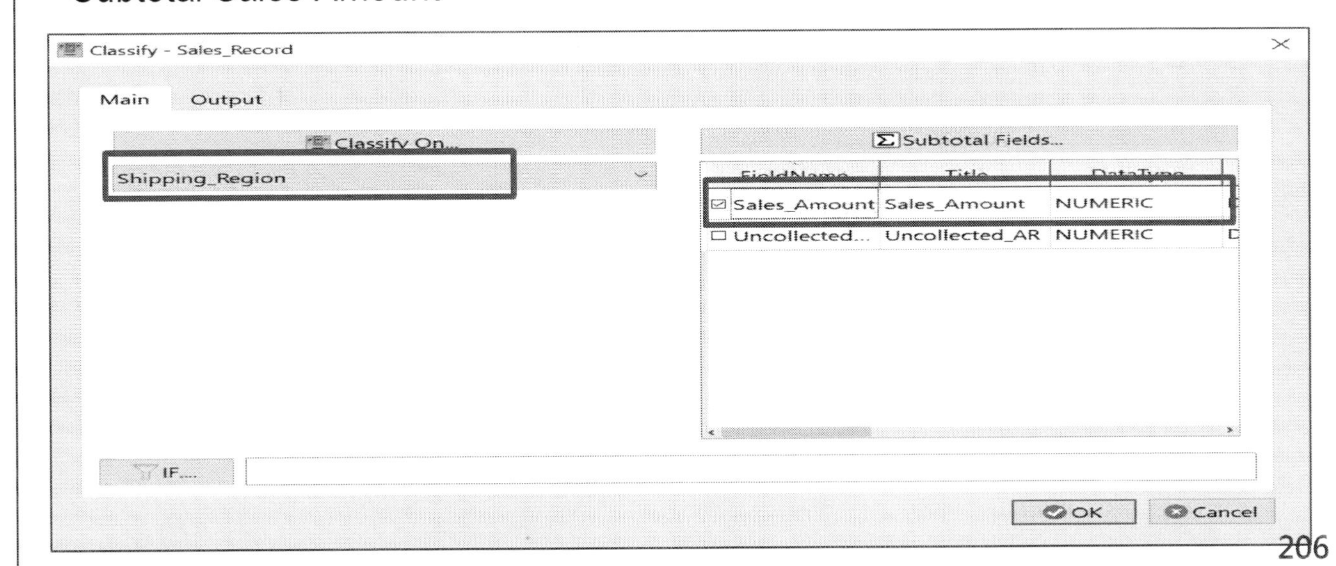

Exercise 5.9
Sales income analysis by region: Which region has the top revenue

- Output to Screen

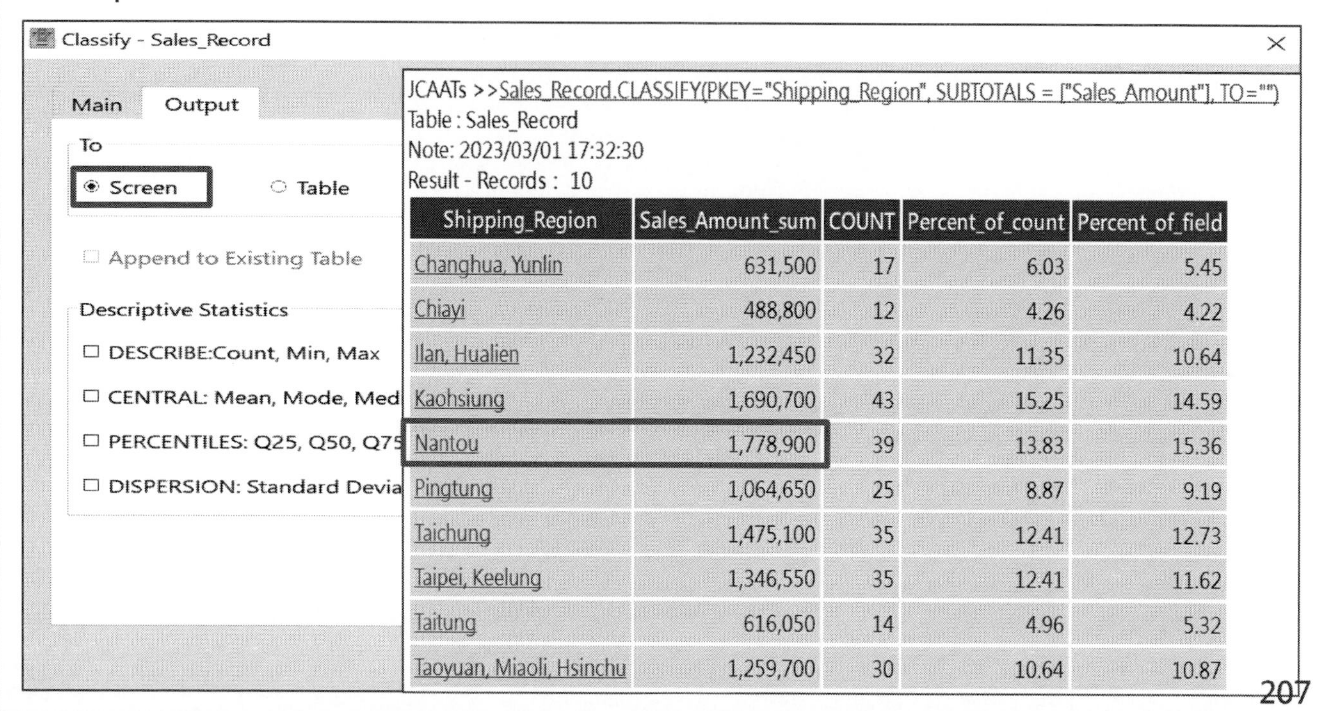

Shipping_Region	Sales_Amount_sum	COUNT	Percent_of_count	Percent_of_field
Changhua, Yunlin	631,500	17	6.03	5.45
Chiayi	488,800	12	4.26	4.22
Ilan, Hualien	1,232,450	32	11.35	10.64
Kaohsiung	1,690,700	43	15.25	14.59
Nantou	1,778,900	39	13.83	15.36
Pingtung	1,064,650	25	8.87	9.19
Taichung	1,475,100	35	12.41	12.73
Taipei, Keelung	1,346,550	35	12.41	11.62
Taitung	616,050	14	4.96	5.32
Taoyuan, Miaoli, Hsinchu	1,259,700	30	10.64	10.87

Exercise 5.9
List of customers whose accumulated transaction amount is greater than 300,000

- Click on " Analyze>Classify "
- Select the field we need to classify on. (Customer_ID)
- Subtotal Sales Amount

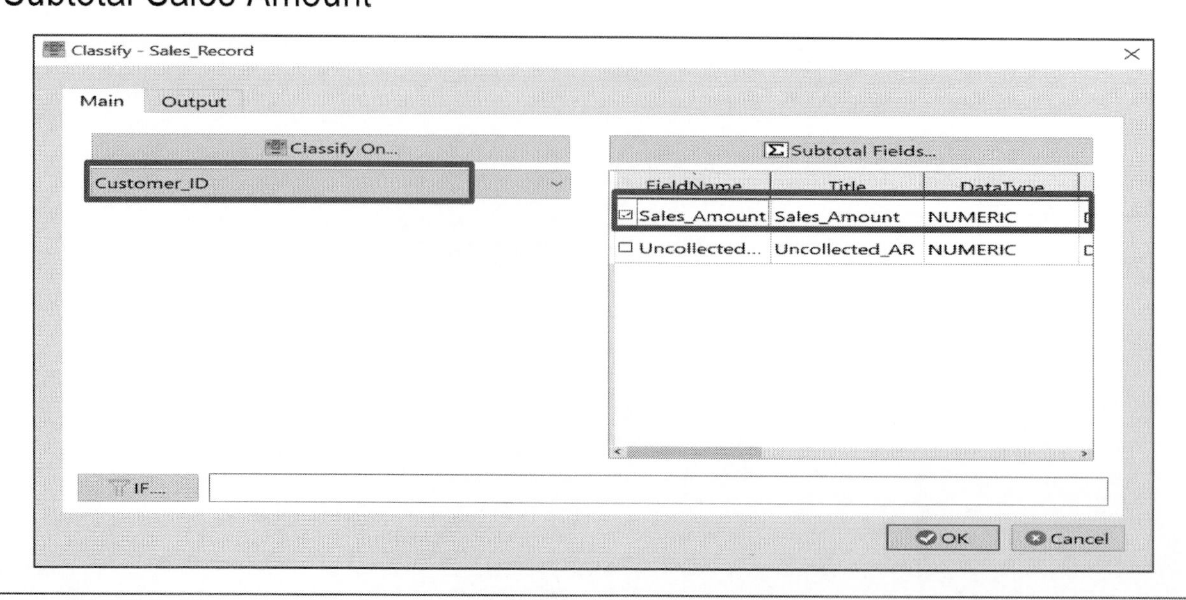

Exercise 5.9
List of customers whose accumulated transaction amount is greater than 300,000

- Output to a new table

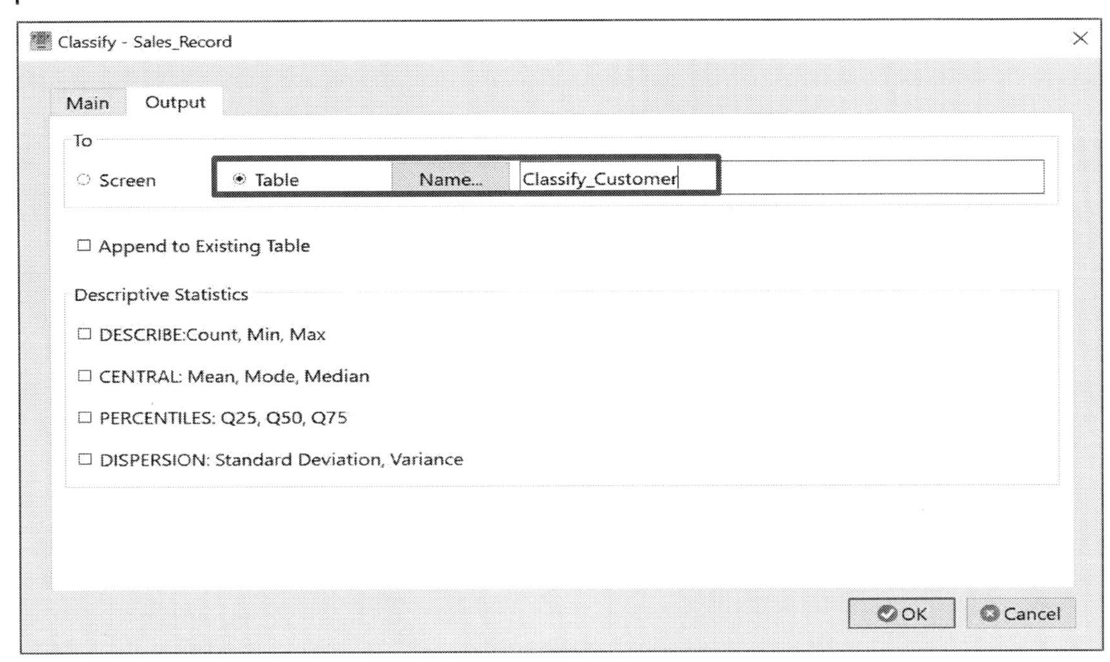

JCAATs-AI Audit Software Copyright © 2023 JACKSOFT.

Exercise 5.9
List of customers whose accumulated transaction amount is greater than 300,000

- Select " Set Filter "

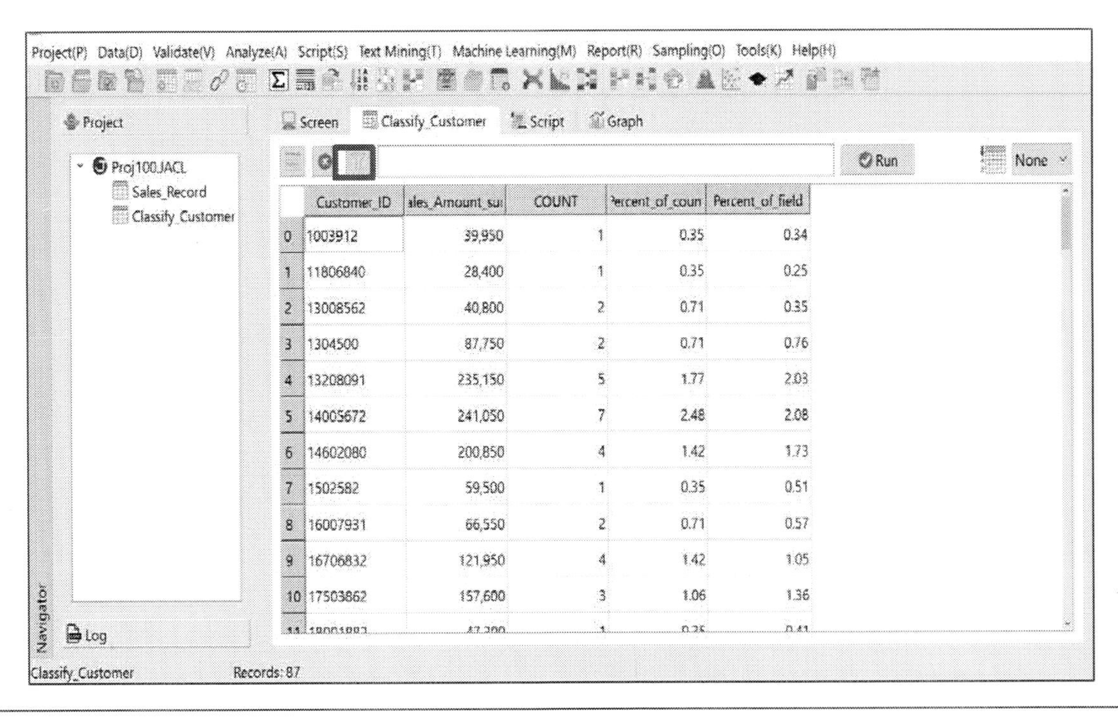

	Customer_ID	ales_Amount_su	COUNT	Percent_of_coun	Percent_of_field
0	1003912	39,950	1	0.35	0.34
1	11806840	28,400	1	0.35	0.25
2	13008562	40,800	2	0.71	0.35
3	1304500	87,750	2	0.71	0.76
4	13208091	235,150	5	1.77	2.03
5	14005672	241,050	7	2.48	2.08
6	14602080	200,850	4	1.42	1.73
7	1502582	59,500	1	0.35	0.51
8	16007931	66,550	2	0.71	0.57
9	16706832	121,950	4	1.42	1.05
10	17503862	157,600	3	1.06	1.36

Exercise 5.9
List of customers whose accumulated transaction amount is greater than 300,000

- Add a Filter Condition:
  Sales_Amount_sum > 300000

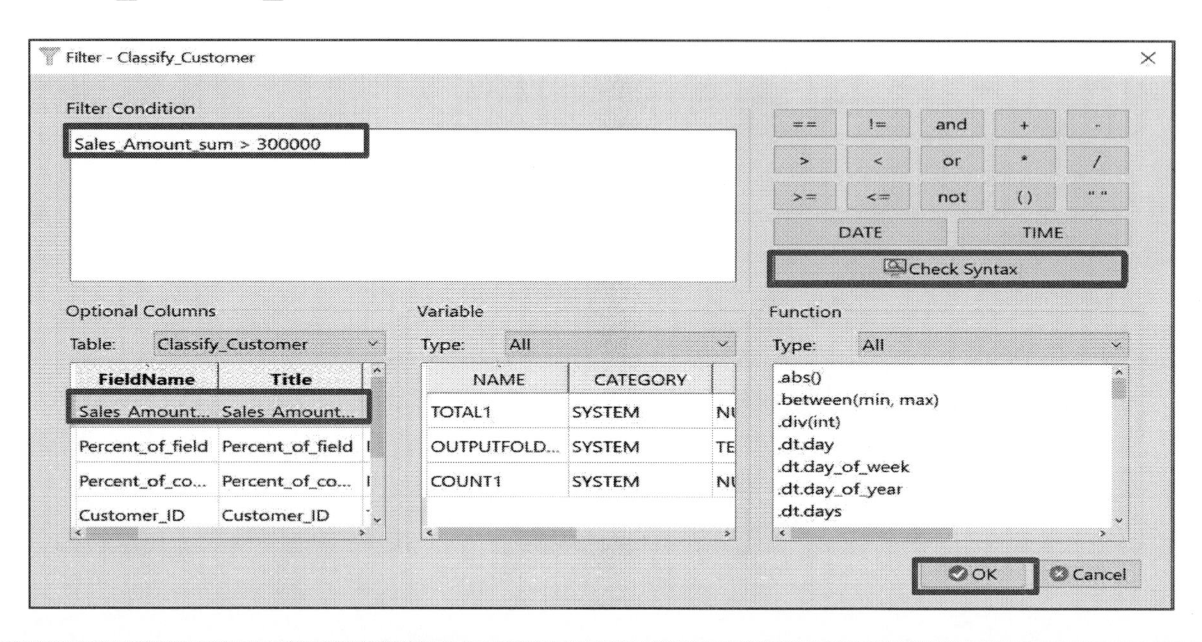

Exercise 5.9
List of customers whose accumulated transaction amount is greater than 300,000

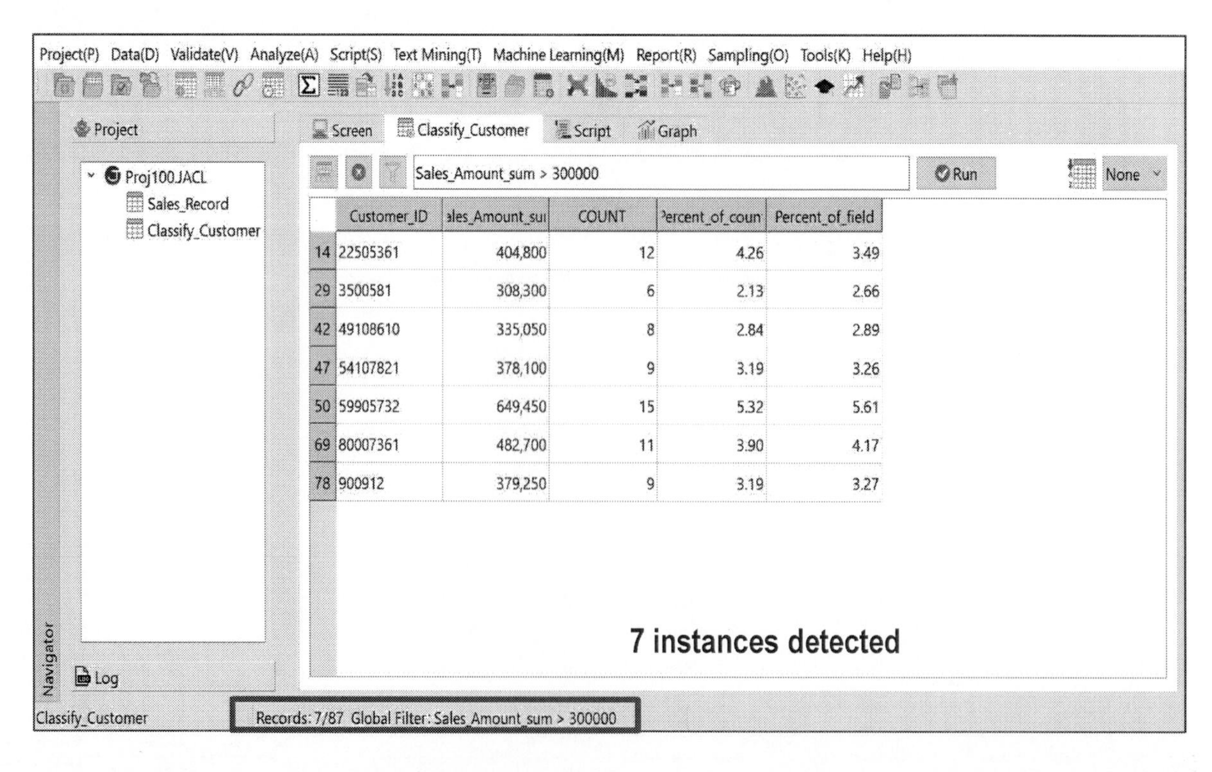

	Customer_ID	ales_Amount_sui	COUNT	Percent_of_coun	Percent_of_field
14	22505361	404,800	12	4.26	3.49
29	3500581	308,300	6	2.13	2.66
42	49108610	335,050	8	2.84	2.89
47	54107821	378,100	9	3.19	3.26
50	59905732	649,450	15	5.32	5.61
69	80007361	482,700	11	3.90	4.17
78	900912	379,250	9	3.19	3.27

Exercise 5.9
List of customers whose accumulated outstanding debt exceeds 200,000

- Click on " Analyze>Classify "
- Select the field we need to classify on. (Customer_ID)
- Subtotal Uncollected_AR

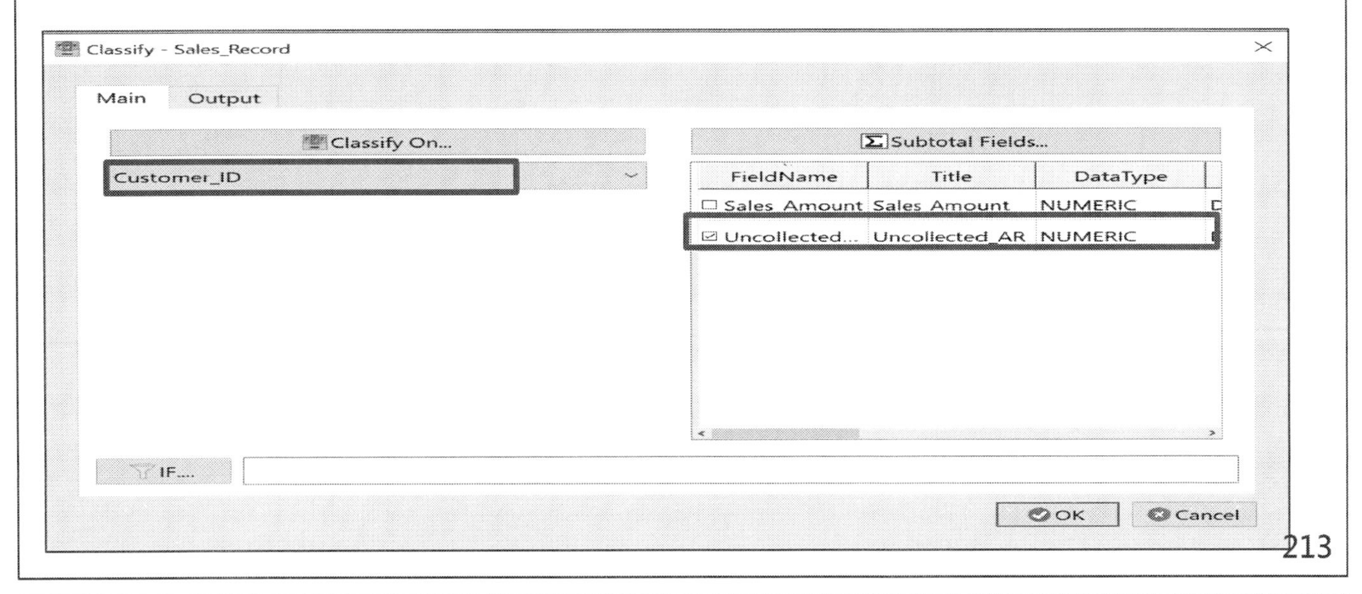

Exercise 5.9
List of customers whose accumulated outstanding debt exceeds 200,000

- Output to a new table

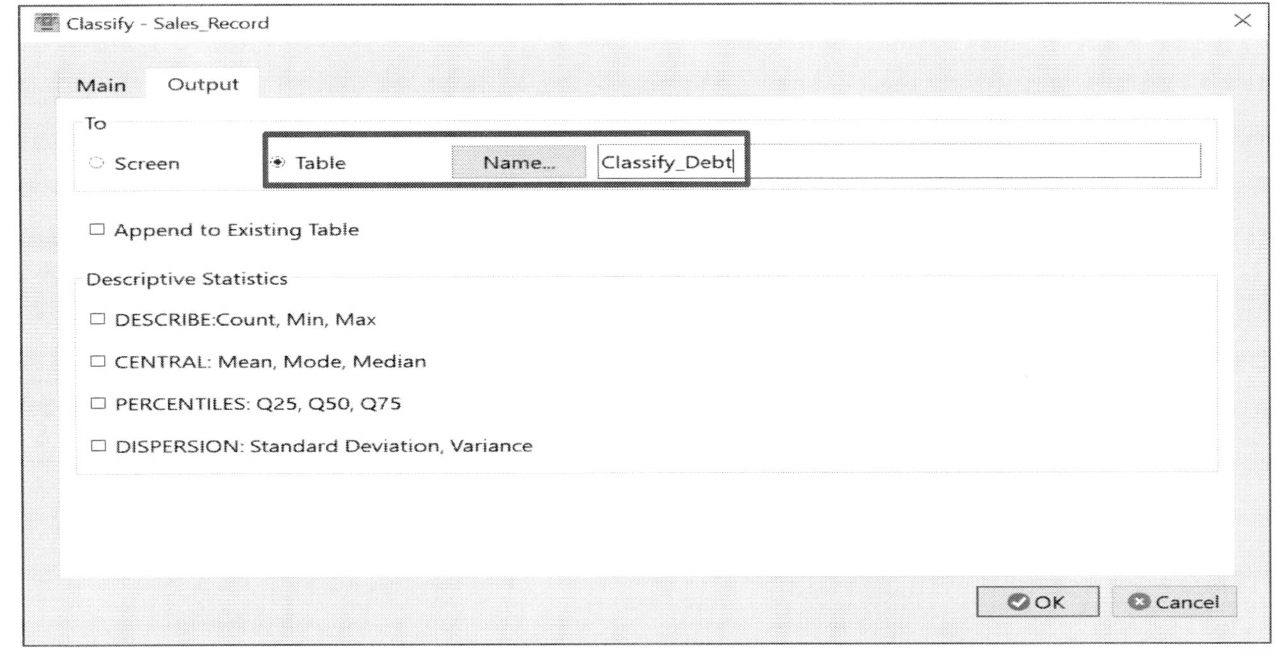

Exercise 5.9
List of customers whose accumulated outstanding debt exceeds 200,000

- Select " Set Filter "

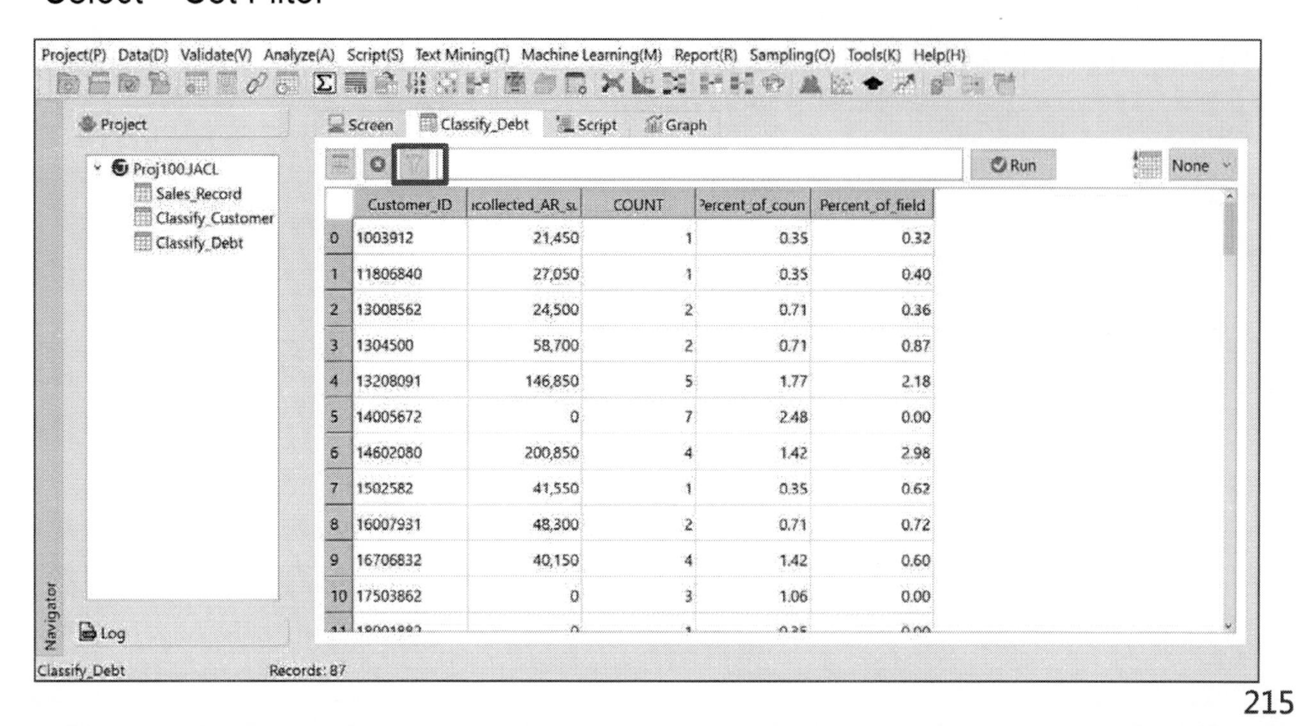

	Customer_ID	ıcollected_AR_sı	COUNT	ʾercent_of_coun	Percent_of_field
0	1003912	21,450	1	0.35	0.32
1	11806840	27,050	1	0.35	0.40
2	13008562	24,500	2	0.71	0.36
3	1304500	58,700	2	0.71	0.87
4	13208091	146,850	5	1.77	2.18
5	14005672	0	7	2.48	0.00
6	14602080	200,850	4	1.42	2.98
7	1502582	41,550	1	0.35	0.62
8	16007931	48,300	2	0.71	0.72
9	16706832	40,150	4	1.42	0.60
10	17503862	0	3	1.06	0.00

Exercise 5.9
List of customers whose accumulated outstanding debt exceeds 200,000

- Add a Filter Condition:
  Uncollected_AR_sum > 200000

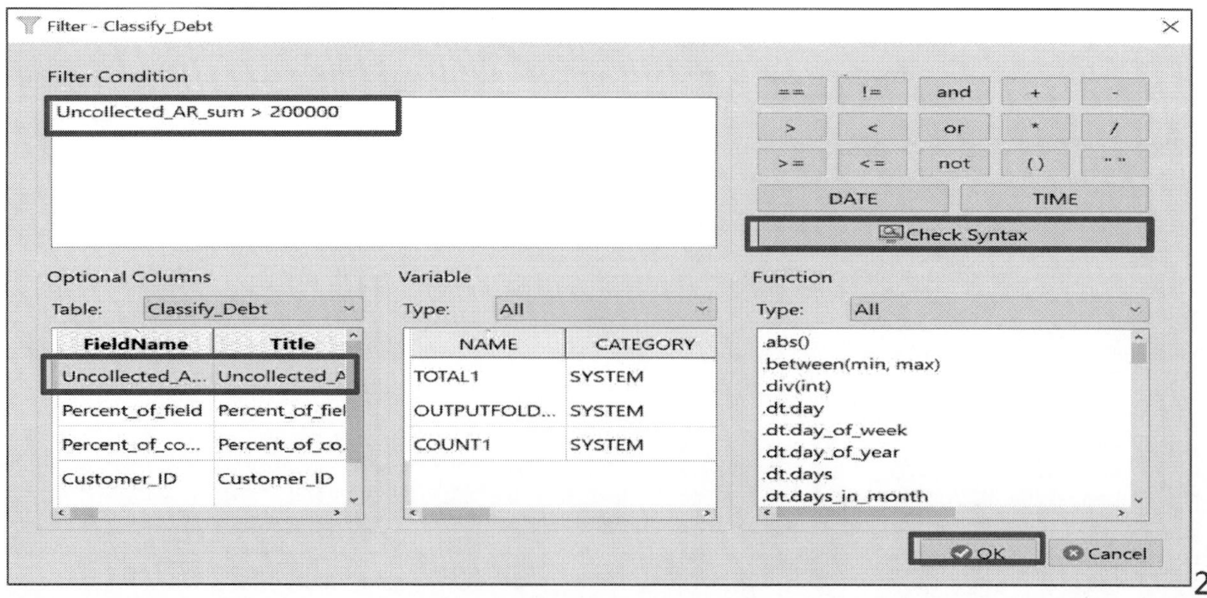

Exercise 5.9

List of customers whose accumulated outstanding debt exceeds 200,000

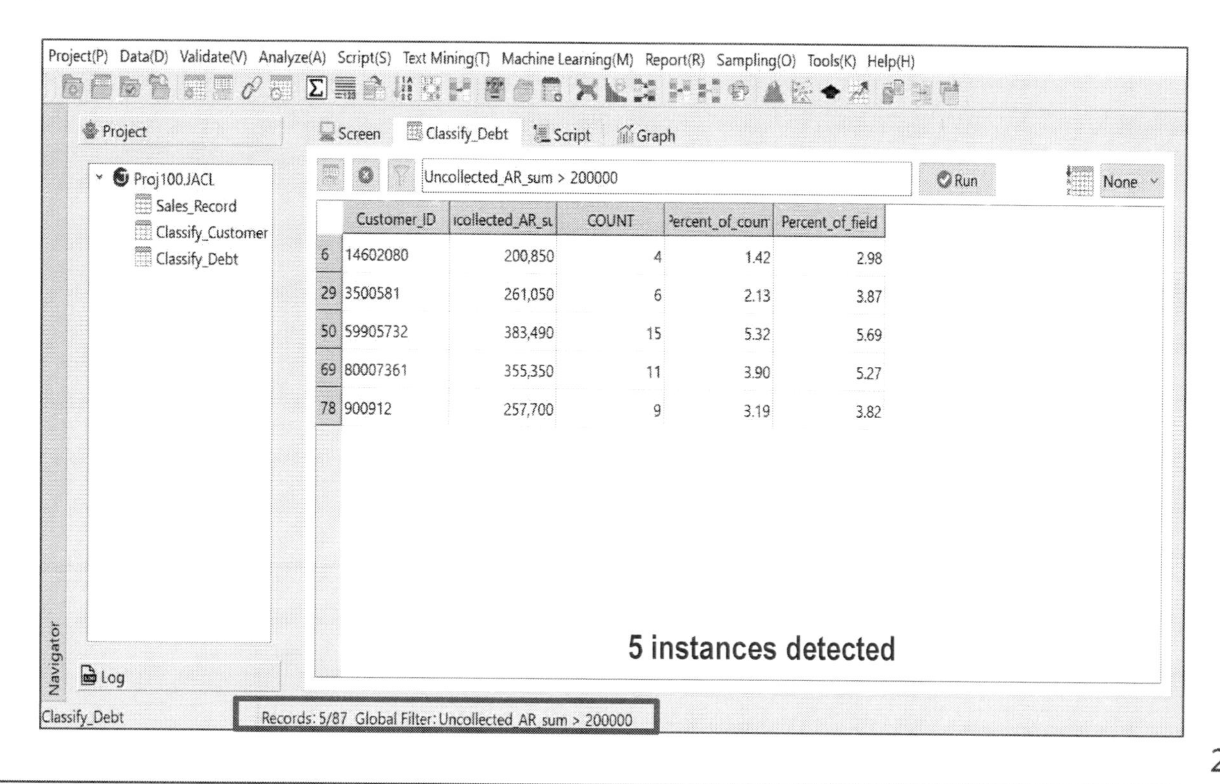

Exercise 5.9

Summarize sales amount by shipment date and analyze whether there is a special concentration of revenue on specific dates

- Click on " Analyze>Summarize "
- Select the field we need to Summarize on.
- Subtotal Sales_Amount

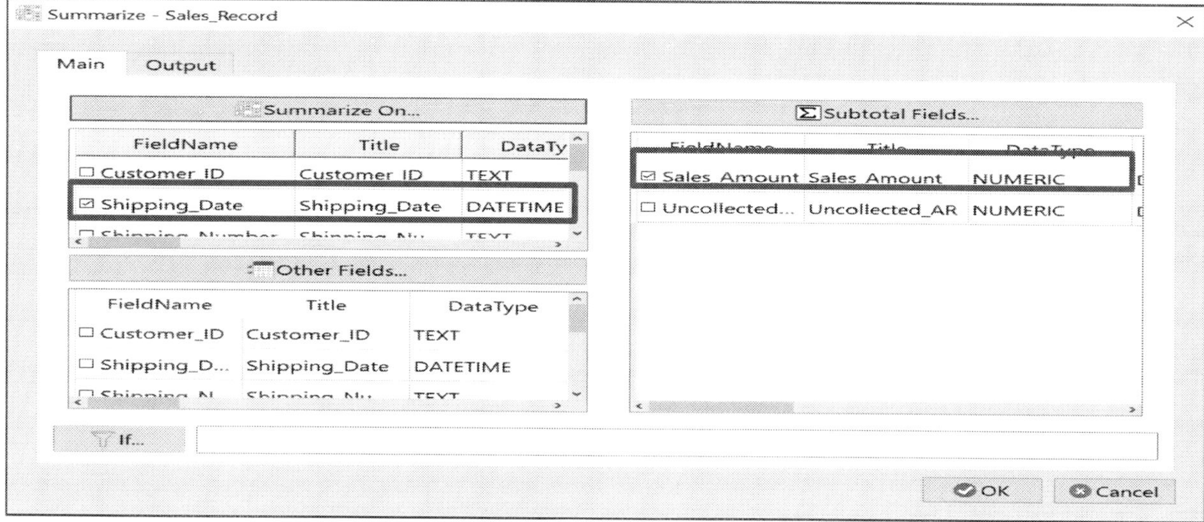

Exercise 5.9
Summarize sales amount by shipment date and analyze whether there is a special concentration of revenue on specific dates

- Output to a new table

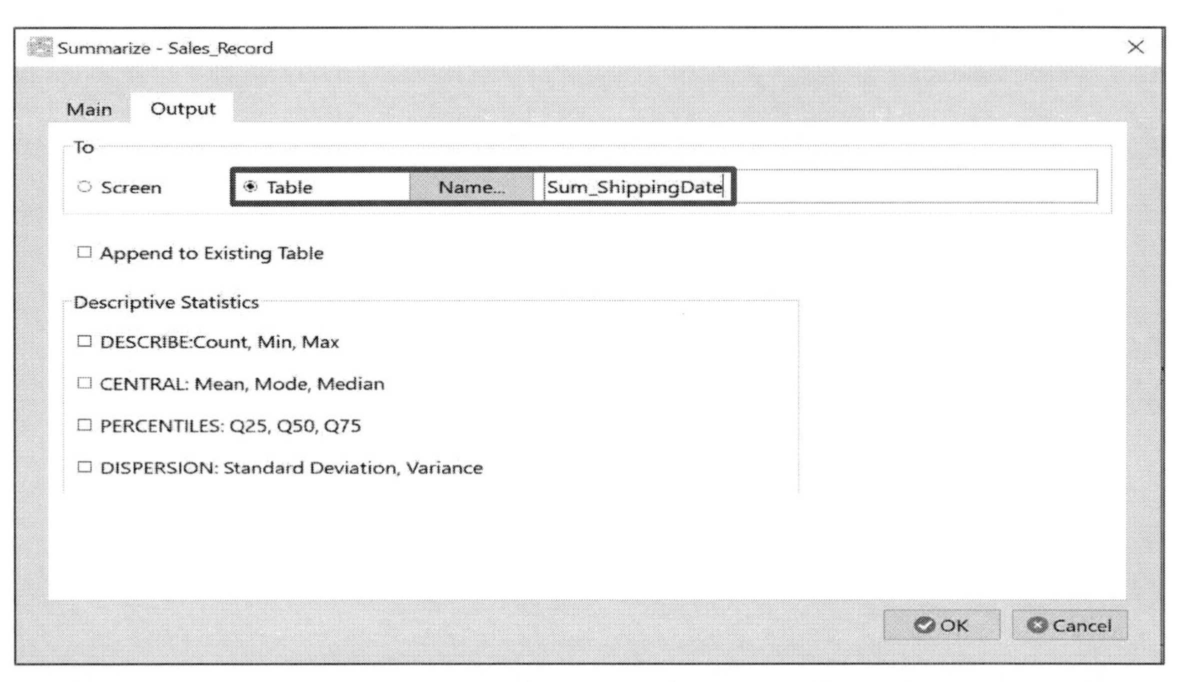

Exercise 5.9
Summarize sales amount by shipment date and analyze whether there is a special concentration of revenue on specific dates

- Add a Filter Condition:

  (Shipping_Date == date(2009-12-31) and COUNT > 10) or

  (Shipping_Date == date(2010-12-31) and COUNT > 10)

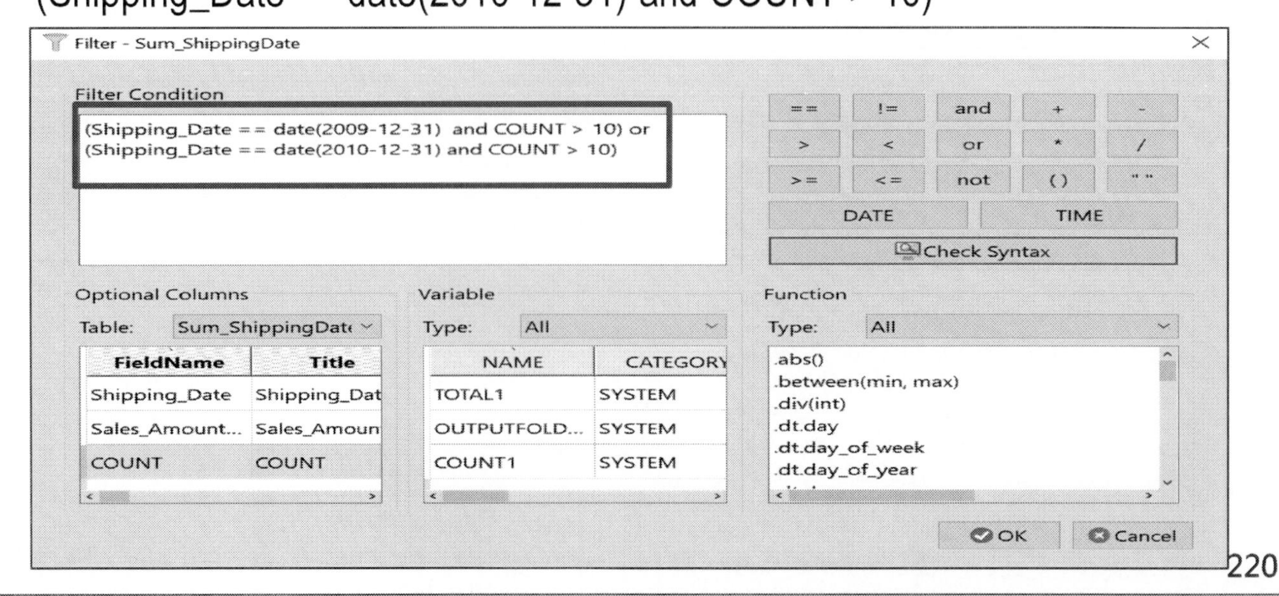

Exercise 5.9
Summarize sales amount by shipment date and analyze whether there is a special concentration of revenue on specific dates

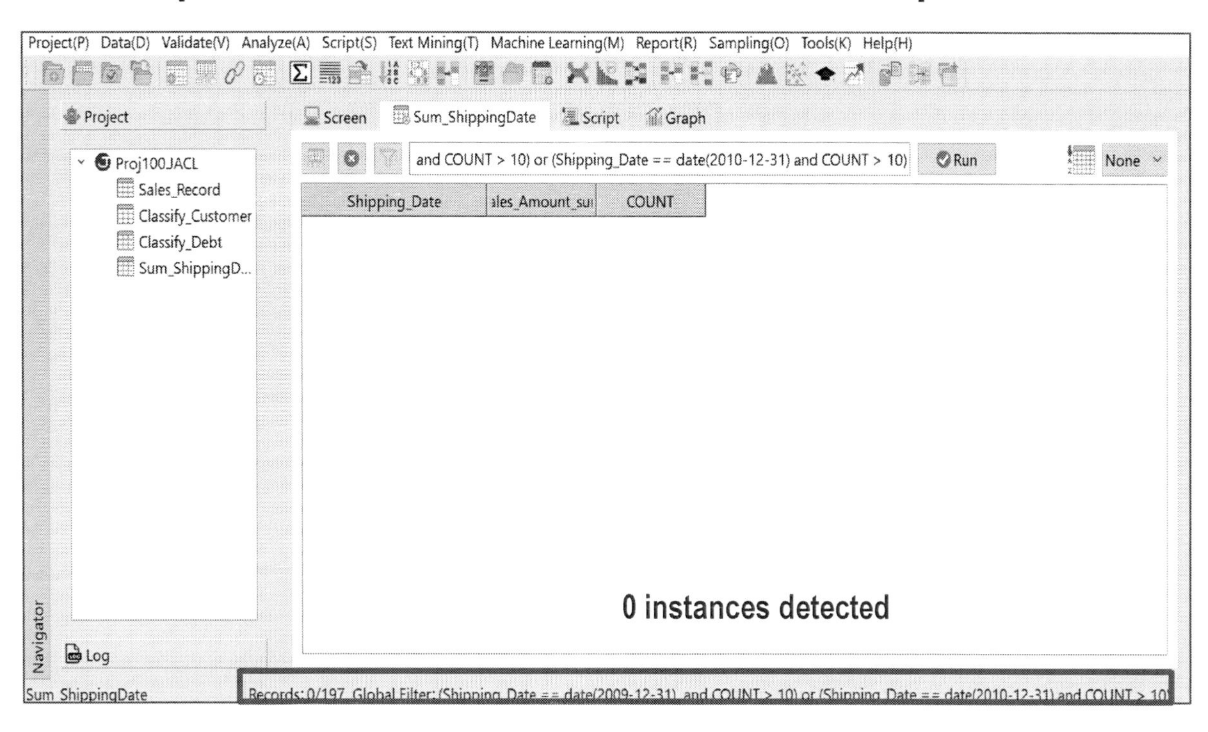

Exercise 5.9
Analysis of Sales Revenue by Region

- Click on " Analyze>Classify "
- Select the field we need to classify on. (Shipping_Region)
- Subtotal Sales_Amount

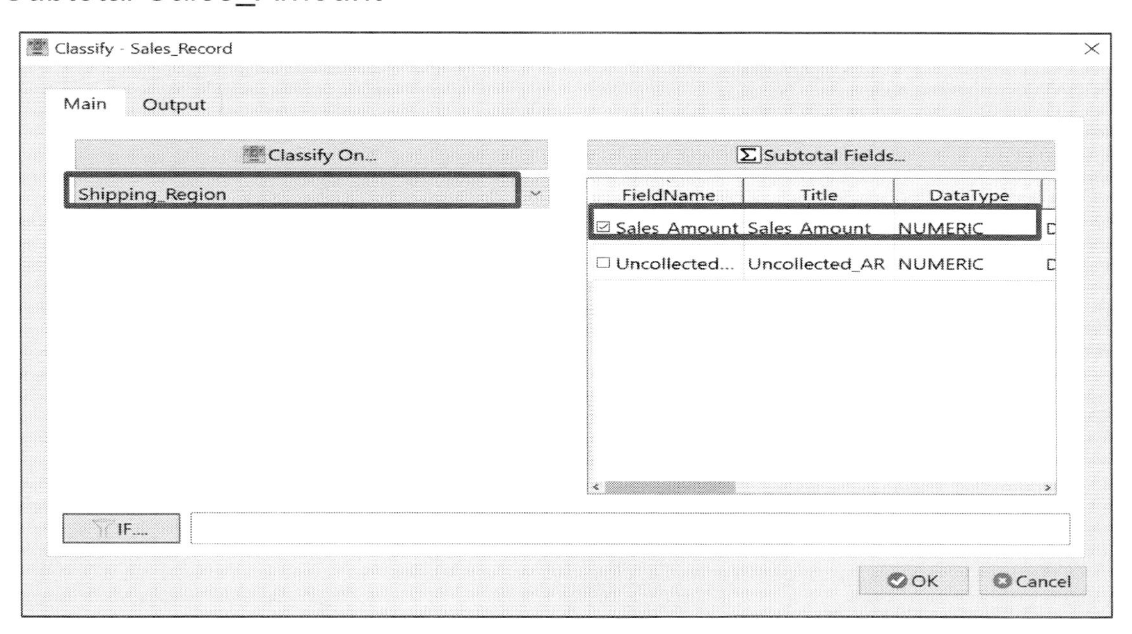

Exercise 5.9
Analysis of Sales Revenue by Region

- Add a Filter Condition:

  (Shipping_Region == "Taipei, Keelung" or Shipping_Region == "Taichung" or Shipping_Region == "Chiayi" or Shipping_Region == "Kaohsiung"

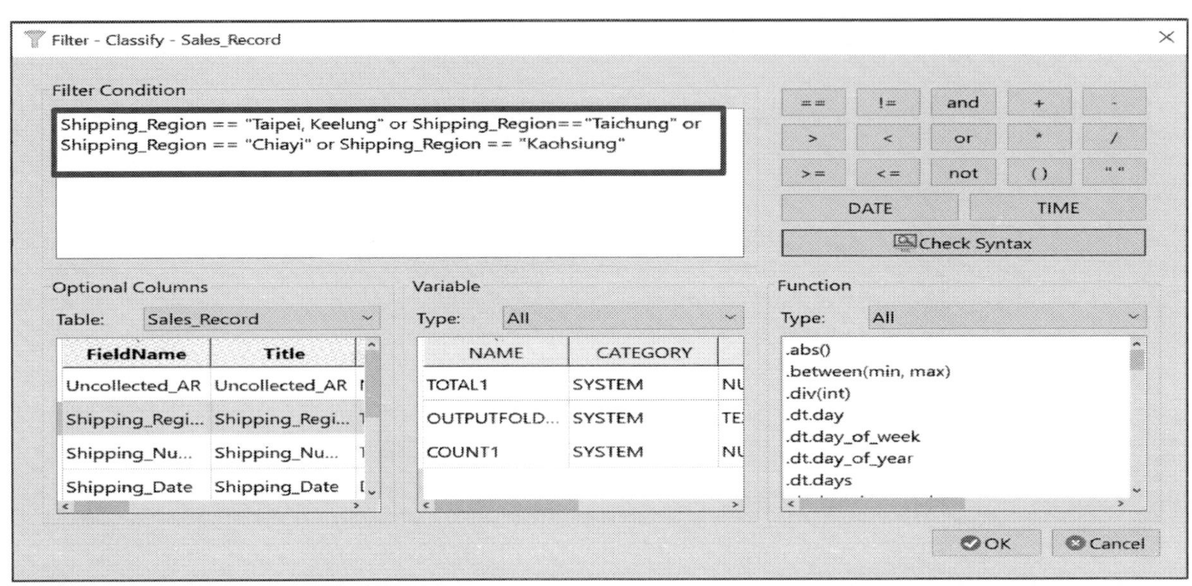

Exercise 5.9
Analysis of Sales Revenue by Region

- Output to Screen

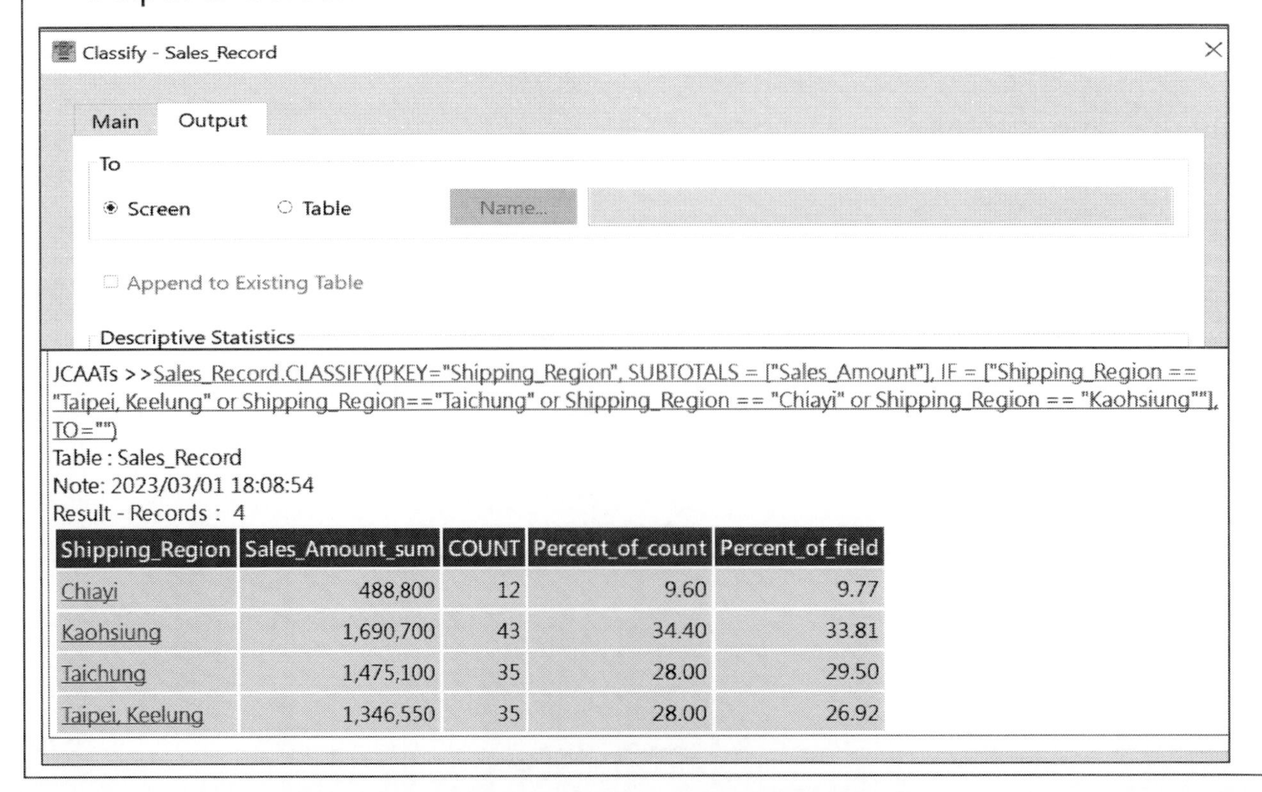

JCAATs >>Sales_Record.CLASSIFY(PKEY="Shipping_Region", SUBTOTALS = ["Sales_Amount"], IF = ["Shipping_Region == "Taipei, Keelung" or Shipping_Region=="Taichung" or Shipping_Region == "Chiayi" or Shipping_Region == "Kaohsiung""], TO="")
Table : Sales_Record
Note: 2023/03/01 18:08:54
Result - Records : 4

Shipping_Region	Sales_Amount_sum	COUNT	Percent_of_count	Percent_of_field
Chiayi	488,800	12	9.60	9.77
Kaohsiung	1,690,700	43	34.40	33.81
Taichung	1,475,100	35	28.00	29.50
Taipei, Keelung	1,346,550	35	28.00	26.92

Exercise 5.9
Analysis of Receivables by Region

- Click on " Analyze>Age "
- Select the field we need to age on. (Shipping_Date)
- Subtotal Uncollected_AR
- Define Cutoff Date, aging periods, and filter condition.

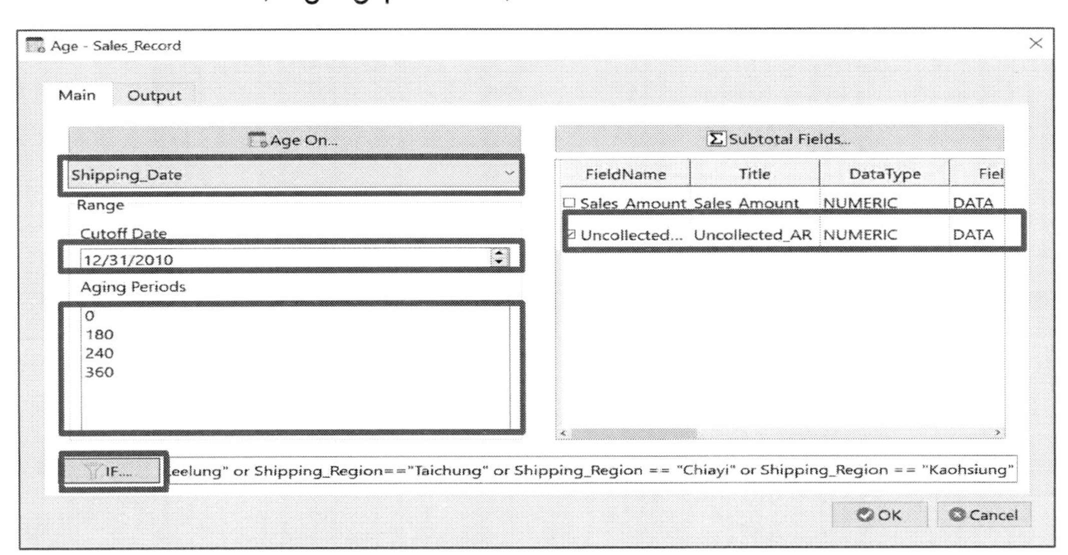

Exercise 5.9
Analysis of Receivables by Region

JCAATs >>Sales_Record.AGE(KEY="Shipping_Date", SUBTOTAL = ["Uncollected_AR"], CUTOFF = ["2010/12/31"], INTERVAL = ["0","181","241","361"], IF = ["Shipping_Region == "Taipei, Keelung" or Shipping_Region=="Taichung" or Shipping_Region == "Chiayi" or Shipping_Region == "Kaohsiung" "], TO="")
Table : Sales_Record
Note: 2023/03/01 18:12:41
Result - Records : 6

Days	Uncollected_AR_sum	COUNT	Percent_of_count	Percent_of_field
< 0	0	0	0.00	0.00
0 ~ 180	987,050	59	47.20	37.68
181 ~ 240	500,000	20	16.00	19.09
241 ~ 360	623,840	26	20.80	23.82
>= 361	508,500	20	16.00	19.41
NaT	0	0	0.00	0.00

Chapter 6 - Exercise Questions

- Exercise 6.1: Please create a new JCAATs project and practice copying from another project data (**ERP_Audit**).

- Exercise 6.2: Please practice using the **Verify command** to test if there are any abnormalities in the project data table. Export the abnormal information for later audit use, and use the **Locate command** to understand the abnormal information.

- Exercise 6.3: Please practice using the **Clean command** to purify the problematic data according to the audit requirements.

JCAATs Learning notes：

Python Based Computer-Assisted Audit Techniques (CAATs)

Data Analysis and Smart Audit

Chapter 6 . Validate

Outline:

1. How to Perform Data Validation
2. Validate Command Interface
3. Validate Command List

JCAATs - AI Audit Software

1. How to Perform Data Validation

How to perform data validation?

- JCAATs provides several commands to assist in verifying the analyzed data, ensuring high data quality, and reducing the risk of audit errors.
- JCAATs data validation procedures：

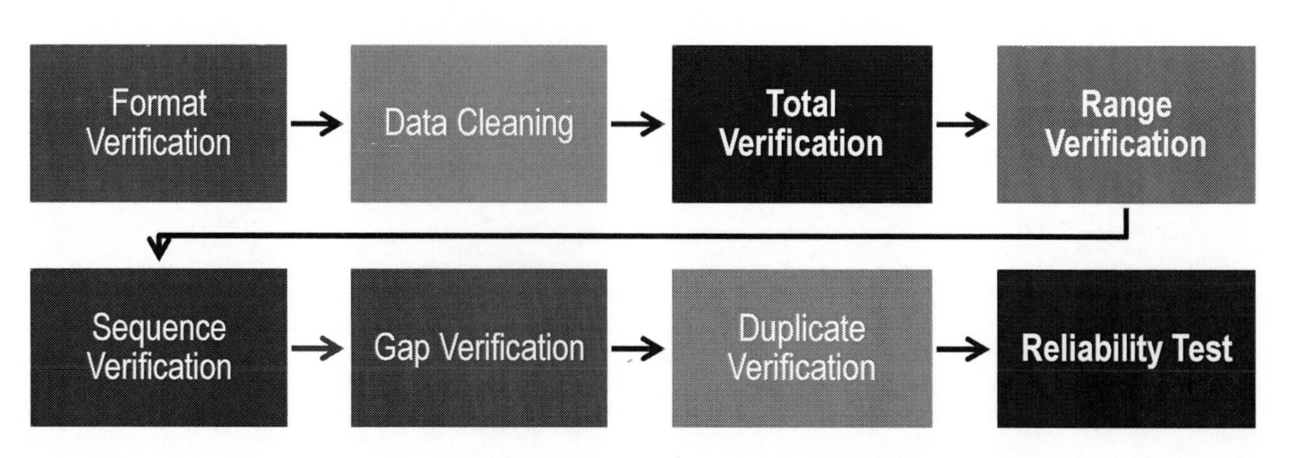

Step 1: Format Verification

- To ensure that a table is valid by checking the following criteria:
  - Text fields should only include text that is meant for visualization.
  - Numeric fields should only include numbers, commas, minus signs, and currency symbols.
  - Datetime fields should only include effective dates.
- JCAATs provides the "**Verify**" command as a way to check the validity of field data. This command can be used to perform various types of format and data validation checks, helping to ensure data accuracy and completeness.

- **If error found:**
  - If the table's format is incorrect, check its validity again after making corrections.
  - If the table content is incorrect, retrieve the data again or try to correct it.
  - If the error still occurs, decide how to deal with the error, such as such as dropping erroneous data or imputing missing values, etc.

Step 2: Data Cleaning

- The primary purpose of data cleaning is to improve the quality and accuracy of the data used in analysis.
  - Providing a purification mechanism to handle missing values in numeric fields.
  - Provide a purification mechanism for missing values in text fields.
  - Provide a purification mechanism for missing values in datetime fields.
- JCAATs can use the "**Clean**" command to purify the analysis data.

- **Tips:**
  - JCAATs provides a greater level of control and flexibility in data analysis compared to ACL and IDEA.
  - Users can choose how to handle gaps or missing values in data, with options to drop them, fill them with the mode value, or leave them unchanged. The system exports a new table for verification purposes, while retaining the original table as evidence, ensuring that data analysis is based on accurate and complete data.

Step 3: Total Validation

- Ensure the data conforms to the summary report's illustration.
- Compare control totals generated by JCAATs with the originally reported totals.
- The control totals can be verified using the following command:
  - **Count** command
  - **Total** command
  - **Profile** command
  - **Statistics** command

- **If the control totals do not match the report,**
  - This usually indicates that the extracted data is not appropriate or does not accurately represent the original data.
  - If there is an excessive amount of data, filters can be used to extract only the required data..
  - If there is insufficient data, the data access process should be repeated to obtain the required amount of data.

Step 4: Range Verification

- Numeric and date data must adhere to defined upper and lower bounds to ensure that they meet the required specifications.
- To check for upper and lower bounds on data use the following commands :
  - **Profile** command
  - **Statistics** command

- **If records do not fall within the specific bounds:**
- If the table contains a large number of records that are outside the desired range, it may be necessary to extract only the relevant records and save them as a new table, ensuring that the data being analyzed is accurate and within the desired range.
- If there is insufficient data, the data access process should be repeated to obtain the required amount of data.

In JCAATs, the Statistics command provides results that include the mode of text data, which allows the analyst to analyze the text boundary conditions in more detail.

Step 5: Sequence Verification

- Check for non-sequential data if the data records required sequence.
- Verify for any non-sequential data that may indicate fraudulent insertions.
- To perform sequence test, you can use:
  - Sequence command
- If a non-sequential item is found in the data:
  - Check the item and determine its validity.
  - Inform the data provider about the non-sequential item.
  - Create a new table to confirm the reason for the non-sequential item. This helps to ensure that the data is accurate and reliable for analysis.

Step 6: Gap Verification

- Verify for any potential gaps or missing values in the data extracted from the source system.
- Check for any irregularities or fraudulent activities in the data drawn from the source system.
- To perform gap test, you can use:
  - Gap command
- If any gaps or missing values are found in the data :
  - Determine the level of impact analysis needed.
  - Notify the data provider of the discovery.

Step 7: Duplicate Verification

- Check for duplicate entries in records or fields to ensure that data is accurate and complete.
- Investigate the possibility of fraudulent insertions in the data to ensure data integrity and security.
- To perform duplicate test, you can use:
  - Duplicate command
- If a duplicate item is found:
  - Examine the findings to determine their validity.
  - Inform the data provider of the issue.
  - Create a new table to confirm the reason for the duplicate.

JCAATs-AI Audit Software
Copyright © 2023 JACKSOFT.

Step 8: Reliability Test

- Select the specific data to check reliability.
- Specifically test against the values resulting from calculations.
- To perform duplicate test, you can use:
  - Locate command
- If a mismatch is found with the original field:
  - the data provider should be informed.
  - find the specific record data for field review and recalculation.
  - If the error is significant and cannot be resolved easily, resetting the data may be necessary.

It is essential to ensure that the calculations and the resulting values are accurate to avoid any potential data issues.

2. Validate Command Interface

Validate Command Interface

- The "validate commands" adhere to standard interface design. There are two primary tabs on the interface, such as
  - **Main** tab: Set main parameters for the command.
  - **Output** tab: Set output parameters for the command.

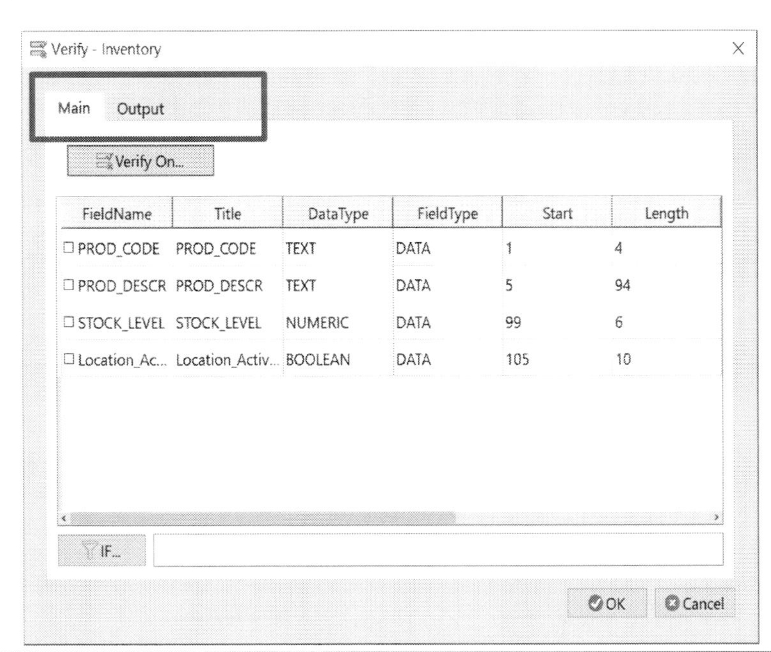

Main tab - Standard

- The standard Main tab includes two main process icon: **On...** and **IF...**
  - **On... :** a dialog to select the required validation fields.
  - **IF... :** a dialog to set your local filter.

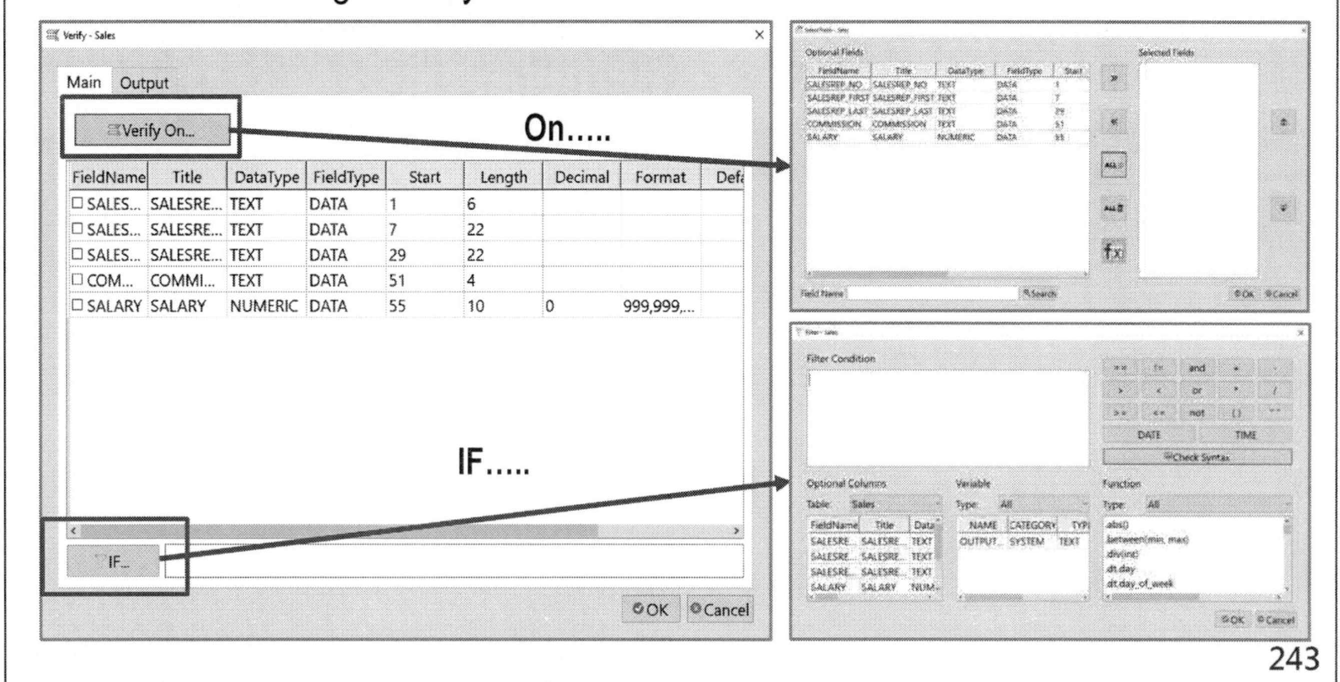

Main tab - Extension

- This interface includes an extension that adds one or two operators to the standard Main tab interface.
- These operators include other affiliated functions, which are often in the form of a checkbox or input parameter (such as **Statistics** command). In some cases, additional execution options are available through button functions (such as **Clean** command).

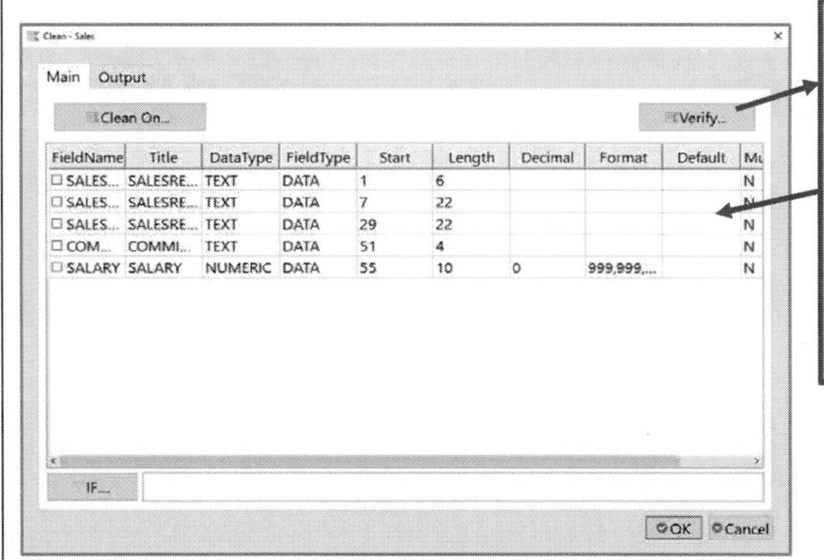

The accessibility features allow for direct execution of the **Verify** command, exporting of fields with errors, and displaying them in a selection table. These features can significantly improve your productivity and efficiency while using the application.

Main tab - Two Columns with Single Selection

- This interface is divided into two columns. The right column supports a single field selection with some of the available operators, while the left column supports selection of multiple fields.

Main tab - Two Columns with Multiple Selections

- This interface has two columns, both of which support multiple selections with some of the available operators. The settings interface includes other affiliated functions, often in the form of a checkbox or input parameter (such as **Sequence**, and **Duplicate**).

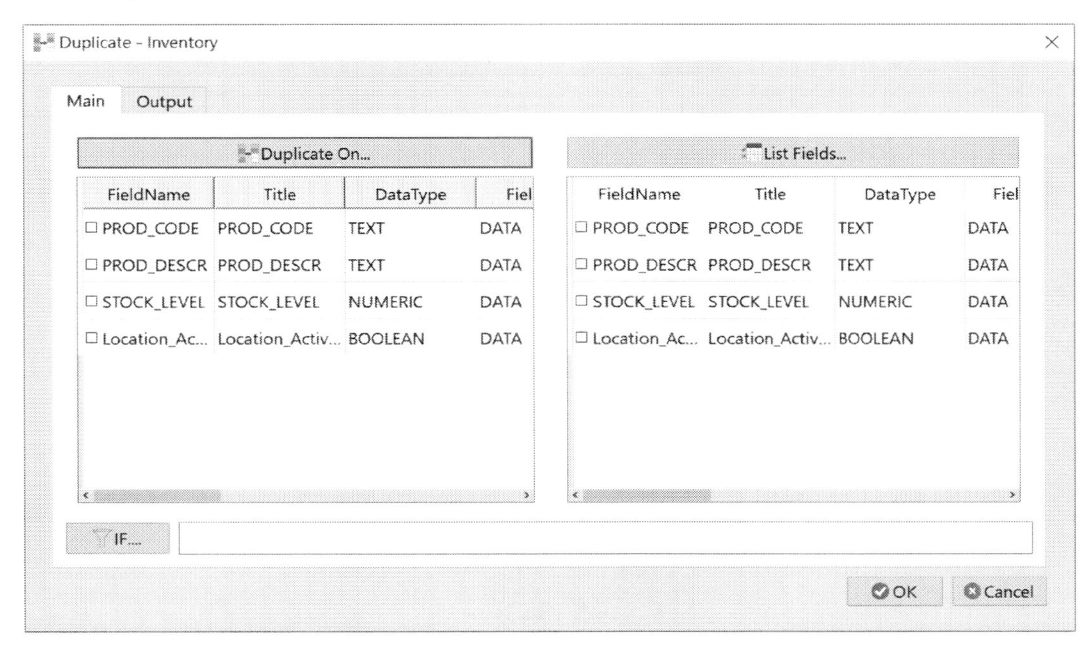

Output tab - Standard

- A significant difference between JCAATs and other audit software is that in JCAATs, all command analysis results can be easily exported to a table.
- This feature is particularly beneficial for complex audits or when multiple analyses need to be performed on the same data set, as it allows users to perform more detailed analysis on the exported data.

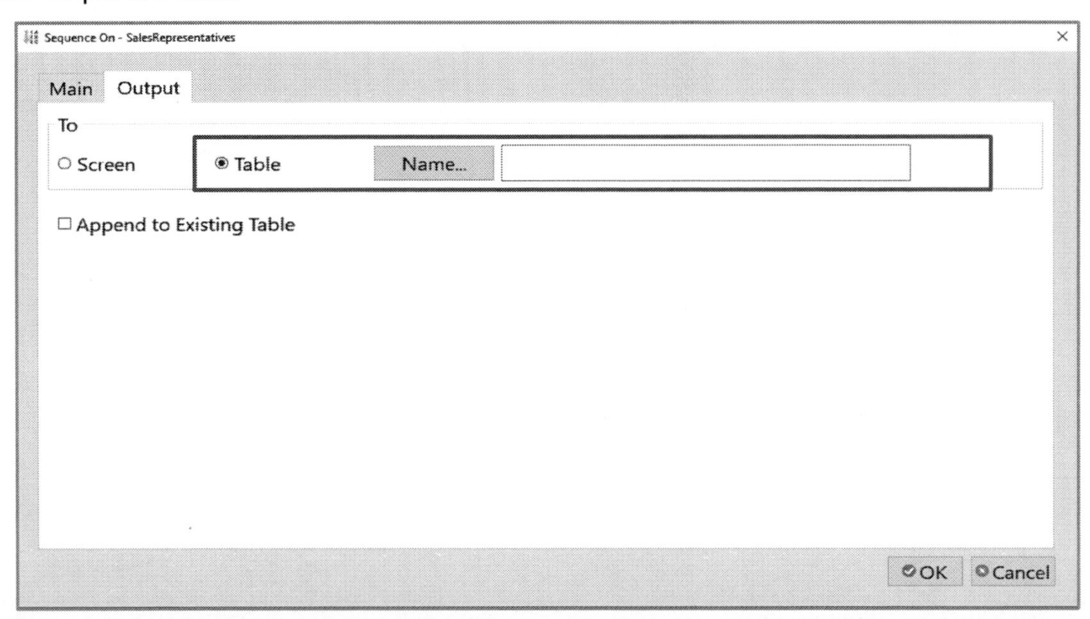

Output tab - with Descriptive Statistics

- In JCAATs, if both the key field and the command result are numeric, users have the option to export an output result that includes **descriptive statistics**. This feature is particularly useful for conducting more detailed data exploration and analysis.

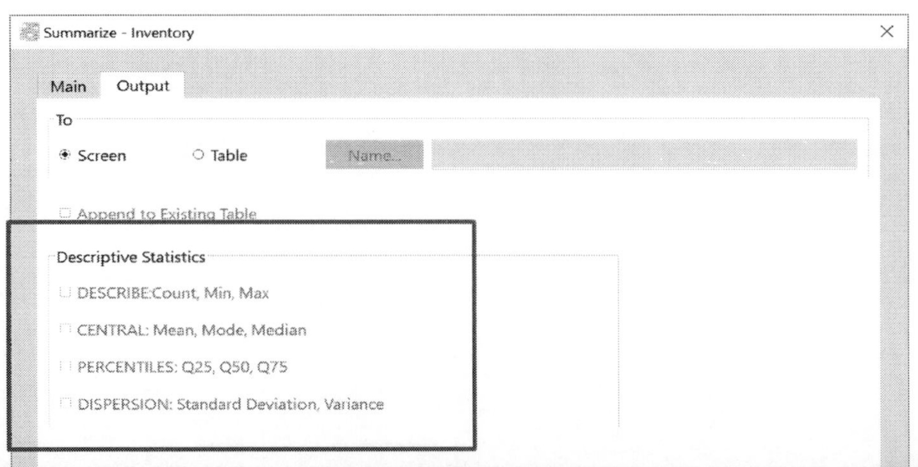

Descriptive statistics include 4 categories with 11 indicators.

Output tab - with Missing Value Imputation

- Missing value imputation is a technique used in data analysis to fill in missing data points with estimated values. It is often used when data is incomplete or missing due to various reasons, such as data entry errors, fraud etc.

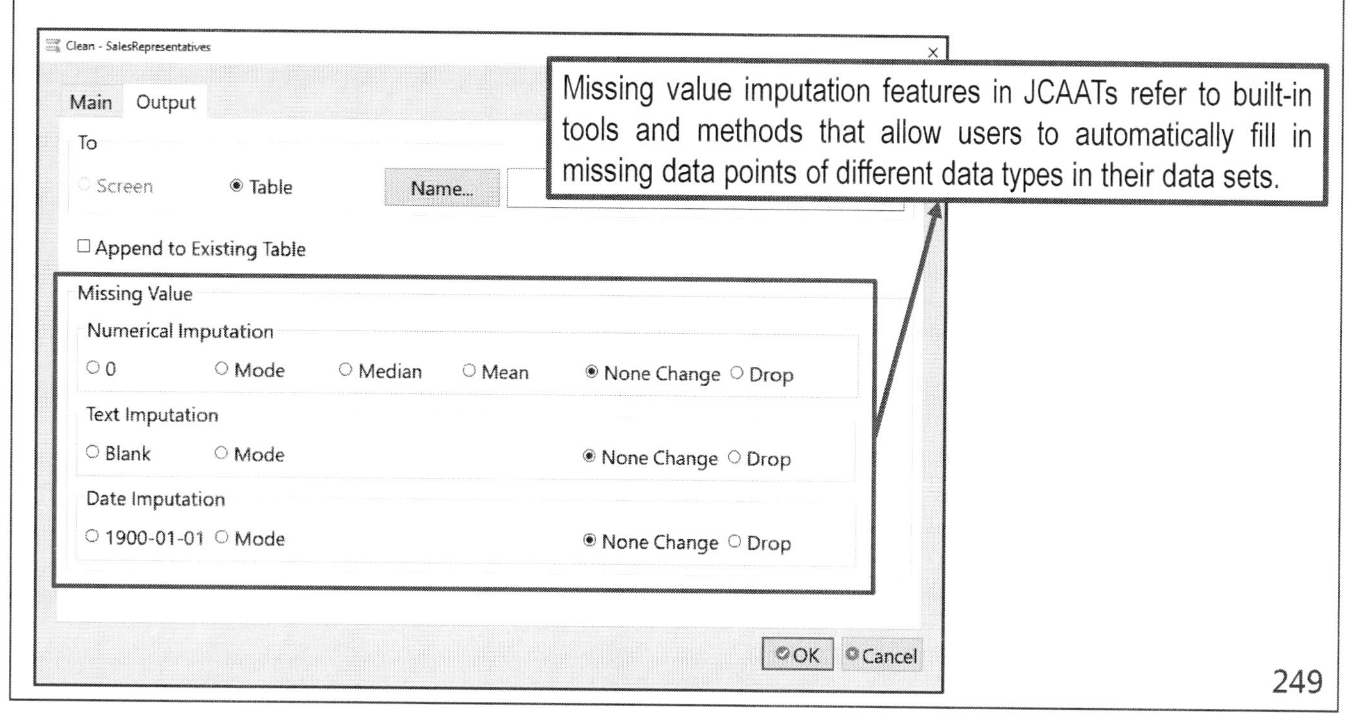

The List of Validate Commands

Test Procedure	Description	Command	Data Type
Confirm & Discover	Obtaining general information from the fields.	Verify	Text, Numeric ,Datetime
		Count	Record-based
		Total	Numeric
		Profile	Numeric
		Statistics	Text, Numeric ,Datetime
Sequence	Testing the sequence of records in the table	Sequence	Text, Numeric ,Datetime
Integrity	Testing for gaps in records within a table, helping users identify any missing or incomplete data.	Gap	Text, Numeric ,Datetime
Uniqueness	Determining whether records contain duplicate data and identifying which specific fields contain unique values.	Duplicate	Text, Numeric ,Datetime
Positioning	Assigning a record location to a specific table.	Locate	Records
Data Cleaning	Helping to fill errors in fields and ensure that the resulting analysis table is complete and accurate.	Clean	Text, Numeric ,Datetime

Chapter 6 - Exercise Questions

( ) 6-1 Which JCAATs command is used to obtain the maximum and minimum of a numeric field?

Ⅰ. Count
Ⅱ. Profile
Ⅲ. Statistic
Ⅳ. Summarize
Ⅴ. Index

a) I and II
b) II, and III
c) I, II, III, and VI
d) I, II, III, VI, and V
e) I, II, III, and V

Chapter 6 - Exercise Questions

( ) 6-2 Which of the following types of data damage are included in the JCAATs Verify command? Choose the correct option(s).

Ⅰ. The table contains a data source.
Ⅱ. Character fields contain only valid characters, and no unprintable characters are present.
Ⅲ. Numeric fields contain only valid numeric data. In addition to numbers, numeric fields can contain one preceding plus sign or minus sign and one decimal point.
Ⅳ. Datetime fields contain valid dates, datetimes, or times.
Ⅴ. Data records are in sequence based on a key field.

a) I, II, III, and IV
b) II, III, and IV
c) I, III, IV, and V
d) I, II, and IV
e) All of the above

Chapter 6 - Exercise Questions

( ) 6-3 What test should be performed in JCAATs to confirm that each staff member has only one salary payment record?

a) Data availability validity
b) Total amount test
c) Date range and availability validity
d) Missing item test
e) Duplicate item test

( ) 6-4 In JCAATs, which command can be used to select a missing item or range in a data file?

a) Classify command
b) Stratify command
c) Count command
d) Statistics command
e) Gap command

Chapter 6 - Exercise Questions

( ) 6-5 Which of the following statements about running commands on JCAATs is correct?

a) Duplicate can subtotal numeric fields.
b) Age can analyze text fields.
c) Summarize can subtotal numeric fields by datetime fields.
d) Classify can select more than one key field.
e) All of the above.

( ) 6-6 Which of the following JCAATs commands can be used to summarize comparisons based on different numerical data ranges?

a) Statistic
b) Stratify
c) Verify
d) Duplicate
e) Sort

Chapter 6 - Exercise Questions

( ) 6-7 In the 2000s, accounts receivable information is detailed in a file. You want to select a due date from March to June, and the accounts receivable amount is above 3,000 NTD. Which condition should you use for the filter?

a) Due_Date.between('2000-03-01','2000-03-31') and Due_Date.between('2000-06-01','2000-06-30') and Amount > 3000
b) Due_Date.between('2000-03-01','2000-03-31') or (Due_Date.between('2000-06-01','2000-06-30') and Amount > 3000)
c) (Due_Date.between('2000-03-01','2000-03-31') or Due_Date.between('2000-06-01','2000-06-30')) and Amount > 3000
d) Due_Date.between('2000-03-01','2000-03-31') and Due_Date.between('2000-06-01','2000-06-30') or Amount > 3000
e) No of the above

JCAATs Learning notes：

JCAATs - AI Audit Software

Exercise 6.1

Please create a new JCAATs project file and practice copying the template project data (ERP_Audit) for later verification use.

Exercise 6.1
Copy the template project data table (ERP_Audit) to the new project.

- Create a new project ERP_Audit
- Select "Data>Copy from another Project"

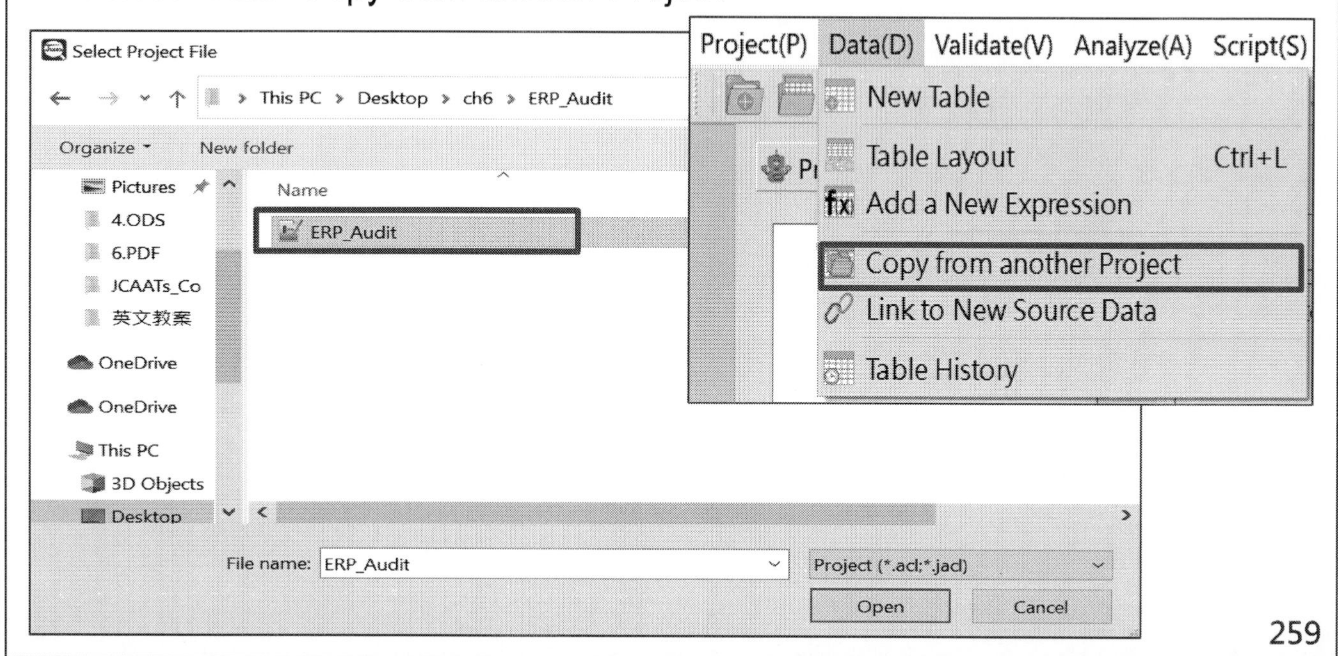

Exercise 6.1
Copy Template Project Data Sheet > Select Table.

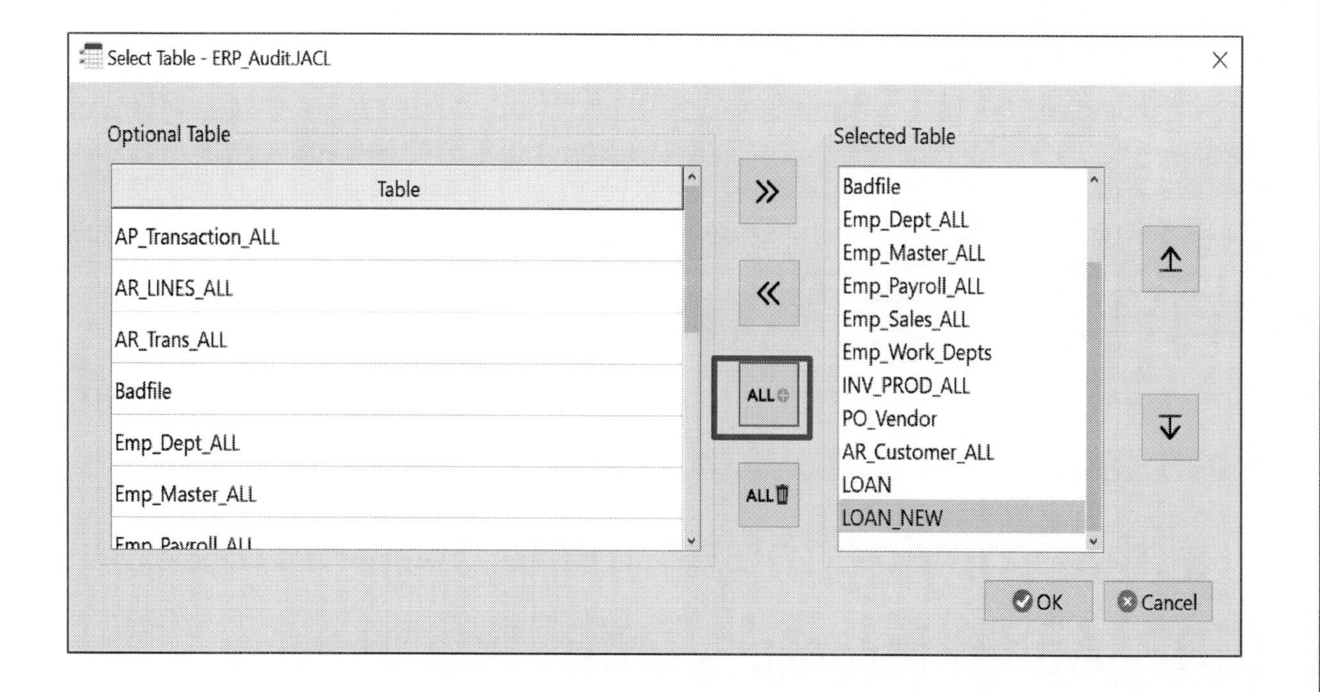

Exercise 6.1
Copy Template Project Table > Finish importing the Table.

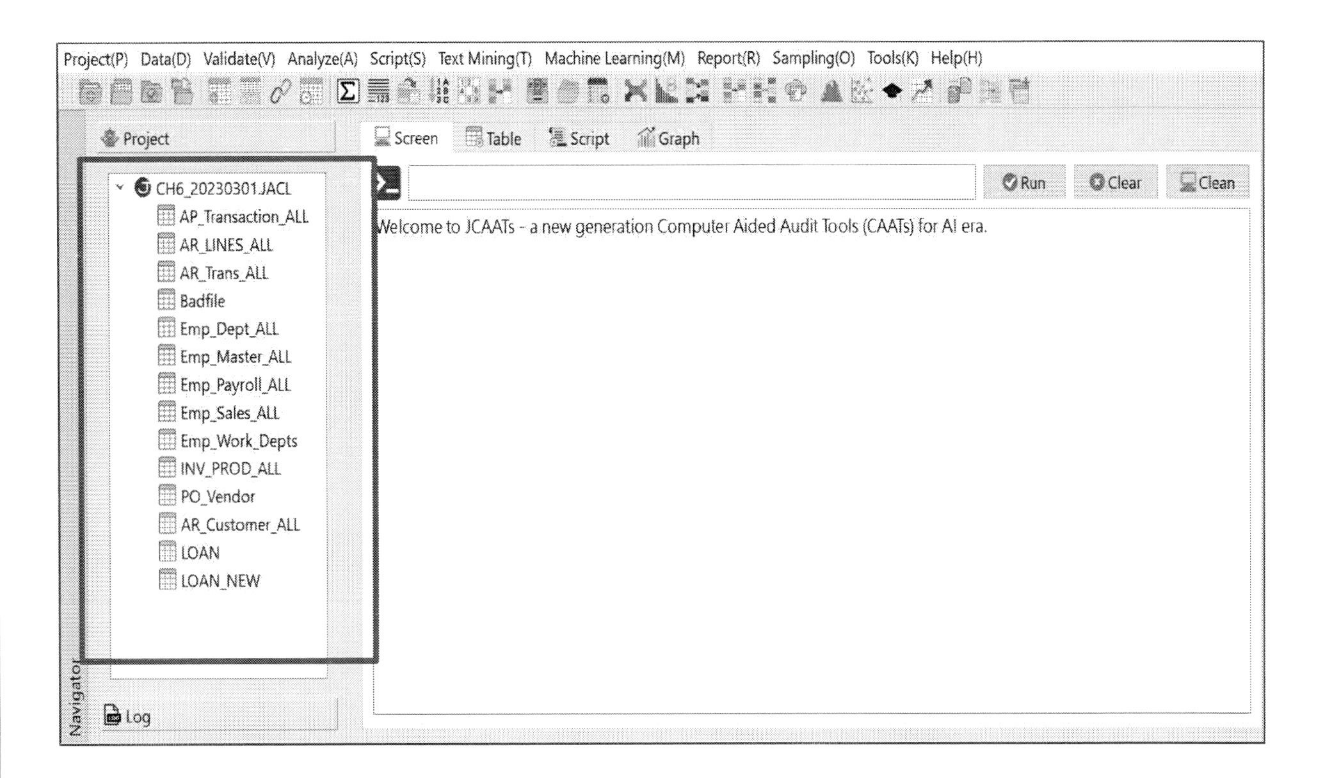

Exercise 6.1
Data>Link to New Source Data

- Select "AP_Transaction_All" Table
- Link to Source Data （*.FIL）

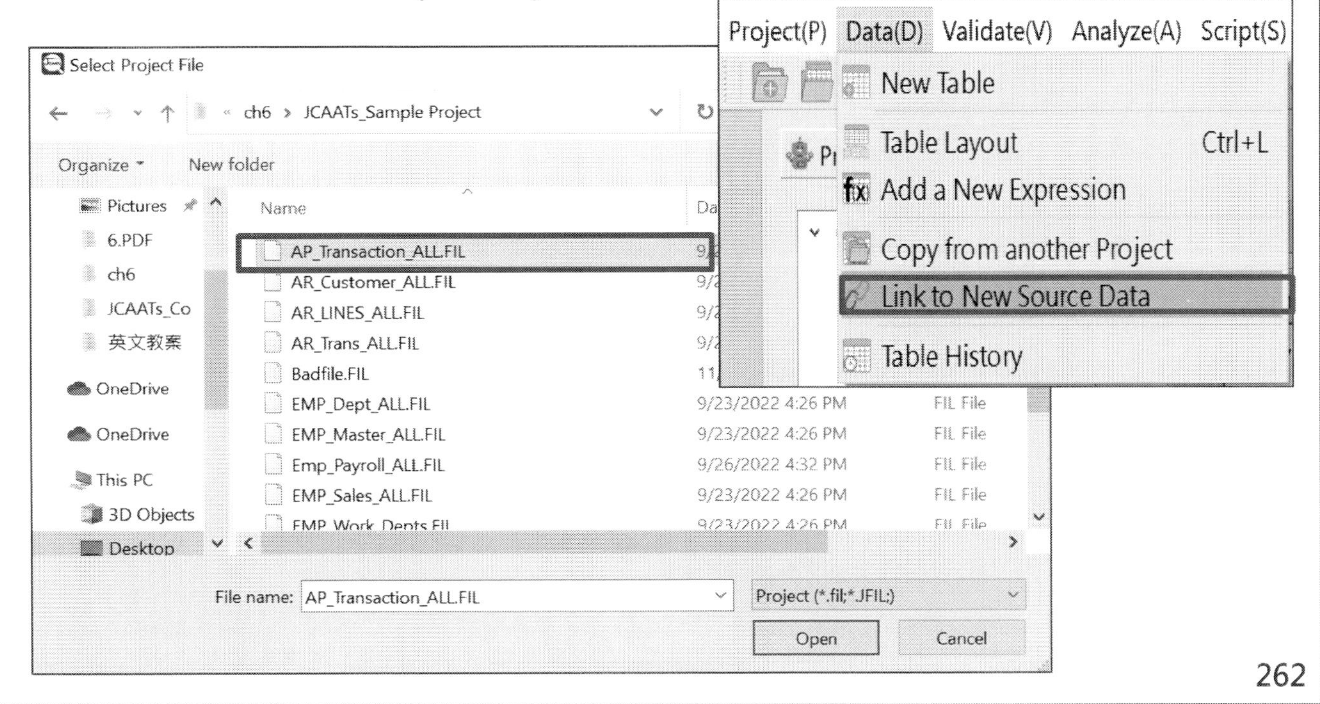

Exercise 6.1
Link to new data source > Complete linking to data source

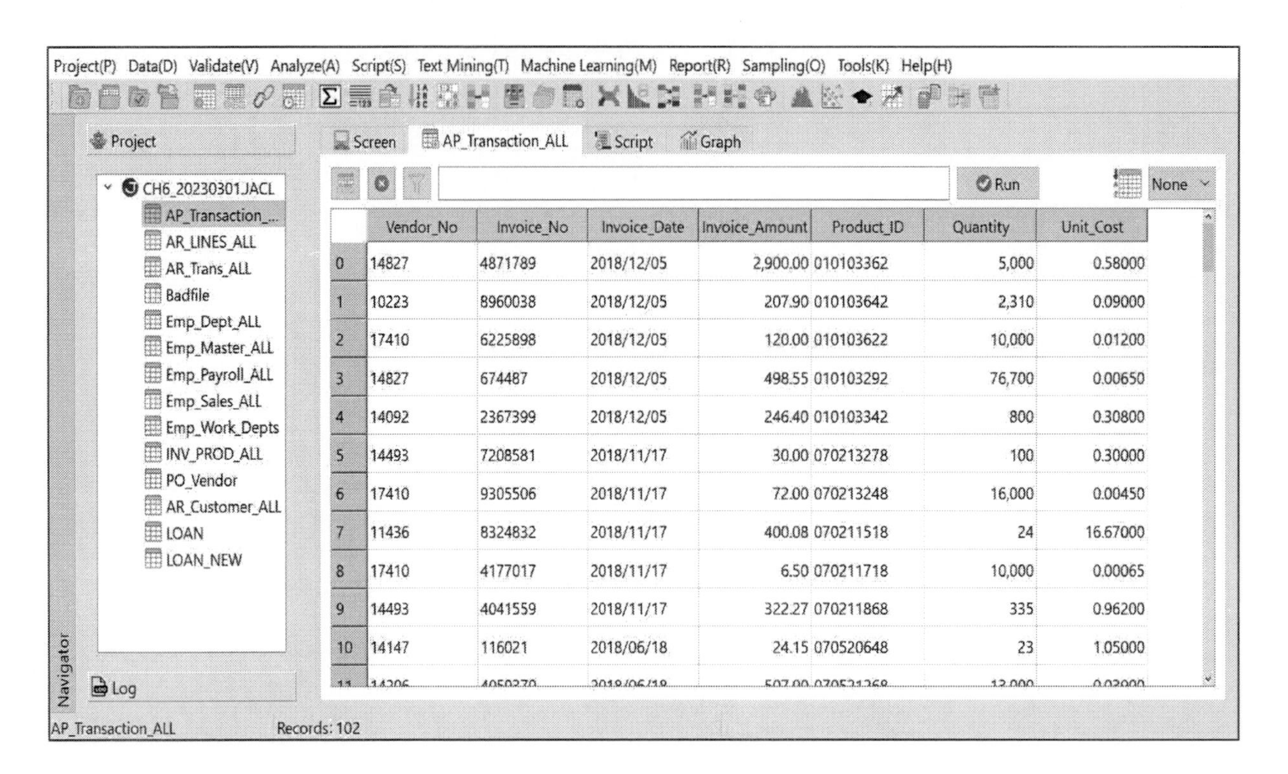

JCAATs Learning notes：

JCAATs - AI Audit Software

Exercise 6.12

Exercise 6.12: A Case Study

1. Please create a new project file named "AR_Audit". (5 points)
2. Please import accounts receivable transaction details data according to the following schema. (15 points)

Field No.	Length	Name	Note	Type	Field No.	Length	Name	Note	Type
1	3	Seq	Transaction No	C	4	10	DueDate	Due Date	D
2	6	CustNo	Customer No	C	5	10	Goods	Item Name	C
3	10	InvDate	Invoice Date	D	6	10	TransAmt	Amount	N

3. Verification of Data Accuracy (20 points)

□ None, there is no error in the data.

□ Yes, there is an error in the row and column and the reason:

Exercise 6.12: A Case Study

4. Number of Records (10 points)
   i. In accounts receivable, the number of transactions for selling cookies is ______________________________
   ii. In accounts receivable, the number of transactions for selling cookies with a transaction amount greater than 200 is

   ____
5. Sum Calculation (10 points)
   i. The total transaction amount for toilet paper in the accounts receivable transaction details is

   ____

Exercise 6.12: A Case Study

6. Please confirm whether there are any missing serial numbers in the accounts receivable transaction details (10 points).
   - □ No, there are no errors in the data.
   - □ Yes, there are a total of error records which are respectively:______________________________
7. Please confirm whether there are any duplicate transaction serial numbers in the accounts receivable transaction (10 points).
   - □ No, there is no such situation in the data.
   - □ Yes, the repeated transaction serial number is ________________

Exercise 6.12: A Case Study

8. Total transaction amount for each product in the accounts receivable transaction details (10 points).
   i. The highest number of products sold is .
   ii. The percentage of the lowest transaction amount to the total amount is ______________________
9. Based on the transaction amount levels of 0, 500, 800, 1000, and 10000 (10 points).
   i. The level with the highest percentage of transactions to the total number is .
   ii. The level with the lowest percentage of transaction amount to the total amount is .

JCAATs-AI Audit Software

Exercise 6.12: A Case Study

10. Please confirm whether there are any duplicate transaction serial numbers in the accounts receivable transaction (10 points).
    □ No, there is no such situation in the data.
    □ Yes, the repeated transaction serial number is
    ____
11. Total transaction amount for each product in the accounts receivable transaction details (10 points).
    i. The highest number of products sold is
    ____
    ii. The percentage of the lowest transaction amount to the total amount is ______________________

Exercise 6.12: A Case Study

12. Based on the transaction amount levels of 0, 500, 800, 1000, and 10000 (10 points).
    i. The level with the highest percentage of transactions to the total number is ______________________________
    ii. The level with the lowest percentage of transaction amount to the total amount is ______________________________

JCAATs-AI Audit Software
Copyright © 2023 JACKSOFT.

Exercise 6.12
Please create a new project file named "AR_Audit"

1. Create a new folder.
2. Click JCAATs-AI audit software.
3. Click "Project" > "Select New Project."
4. Define a project name.
5. Save.

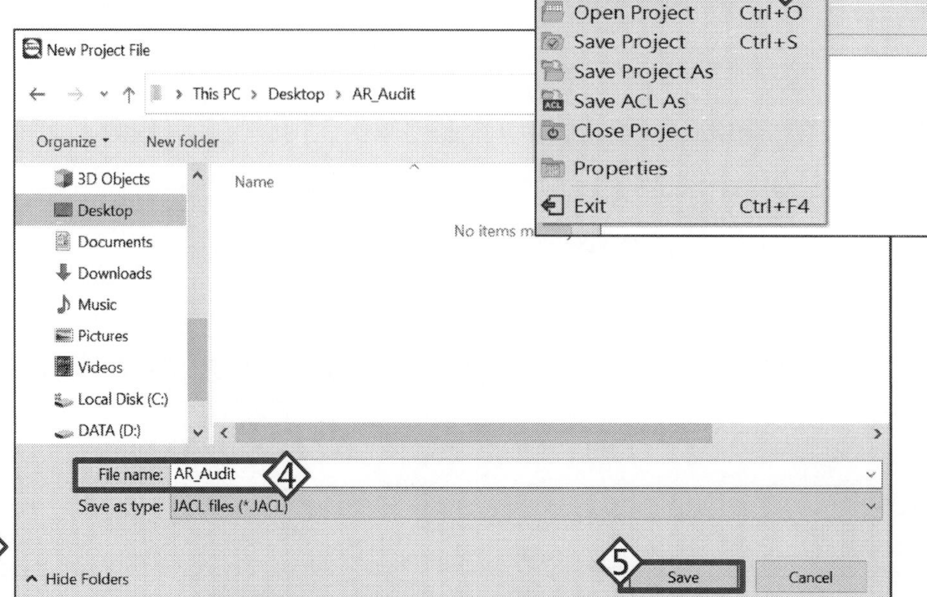

Exercise 6.12
Please import accounts receivable transaction details data according to the following schema.

- Select "Data > New Table".
- After selecting the data source platform as "File", click "Next".

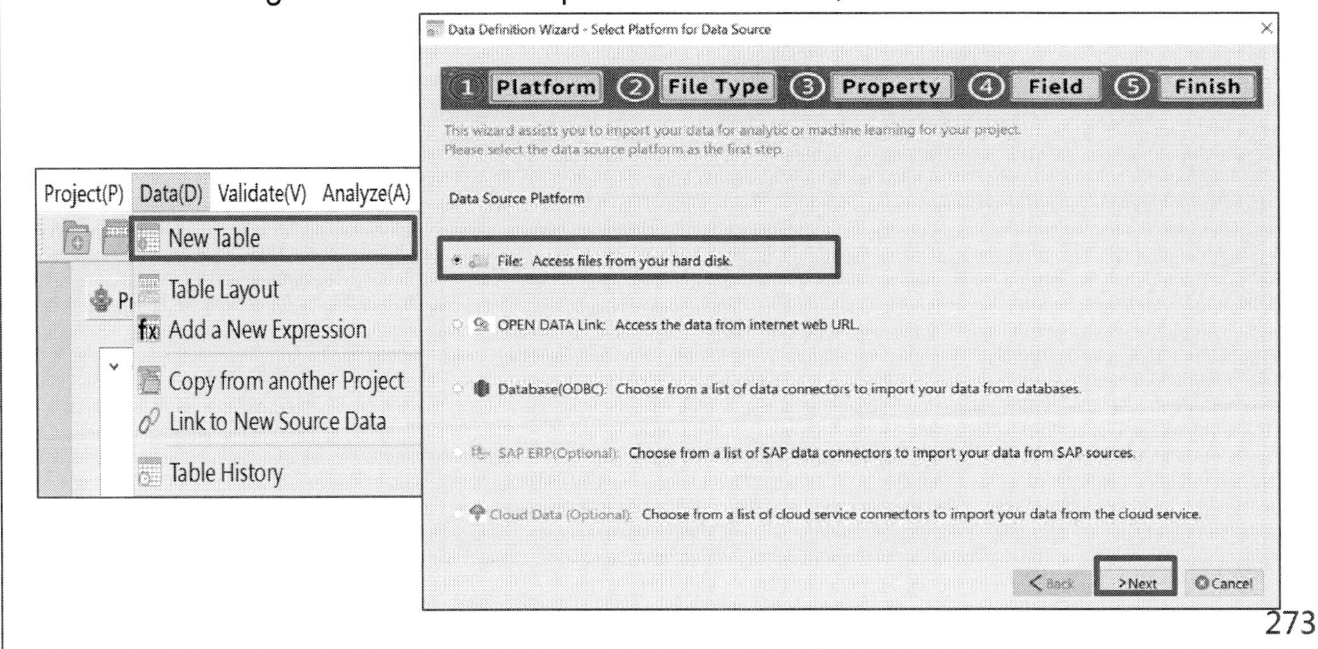

Exercise 6.12
Please import accounts receivable transaction details data according to the following schema.

- Select AR_Transaction_Lines to import.
- Click "Open" to proceed.

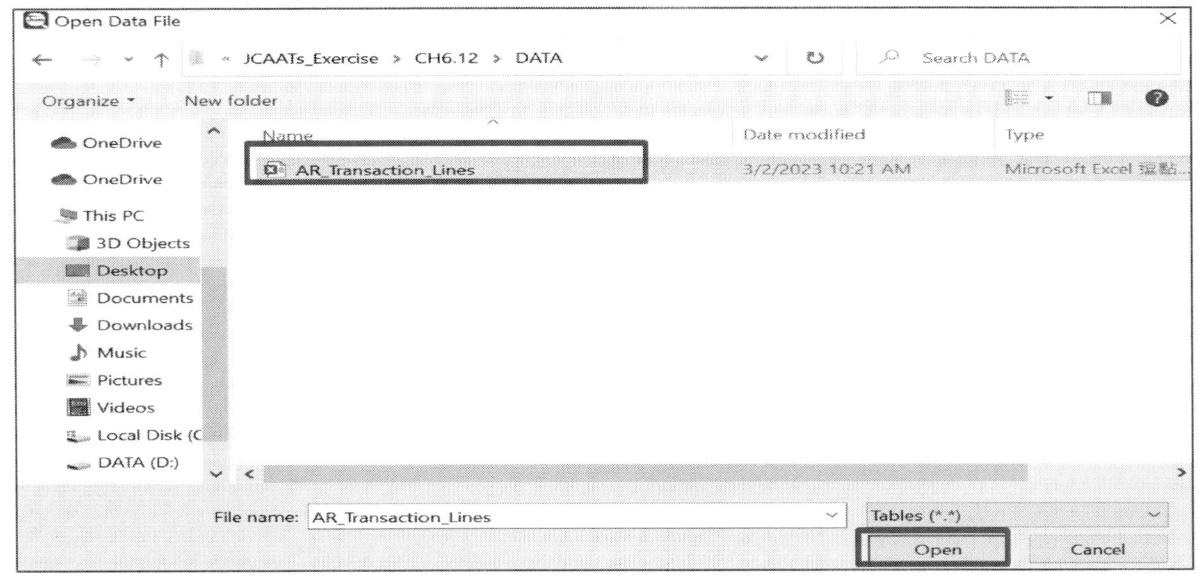

Exercise 6.12
Please import accounts receivable transaction details data according to the following schema.

- JCAATs will automatically detect the file format to ensure accuracy. If there are no errors, select "Next ".

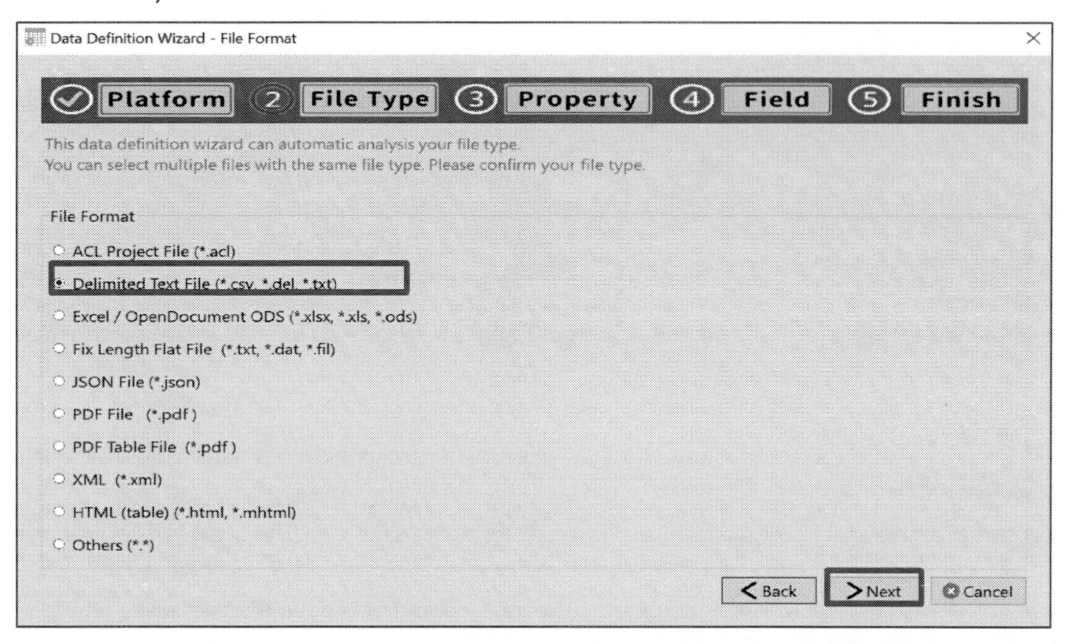

JCAATs-AI Audit Software

Exercise 6.12
Please import accounts receivable transaction details data according to the following schema.

- Ensure whether the first row is the field name or not, or specify the starting line. If there are no errors, select "Next".

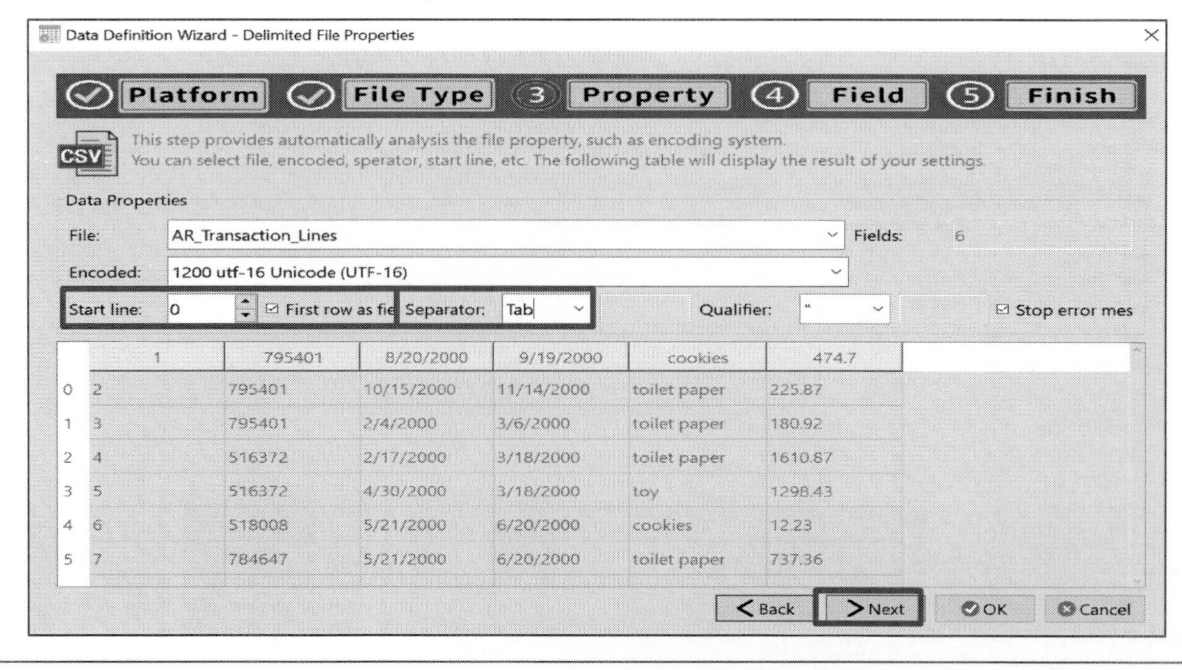

Exercise 6.12

Please import accounts receivable transaction details data according to the following schema.

- After completing the settings, select "Next" to proceed.

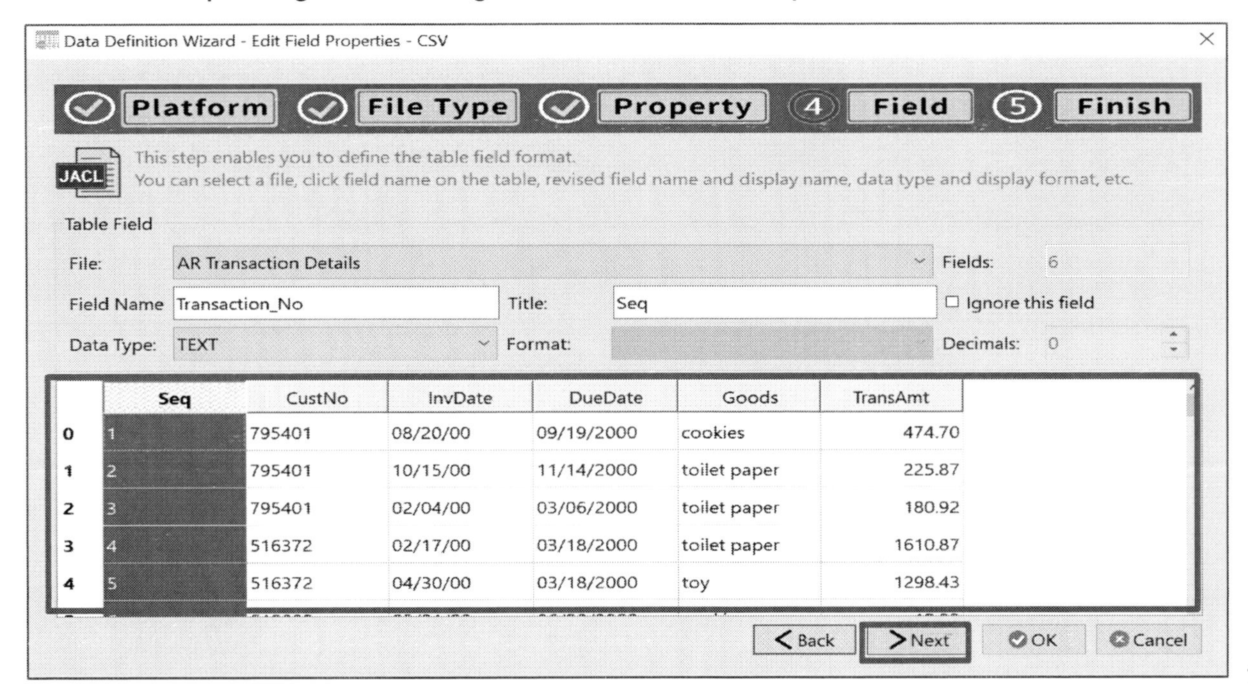

Exercise 6.12

Please import accounts receivable transaction details data according to the following schema.

- The default data file path is the project folder, but it can be modified as needed. After ensuring that the path and information are correct, select "Done".

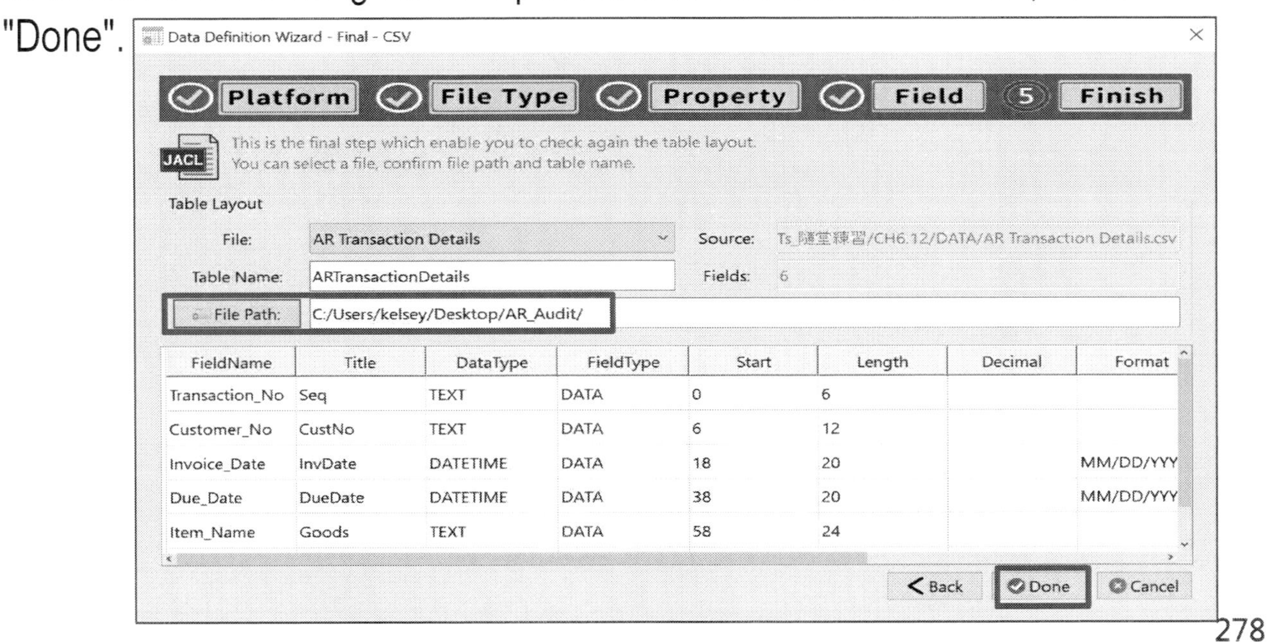

Exercise 6.12
Please import accounts receivable transaction details data according to the following schema.

- After the import progress is completed, we can see that the data table has been successfully imported.

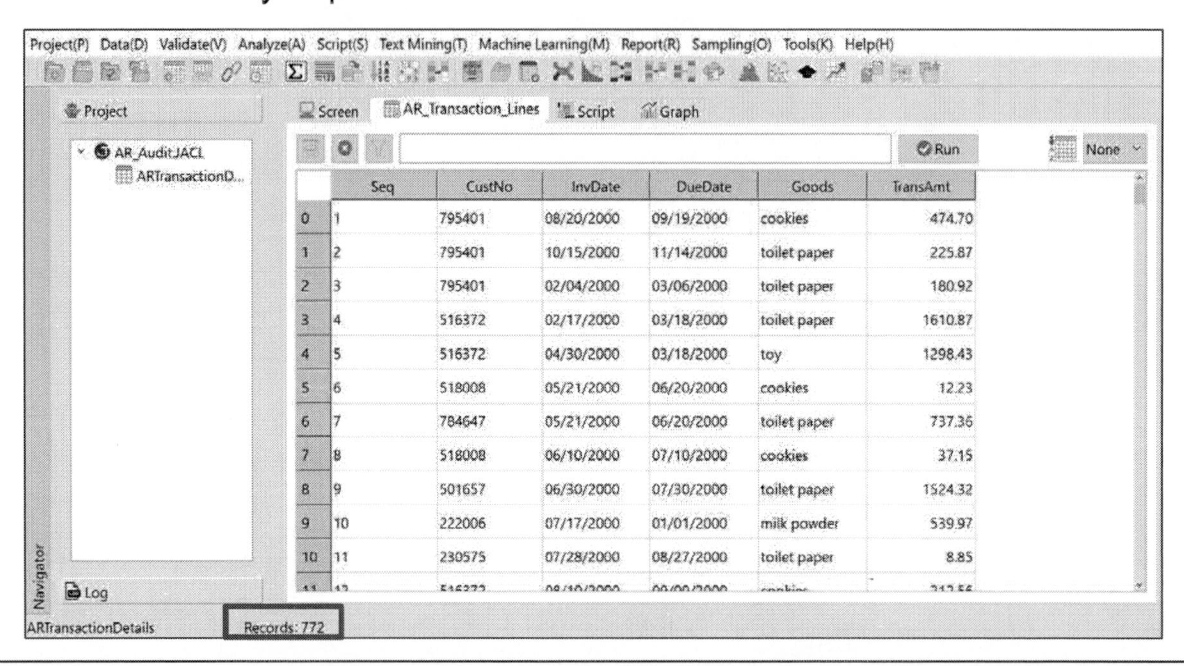

Exercise 6.12
Verification of Data Accuracy

- Open the " AR_Transcation_Lines" Table
- Click on "Validate>Verify"

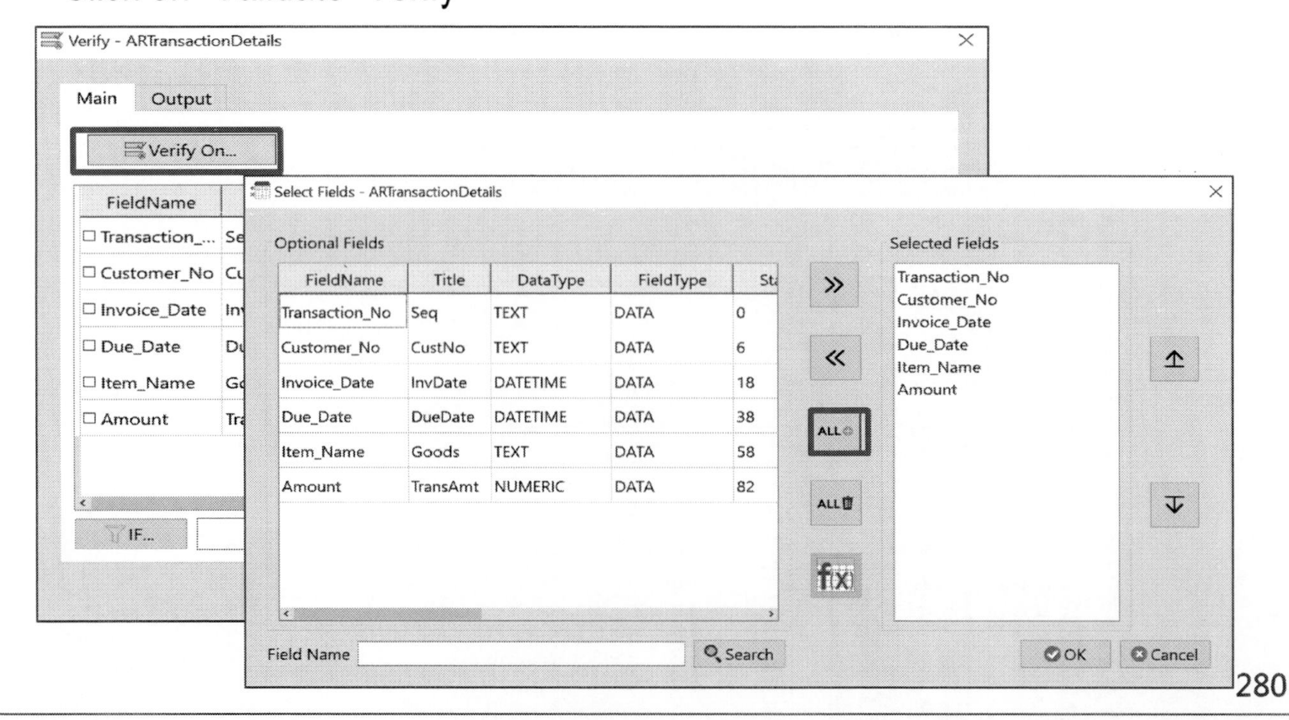

Exercise 6.12
Verification of Data Accuracy

- Output to Screen.

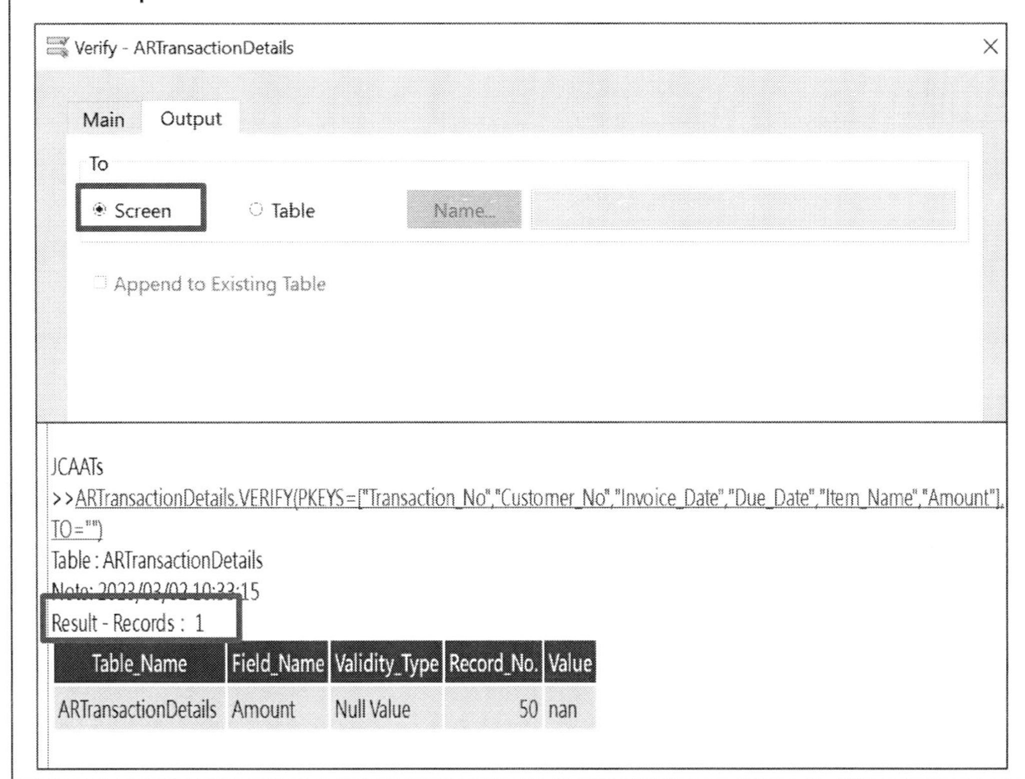

Table_Name	Field_Name	Validity_Type	Record_No.	Value
ARTransactionDetails	Amount	Null Value	50	nan

Exercise 6.12
In accounts receivable, the number of transactions for selling cookies is

- Click on " Analyze>Classify "

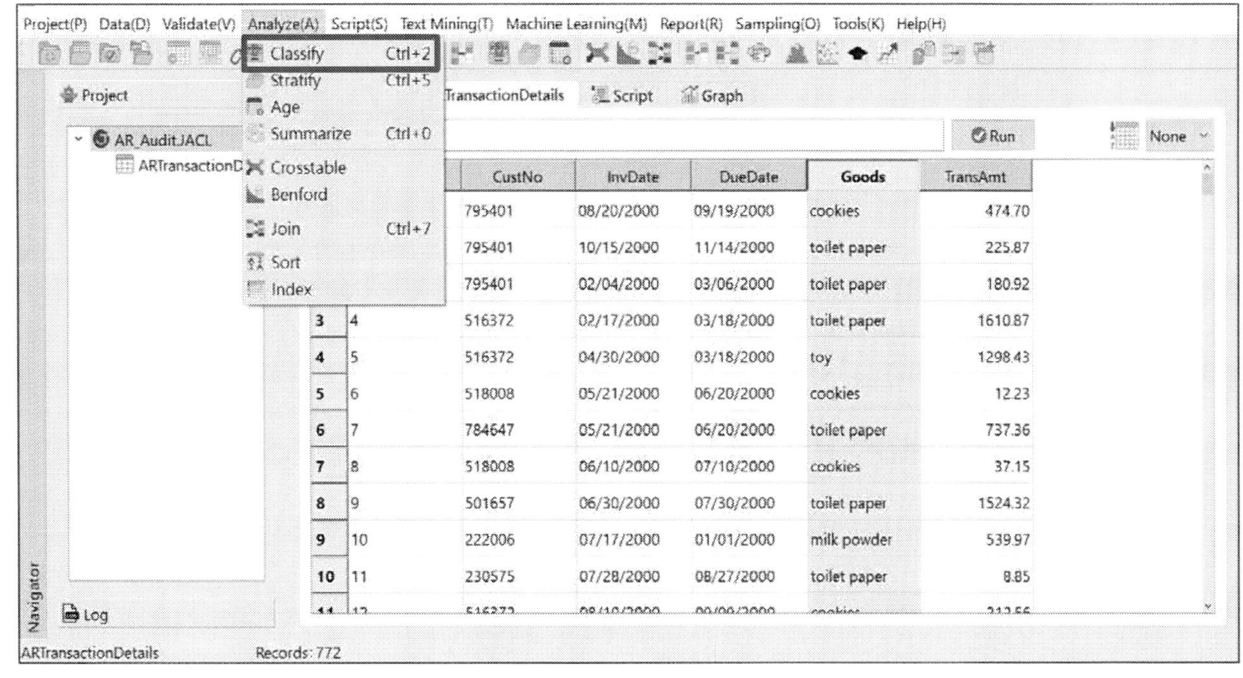

		CustNo	InvDate	DueDate	Goods	TransAmt
		795401	08/20/2000	09/19/2000	cookies	474.70
		795401	10/15/2000	11/14/2000	toilet paper	225.87
		795401	02/04/2000	03/06/2000	toilet paper	180.92
3	4	516372	02/17/2000	03/18/2000	toilet paper	1610.87
4	5	516372	04/30/2000	03/18/2000	toy	1298.43
5	6	518008	05/21/2000	06/20/2000	cookies	12.23
6	7	784647	05/21/2000	06/20/2000	toilet paper	737.36
7	8	518008	06/10/2000	07/10/2000	cookies	37.15
8	9	501657	06/30/2000	07/30/2000	toilet paper	1524.32
9	10	222006	07/17/2000	01/01/2000	milk powder	539.97
10	11	230575	07/28/2000	08/27/2000	toilet paper	8.85

Exercise 6.12

In accounts receivable, the number of transactions for selling cookies is

- Select the field to classify. (Item_Name)
- Subtotal Amount

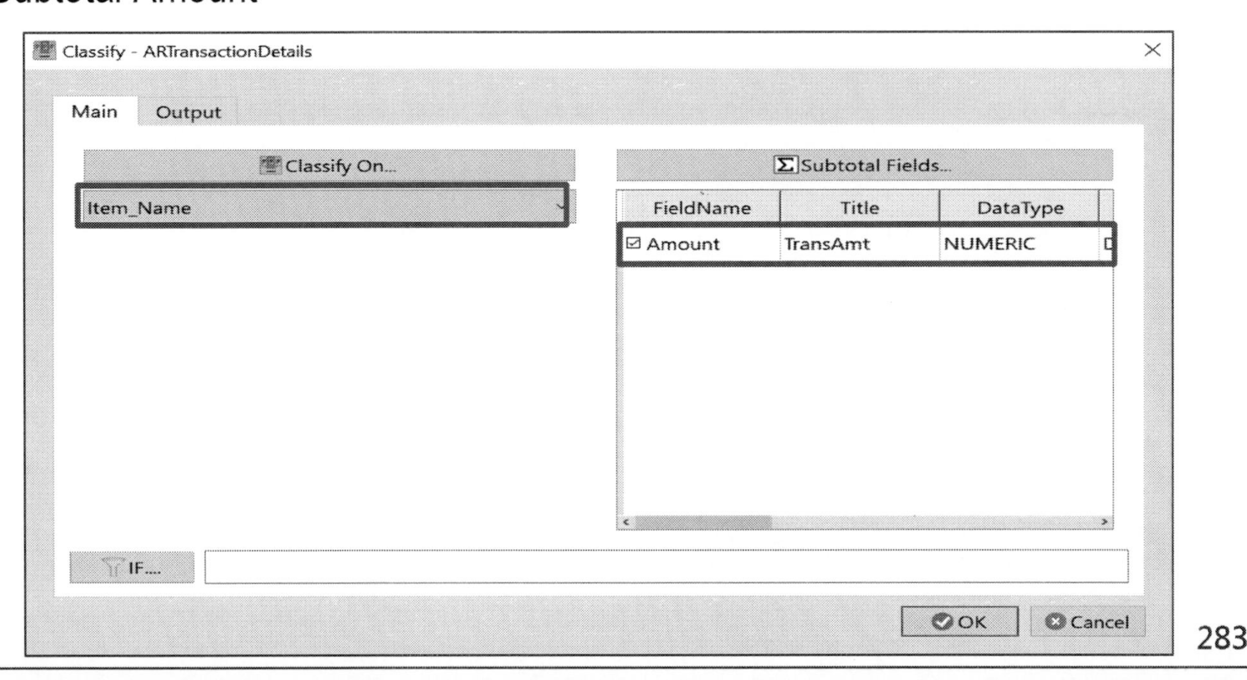

Exercise 6.12

In accounts receivable, the number of transactions for selling cookies is

- Output to Screen

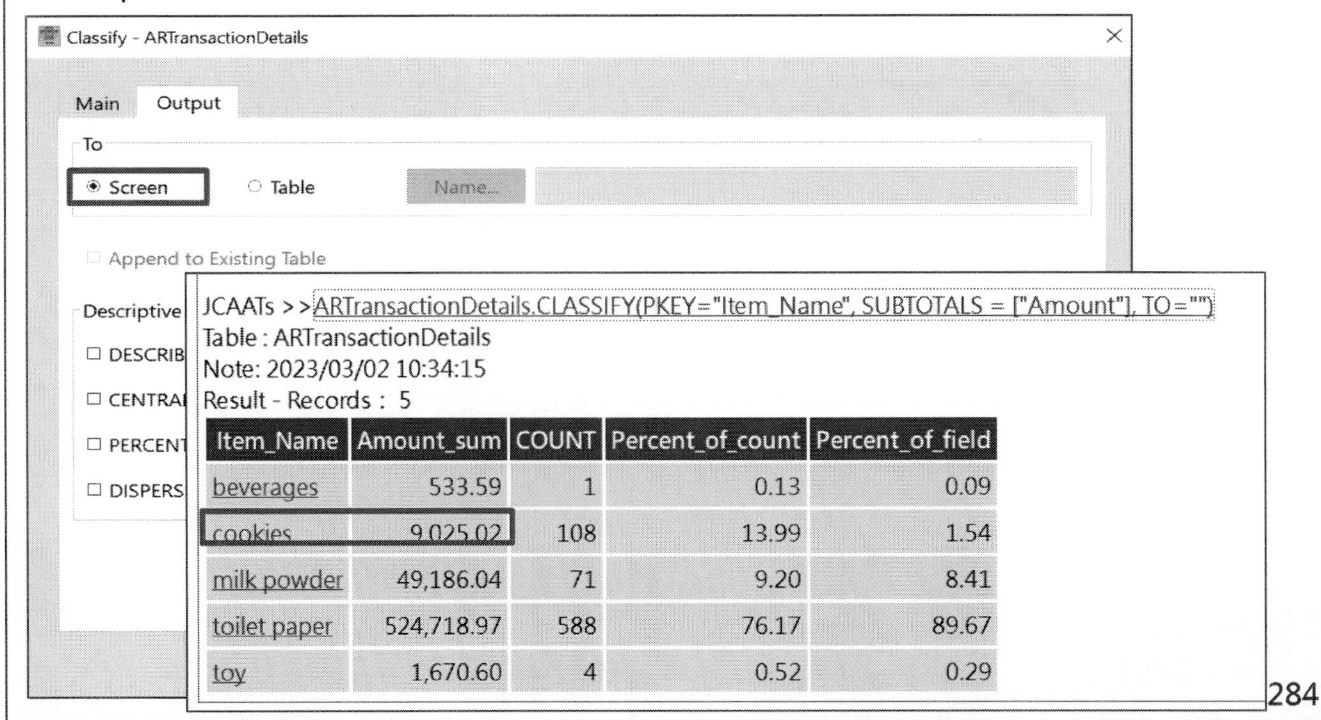

JCAATs >>ARTransactionDetails.CLASSIFY(PKEY="Item_Name", SUBTOTALS = ["Amount"], TO="")
Table : ARTransactionDetails
Note: 2023/03/02 10:34:15
Result - Records : 5

Item_Name	Amount_sum	COUNT	Percent_of_count	Percent_of_field
beverages	533.59	1	0.13	0.09
cookies	9,025.02	108	13.99	1.54
milk powder	49,186.04	71	9.20	8.41
toilet paper	524,718.97	588	76.17	89.67
toy	1,670.60	4	0.52	0.29

Exercise 6.12
In accounts receivable, the number of transactions for selling cookies with a transaction amount greater than 200 is

- Click on "cookies" to drill down for information.

JCAATs >>ARTransactionDetails.CLASSIFY(PKEY="Item_Name", SUBTOTALS = ["Amount"], TO="")
Table : ARTransactionDetails
Note: 2023/03/02 10:34:15
Result - Records : 5

Item_Name	Amount_sum	COUNT	Percent_of_count	Percent_of_field
beverages	533.59	1	0.13	0.09
cookies	9,025.02	108	13.99	1.54
milk powder	49,186.04	71	9.20	8.41
toilet paper	524,718.97	588	76.17	89.67
toy	1,670.60	4	0.52	0.29

Exercise 6.12
In accounts receivable, the number of transactions for selling cookies with a transaction amount greater than 200 is

- Select "Set Filter"

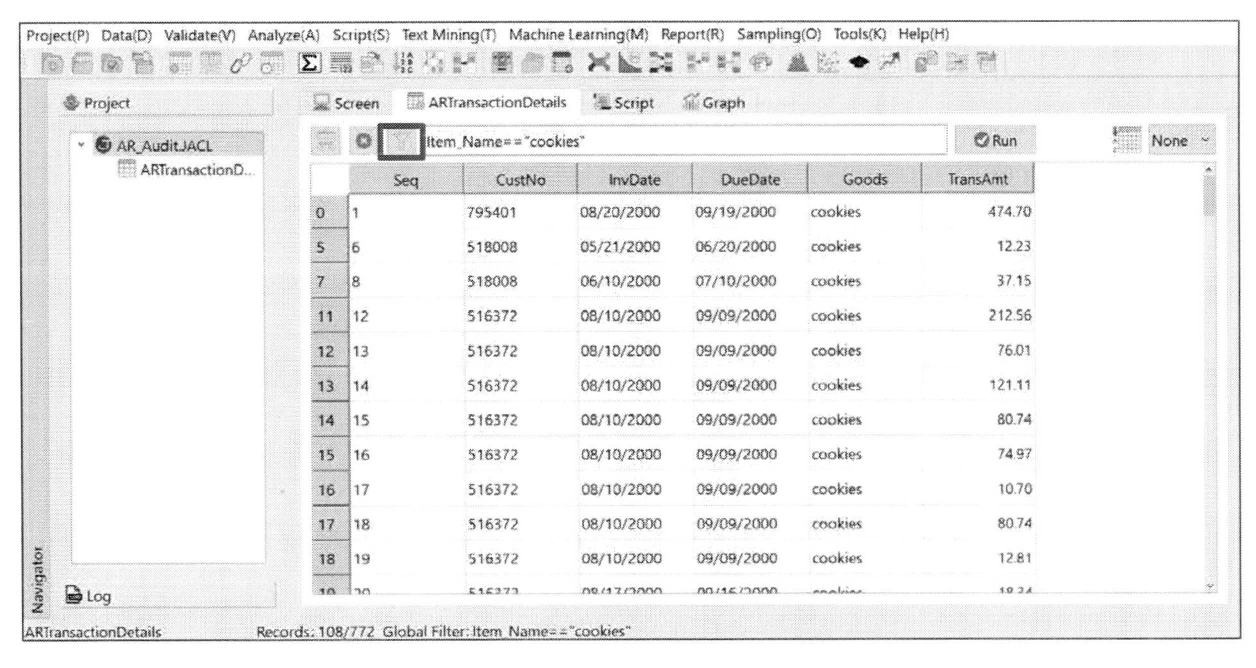

Exercise 6.12

In accounts receivable, the number of transactions for selling cookies with a transaction amount greater than 200 is

- Modify Filter Condition:

  Item_Name == "cookies" and Amount > 200

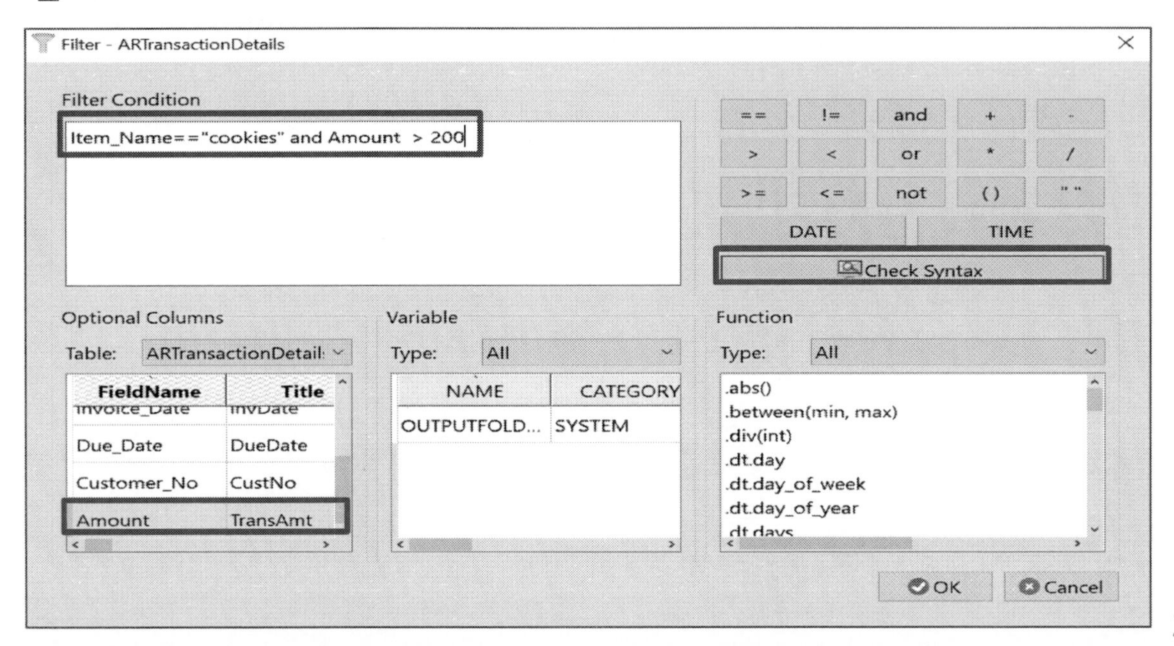

JCAATs-AI Audit Software Copyright © 2023 JACKSOFT.

Exercise 6.12

In accounts receivable, the number of transactions for selling cookies with a transaction amount greater than 200 is

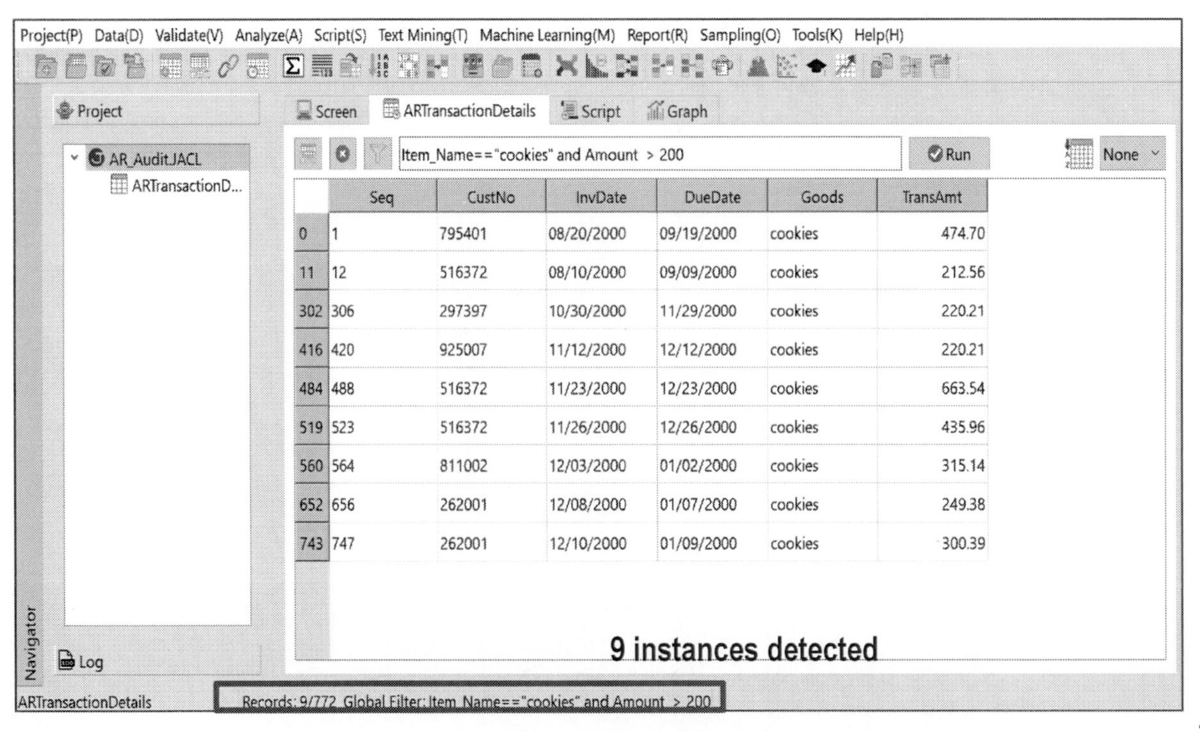

	Seq	CustNo	InvDate	DueDate	Goods	TransAmt
0	1	795401	08/20/2000	09/19/2000	cookies	474.70
11	12	516372	08/10/2000	09/09/2000	cookies	212.56
302	306	297397	10/30/2000	11/29/2000	cookies	220.21
416	420	925007	11/12/2000	12/12/2000	cookies	220.21
484	488	516372	11/23/2000	12/23/2000	cookies	663.54
519	523	516372	11/26/2000	12/26/2000	cookies	435.96
560	564	811002	12/03/2000	01/02/2000	cookies	315.14
652	656	262001	12/08/2000	01/07/2000	cookies	249.38
743	747	262001	12/10/2000	01/09/2000	cookies	300.39

Exercise 6.12

The total transaction amount for toilet paper in the accounts receivable transaction details is

JCAATs >>ARTransactionDetails.CLASSIFY(PKEY="Item_Name", SUBTOTALS = ["Amount"], TO="")
Table : ARTransactionDetails
Note: 2023/03/02 10:34:15
Result - Records : 5

Item_Name	Amount_sum	COUNT	Percent_of_count	Percent_of_field
beverages	533.59	1	0.13	0.09
cookies	9,025.02	108	13.99	1.54
milk powder	49,186.04	71	9.20	8.41
toilet paper	524,718.97	588	76.17	89.67
toy	1,670.60	4	0.52	0.29

Exercise 6.12

Please confirm whether there are any missing serial numbers in the accounts receivable transaction details

- Click on " Validate>Gap "

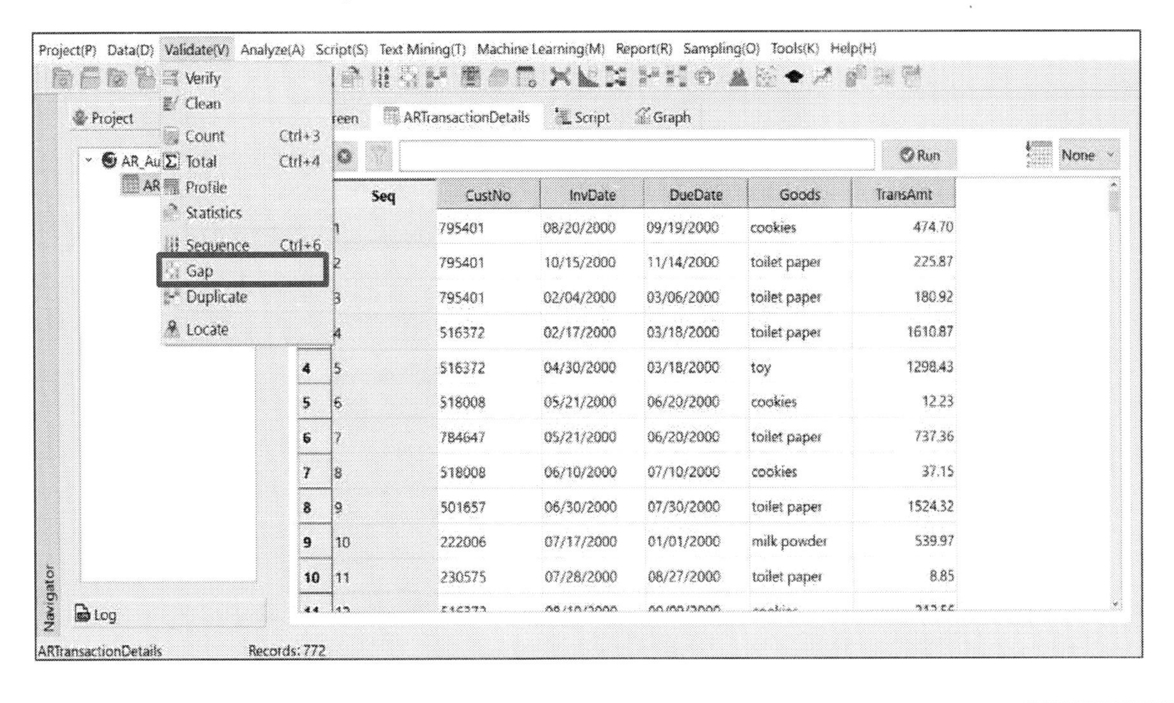

Exercise 6.12

Please confirm whether there are any missing serial numbers in the accounts receivable transaction details

- Select the field to Gap on and choose the output type as List Gap Ranges.

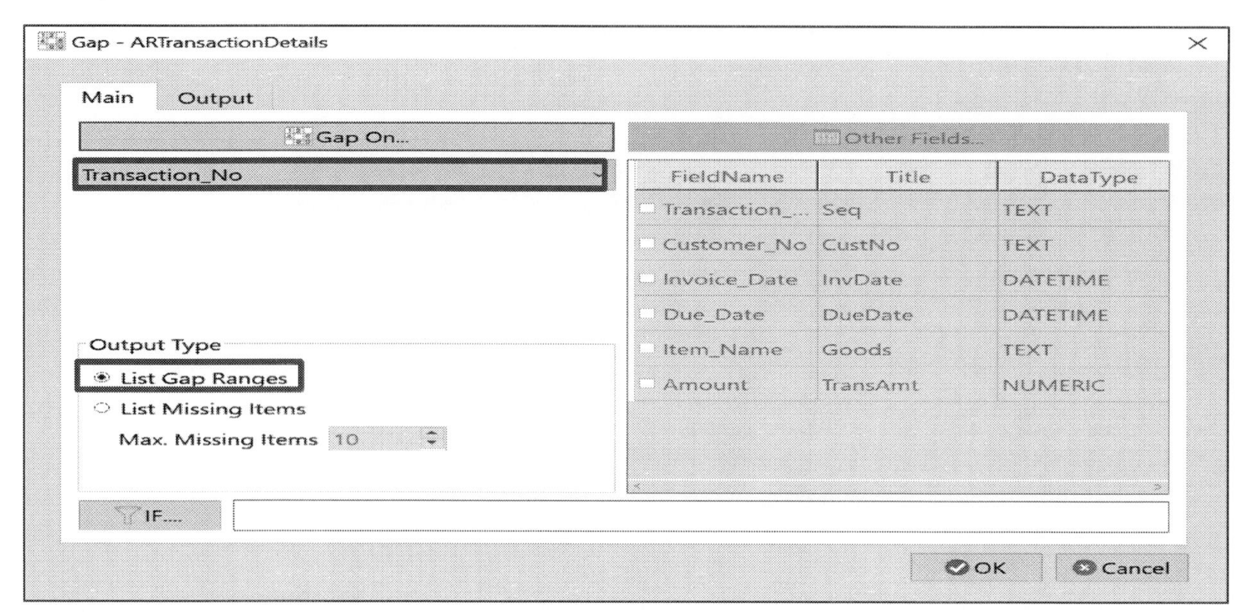

JCAATs-AI Audit Software
Copyright © 2023 JACKSOFT.

Exercise 6.12

Please confirm whether there are any missing serial numbers in the accounts receivable transaction details

- Output to Screen

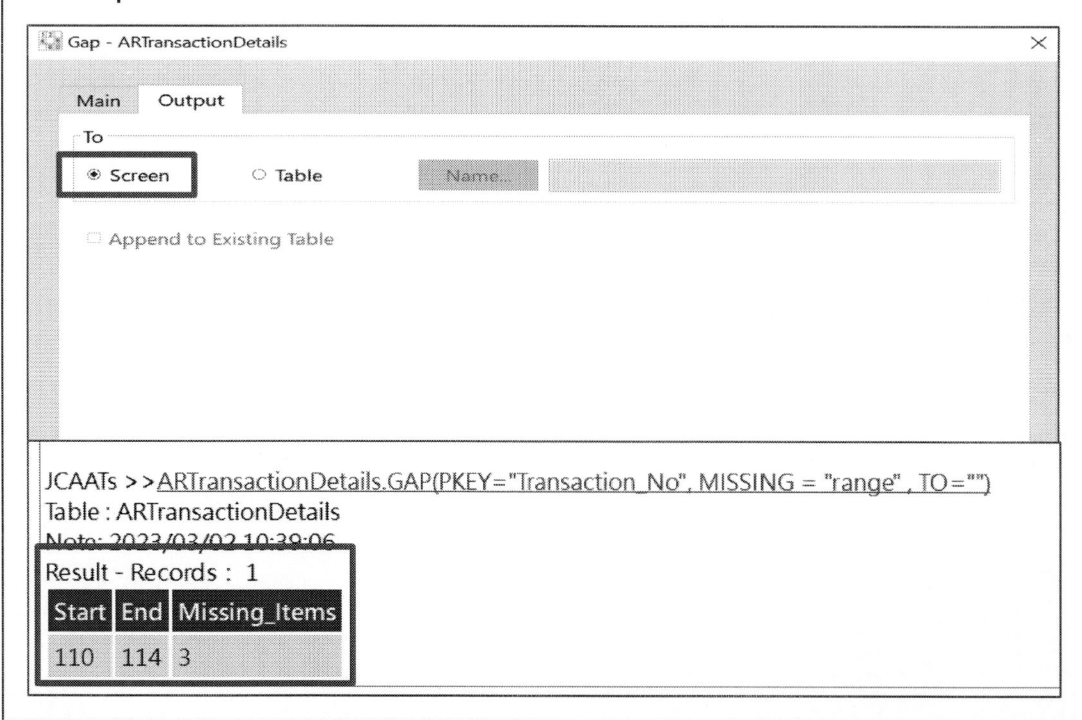

Exercise 6.12
Please confirm whether there are any duplicate transaction serial numbers in the accounts receivable transaction

- Click on " Validate>Duplicate "

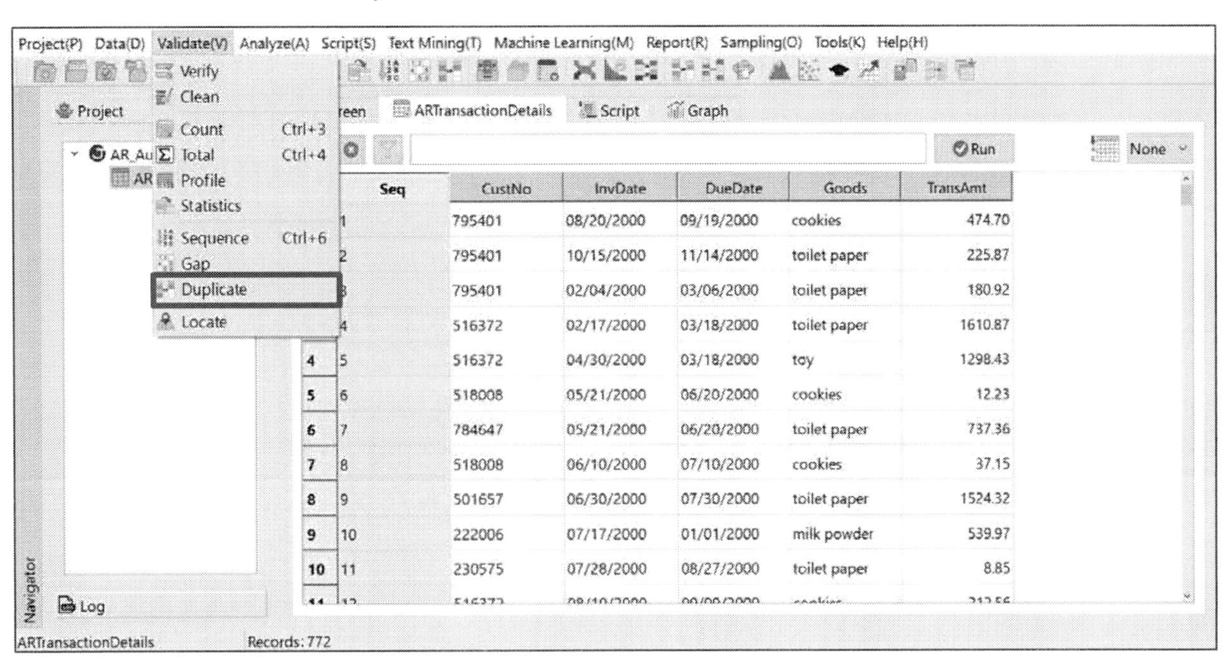

Exercise 6.12
Please confirm whether there are any duplicate transaction serial numbers in the accounts receivable transaction

- Select the field we need to Duplicate on.

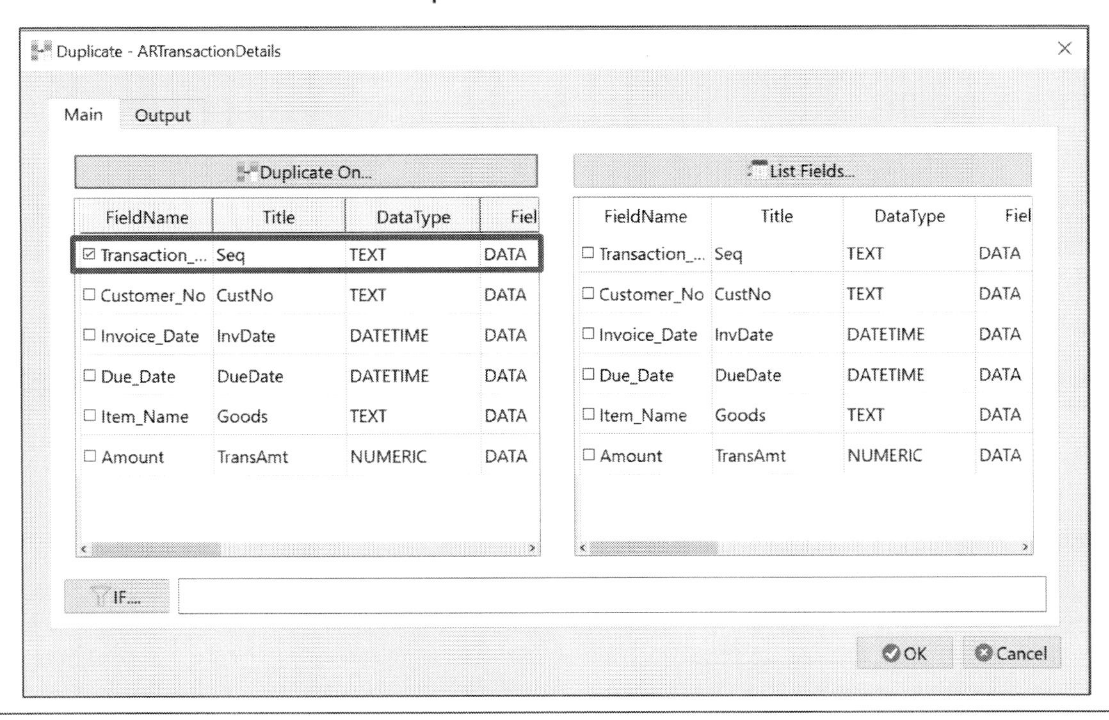

Exercise 6.12

Please confirm whether there are any duplicate transaction serial numbers in the accounts receivable transaction

- Output to Screen

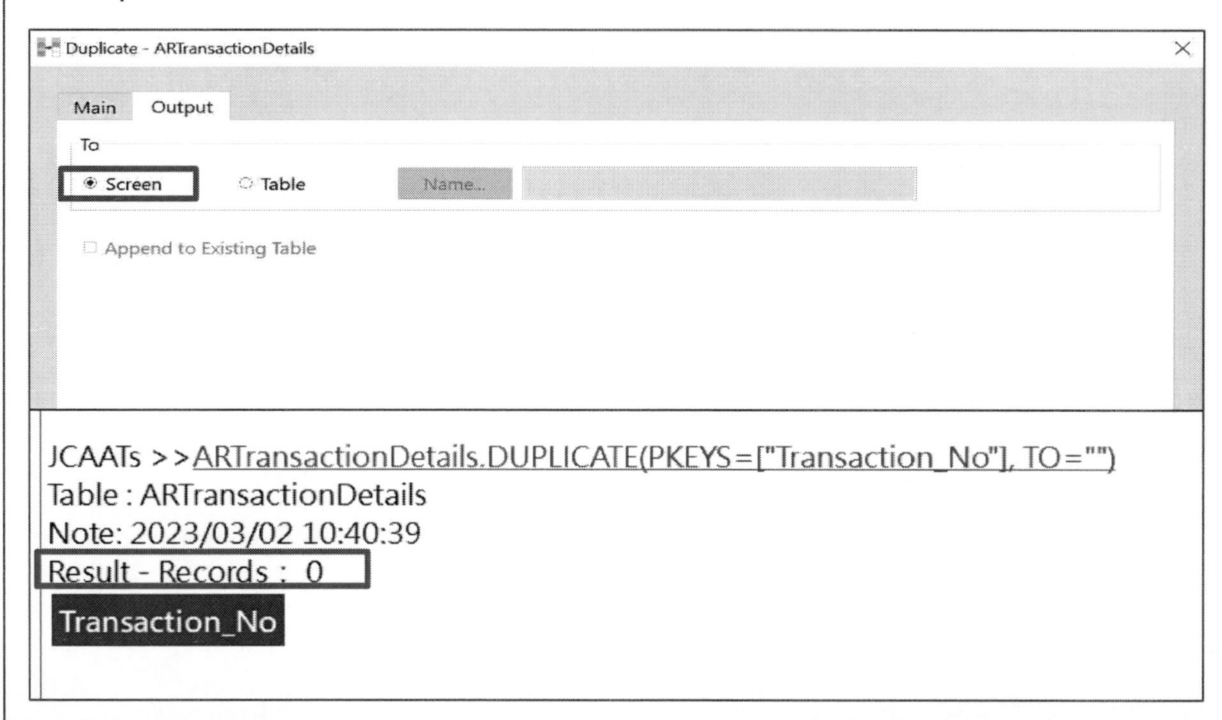

Exercise 6.12:

Total transaction amount for each product in the accounts receivable transaction details:

The highest number of products sold is?

- Select the field we need to Classify
- Subtotal Amount

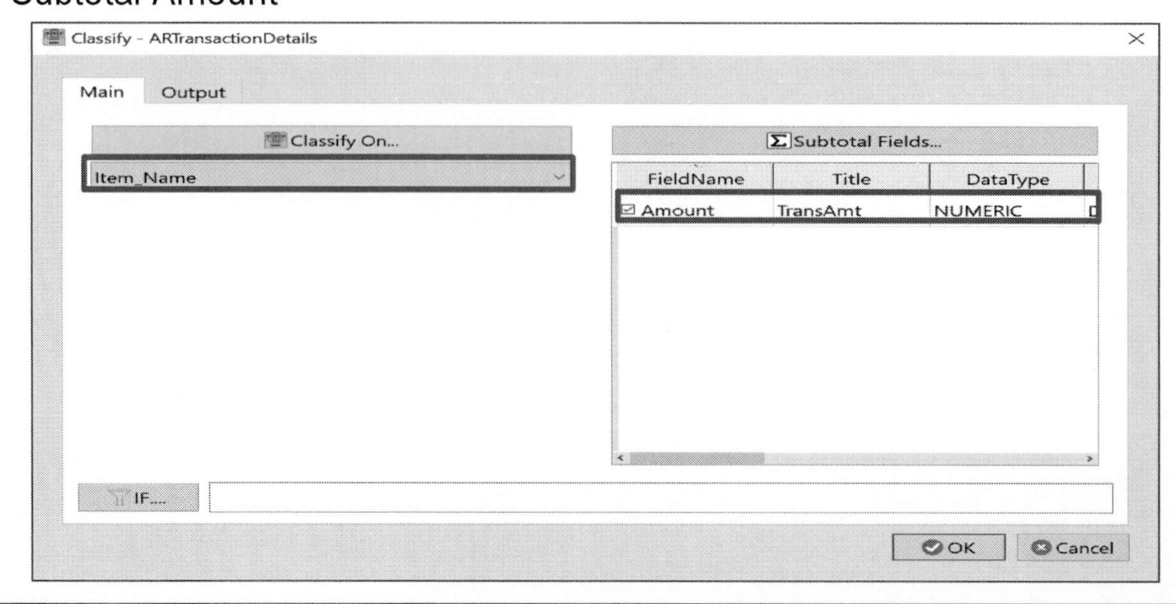

Exercise 6.12
Total transaction amount for each product in the accounts receivable transaction details: The highest number of products sold is?

- Output to a new table

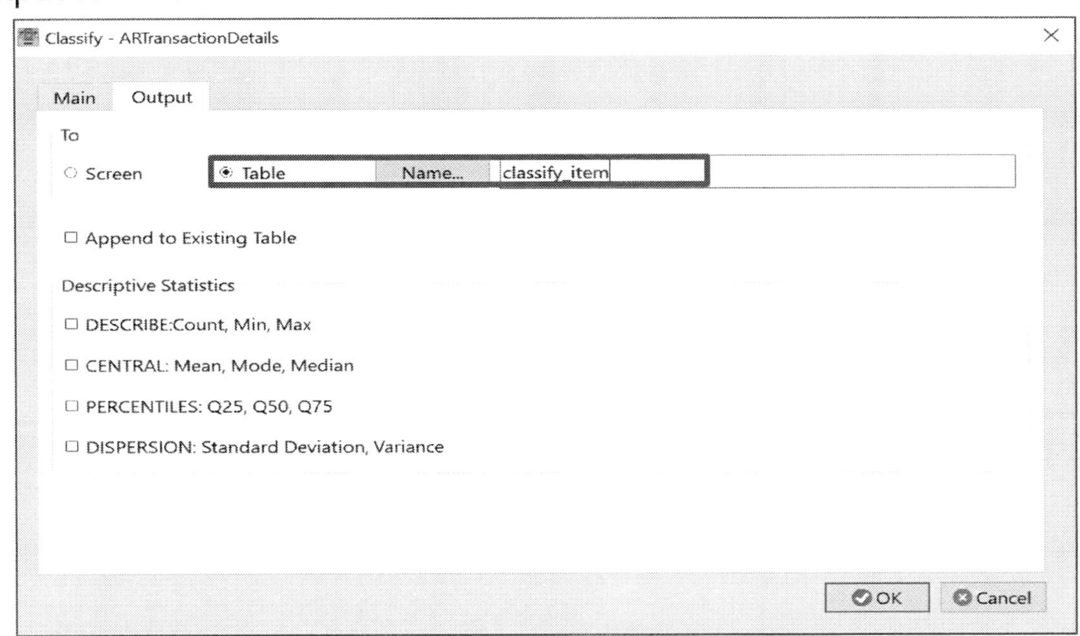

JCAATs-AI Audit Software Copyright © 2023 JACKSOFT.

Exercise 6.12
Total transaction amount for each product in the accounts receivable transaction details: The highest number of products sold is?

- Right-click on the field to sort the column in descending order.

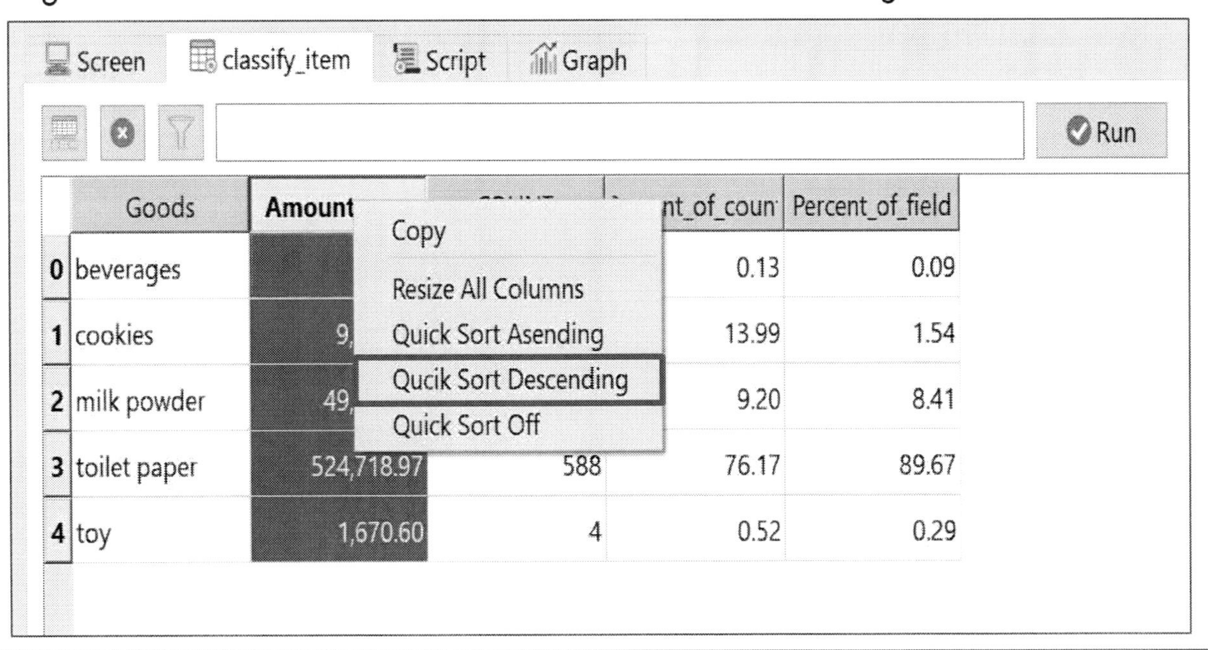

Exercise 6.12

Total transaction amount for each product in the accounts receivable transaction details:

The highest number of products sold is?

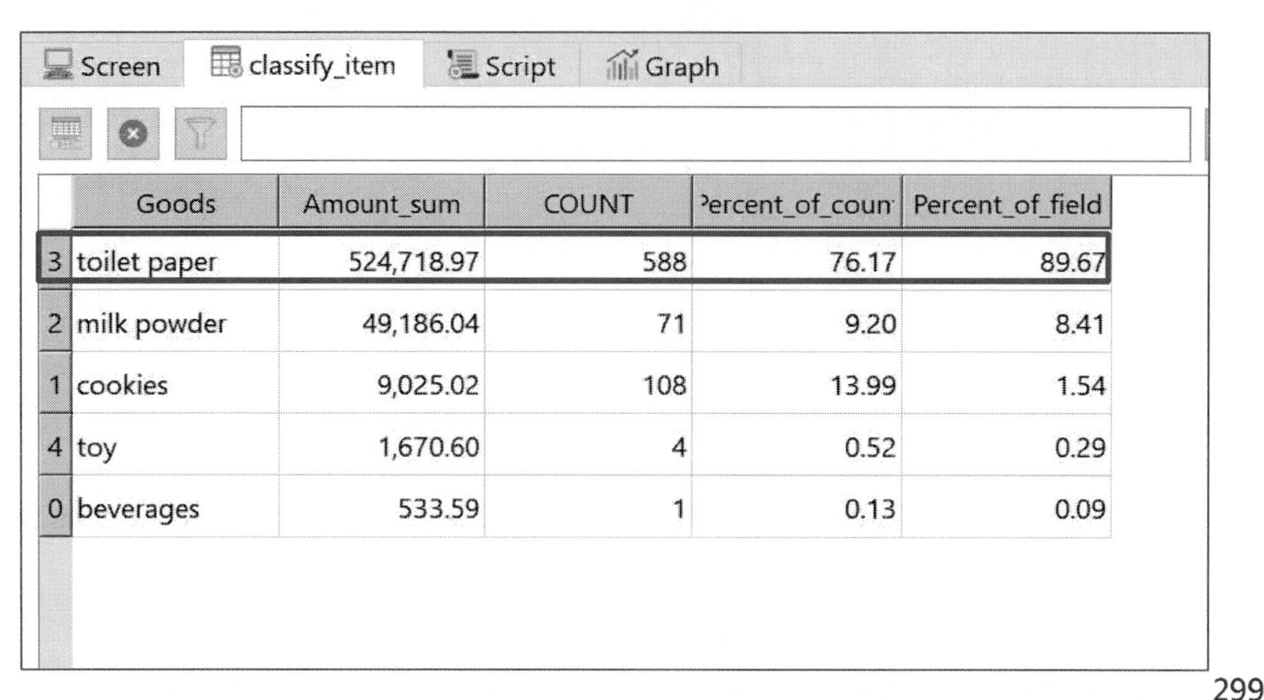

	Goods	Amount_sum	COUNT	Percent_of_coun	Percent_of_field
3	toilet paper	524,718.97	588	76.17	89.67
2	milk powder	49,186.04	71	9.20	8.41
1	cookies	9,025.02	108	13.99	1.54
4	toy	1,670.60	4	0.52	0.29
0	beverages	533.59	1	0.13	0.09

JCAATs-AI Audit Software Copyright © 2023 JACKSOFT.

Exercise 6.12

Total transaction amount for each product in the accounts receivable transaction details:

The percentage of the lowest transaction amount to the total amount is?

- Right-click on the field to sort the column in Ascending order.

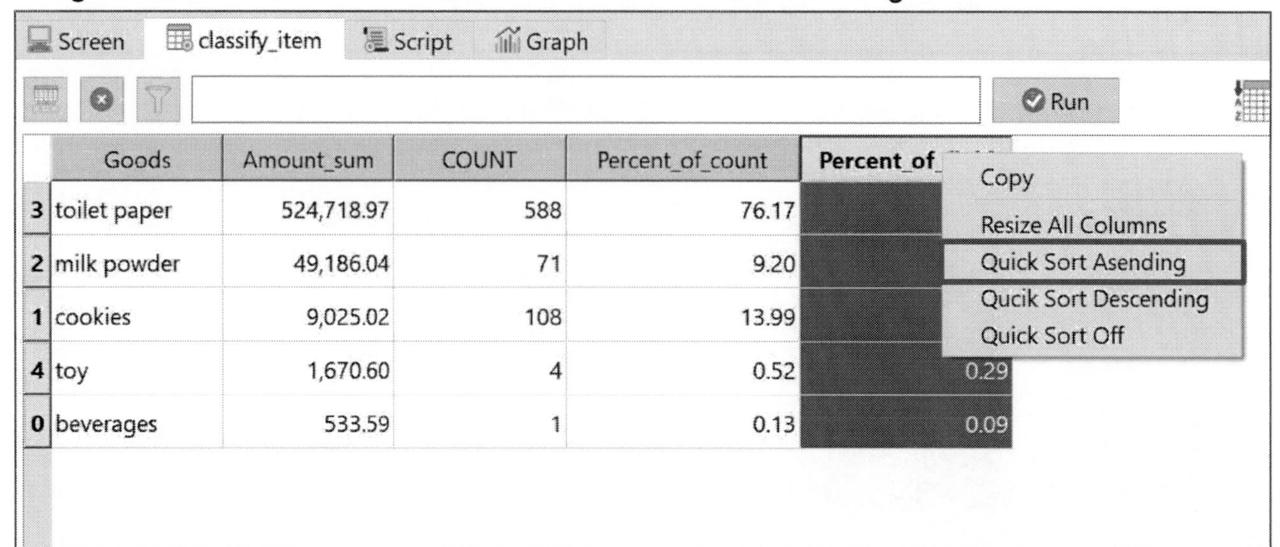

	Goods	Amount_sum	COUNT	Percent_of_count	Percent_of_
3	toilet paper	524,718.97	588	76.17	
2	milk powder	49,186.04	71	9.20	
1	cookies	9,025.02	108	13.99	
4	toy	1,670.60	4	0.52	0.29
0	beverages	533.59	1	0.13	0.09

Exercise 6.12

Total transaction amount for each product in the accounts receivable transaction details:

The percentage of the lowest transaction amount to the total amount is?

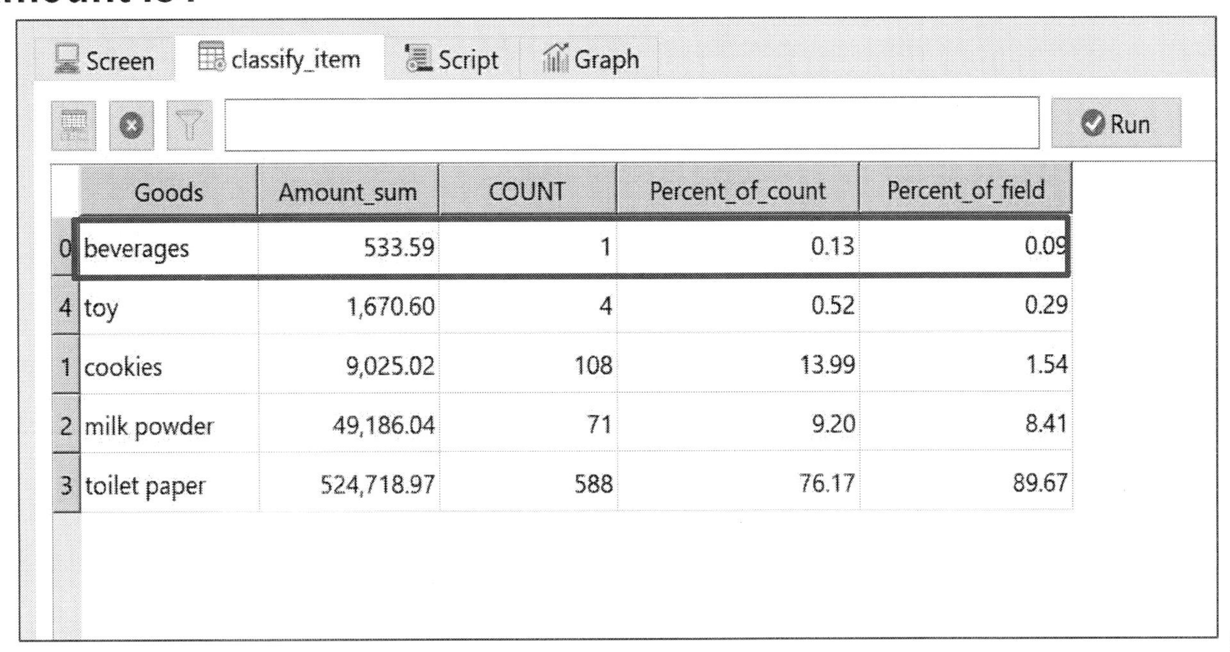

	Goods	Amount_sum	COUNT	Percent_of_count	Percent_of_field
0	beverages	533.59	1	0.13	0.09
4	toy	1,670.60	4	0.52	0.29
1	cookies	9,025.02	108	13.99	1.54
2	milk powder	49,186.04	71	9.20	8.41
3	toilet paper	524,718.97	588	76.17	89.67

Exercise 6.12

Based on the transaction amount levels of 0, 500, 800, 1000, and 10000, the level with the highest percentage of transactions to the total number is?

- Click on " Analyze>Stratify "

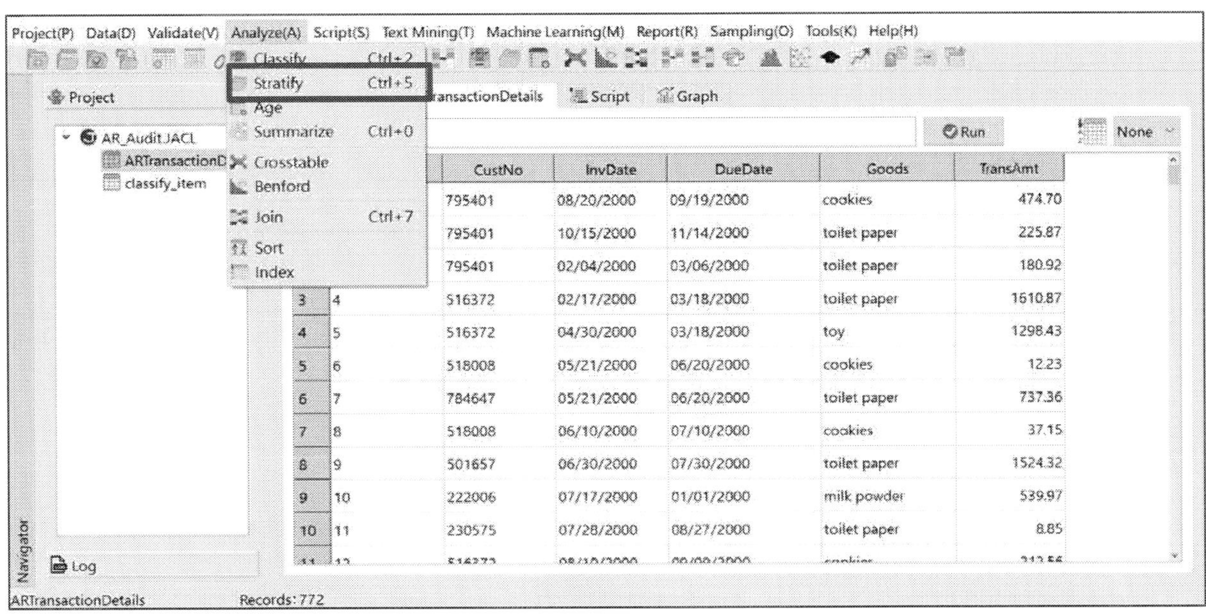

Exercise 6.12

Based on the transaction amount levels of 0, 500, 800, 1000, and 10000, the level with the highest percentage of transactions to the total number is?

- Select the field to stratify, subtotal Amount
- Define Interval Size as custom

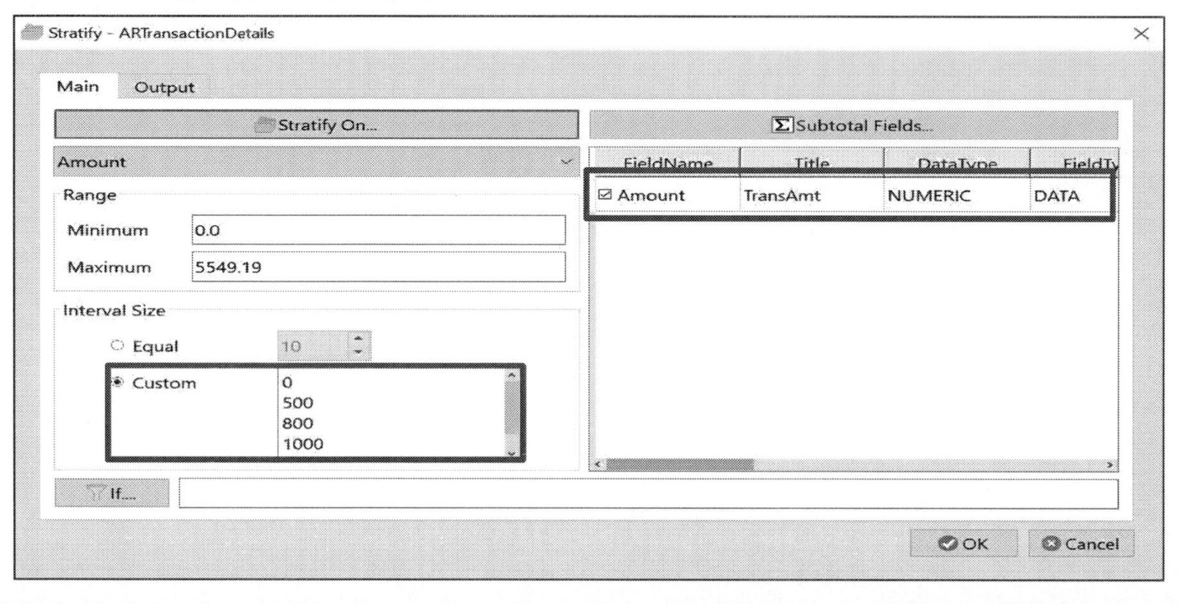

Exercise 6.12

Based on the transaction amount levels of 0, 500, 800, 1000, and 10000, the level with the highest percentage of transactions to the total number is?

- Output to a new table

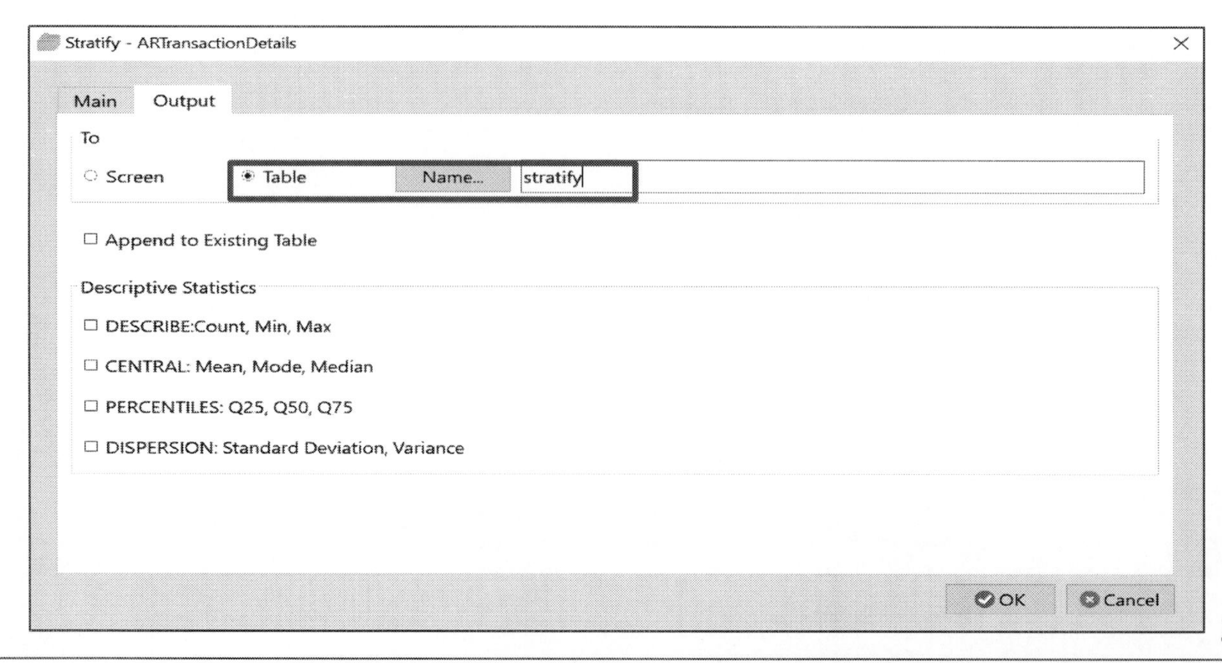

Exercise 6.12

Based on the transaction amount levels of 0, 500, 800, 1000, and 10000, the level with the highest percentage of transactions to the total number is?

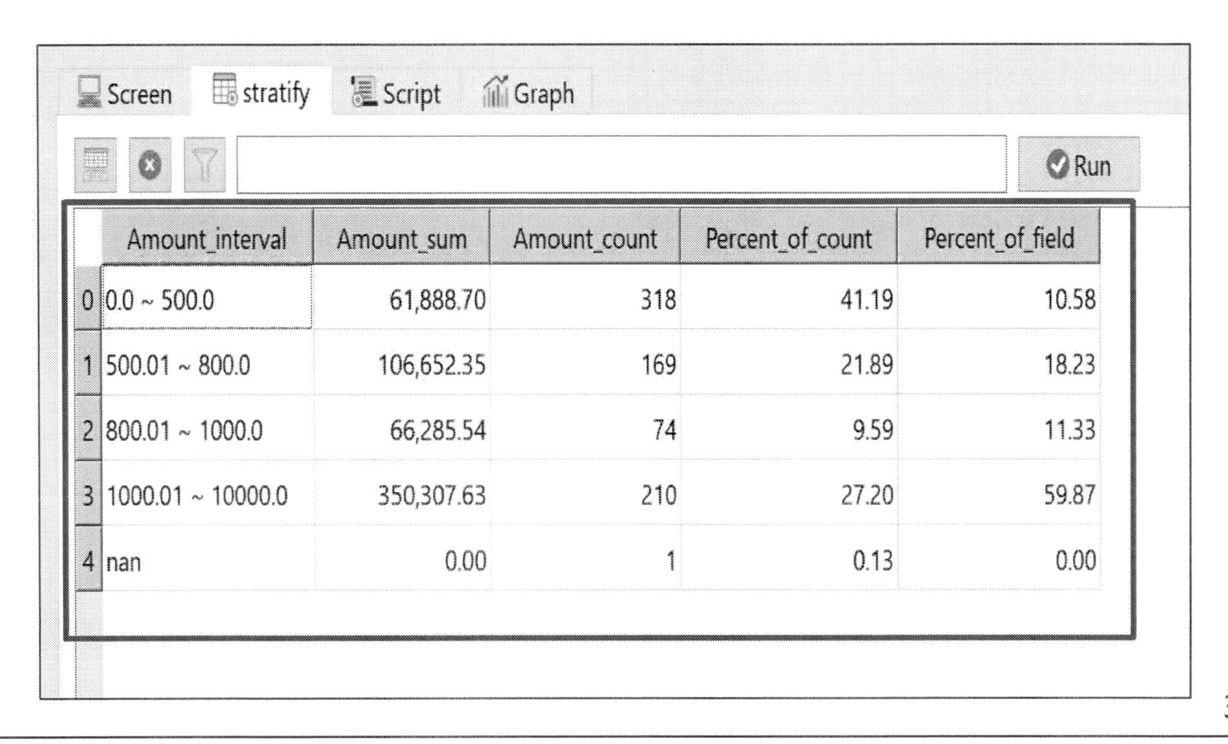

	Amount_interval	Amount_sum	Amount_count	Percent_of_count	Percent_of_field
0	0.0 ~ 500.0	61,888.70	318	41.19	10.58
1	500.01 ~ 800.0	106,652.35	169	21.89	18.23
2	800.01 ~ 1000.0	66,285.54	74	9.59	11.33
3	1000.01 ~ 10000.0	350,307.63	210	27.20	59.87
4	nan	0.00	1	0.13	0.00

JCAATs-AI Audit Software
Copyright © 2023 JACKSOFT.

Exercise 6.12

Based on the transaction amount levels of 0, 500, 800, 1000, and 10000, the level with the highest percentage of transactions to the total number is?

- Right-click on the field to sort the column in Descending order.

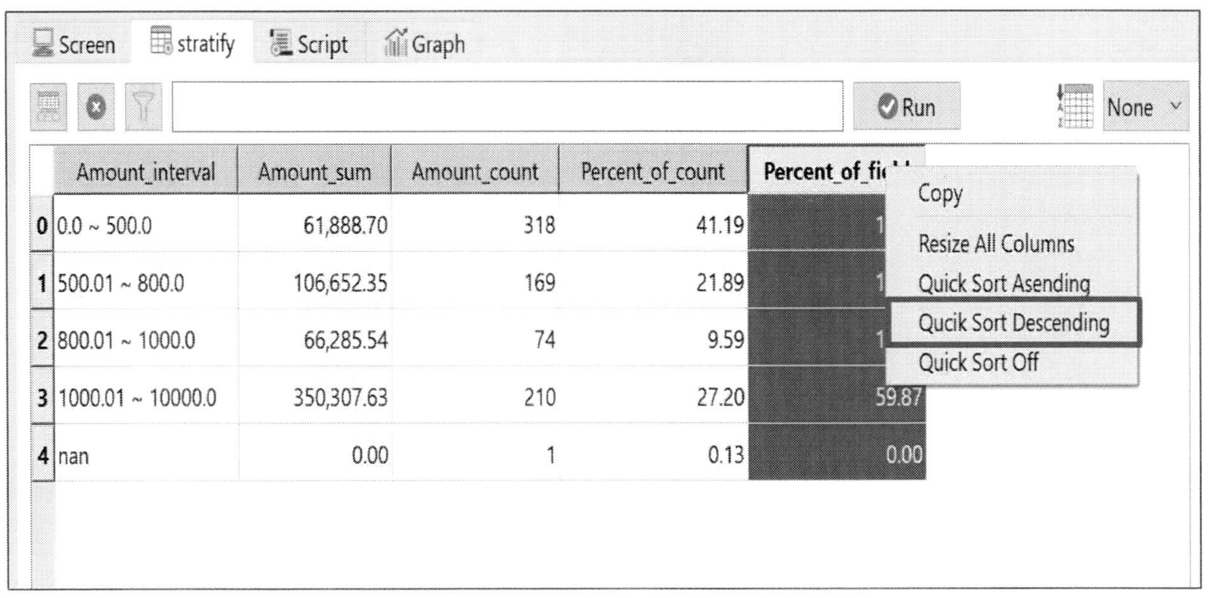

	Amount_interval	Amount_sum	Amount_count	Percent_of_count	Percent_of_fi...
0	0.0 ~ 500.0	61,888.70	318	41.19	1
1	500.01 ~ 800.0	106,652.35	169	21.89	1
2	800.01 ~ 1000.0	66,285.54	74	9.59	1
3	1000.01 ~ 10000.0	350,307.63	210	27.20	59.87
4	nan	0.00	1	0.13	0.00

Exercise 6.12

Based on the transaction amount levels of 0, 500, 800, 1000, and 10000, the level with the highest percentage of transactions to the total number is?

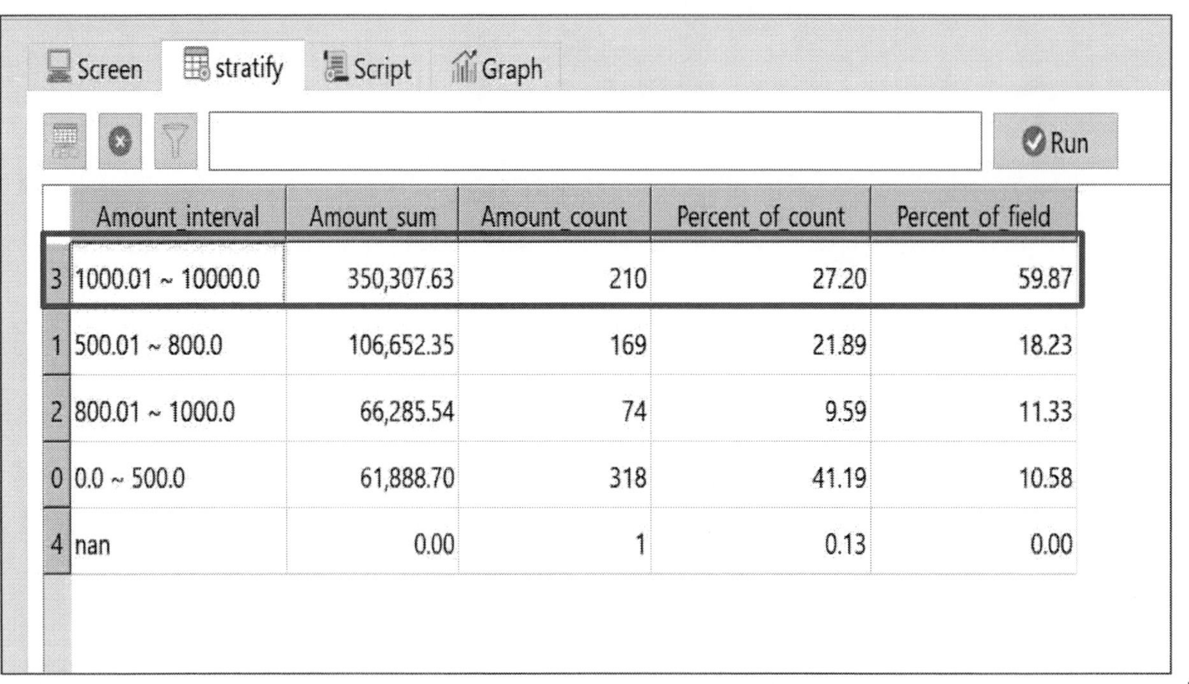

	Amount_interval	Amount_sum	Amount_count	Percent_of_count	Percent_of_field
3	1000.01 ~ 10000.0	350,307.63	210	27.20	59.87
1	500.01 ~ 800.0	106,652.35	169	21.89	18.23
2	800.01 ~ 1000.0	66,285.54	74	9.59	11.33
0	0.0 ~ 500.0	61,888.70	318	41.19	10.58
4	nan	0.00	1	0.13	0.00

Exercise 6.12

Based on the transaction amount levels of 0, 500, 800, 1000, and 10000, the level with the lowest percentage of transaction amount to the total amount is?

- Right-click on the field to sort the column in Ascending order.

Screen stratify Script Graph Run None

	Amount_interval	Amount_sum	Amount_count	Percent_of_count	Percent_of_fi...
0	0.0 ~ 500.0	61,888.70	318	41.19	1
1	500.01 ~ 800.0	106,652.35	169	21.89	1
2	800.01 ~ 1000.0	66,285.54	74	9.59	1
3	1000.01 ~ 10000.0	350,307.63	210	27.20	59.87
4	nan	0.00	1	0.13	0.00

Copy
Resize All Columns
Quick Sort Asending
Qucik Sort Descending
Quick Sort Off

Exercise 6.12

Based on the transaction amount levels of 0, 500, 800, 1000, and 10000, the level with the lowest percentage of transaction amount to the total amount is?

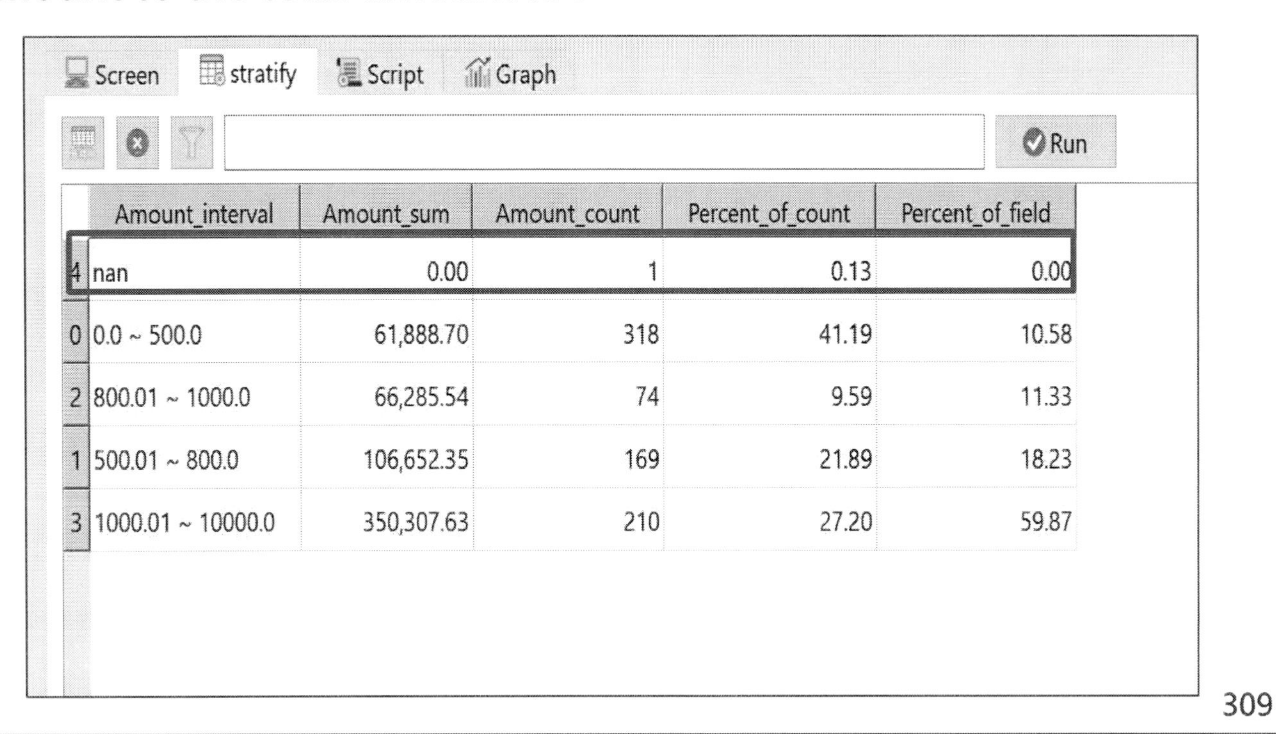

	Amount_interval	Amount_sum	Amount_count	Percent_of_count	Percent_of_field
4	nan	0.00	1	0.13	0.00
0	0.0 ~ 500.0	61,888.70	318	41.19	10.58
2	800.01 ~ 1000.0	66,285.54	74	9.59	11.33
1	500.01 ~ 800.0	106,652.35	169	21.89	18.23
3	1000.01 ~ 10000.0	350,307.63	210	27.20	59.87

JCAATs Learning notes：

JCAATs - AI Audit Software

Python Based Computer-Assisted Audit Techniques (CAATs)

Data Analysis and Smart Audit

Chapter 7 . Analysis

Outline:

1. Data Exploration
   - Classify, Stratify, Age and Summarize commands
2. Analysis Review
   - Crosstable and Benford commands
3. Data Join
   - Join command
4. Data Order
   - Sort and Index commands

JCAATs - AI Audit Software

1. Data Exploration

Data Exploration:

- Data exploration refers to the process of systematically classifying and exploring data to gain insights and identify trends and patterns.
- It involves organizing data into different categories based on their types and characteristics and analyzing them to extract meaningful information.

- Four commands of data exploration:
  - **Classify** (Single Field)
  - **Summarize**(Multiple Fields)
  - **Stratify**(Single Numeric Field)
  - **Age**(Single Datetime Field)

In JCAATs, multiple fields can be mixed for exploration by category, and information can be explored beyond the subtotal value, including descriptive statistics. This allows users to gain a deeper understanding of the data and identify unusual items or anomalies that may require further investigation.

Interface for the Commands

- This type of “analysis commands” adhere to standard interface design. There are two primary tabs on the interface, such as
  - **Main** tab: Set main parameters for the command.
  - **Output** tab: Set output parameters for the command.

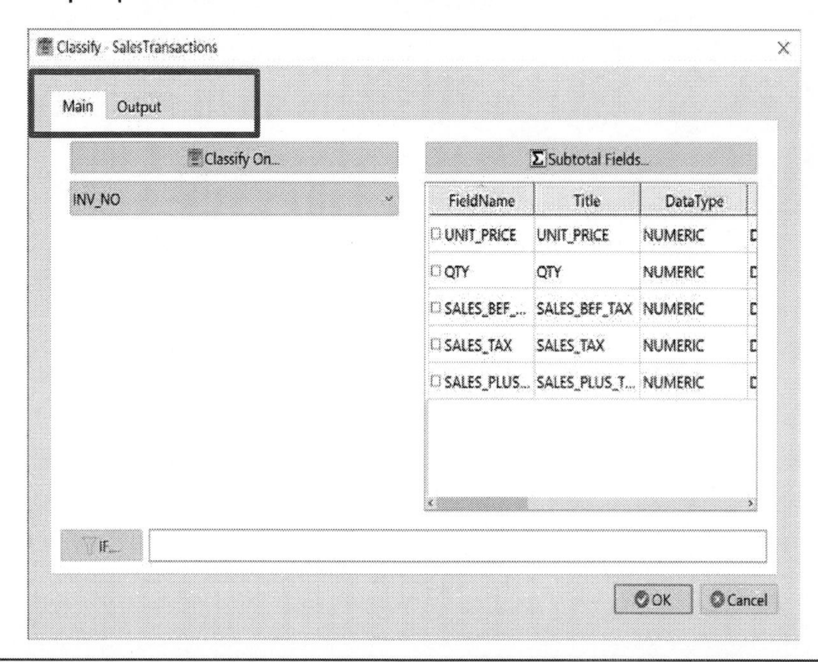

Main tab - Two Columns with Single Selection

- This interface for Classify, Stratify and Age commands are divided into two columns. The right column supports a single field selection with some of the available operators, while the left column supports selection of multiple numeric fields to calculate their subtotal.

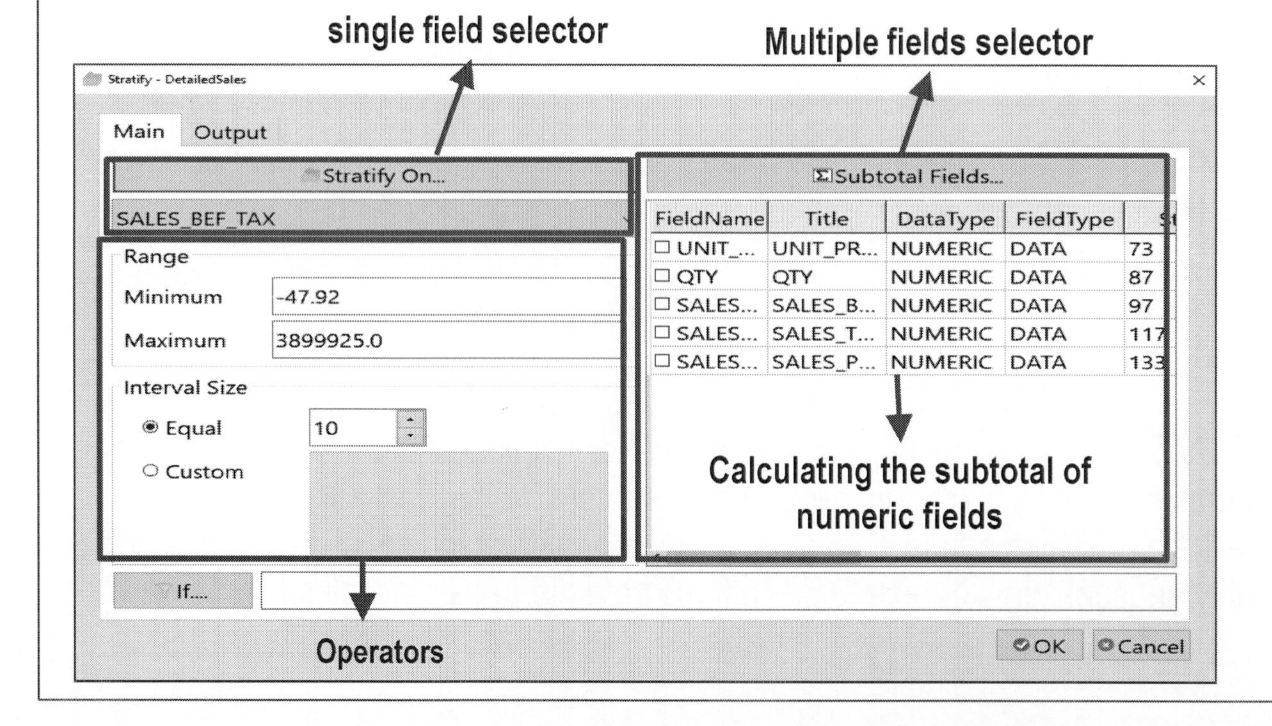

Output tab - with descriptive statistics

- If users select subtotal fields, users have the option to export an output result that includes **descriptive statistics**. This feature is particularly useful for conducting more detailed data exploration and analysis. If users do not select any subtotal fields, the result will based on the grouping count and the description statistics area is disabled.

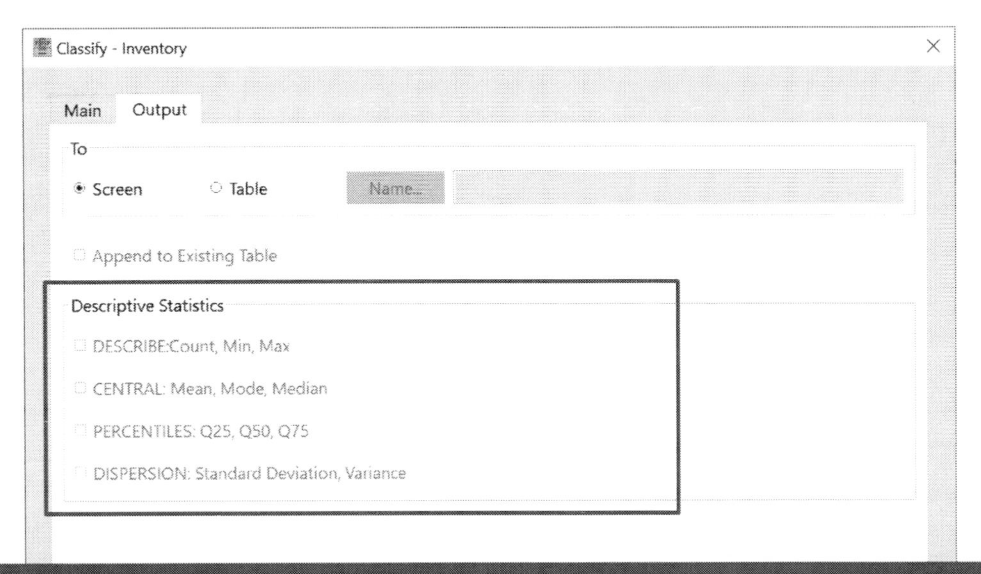

Descriptive statistics include 4 categories with 11 indicators.

Main tab - Three Tables

- The Summarize command can select multiple fields (such as, text, datetime, numeric and others) as key fields for grouping calculation.
- Other Fields: The field is displayed with the first record data of this category.

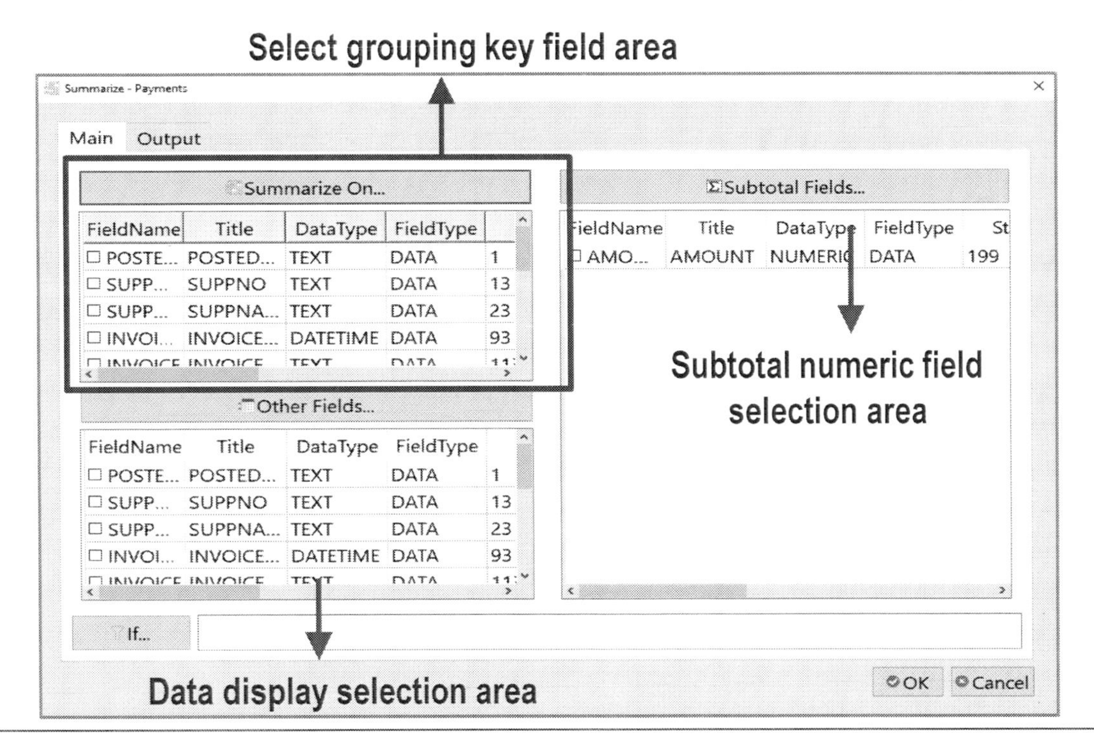

Result Screen

- If users specify to view the resulting data on the screen, the screen will display a grouping table, which organizes the information into relevant categories or groups.

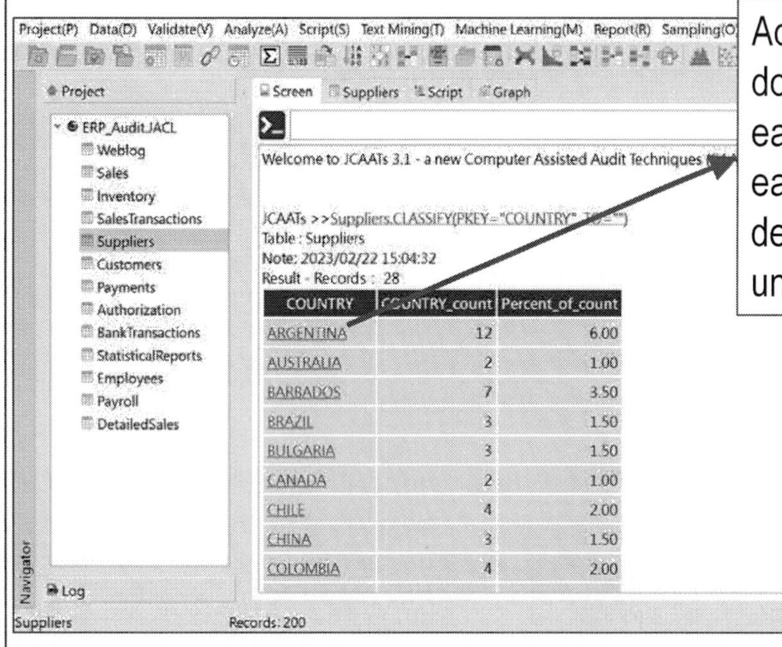

COUNTRY	COUNTRY_count	Percent_of_count
ARGENTINA	12	6.00
AUSTRALIA	2	1.00
BARBADOS	7	3.50
BRAZIL	3	1.50
BULGARIA	3	1.50
CANADA	2	1.00
CHILE	4	2.00
CHINA	3	1.50
COLOMBIA	4	2.00

Additionally, the table features a drill-down function, which allows users to easily access more detailed data for each category or group, providing a deeper level of insight and understanding.

Result Graph

- The exploration commands (except Summarize) provides a bar chart result graph with the key field as the X-axis. Users can zoom in and out, and they can also click on the tiles to view detailed information and discuss it in depth.

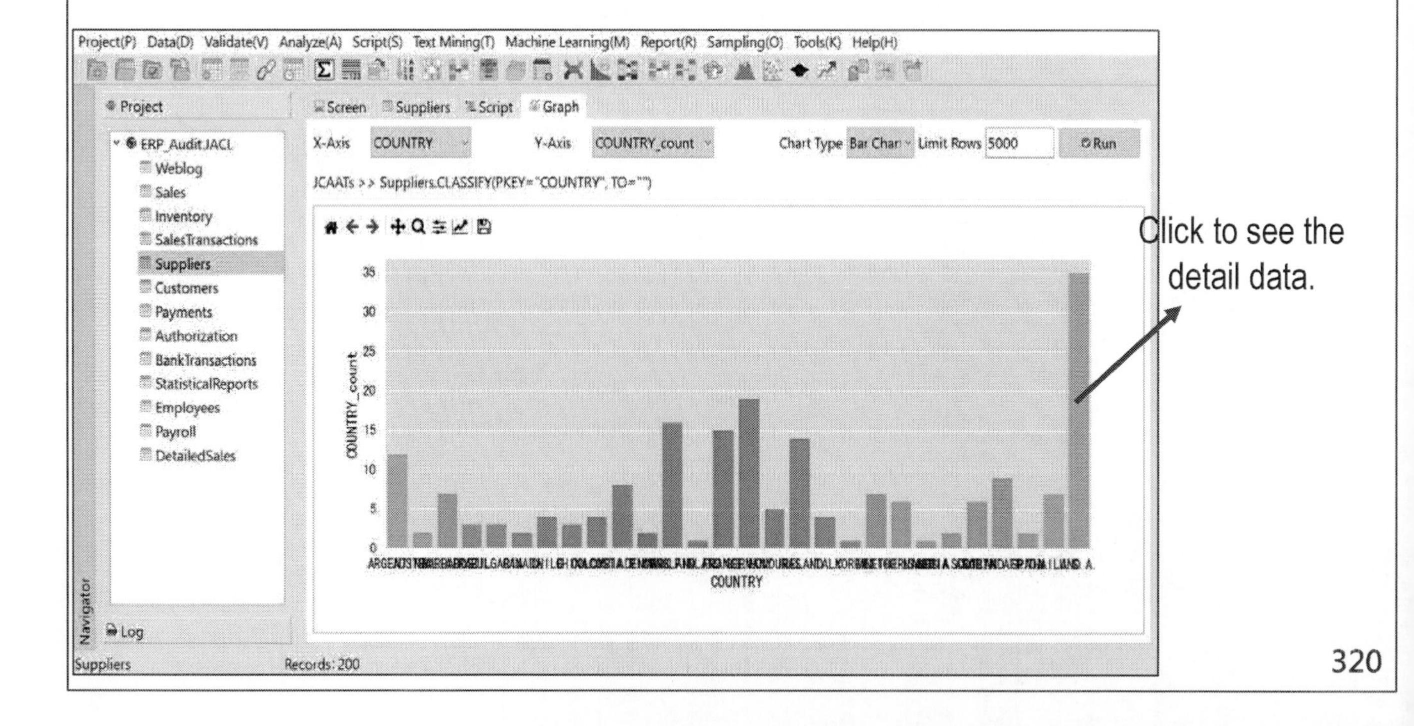

JCAATs - AI Audit Software

2. Analytical Review

Analytical Review

- Analytical review is a process used in auditing, accounting, and other fields to analyze and review data in a systematic and structured manner.
- This process involves using a variety of tools and techniques to identify patterns, anomalies, and trends within the data, which can then be used to draw conclusions or make decisions.
- Two commands of analytical review:
  - **Benford** (Single Numeric Field)
  - **Crosstable** (Two Text Fields)

Main tab - Crosstable

- Cross-table is a technique used to analyze the relationship between two text fields.
- It can select multiple numeric fields to calculate subtotals. If no any subtotal field is selected, the count will be the result.

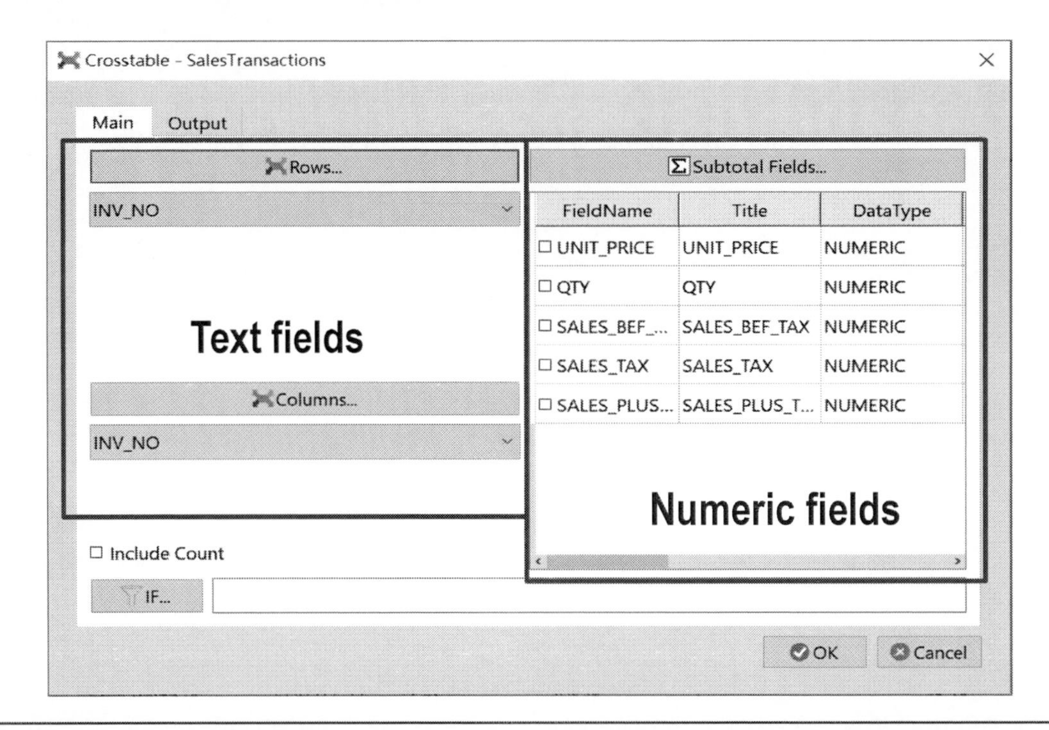

Result Screen of Crosstable

- Cross-table technique involves creating a table that displays the frequency distribution of one variable relative to another, which can help to identify patterns or correlations between the variables.

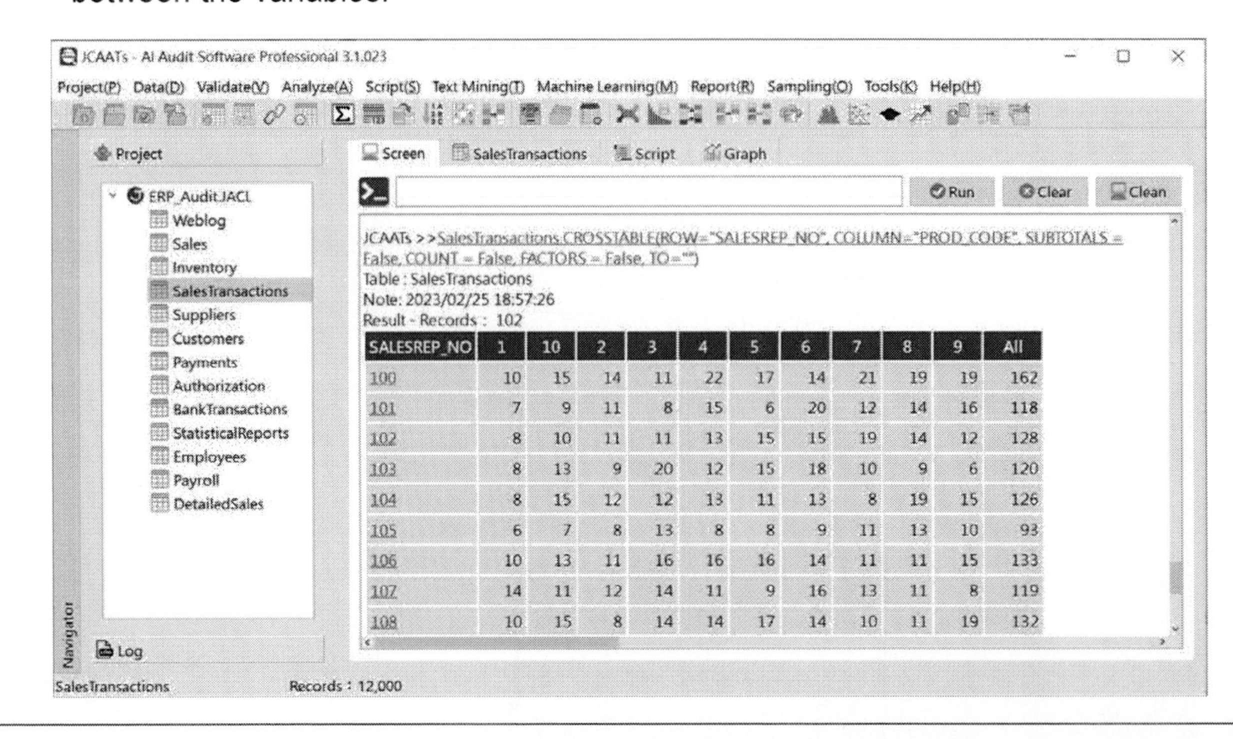

SALESREP_NO	1	10	2	3	4	5	6	7	8	9	All
100	10	15	14	11	22	17	14	21	19	19	162
101	7	9	11	8	15	6	20	12	14	16	118
102	8	10	11	11	13	15	15	19	14	12	128
103	8	13	9	20	12	15	18	10	9	6	120
104	8	15	12	12	13	11	13	8	19	15	126
105	6	7	8	13	8	8	9	11	13	10	93
106	10	13	11	16	16	16	14	11	11	15	133
107	14	11	12	14	11	9	16	13	11	8	119
108	10	15	8	14	14	17	14	10	11	19	132

Result Graph of Crosstable

- Cross-table Graph can be used to identify trends, uncover hidden relationships, and test hypotheses, making it a valuable tool for data analysis in a wide range of fields.

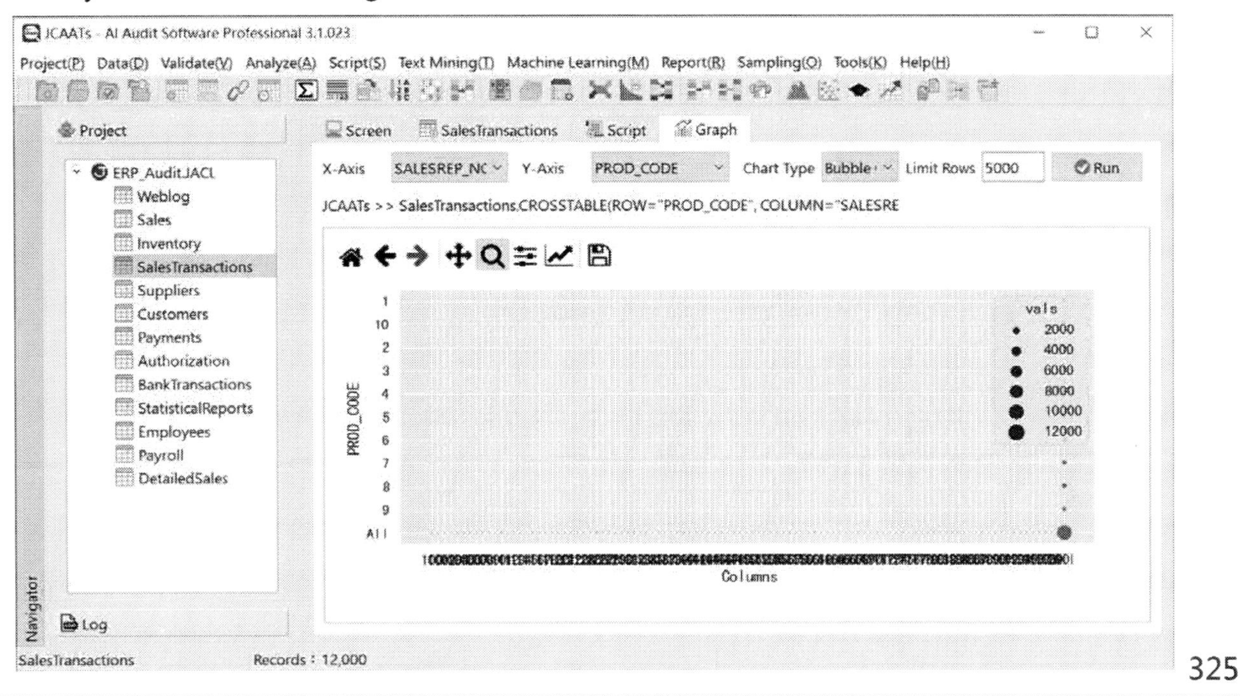

Benford

- Among the numbers 1 to 9, the digit "1" has the highest frequency of appearance, which is about 30%. The digit "2" comes in second, and the frequency decreases as the digit value increases.

$$P(d) = \log_{10}(d+1) - \log_{10}(d) = \log_{10}\left(\frac{d+1}{d}\right) = \log_{10}\left(1 + \frac{1}{d}\right).$$

Digit	Probability
1	30.1%
2	17.6%
3	12.5%
4	9.7%
5	7.9%
6	6.7%
7	5.8%
8	5.1%
9	4.6%

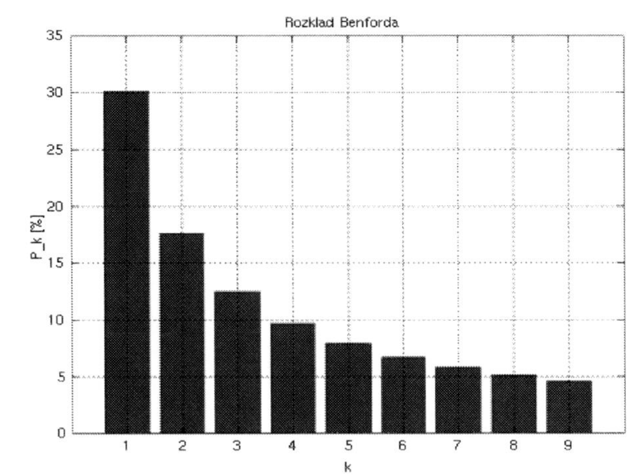

Source: https://en.wikipedia.org/wiki/Benford%27s_law

Main tab - Benford

- JCAATs can select the starting position and number of digits to make Benford analysis more resilient.
- Users can flexibly set the upper and lower boundary range with different confidence rate for test verification, which makes the check results more reliable.

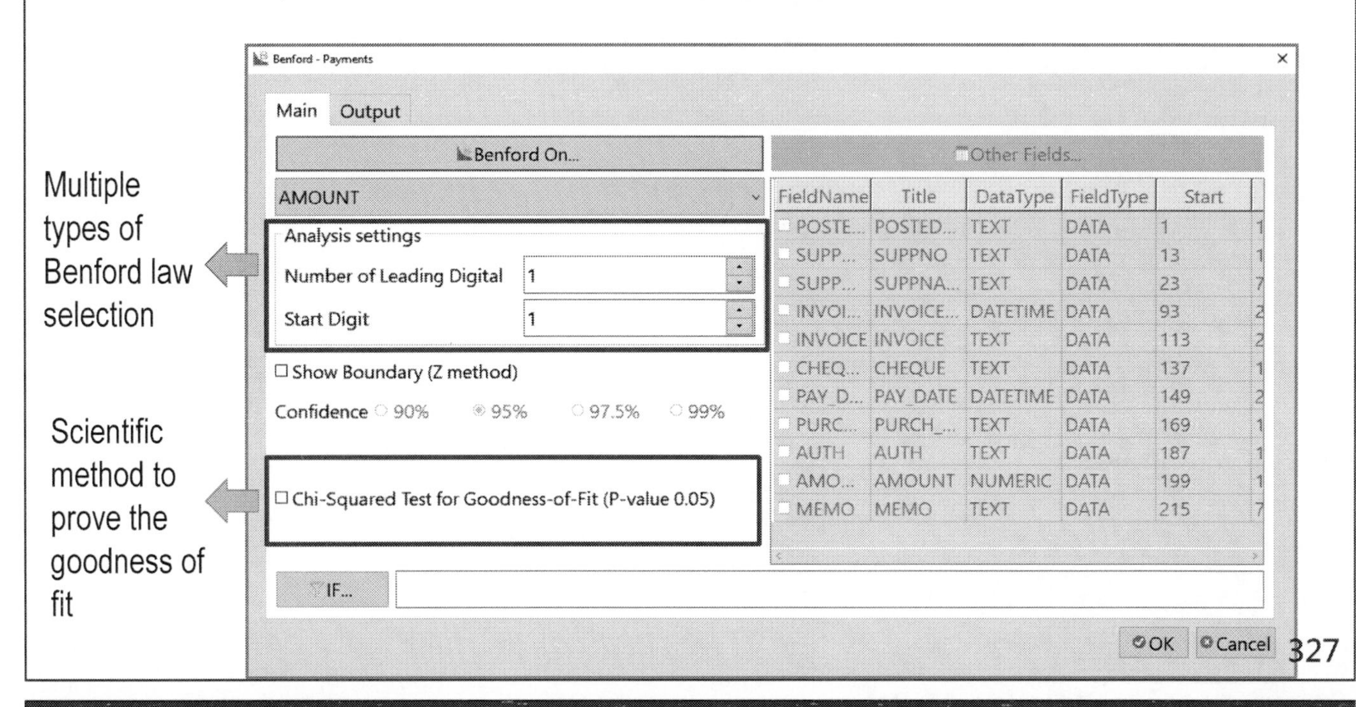

Result Screen of Benford

- Benford's Law has a wide range of applications, including forensic accounting, fraud detection, and data validation. By examining the first digits of a dataset and comparing them to the expected distribution under Benford's Law, analysts can identify potential anomalies or errors in the data.

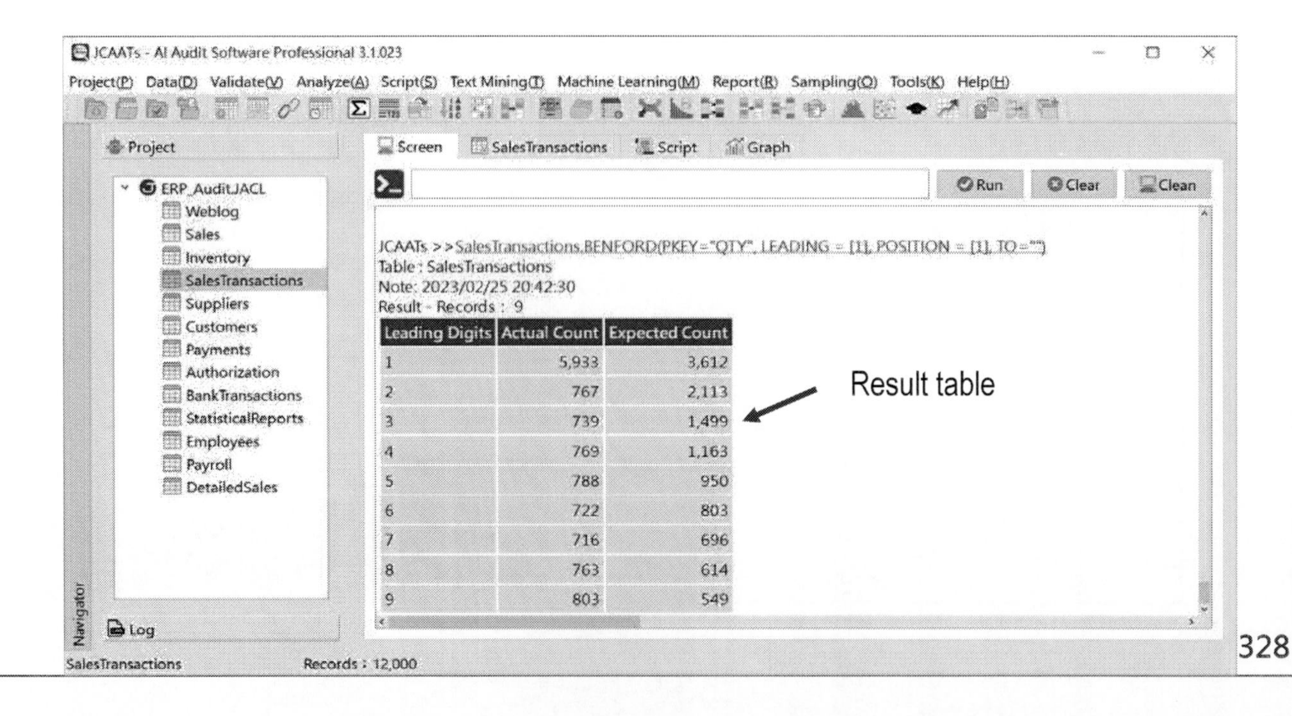

Leading Digits	Actual Count	Expected Count
1	5,933	3,612
2	767	2,113
3	739	1,499
4	769	1,163
5	788	950
6	722	803
7	716	696
8	763	614
9	803	549

Result Graph of Benford

- This benford's law pattern can be observed in a wide range of datasets, including financial data, population numbers, physical constants, and even the lengths of rivers.

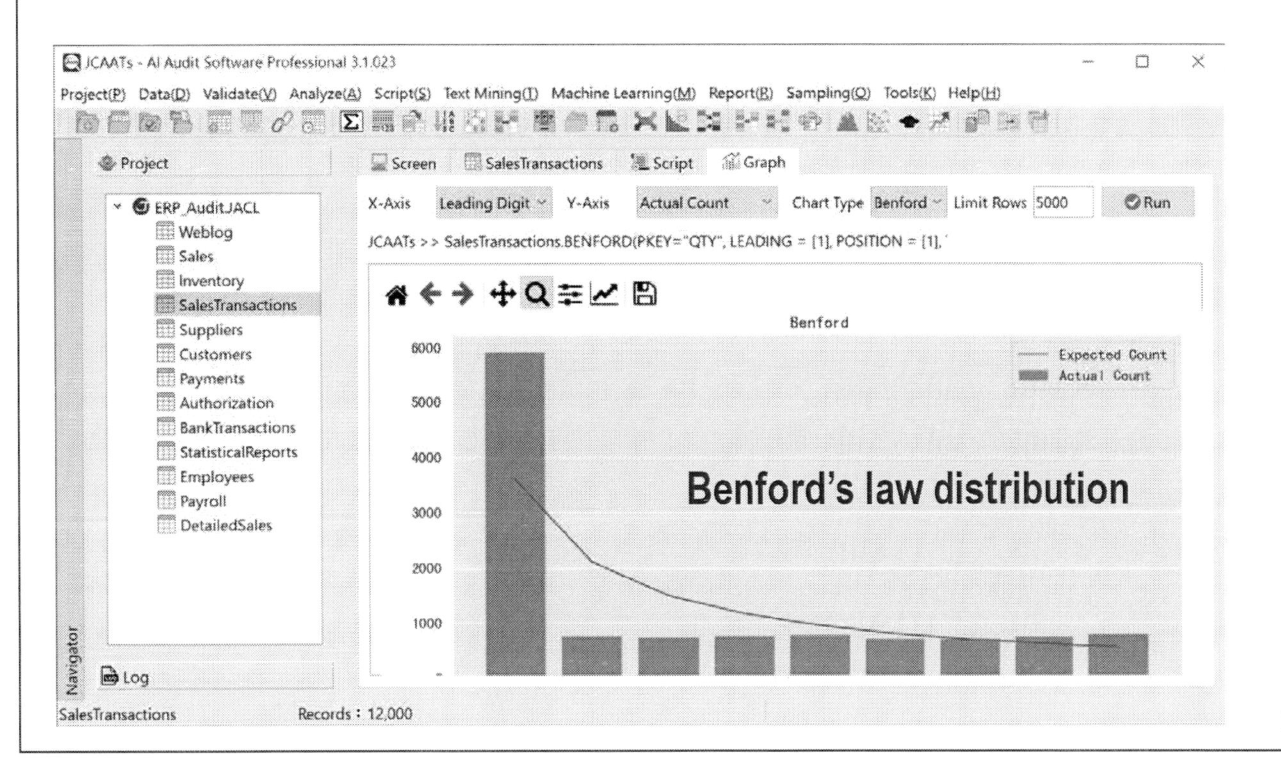

JCAATs - AI Audit Software

Data Join

- Data join is a common operation in data analysis and is often performed using specialized software such as audit software or data manipulation tools.
- The purpose of data join is to consolidate information from different sources into a single dataset for analysis, reporting, or other purposes.
- Data join involves identifying a common variable or set of variables that can be used to match the records in the datasets being merged.
- For example, if one dataset contains customer information and another dataset contains sales data, the common variable might be a customer ID or email address.
- There are several types of join operations in database, including inner join, left join, right join, and outer join, each of which specifies how to handle records that do not have a match in the other dataset.

Join Command

In JCAATs, users can use the **Join** command to combine two tables with the same key-value field, and then export the resulting third table.

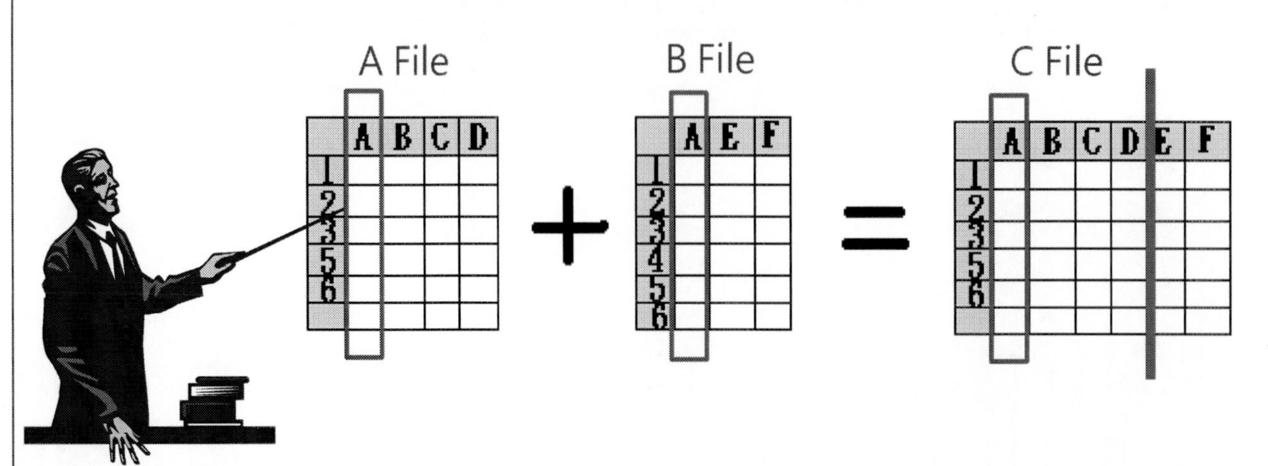

How to perform Join command?

- The JOIN command is used to **combine two tables into a new data table** based on **the key fields** and condition.
- When performing a join operation, two tables are involved. The first table that was opened is called the **primary** table, and the second is called the **secondary** table.
  - **While using the JOIN command, it is important to determine which table is the primary and which is the secondary.**
- The JOIN command can combine two tables into the third table. Any two tables that are ready to relate or connect must have an identifying feature field. This field is called the **key-value** field.

Six Types of Join

- Type 1 : Matched Primary with the first Secondary
- Type 2 : Matched All Primary with the first Secondary
- Type 3 : Matched All Secondary with the first Primary
- Type 4 : Matched All Primary and Secondary with the first
- Type 5 : Unmatched Primary
- Type 6 : Many to Many

Six Types of Join Icons

	JCAATs
1	Matched Primary with the first Secondary
2	Matched All Primary with the first Secondary
3	Matched All Secondary with the first Primary
4	Matched All Primary and Secondary with the first
5	Unmatch Primary
6	Many to Many

The Procedure of Join

1. Determine the purpose of the JOIN operation.
2. Identify the Primary and Secondary tables to be joined.
3. The files to be joined must belong to the same JCAATS project.
4. Both files need to have a common feature field/key-value field (e.g., employee ID, ID number).
5. Data type and length of feature field must to be consistent
6. Choose the join category:
   - **A. Matched Primary with the first Secondary**
   - **B. Matched All Primary with the first Secondary**
   - **C. Matched All Secondary with the first Primary**
   - **D. Matched All Primary and Secondary with the first**
   - **E. Unmatched Primary**
   - **F. Many to Many**

The Process Steps of Join Command

- To use Join command:
  1. Open the Join command
  2. Select the primary table
  3. Select the secondary table
  4. Select the key field for both the primary and secondary tables
  5. Select which fields to include in the resulting table from the primary and secondary tables.
  6. Use filters to specify the data to be included. (selectivity)
  7. Select the Join execution type
  8. Name the resulting table.

Examples of Join

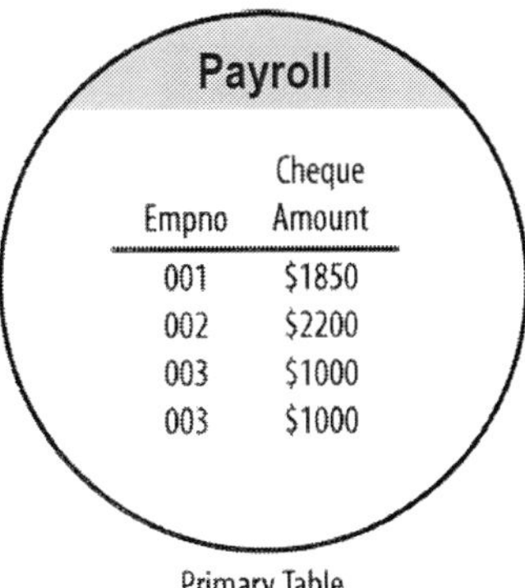
Payroll

Empno	Cheque Amount
001	$1850
002	$2200
003	$1000
003	$1000

Primary Table

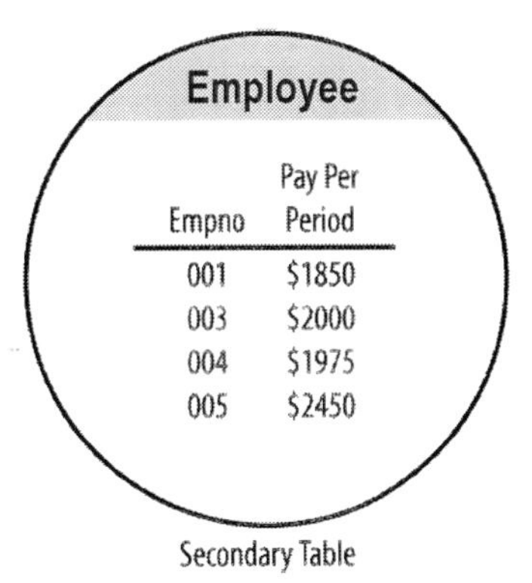
Employee

Empno	Pay Per Period
001	$1850
003	$2000
004	$1975
005	$2450

Secondary Table

5 **Unmatched Primary**

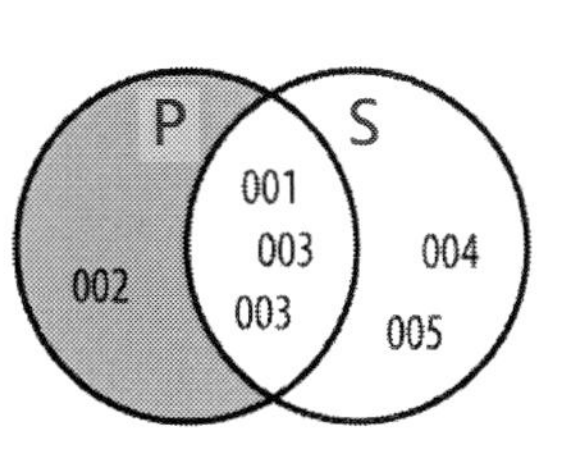

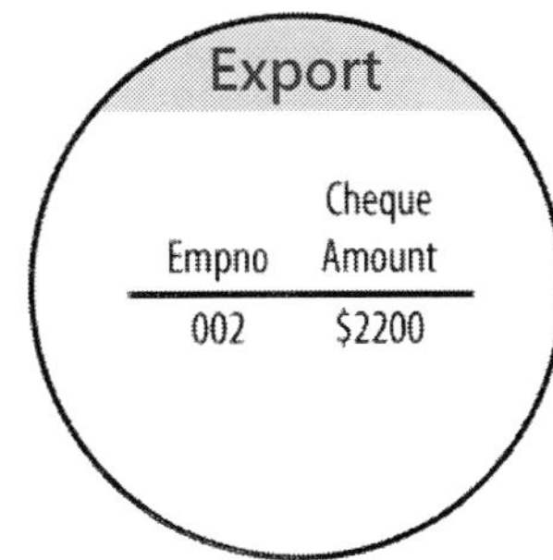
Export

Empno	Cheque Amount
002	$2200

Matched Primary with the first Secondary

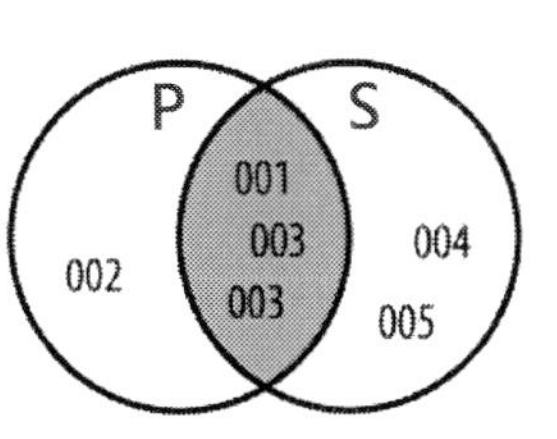

Export

Empno	Cheque Amount	Pay Per Period
001	$1850	$1850
003	$1000	$2000
003	$1000	$2000

Examples of Join

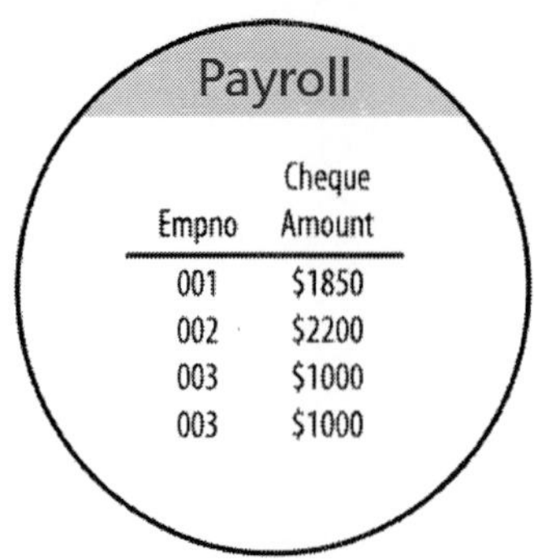

Payroll

Empno	Cheque Amount
001	$1850
002	$2200
003	$1000
003	$1000

Primary Table

Employee

Empno	Pay Per Period
001	$1850
003	$2000
004	$1975
005	$2450

Secondary Table

Matched All Secondary with the first Primary

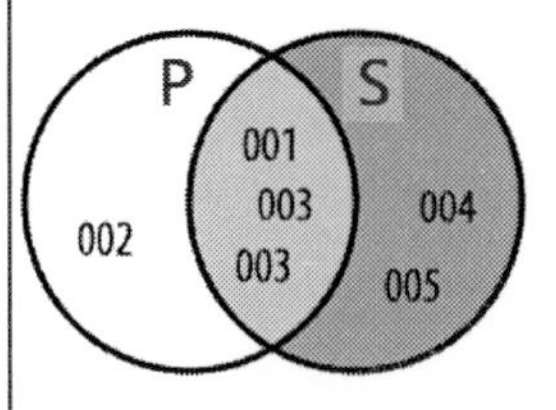

Export

Empno	Cheque Amount	Pay Per Period
001	$1850	$1850
003	$1000	$2000
003	$1000	$2000
004	$0	$1975
005	$0	$2450

Matched All Primary with the first Secondary

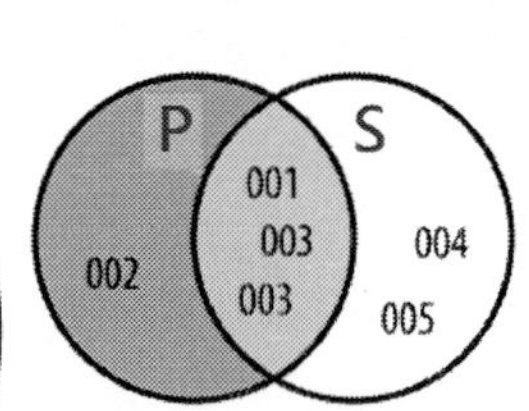

Export

Empno	Cheque Amount	Pay Per Period
001	$1850	$1850
002	$2200	$0
003	$1000	$2000
003	$1000	$2000

Examples of Join

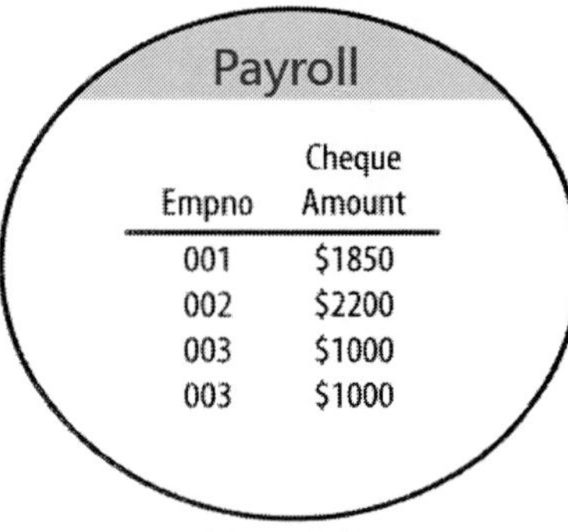

Payroll

Empno	Cheque Amount
001	$1850
002	$2200
003	$1000
003	$1000

Primary Table

Employee

Empno	Pay Per Period
001	$1850
003	$2000
004	$1975
005	$2450

Secondary Table

Matched All Primary and Secondary with the first

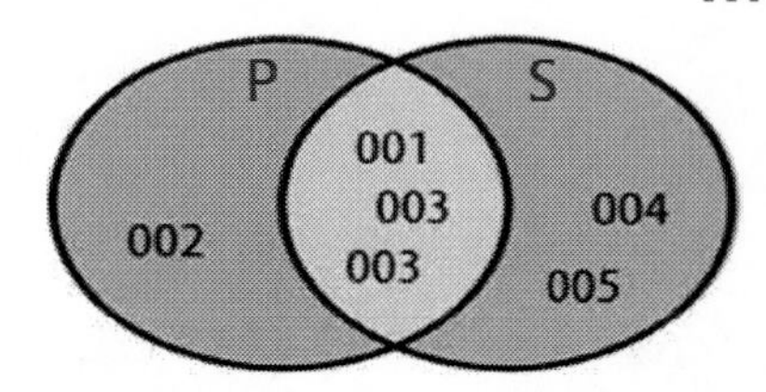

Export

Empno	Cheque Amount	Pay Per Period
001	$1850	$1850
002	$2200	$0
003	$1000	$2000
003	$1000	$2000
004	$0	$1975
005	$0	$2450

Examples of Join

1. Find all match data between the same employee code of payroll and employee table
2. Filter out data with the correct date
3. Check whether the actual payment in the payroll is consistent with the salary recorded in the employee file

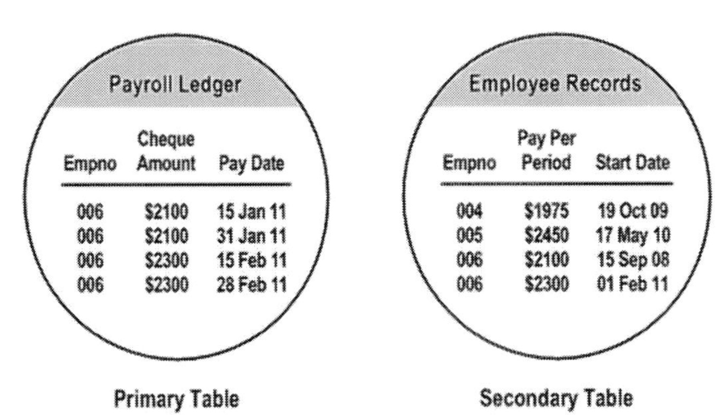

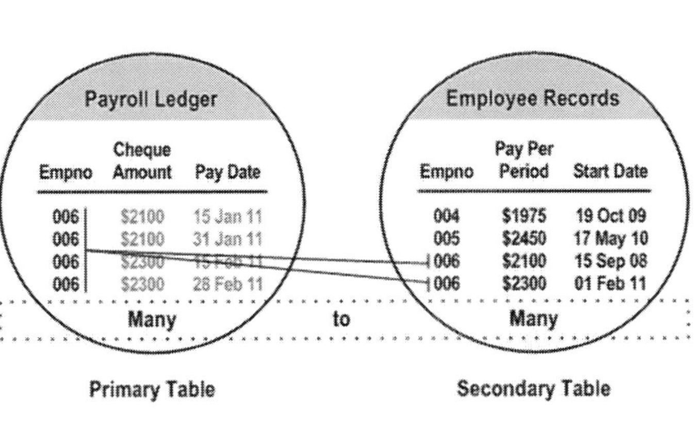

Many-to-Many

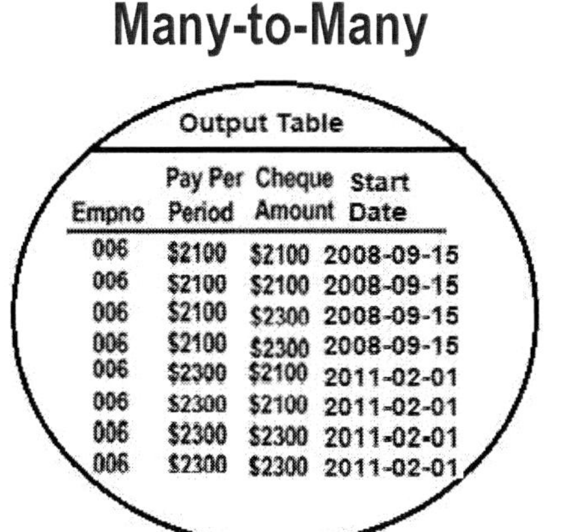

Main tab - Join

- Once the common fields is identified, the datasets can be joined using a join operation, which brings together the matching records from each dataset into a new output dataset.

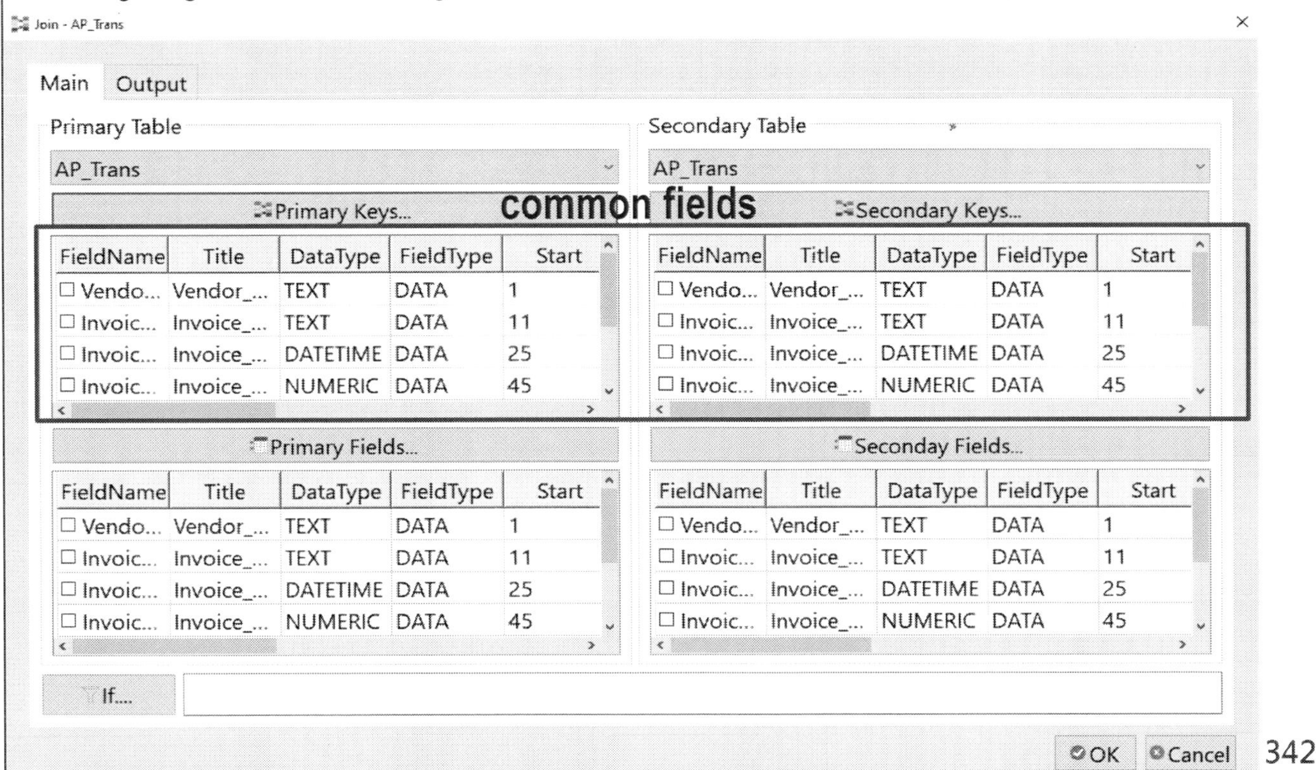

Output tab - Join

- It is important to carefully consider the quality of the data being merged and to ensure that the join operation is performed correctly to avoid introducing errors or bias into the analysis.

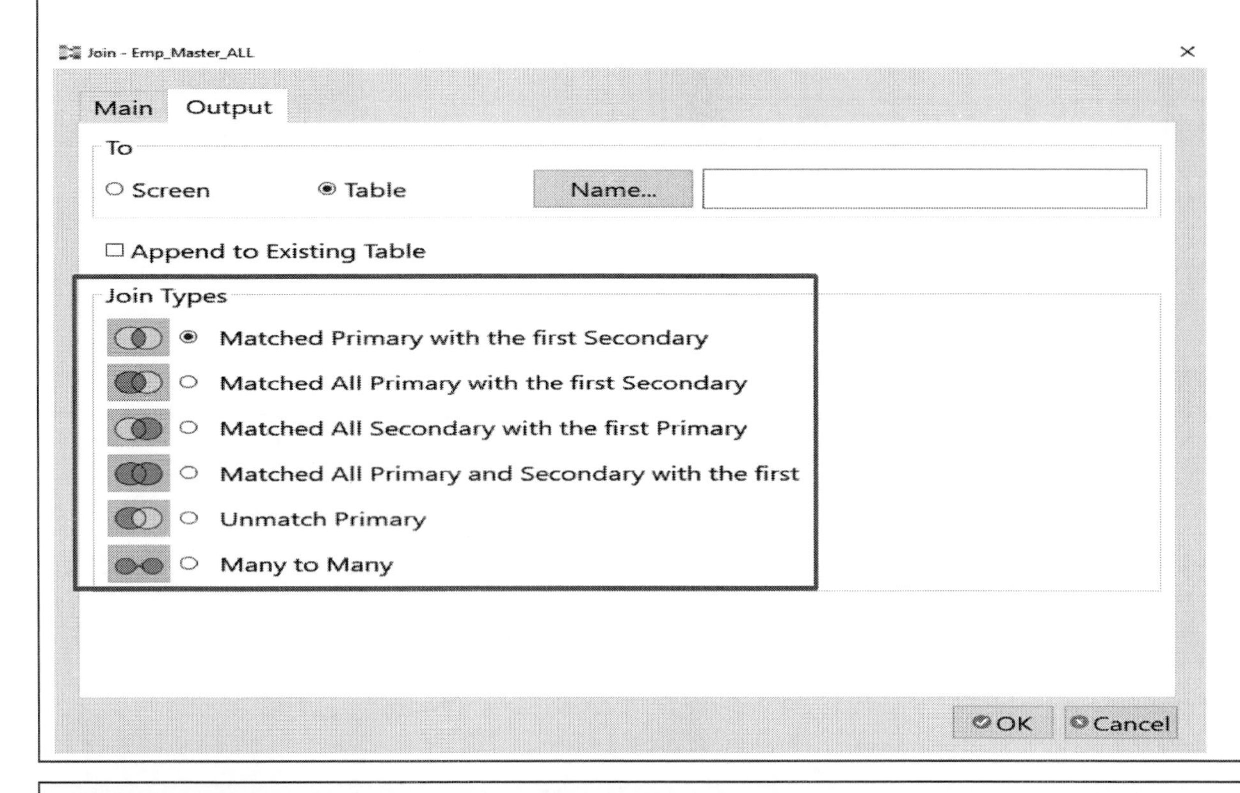

JCAATs - AI Audit Software

4. Data Order

Data Order

- Data sequence refers to the order or arrangement of data elements within a dataset.
- Sorting and indexing are two common operations used to manipulate the sequence of data within a dataset.

- Three commands for data sequence:
  - **Sort**
  - **Index**
  - **Quick Sort**

Both sort and index are important operations in data analysis, as they allow analysts to manipulate the sequence of data within a dataset to better understand patterns and relationships.

Sort

- Sorting is the process of arranging the data elements in a specific order based on one or more fields and generate a new resulting table. For example, a list of sales figures could be sorted in descending order to show the highest sales first.
- The resulting table have the same record structure as the source table.

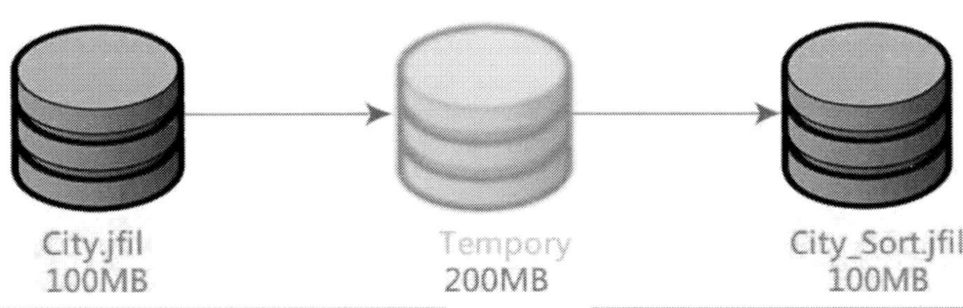

index	Name	City
1	Juan Carlos Pecoroff	Berazategui
2	Lawrence O'Mara	Norton Shores
3	Carmen Bacardi Bolivar	Alajuela
4	Arthur H. Penn	Berlin
5	Simon Allen	London

index	Name	City
1	Carmen Bacardi Bolivar	Alajuela
2	Juan Carlos Pecoroff	Berazategui
3	Arthur H. Penn	Berlin
4	Simon Allen	London
5	Lawrence O'Mara	Norton Shores

Main tab - Sort

- Sorting can be performed in ascending or descending order, and can be based on one or more variables.

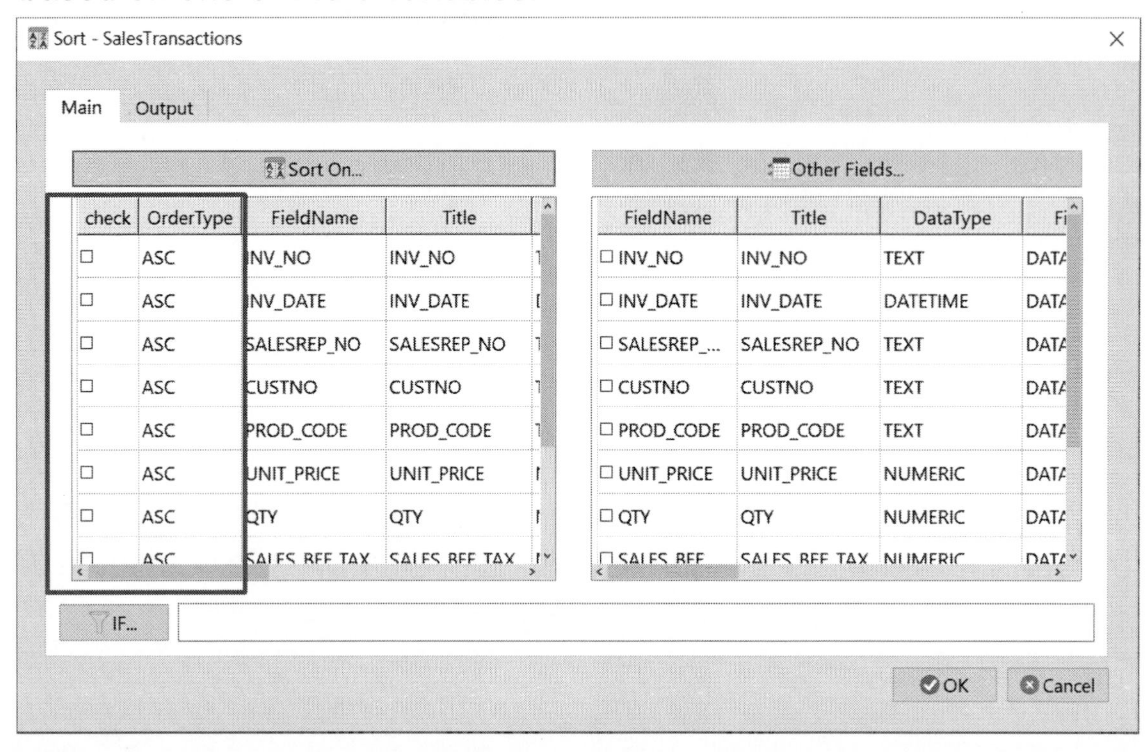

Output tab - Sort

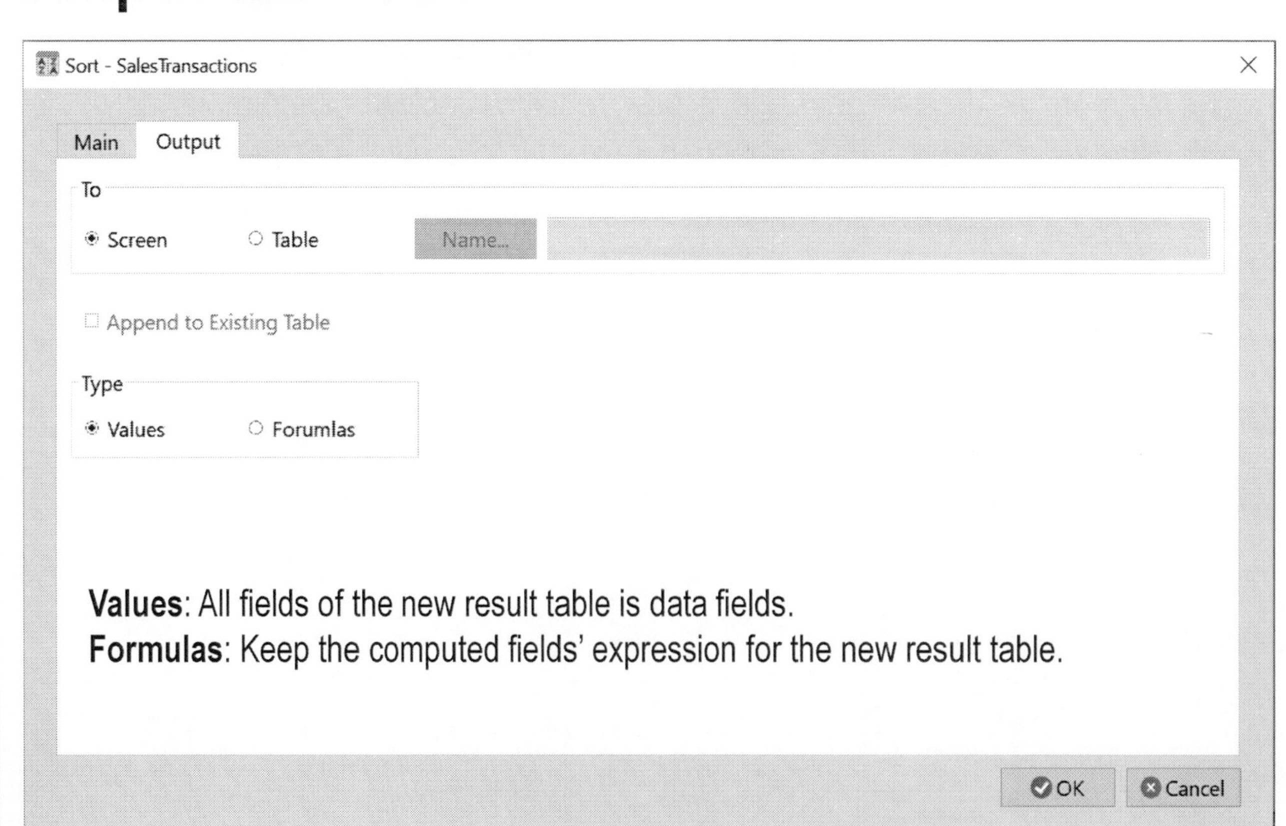

Values: All fields of the new result table is data fields.
Formulas: Keep the computed fields' expression for the new result table.

Index

- Indexing is the process of creating an index that maps the values of a specific fields to their corresponding positions in the dataset.
- An index can be used to quickly locate specific data elements within a dataset based on the values of one or more fields. For example, an index could be created for a list of city names, allowing the analyst to quickly locate the information for a specific city without having to search through the entire dataset.

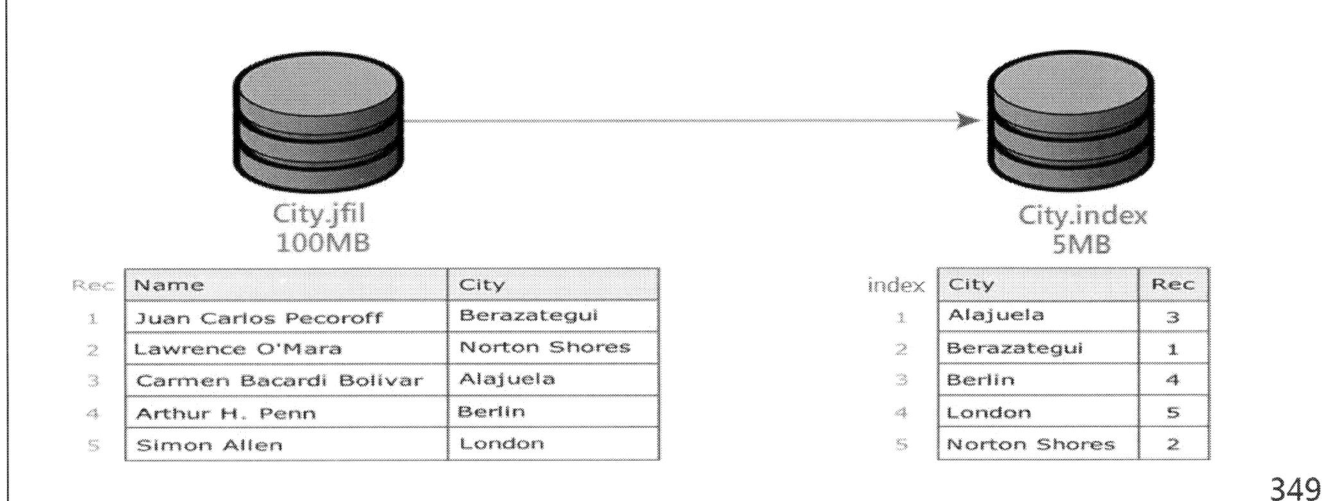

Rec	Name	City
1	Juan Carlos Pecoroff	Berazategui
2	Lawrence O'Mara	Norton Shores
3	Carmen Bacardi Bolivar	Alajuela
4	Arthur H. Penn	Berlin
5	Simon Allen	London

index	City	Rec
1	Alajuela	3
2	Berazategui	1
3	Berlin	4
4	London	5
5	Norton Shores	2

Main tab - Index

- It is important to carefully consider the impact of these operations on the analysis and to ensure that the resulting sequence accurately reflects the underlying data.

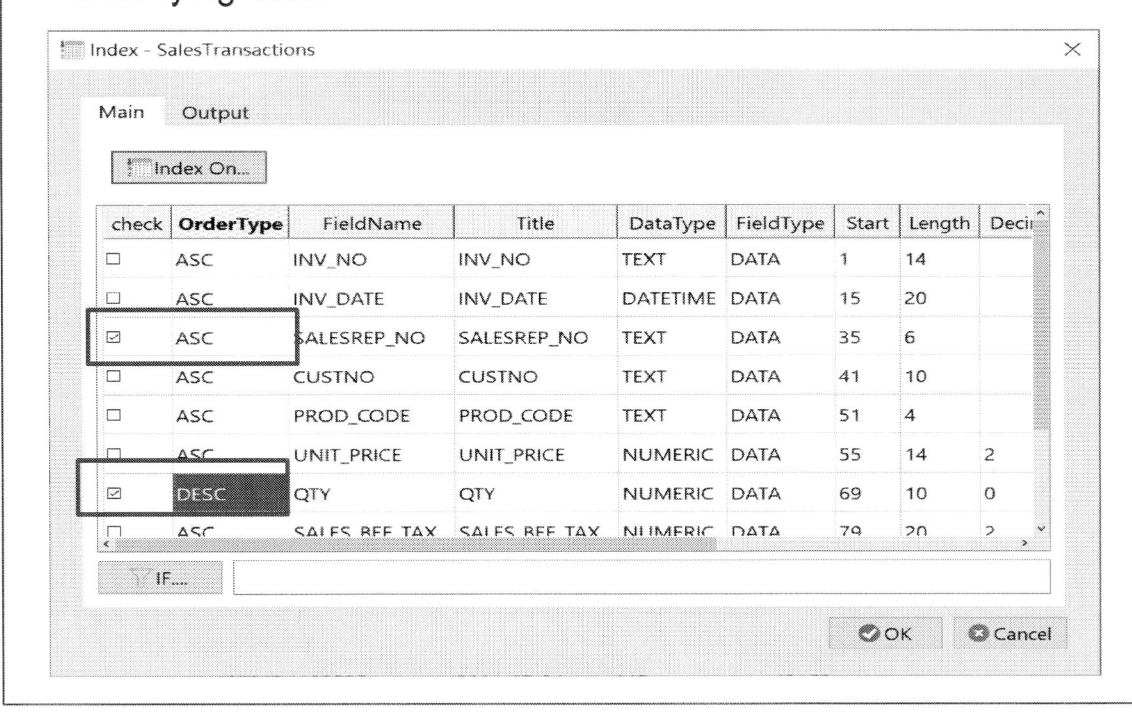

Output tab - Index

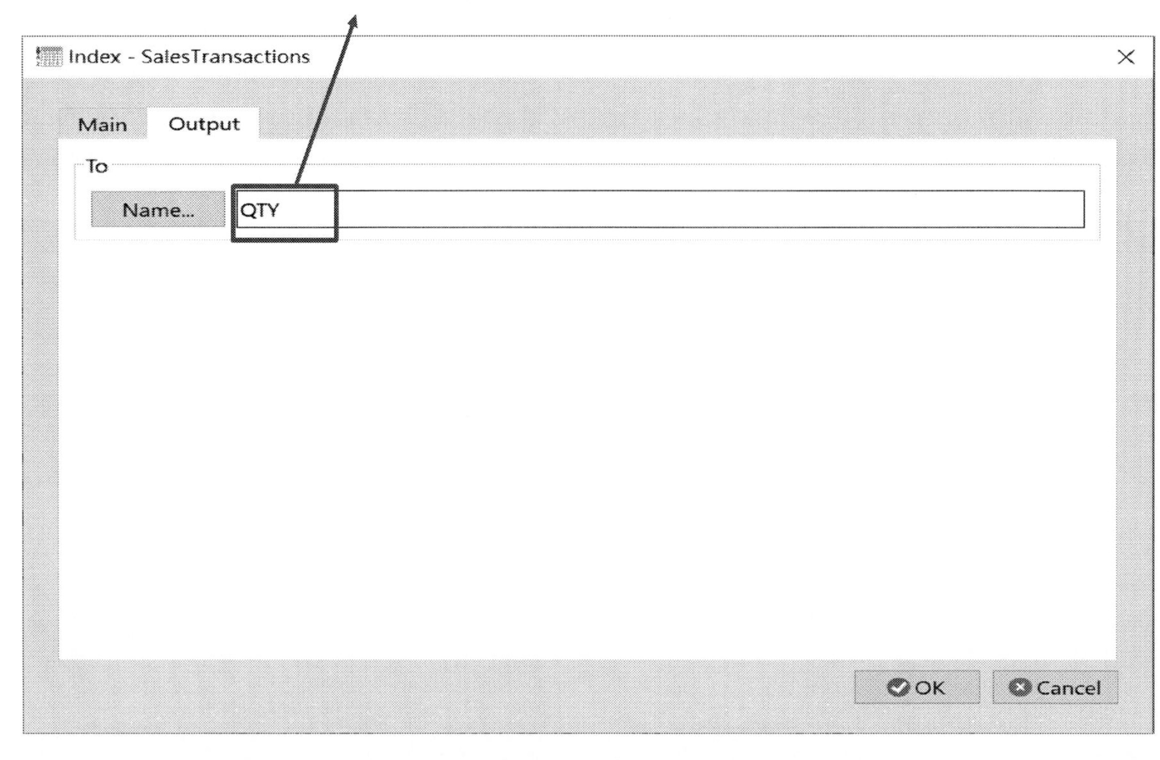

Display the index data

- The data display followed by the index.

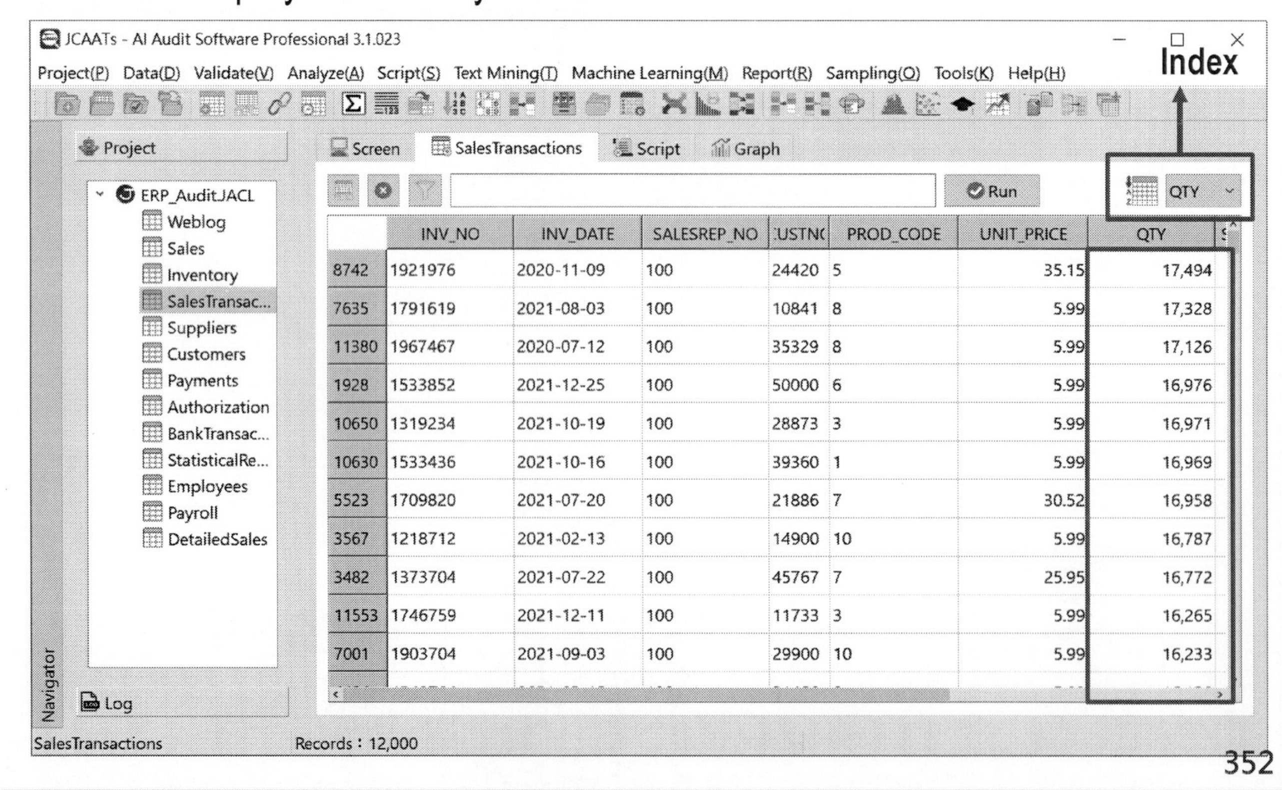

	INV_NO	INV_DATE	SALESREP_NO	:USTN(	PROD_CODE	UNIT_PRICE	QTY
8742	1921976	2020-11-09	100	24420	5	35.15	17,494
7635	1791619	2021-08-03	100	10841	8	5.99	17,328
11380	1967467	2020-07-12	100	35329	8	5.99	17,126
1928	1533852	2021-12-25	100	50000	6	5.99	16,976
10650	1319234	2021-10-19	100	28873	3	5.99	16,971
10630	1533436	2021-10-16	100	39360	1	5.99	16,969
5523	1709820	2021-07-20	100	21886	7	30.52	16,958
3567	1218712	2021-02-13	100	14900	10	5.99	16,787
3482	1373704	2021-07-22	100	45767	7	25.95	16,772
11553	1746759	2021-12-11	100	11733	3	5.99	16,265
7001	1903704	2021-09-03	100	29900	10	5.99	16,233

Quick Sort

- When you right-click on the name of a field or column within a dataset, a menu will typically appear that provides a selection of sorting options, including quick sort ascending or descending.

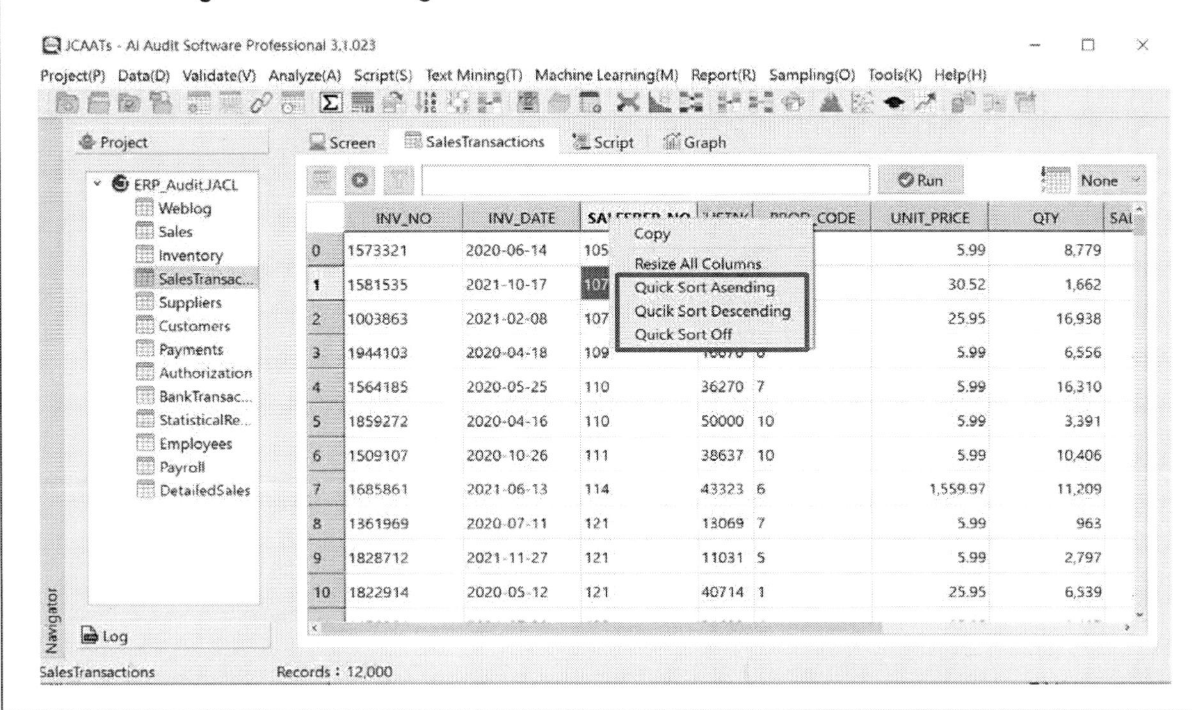

Sort vs Index

Condition	Sort	Index
Speed	Slow	Fast
File Size	Big	Small
Disk Size	Many	Few
All file processing results	Fast	Slow
Find some record processing results	Slow	Fast

Chapter 7 - Exercise Questions

( ) 7-1 Which of the following functions can be used to show null data (Nan/NaT) that a user wants to know?

a) isin([values])
b) str.isalnum()
c) mode()
d) isna()
e) isnull()

( ) 7-2 You have two tables in JCAATs: one containing vendor information and the other containing a government-released company blacklist. How can you determine if your company is listed in the blacklist and therefore at high risk?

a) Use the Merge command to combine the two tables.
b) Use the Extract command to extract relevant information from the two tables.
c) Use the Join command to combine the two tables based on a shared key field.
d) Use the Verify command to check for matches between the two tables.
e) Use the Summarize command to aggregate information from the two tables.

Chapter 7 - Exercise Questions

( ) 7-3 Which command in JCAATs is used to analyze word token and term frequency?

Ⅰ. Keyword
Ⅱ. Sentiment
Ⅲ. Text Cloud
Ⅳ. Benford

a) I, II, and IV
b) II, III, and IV
c) I and IV
d) I and III
e) II and III

Chapter 7 - Exercise Questions

( ) 7-4 To begin an audit project in JCAATs, it is important to verify the date range of the data. Which command should you use to determine the data range in JCAATs?

a) DATE Command, which displays the current date and time settings in the system.
b) PROFILE Command, which displays the user profile information.
c) STATISTICS Command, which provides summary statistic description about the data in the table, including the minimum and maximum date range.
d) AGE Command, which calculates the age of a record based on a specified date.
e) CHART Command, which is used to generate graphical representations of data.

Chapter 7 - Exercise Questions

( ) 7-5 You want to perform an audit on an entire year's worth of data, but the data is spread among 12 monthly transaction tables. Which JCAATs command is best suited to combine these 12 tables into one table?

a) Append
b) Extract/Append
c) Merge
d) Join
e) Combine

( ) 7-6 Which of the following statements about determining sample size is true?

a) As the confidence level increases, the sample size increases.
b) As the standard deviation becomes higher, the sample size gets bigger.
c) As the precision value becomes smaller, the sample size gets larger.
d) The sample size is affected by population expectation.
e) All of the above.

Chapter 7 - Exercise Questions

( ) 7-7 Which JCAATs command is best suited for identifying employees who have exceeded their travel and expense reimbursement limits by comparing their claims to the limits?

a) MERGE
b) EXTRACT
c) JOIN (Matched All primary with the first secondary)
d) VERIFY
e) CROSSTABLE

( ) 7-8 Which JCAATs command is best suited for checking employee records against a vendor listing to identify any instances where employees are also listed as vendors?

a) DOUBLE
b) APPEND
c) JOIN
d) MERGE
e) EVALUATE

Chapter 7 - Exercise Questions

( ) 7-9 Which of the following statements is correct about the Age command in JCAATs?

a) It summarizes the data for a range of numerical amounts.
b) It summarizes the data for a date range period.
c) It summarizes the data according to the front.
d) Before using the command, you need to execute a Duplicate command to get the maximum and minimum.
e) None of the above.

( ) 7-10 Which of the following commands does not have a Subtotal function in JCAATs?

a) Age
b) Summarize
c) Stratify
d) Duplicate
e) Classify

Chapter 7 - Exercise Questions

( ) 7-11 Which one of the JCAATs Join command rules is correct?

a) The two tables need to link to the same project.
b) The two tables need to link with at least one common key connected.
c) The patterns of the two tables' link keys should be the same.
d) The field length of the two tables' link keys should be the same.
e) All of the above

( ) 7-12 Mr. Chen, an auditor, intends to use JCAATs to check whether the customer's account receivable balance has null data, and he wants to report all null data. Which commands can he use in the JCAATs project for this analysis?

a) Clean, Age, Gap
b) Verify, Filter(.isna()), Extract
c) Duplicate, Expression(.isin()), Join
d) Join, Sequence, Expression(.isna())
e) None of the above

JCAATs - AI Audit Software

Exercise 7.17

Exercise 7.17: A Case Study

1. Please create a new project file with the project name Loan_Audit.

2. Please import loan detail file data according to the provided file format schema below.

Field No.	Length	Name	Type	Field No.	Length	Name	Type
1	4	Bank_Code	C	4	10	Amount	N
2	7	Loan_NO	C	5	8	Loan_Date	D
3	10	Cust_ID	C	6	5	Bank_Teller_No	C

Exercise 7.17: A Case Study

3. Please import the customer master data according to the file format (Schema) provided below.

Field No.	Length	Name	Type	Field No.	Length	Name	Type
1	10	Cust_ID	C	5	1	Dishonored_Account	C
2	3	Cust_Name	C	6	1	Suspected_Fraud	C
3	8	BirthDate	D	7	10	Month_Income	N
4	10	Phone	C	8	10	Credit_Limit	N

Exercise 7.17: A Case Study

4. Verify data
   i. Verify whether there are errors in the loan details file.
      □ No, the data is correct.
      □ Yes, there is an error in the row and column and the reason:
      ______________.

   ii. Verify whether there are errors in the customer master file.
      □ No, the data is correct.
      □ Yes, there is an error in the row and column and the reason:
      ______________.

Exercise 7.17: A Case Study

5. Validate the data (the fourth verification error does not need to be considered for the following questions).
   1) The record number of loans in the 1st quarter of 2016 is _______ .
   2) The record number of loans in the 2nd quarter of 2016 is _______ .
   3) The total amount of loans in the 3rd quarter of 2016 is _________ .
   4) The total amount of loans in the 4th quarter of 2016 is _________ .
   5) The average loan amount in the first half of 2016 is _________ .
   6) The average loan amount in the second half of 2016 is _________ .

Exercise 7.17: A Case Study

7) Please check whether there are any missing loan numbers.
   □ No, there is no missing information.
   □ Yes, please list the missing information.
8) Please check whether there are any duplicated loan numbers.
   □ No, there is no duplicated information.
   □ Yes, the duplicated information is: .
9) Please check whether there are any errors in the order of loan numbers.
   □ No, there is no error in the data.
   □ Yes, the duplicated information is: .
10) Please check whether there are any duplicated or suspected duplicated customer master files.
   □ No, there is no duplicated information.
   □ Yes, the duplicated information is:

Exercise 7.17: A Case Study

6. Analyze data
   1) Please divide the loan amount of this year (2016) into 6 tiers. Which branch has the highest total loan amount in the tier with the largest amount?
   2) Please divide the loan amount of this year (2016) into tiers of 0, 650,000, 800,000, 900,000, and 1,500,000. Which branch has the highest concentration of loans in the tier with the lowest total amount?
   3) As of the end of November 2016, the total amount of loans that have been overdue for more than six months (183 days) is .
   4) The employee with the highest total amount of loans disbursed is .
   5) The employee with the highest total amount of daily loans is .

Exercise 7.17: A Case Study

6) Have loans been granted to non-bank customers?
   □ No, there is no such situation.
   □ Yes, please list the loan number and loan amount.
7) Have loans been granted to rejected or suspected fraudulent customers?
   □ No, there is no such situation.
   □ Yes, the loan number, loan amount, and customer name are:
   ______________________________.
8) Have loans been granted to customers with annual income (calculated based on two months' income) below 300,000?
   □ No, there is no such situation.
   □ Yes, the customer ID and loan officer code are: .
9) Whether there is a situation of lending exceeding the customer's credit limit.
   □ No, there is no such situation.
   □ Yes, the customer ID and the loan officer code are respectively:

Exercise 7.17
Please create a new project file with the project name Loan_Audit.

1. Create a new folder.
2. Click JCAATs-AI audit software.
3. Click "Project" > "Select New Project."
4. Define a project name.
5. Save.

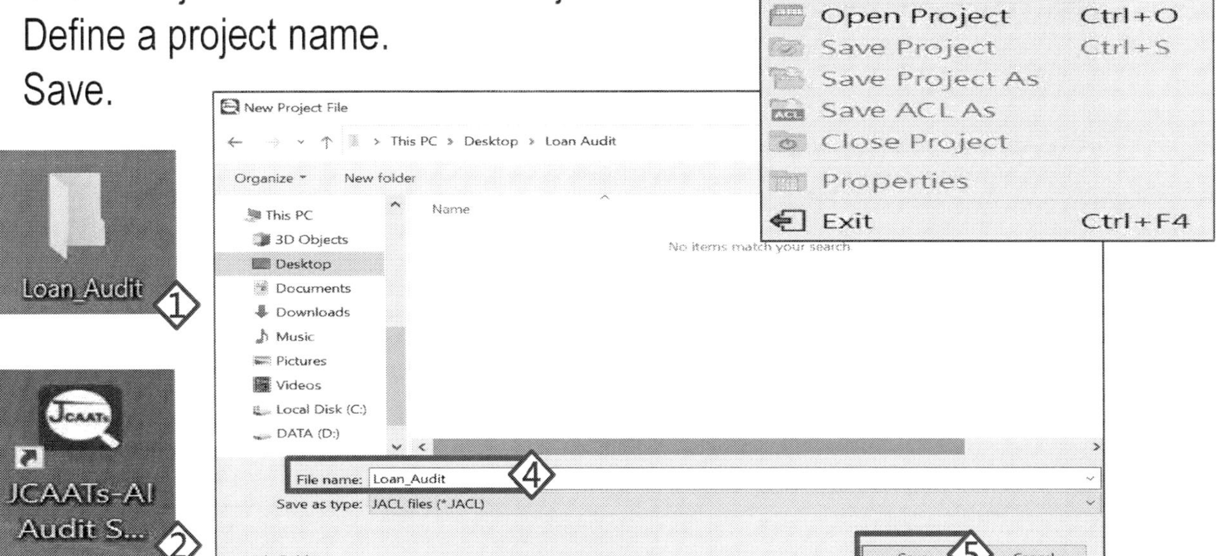

Exercise 7.17
Please import loan detail file data according to the provided file format schema below.

- Select "Data" > "New Table.
- After selecting the data source platform as "File, " click "Next".

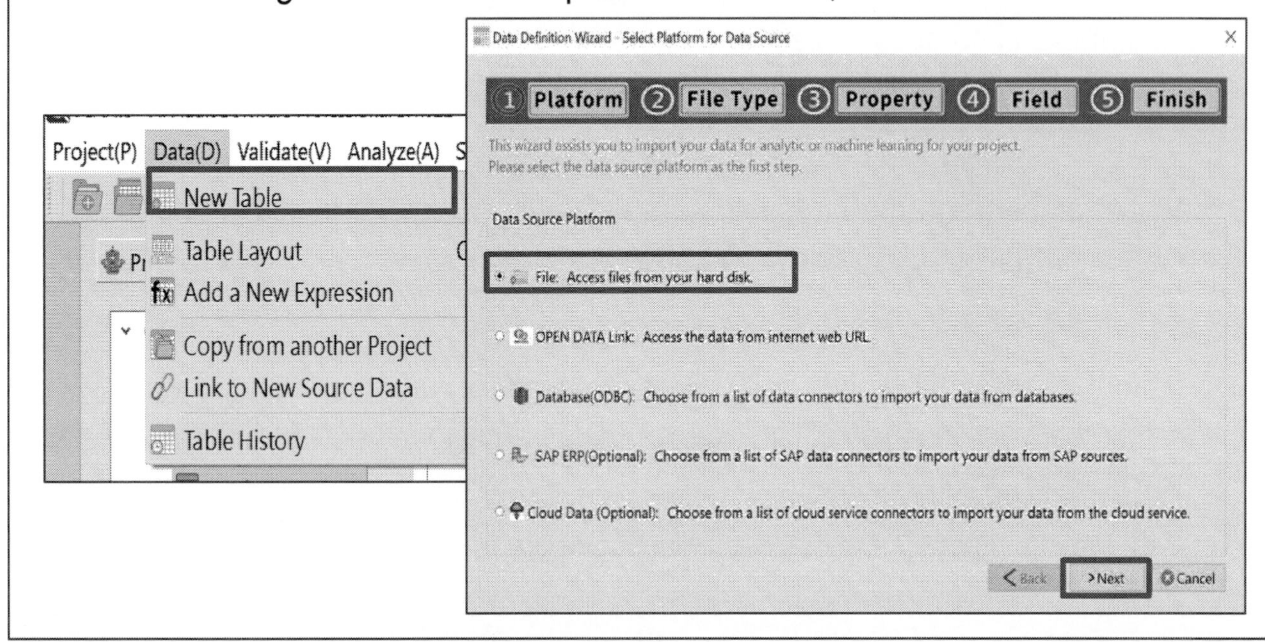

JCAATs-AI Audit Software Copyright © 2023 JACKSOFT.

Exercise 7.17
Please import loan detail file data according to the provided file format schema below.

- Select Loan_Details to import.
- Click "Open" to proceed.

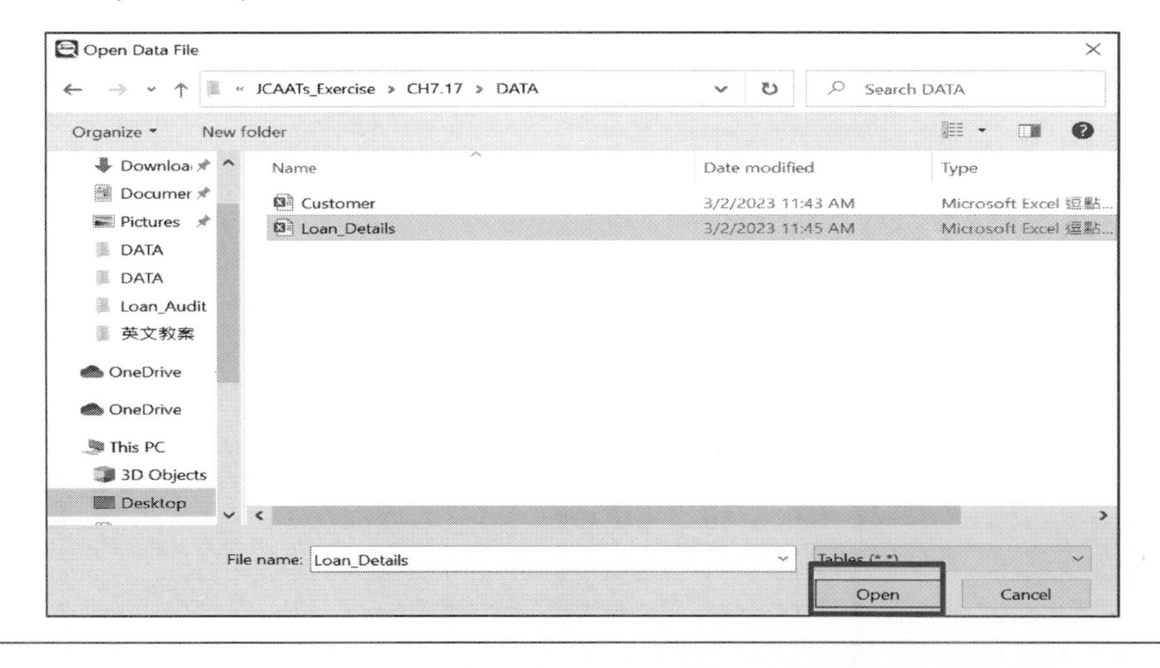

Exercise 7.17
Please import loan detail file data according to the provided file format schema below.

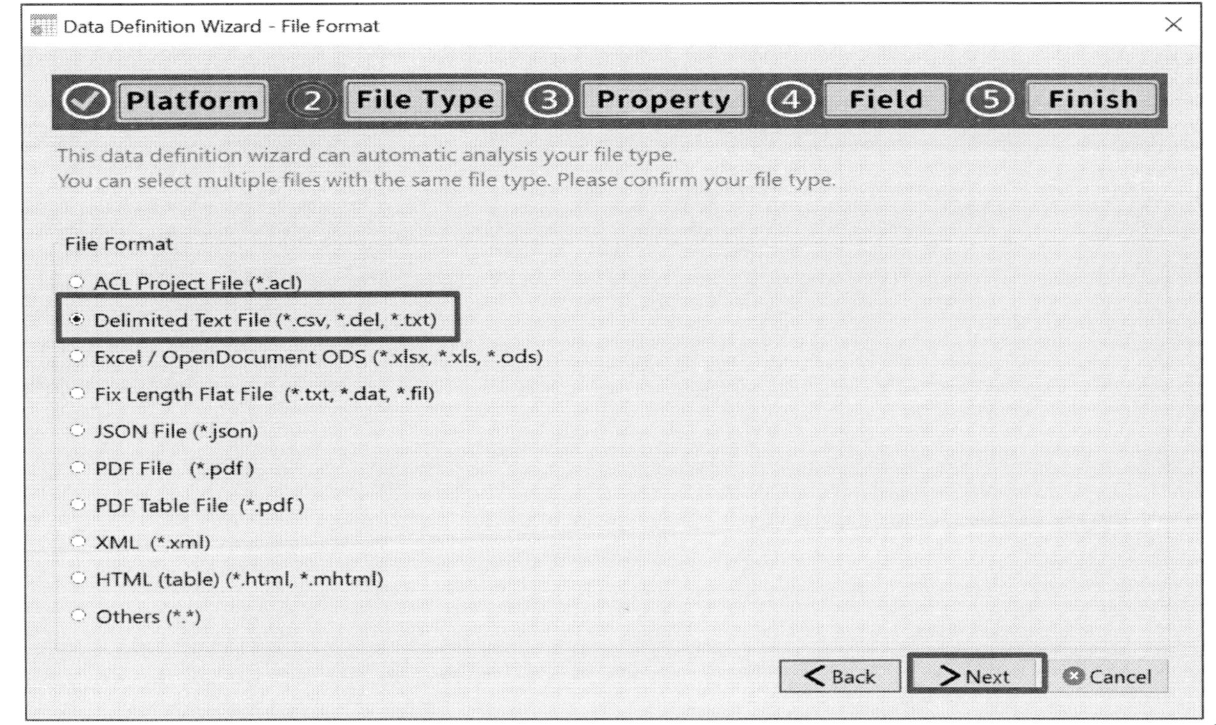

Exercise 7.17
Please import loan detail file data according to the provided file format schema below.

- Set the data file separator as "Tab"

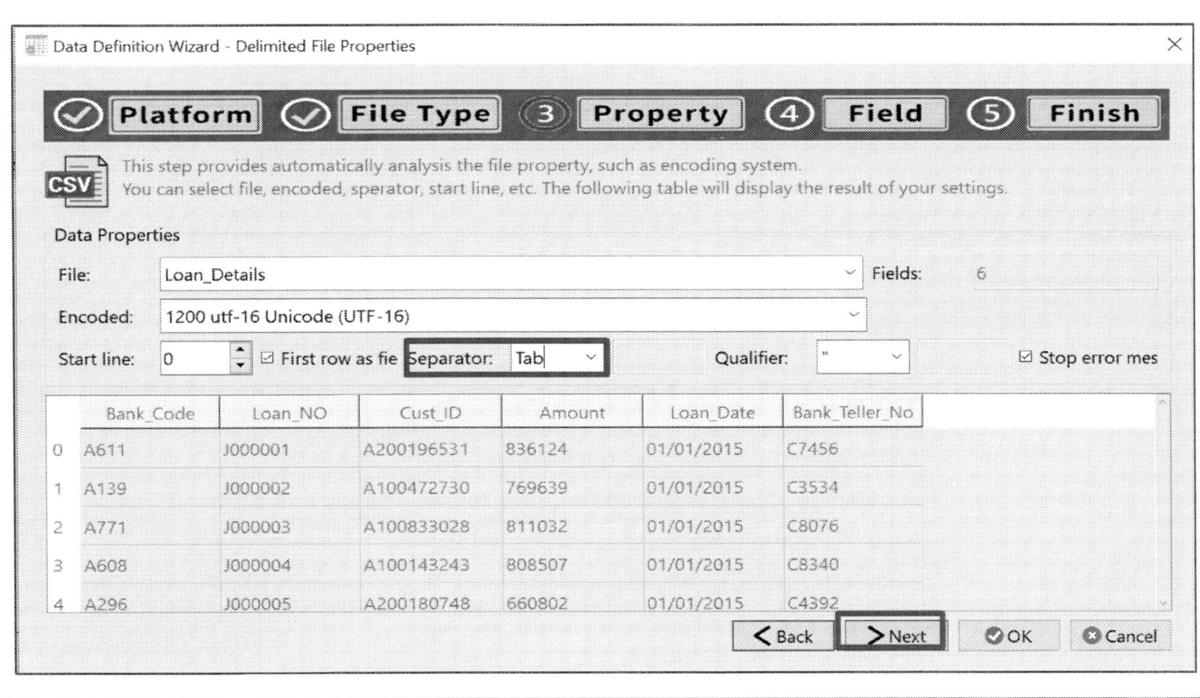

	Bank_Code	Loan_NO	Cust_ID	Amount	Loan_Date	Bank_Teller_No
0	A611	J000001	A200196531	836124	01/01/2015	C7456
1	A139	J000002	A100472730	769639	01/01/2015	C3534
2	A771	J000003	A100833028	811032	01/01/2015	C8076
3	A608	J000004	A100143243	808507	01/01/2015	C8340
4	A296	J000005	A200180748	660802	01/01/2015	C4392

Exercise 7.17

Please import loan detail file data according to the provided file format schema below.

- Define the field name , display name , data type and data format
- Select "Next".

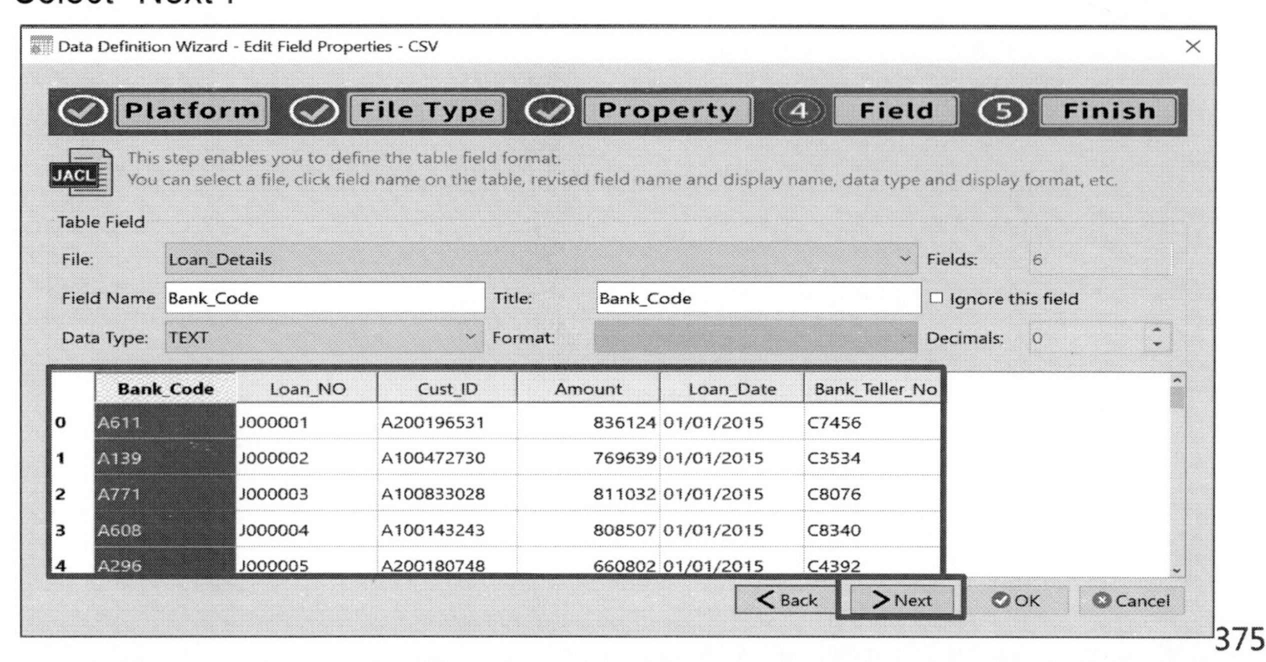

Exercise 7.17

Please import loan detail file data according to the provided file format schema below.

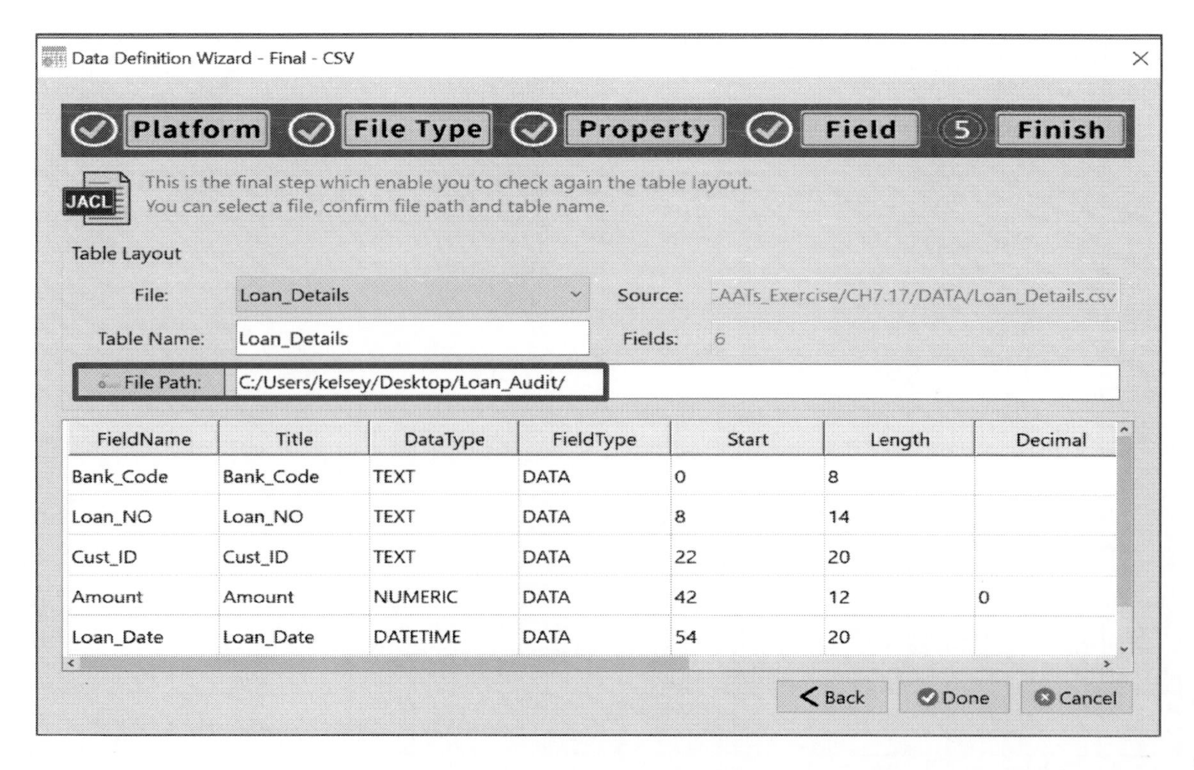

Exercise 7.17

Please import loan detail file data according to the provided file format schema below.

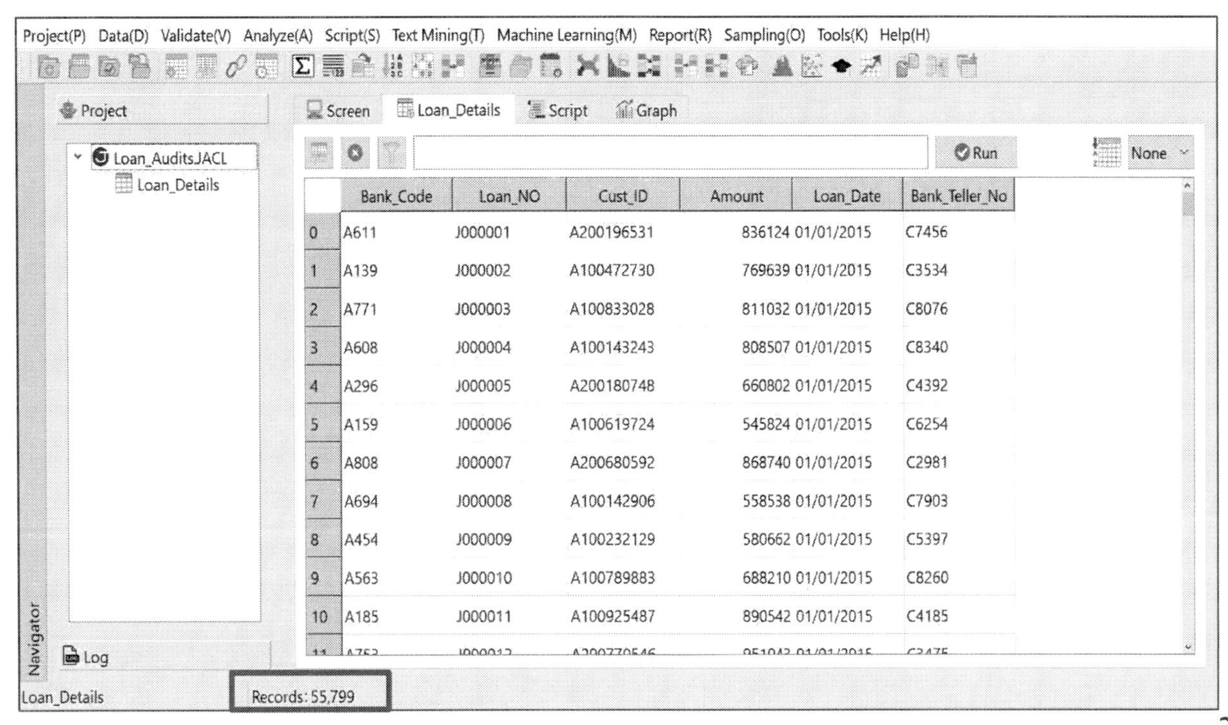

	Bank_Code	Loan_NO	Cust_ID	Amount	Loan_Date	Bank_Teller_No
0	A611	J000001	A200196531	836124	01/01/2015	C7456
1	A139	J000002	A100472730	769639	01/01/2015	C3534
2	A771	J000003	A100833028	811032	01/01/2015	C8076
3	A608	J000004	A100143243	808507	01/01/2015	C8340
4	A296	J000005	A200180748	660802	01/01/2015	C4392
5	A159	J000006	A100619724	545824	01/01/2015	C6254
6	A808	J000007	A200680592	868740	01/01/2015	C2981
7	A694	J000008	A100142906	558538	01/01/2015	C7903
8	A454	J000009	A100232129	580662	01/01/2015	C5397
9	A563	J000010	A100789883	688210	01/01/2015	C8260
10	A185	J000011	A100925487	890542	01/01/2015	C4185

JCAATs-AI Audit Software
Copyright © 2023 JACKSOFT.

Exercise 7.17

Please import the customer master data according to the file format (Schema) provided below.

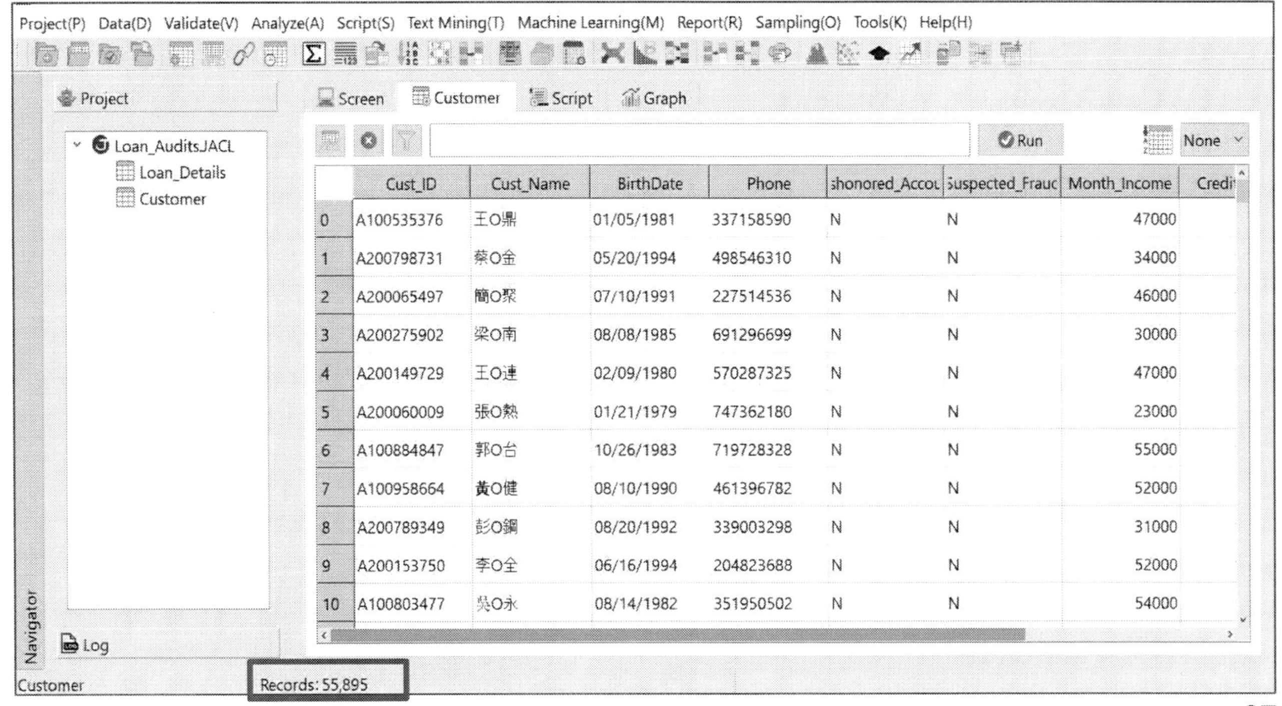

	Cust_ID	Cust_Name	BirthDate	Phone	shonored_Accou	Suspected_Frauc	Month_Income	Credi
0	A100535376	王O鼎	01/05/1981	337158590	N	N	47000	
1	A200798731	蔡O金	05/20/1994	498546310	N	N	34000	
2	A200065497	簡O聚	07/10/1991	227514536	N	N	46000	
3	A200275902	梁O南	08/08/1985	691296699	N	N	30000	
4	A200149729	王O連	02/09/1980	570287325	N	N	47000	
5	A200060009	張O熱	01/21/1979	747362180	N	N	23000	
6	A100884847	郭O台	10/26/1983	719728328	N	N	55000	
7	A100958664	黃O健	08/10/1990	461396782	N	N	52000	
8	A200789349	彭O鋼	08/20/1992	339003298	N	N	31000	
9	A200153750	李O全	06/16/1994	204823688	N	N	52000	
10	A100803477	吳O永	08/14/1982	351950502	N	N	54000	

Exercise 7.17

Verify whether there are errors in the loan details file.

- Open data set we need (Loan_Details)
- Select "Validate>Verify"

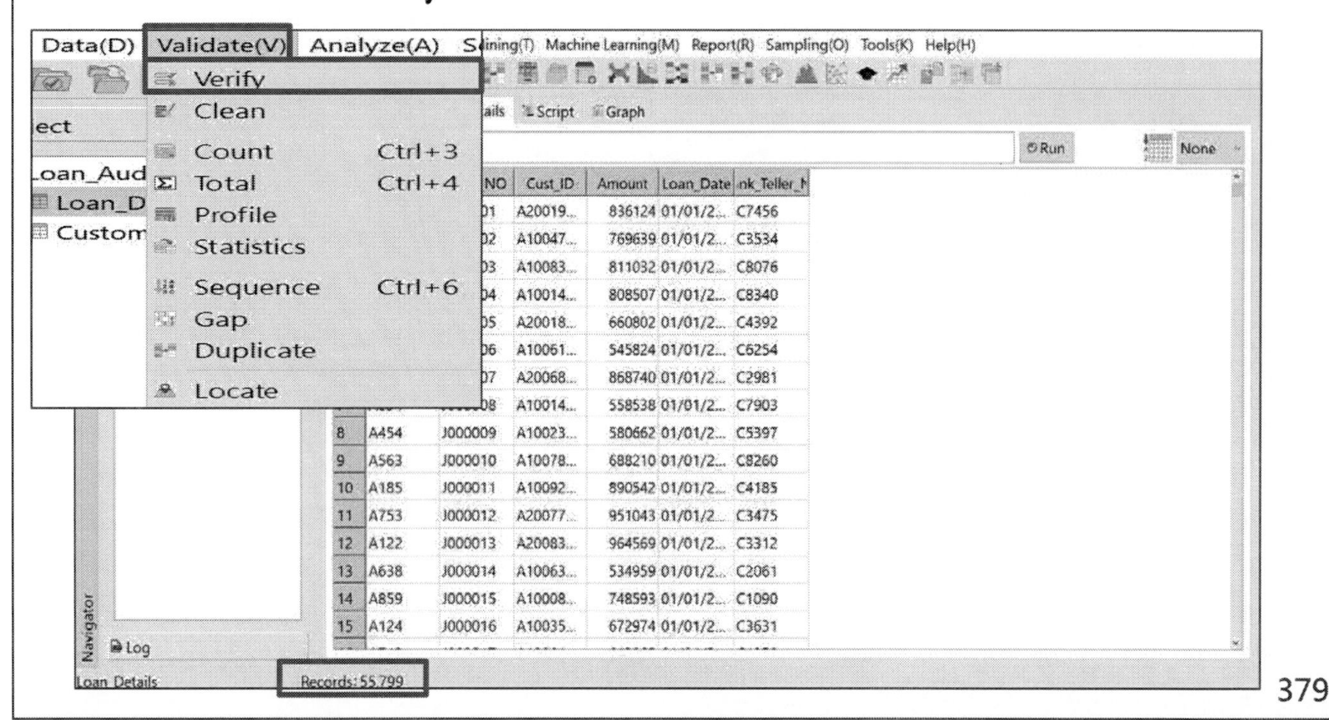

JCAATs-AI Audit Software
Copyright © 2023 JACKSOFT.

Exercise 7.17

Verify whether there are errors in the loan details file.

- Select "Verify On" and choose "Add All" for selected fields.

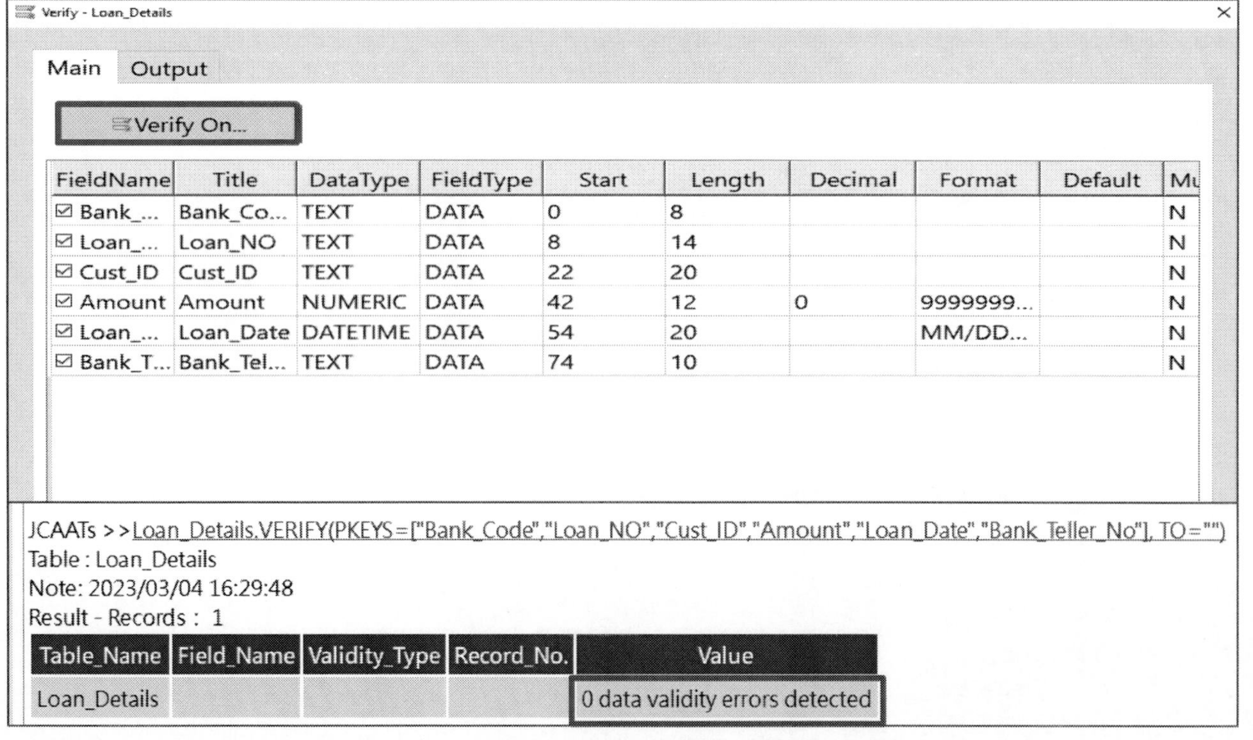

Exercise 7.17

Verify whether there are errors in the customer master file.

- Select "Verify On" and choose "Add All" for selected fields.

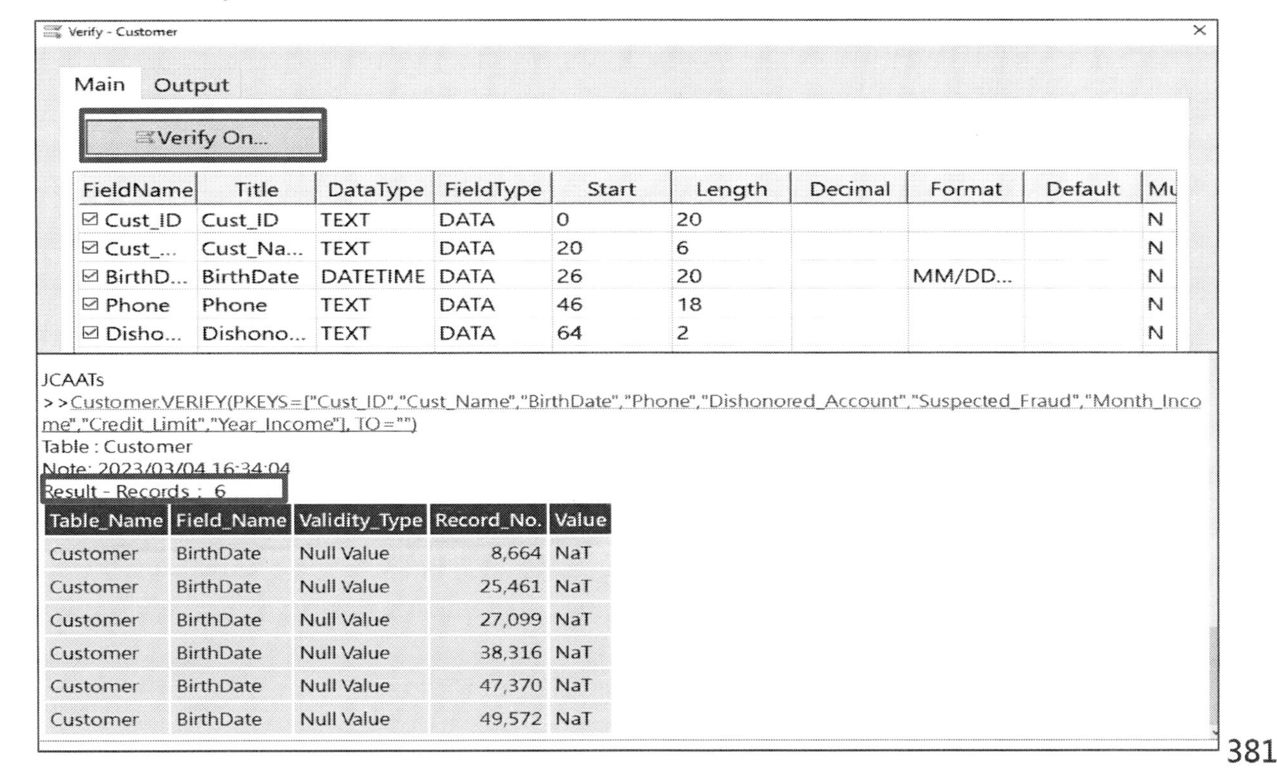

Exercise 7.17

Validate the data: The record number of loans in the 1st quarter of 2016 is?

- Open data set we need (Loan_Details)
- Select "Validate>Count"

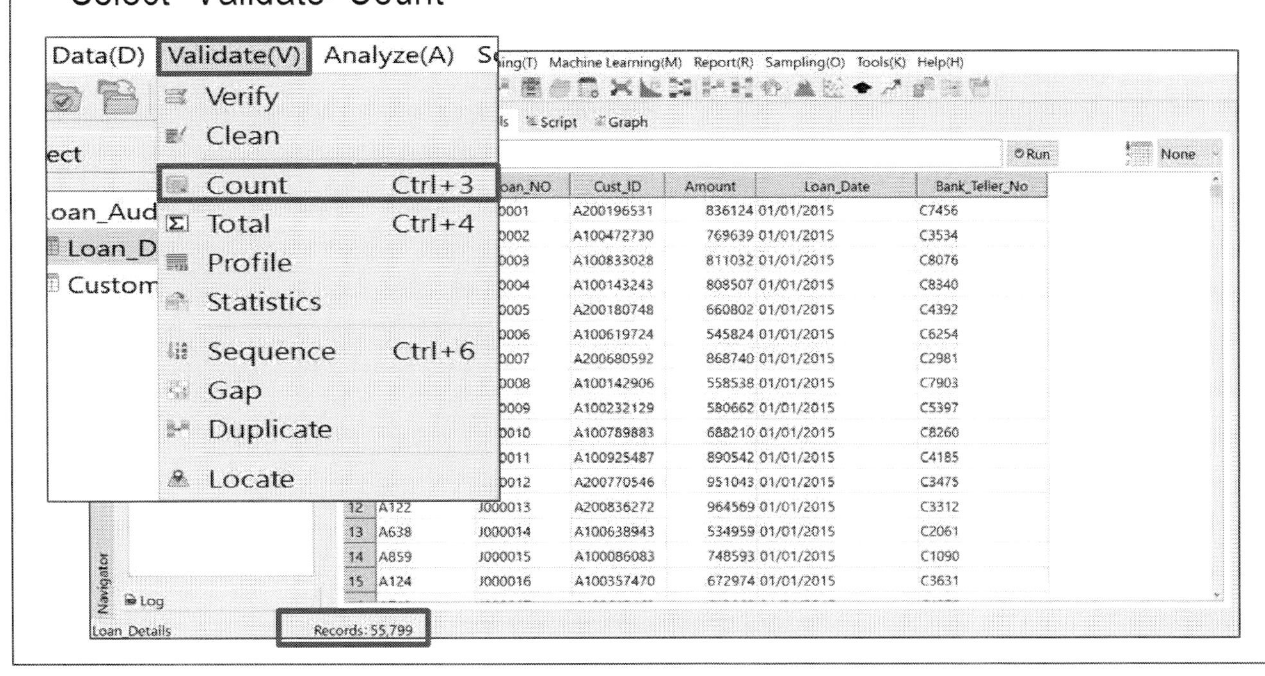

Exercise 7.17

Validate the data: The record number of loans in the 1st quarter of 2016 is?

- Add a Filter Condition:

  Loan_Date.between(date(2016-01-01), date(2016-03-31))

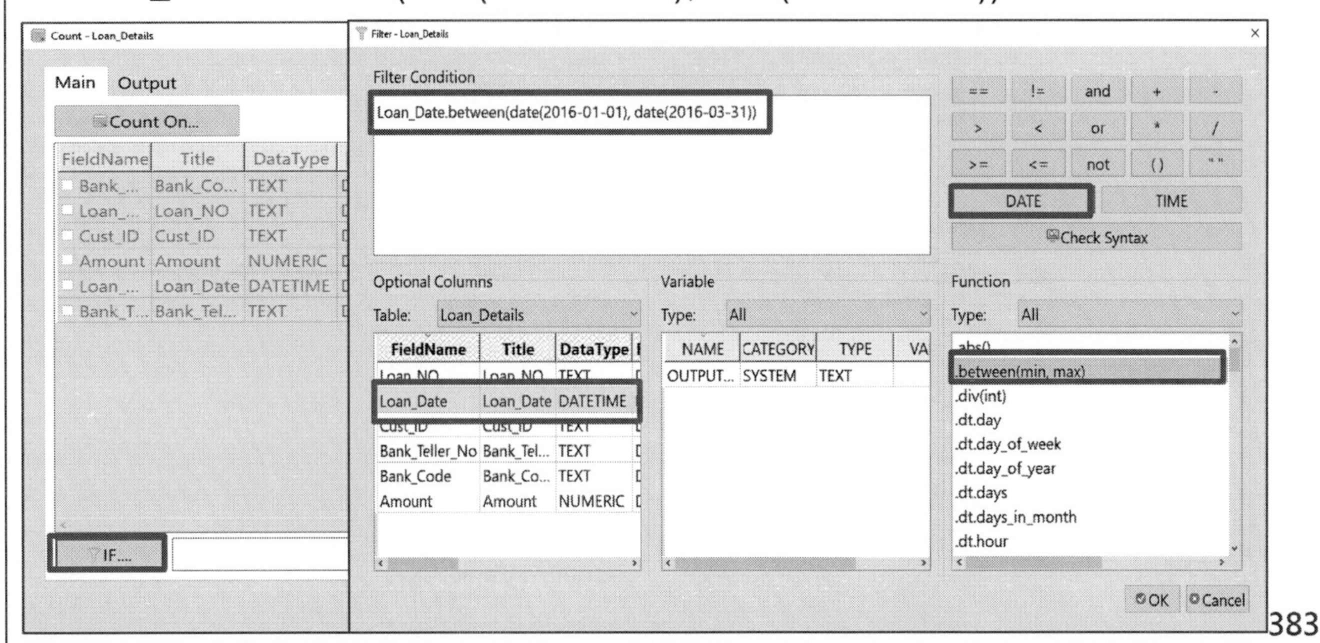

Exercise 7.17

Validate the data: The record number of loans in the 1st quarter of 2016 is?

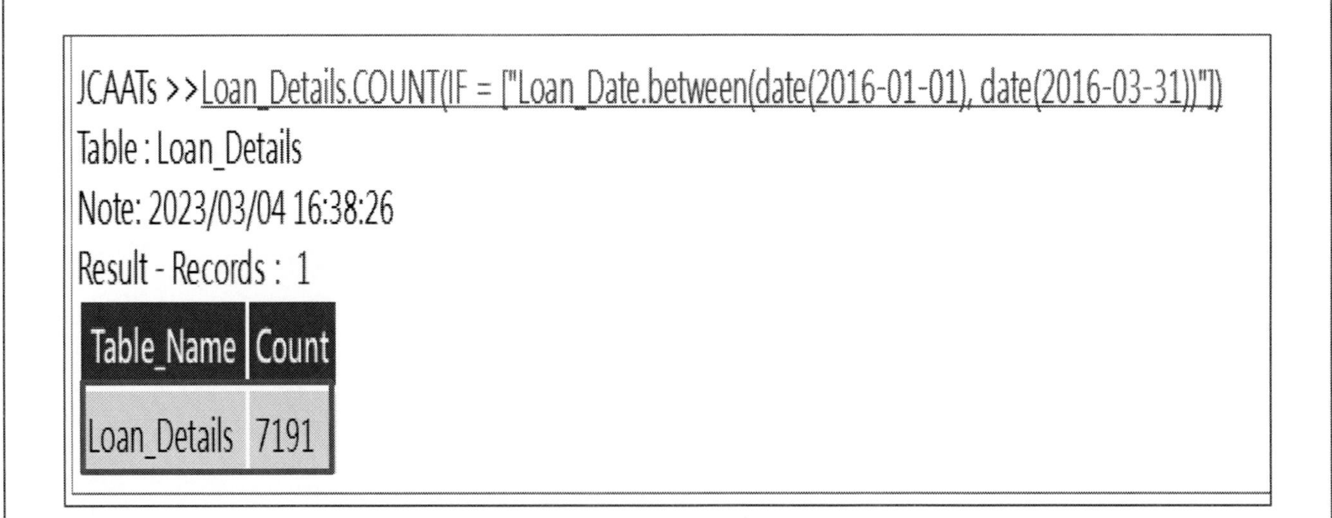

Exercise 7.17
Validate the data: The record number of loans in the 2nd quarter of 2016 is

- Add a Filter Condition:
  Loan_Date.between(date(2016-04-01), date(2016-06-30))

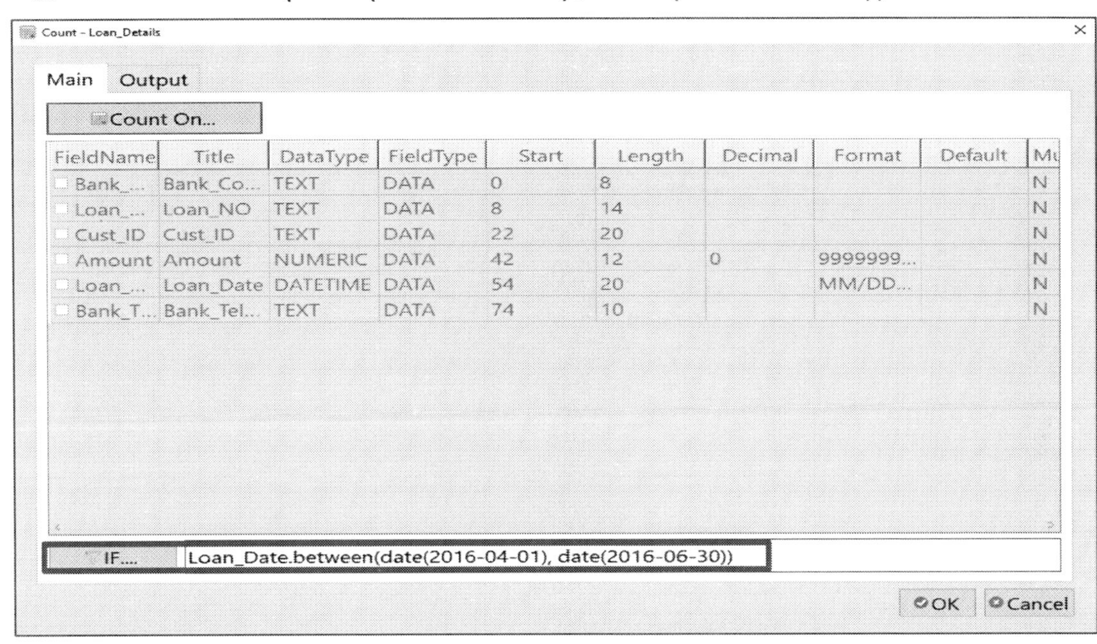

FieldName	Title	DataType	FieldType	Start	Length	Decimal	Format	Default	Mu
Bank_...	Bank_Co...	TEXT	DATA	0	8				N
Loan_...	Loan_NO	TEXT	DATA	8	14				N
Cust_ID	Cust_ID	TEXT	DATA	22	20				N
Amount	Amount	NUMERIC	DATA	42	12	0	9999999...		N
Loan_...	Loan_Date	DATETIME	DATA	54	20		MM/DD...		N
Bank_T...	Bank_Tel...	TEXT	DATA	74	10				N

Exercise 7.17
Validate the data: The record number of loans in the 2nd quarter of 2016 is

JCAATs >>Loan_Details.COUNT(IF = ["Loan_Date.between(date(2016-04-01), date(2016-06-30)) "])

Table : Loan_Details

Note: 2023/03/04 16:39:53

Result - Records : 1

Table_Name	Count
Loan_Details	7287

Exercise 7.17
Validate the data: The total amount of loans in the 3rd quarter of 2016 is

- Open data set we need (Loan_Details)
- Select "Validate>Total"

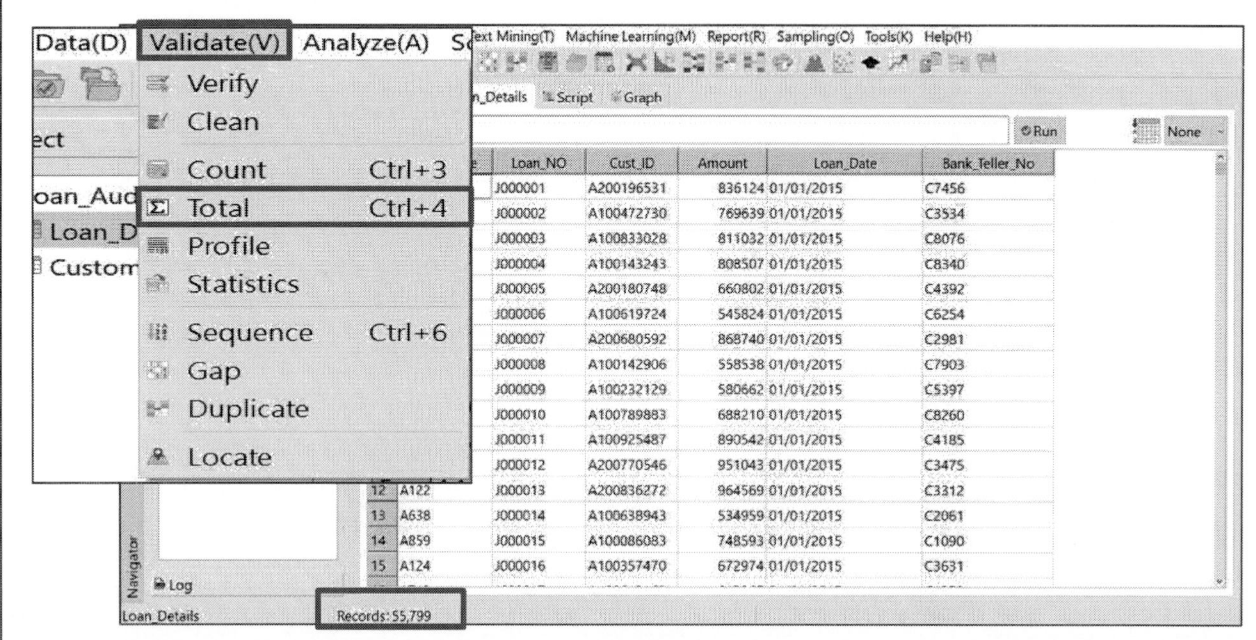

JCAATs-AI Audit Software Copyright © 2023 JACKSOFT.

Exercise 7.17
Validate the data: The total amount of loans in the 3rd quarter of 2016 is

- Add a Filter Condition:

  Loan_Date.between(date(2016-07-01), date(2016-09-30))

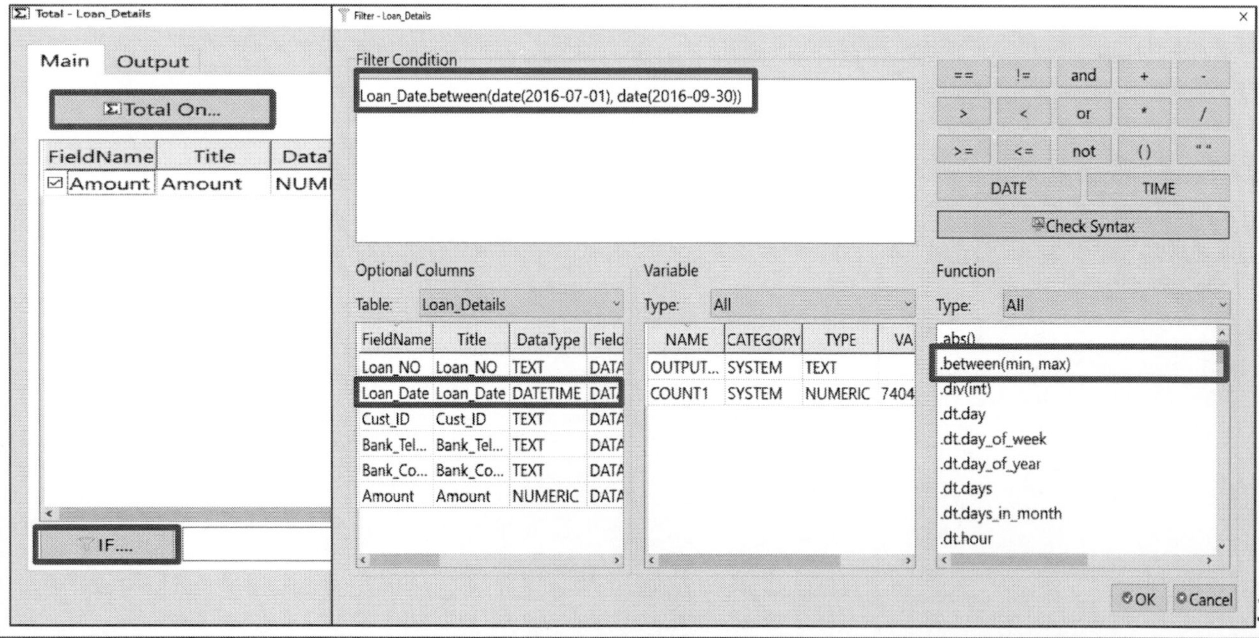

Exercise 7.17
Validate the data: The total amount of loans in the 3rd quarter of 2016 is

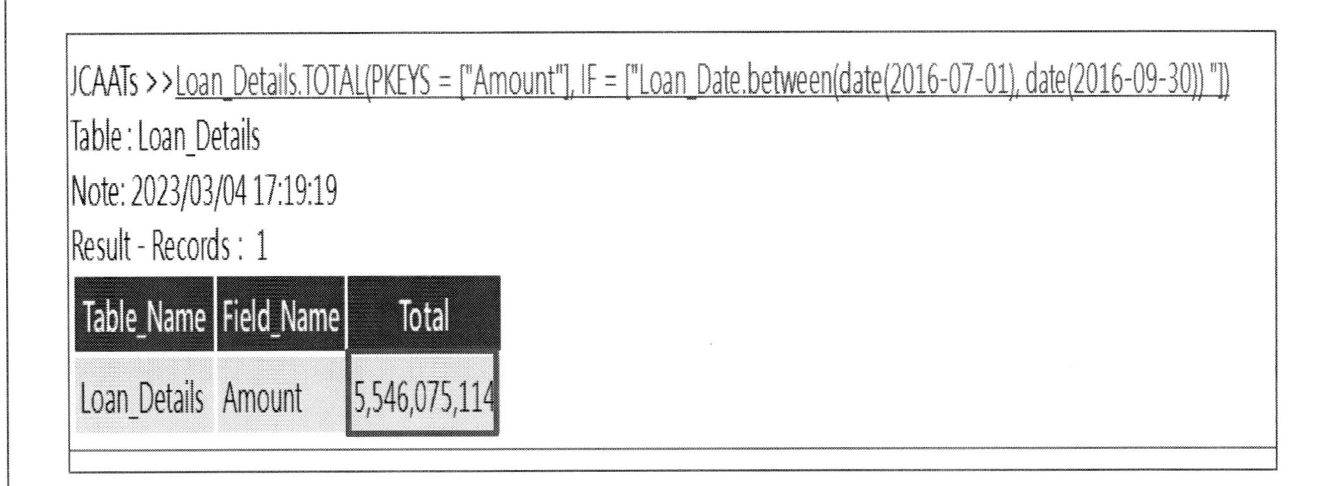

JCAATs >>Loan_Details.TOTAL(PKEYS = ["Amount"], IF = ["Loan_Date.between(date(2016-07-01), date(2016-09-30)) "])
Table : Loan_Details
Note: 2023/03/04 17:19:19
Result - Records : 1

Table_Name	Field_Name	Total
Loan_Details	Amount	5,546,075,114

Exercise 7.17
Validate the data: The total amount of loans in the 4th quarter of 2016 is

- Add a Filter Condition:
  Loan_Date.between(date(2016-10-01), date(2016-12-31))

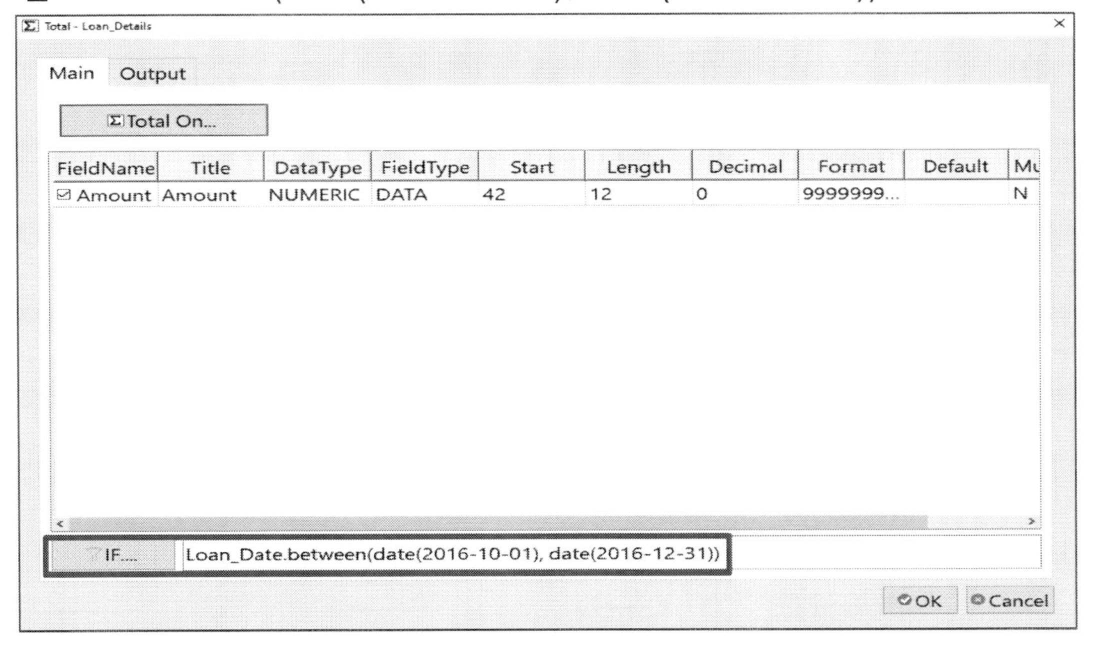

Exercise 7.17
Validate the data: The total amount of loans in the 4th quarter of 2016 is

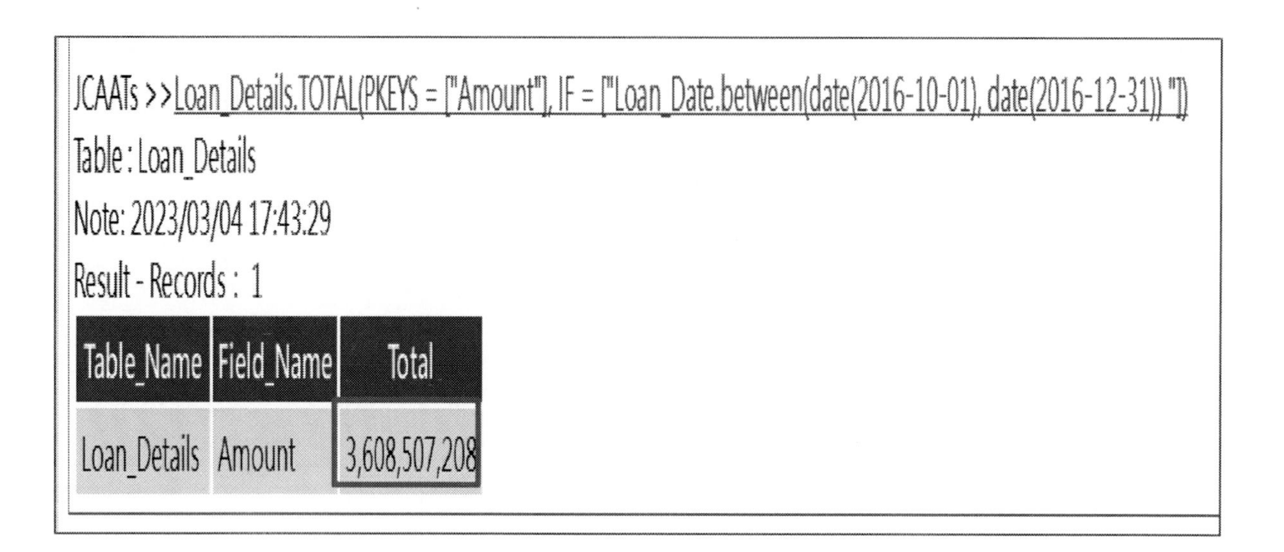
JCAATs >>Loan_Details.TOTAL(PKEYS = ["Amount"], IF = ["Loan_Date.between(date(2016-10-01), date(2016-12-31)) "])
Table : Loan_Details
Note: 2023/03/04 17:43:29
Result - Records : 1

Table_Name	Field_Name	Total
Loan_Details	Amount	3,608,507,208

JCAATs-AI Audit Software
Copyright © 2023 JACKSOFT.

Exercise 7.17
Validate the data: The average loan amount in the first half of 2016 is?

- Open data set we need (Loan_Details)
- Select "Validate>Profile"

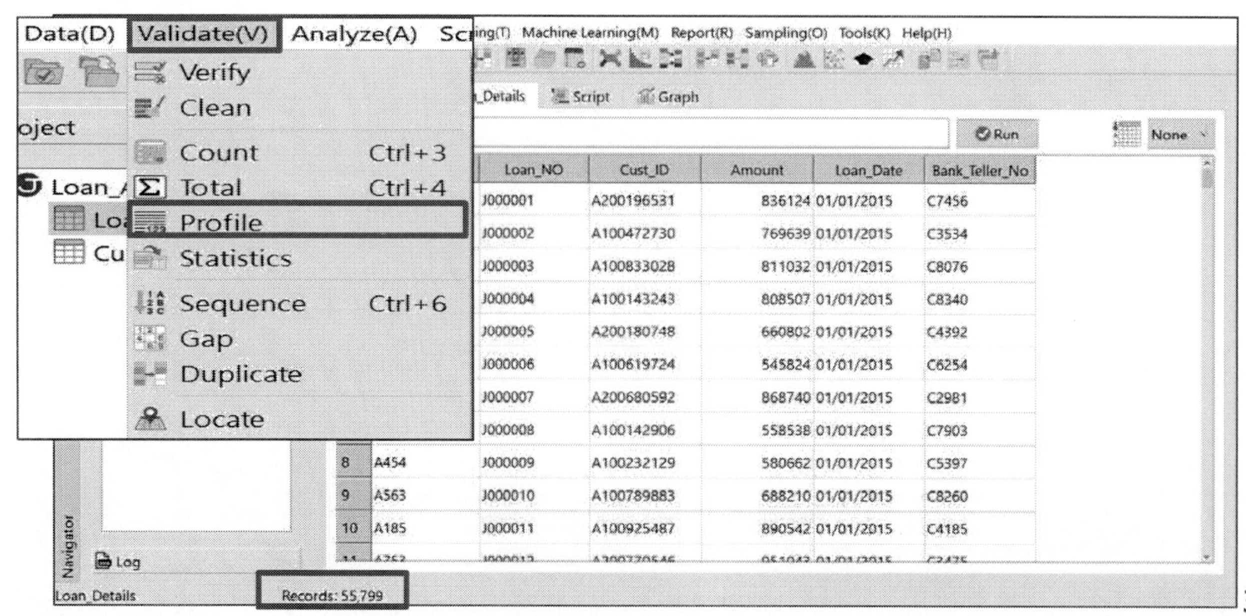

Exercise 7.17
Validate the data: The average loan amount in the first half of 2016 is?

- Add a Filter Condition:

  Loan_Date.between(date(2016-01-01), date(2016-06-30))

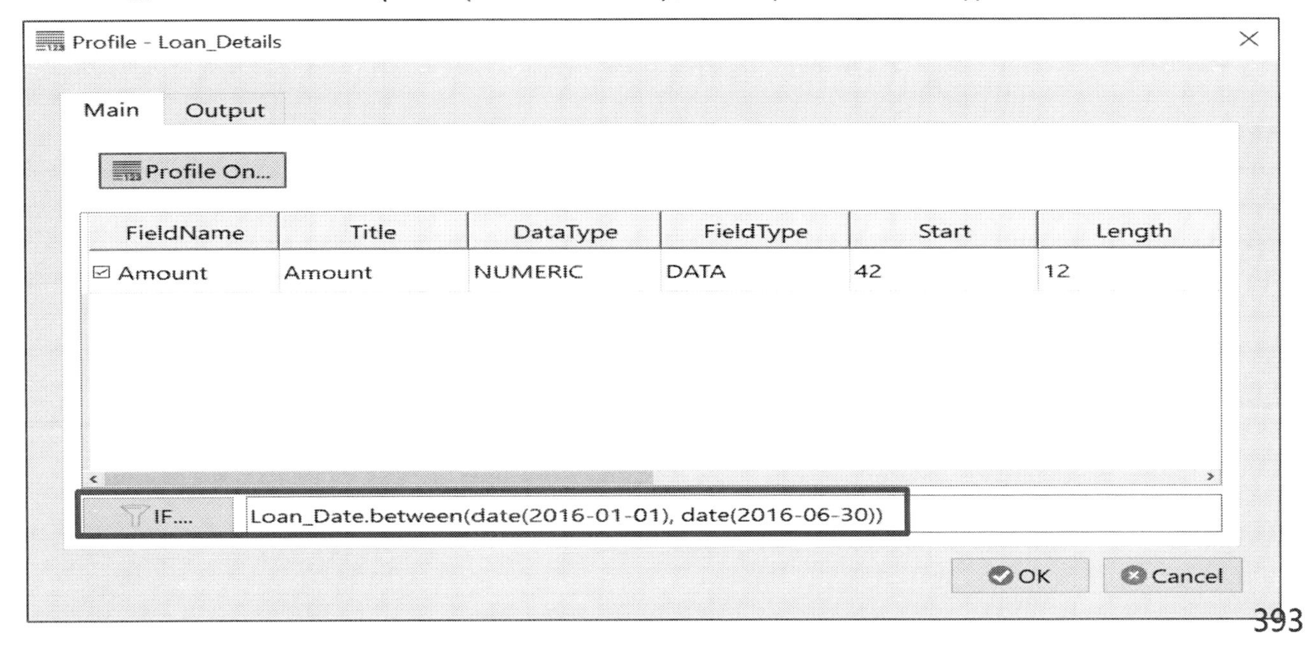

JCAATs-AI Audit Software

Exercise 7.17
Validate the data: The average loan amount in the first half of 2016 is?

- Select a Descriptive Statistics:

  CENTRAL: Mean, Mode, Median

Exercise 7.17
Validate the data: The average loan amount in the first half of 2016 is?

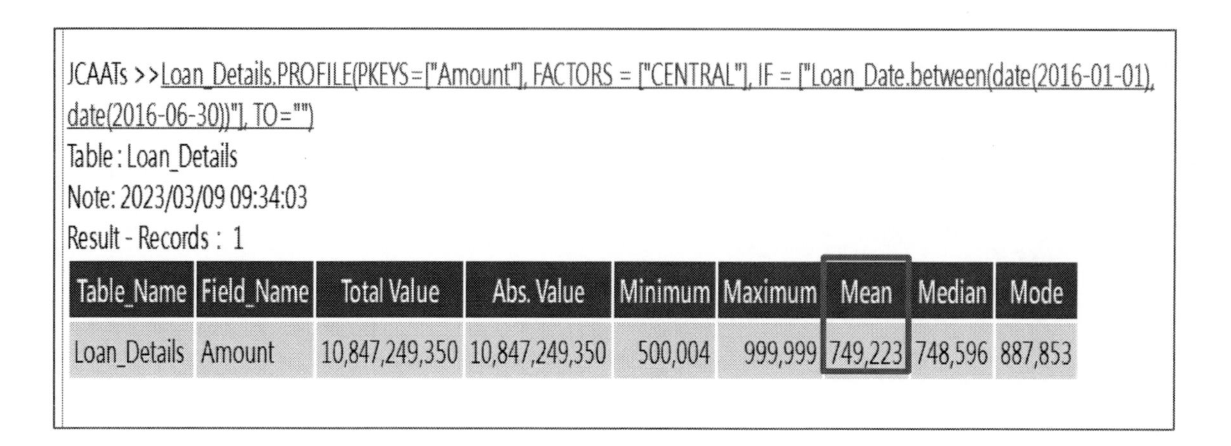

JCAATs >>Loan_Details.PROFILE(PKEYS=["Amount"], FACTORS = ["CENTRAL"], IF = ["Loan_Date.between(date(2016-01-01), date(2016-06-30))"], TO="")
Table : Loan_Details
Note: 2023/03/09 09:34:03
Result - Records : 1

Table_Name	Field_Name	Total Value	Abs. Value	Minimum	Maximum	Mean	Median	Mode
Loan_Details	Amount	10,847,249,350	10,847,249,350	500,004	999,999	749,223	748,596	887,853

Exercise 7.17
Validate the data: The average loan amount in the second half of 2016 is?

- Add a Filter Condition:
  Loan_Date.between(date(2016-07-01), date(2016-12-31))

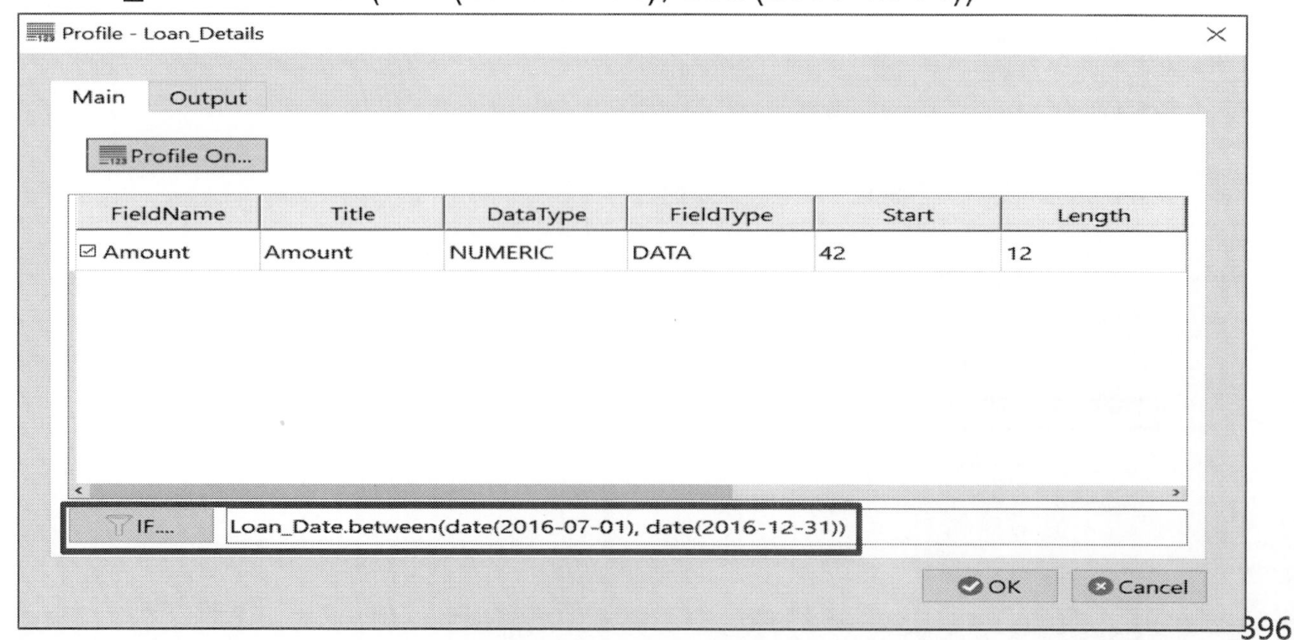

Exercise 7.17
Validate the data: The average loan amount in the second half of 2016 is?

- Select a Descriptive Statistics:
  CENTRAL: Mean, Mode, Median

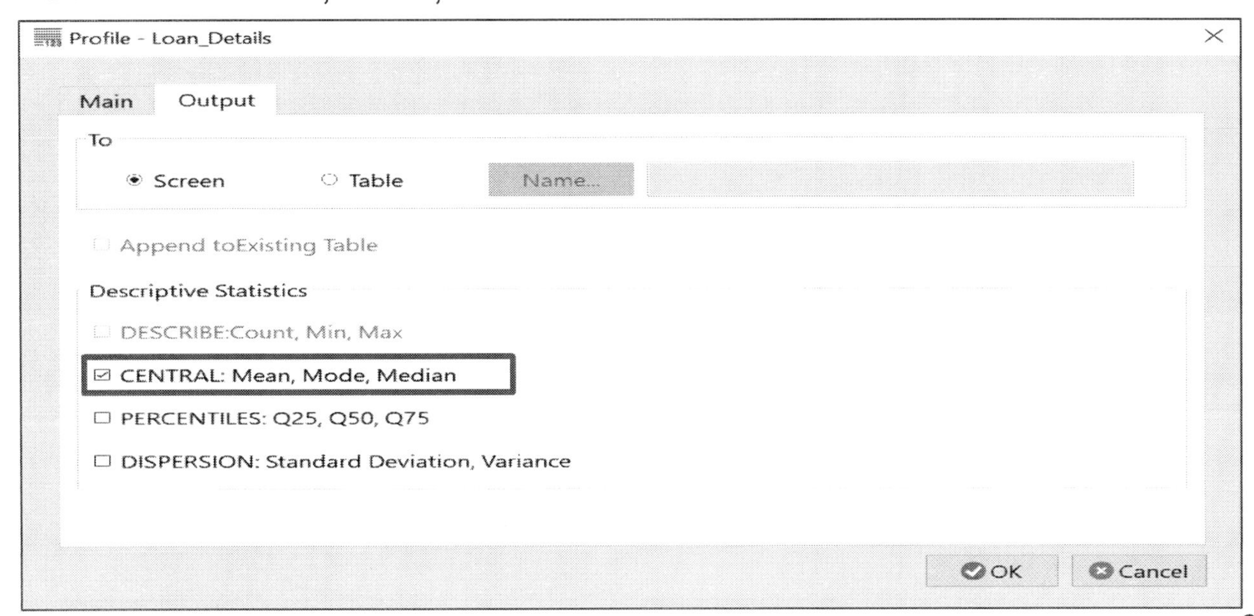

Exercise 7.17
Validate the data: The average loan amount in the second half of 2016 is?

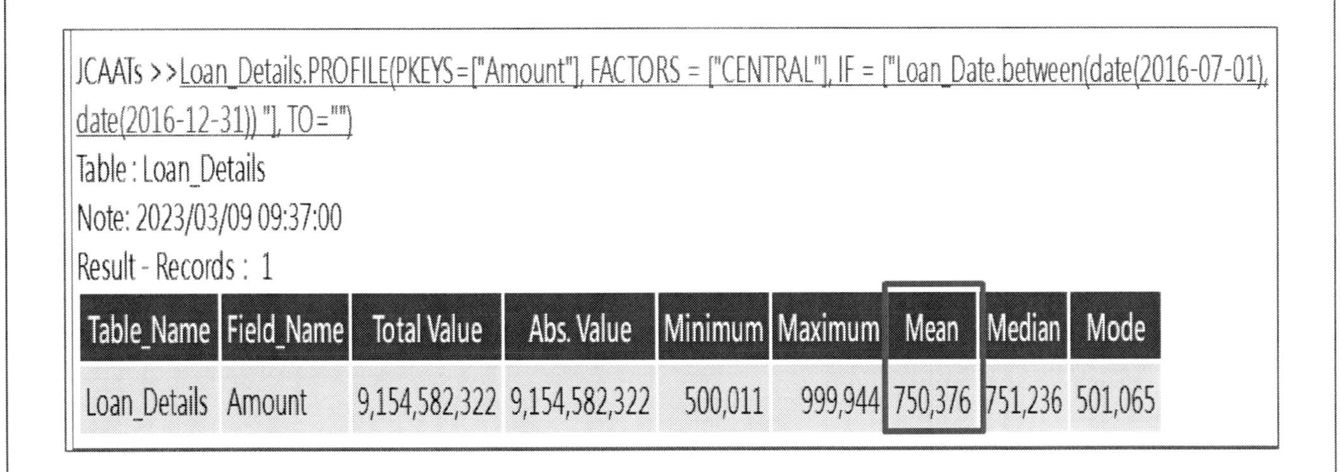

JCAATs >>Loan_Details.PROFILE(PKEYS=["Amount"], FACTORS = ["CENTRAL"], IF = ["Loan_Date.between(date(2016-07-01), date(2016-12-31)) "], TO="")
Table : Loan_Details
Note: 2023/03/09 09:37:00
Result - Records : 1

Table_Name	Field_Name	Total Value	Abs. Value	Minimum	Maximum	Mean	Median	Mode
Loan_Details	Amount	9,154,582,322	9,154,582,322	500,011	999,944	750,376	751,236	501,065

Exercise 7.17
Please check whether there are any missing loan numbers.

- Open data set we need (Loan_Details)
- Select "Validate>Gap", choose Output Type as List Missing Items

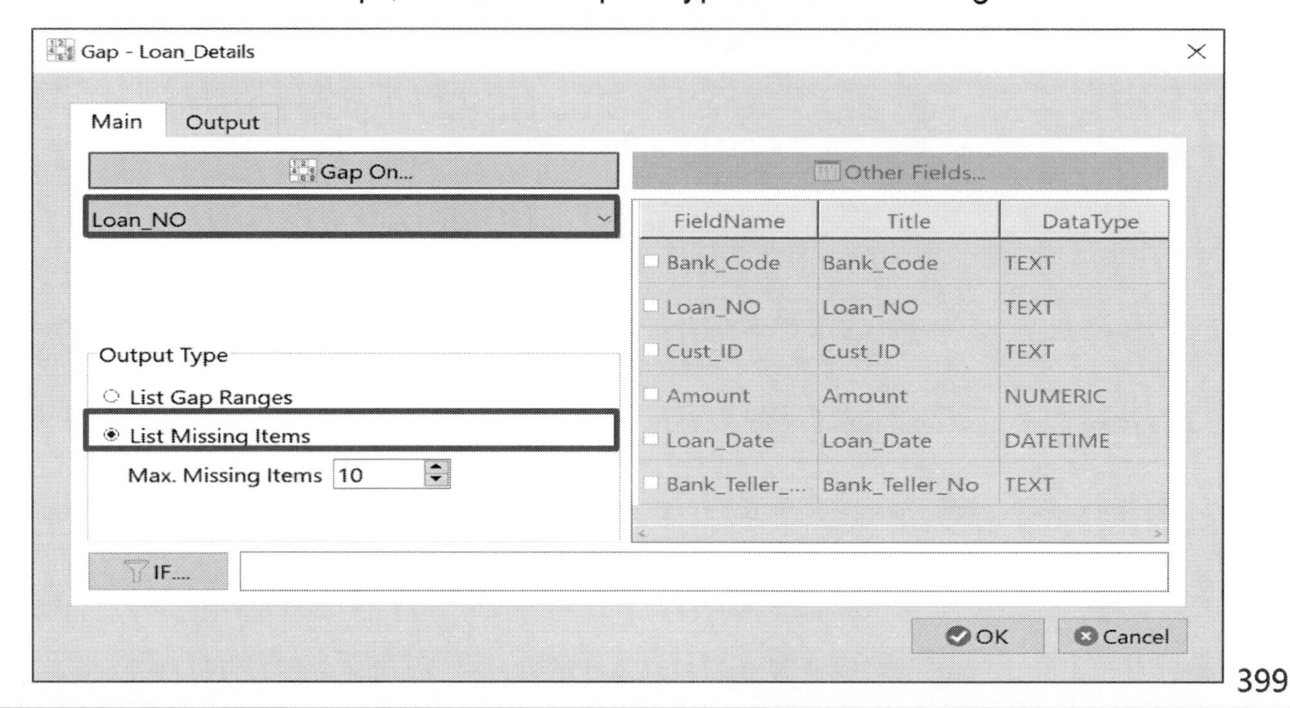

Exercise 7.17
Please check whether there are any missing loan numbers.

JCAATs >>Loan_Details.GAP(PKEY="Loan_NO", MAX = 10, MISSING = "item" , TO="")
Table : Loan_Details
Note: 2023/03/09 09:38:12
Result - Records : 7

Loan_NO
J15062
J15091
J15143
J15242
J15243
J15317
J15318

Exercise 7.17
Please check whether there are any duplicated loan numbers.

- Open data set we need (Loan_Details)
- Select "Validate>Duplicate", choose the field "Loan_No" to analysis

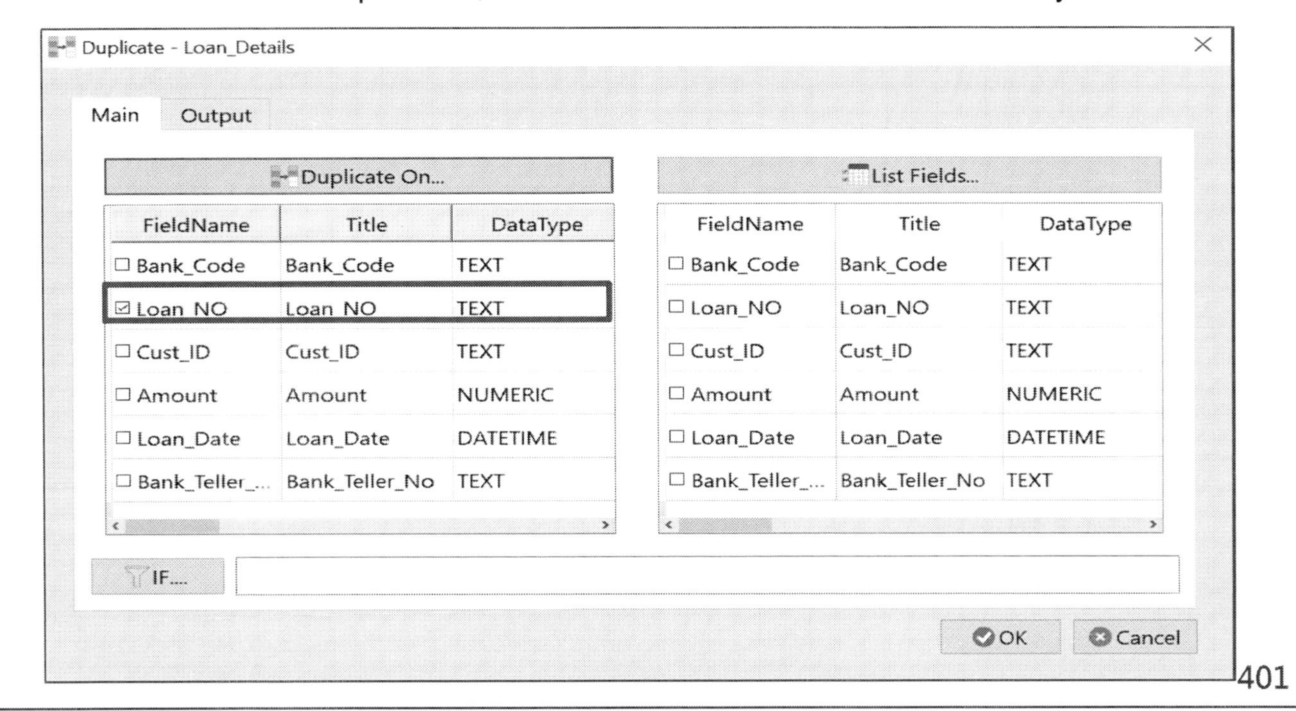

Exercise 7.17
Please check whether there are any duplicated loan numbers.

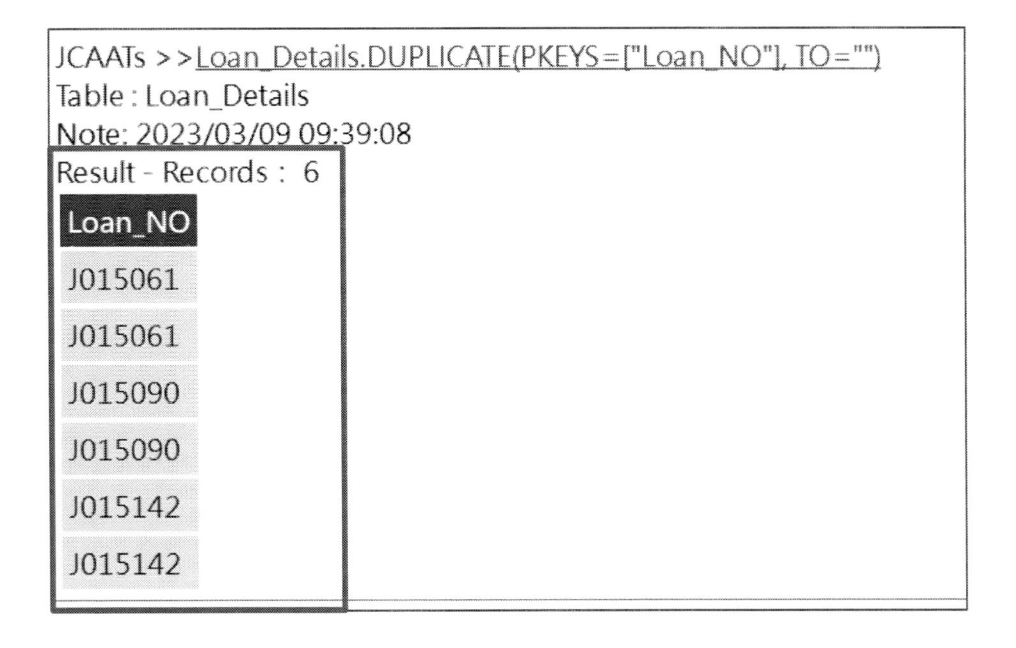

Exercise 7.17

Please check whether there are any errors in the order of loan numbers.

- Open data set we need (Loan_Details)
- Select "Validate>Sequence", choose the field "Loan_No" to analysis

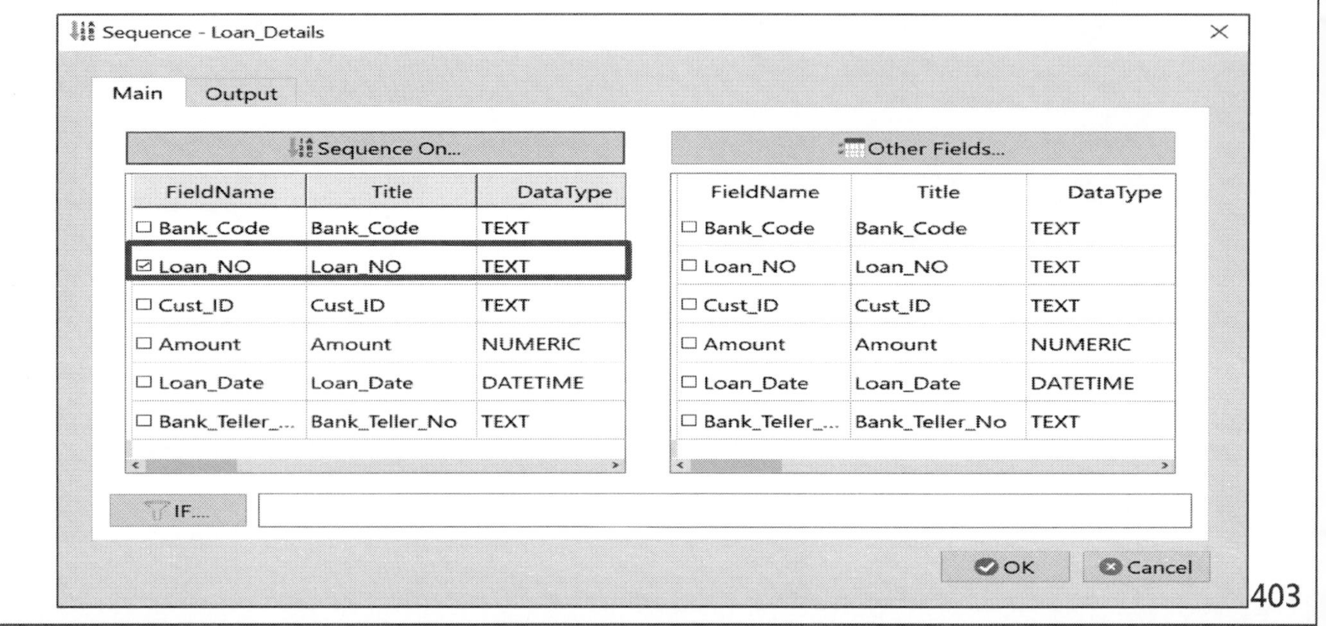

JCAATs-AI Audit Software
Copyright © 2023 JACKSOFT.

Exercise 7.17

Please check whether there are any errors in the order of loan numbers.

JCAATs >>Loan_Details.SEQUENCE(PKEYS=["Loan_NO"], FIELDS = [])
Table : Loan_Details
Note: 2023/03/09 09:39:56
Result - Records : 2

Loan_NO
J014949
J014981

Exercise 7.17

Please check whether there are any duplicated or suspected duplicated customer master files.

- Open data set we need (Customer)
- Select "Validate>Duplicate", choose the field "Cust_ID" to analysis

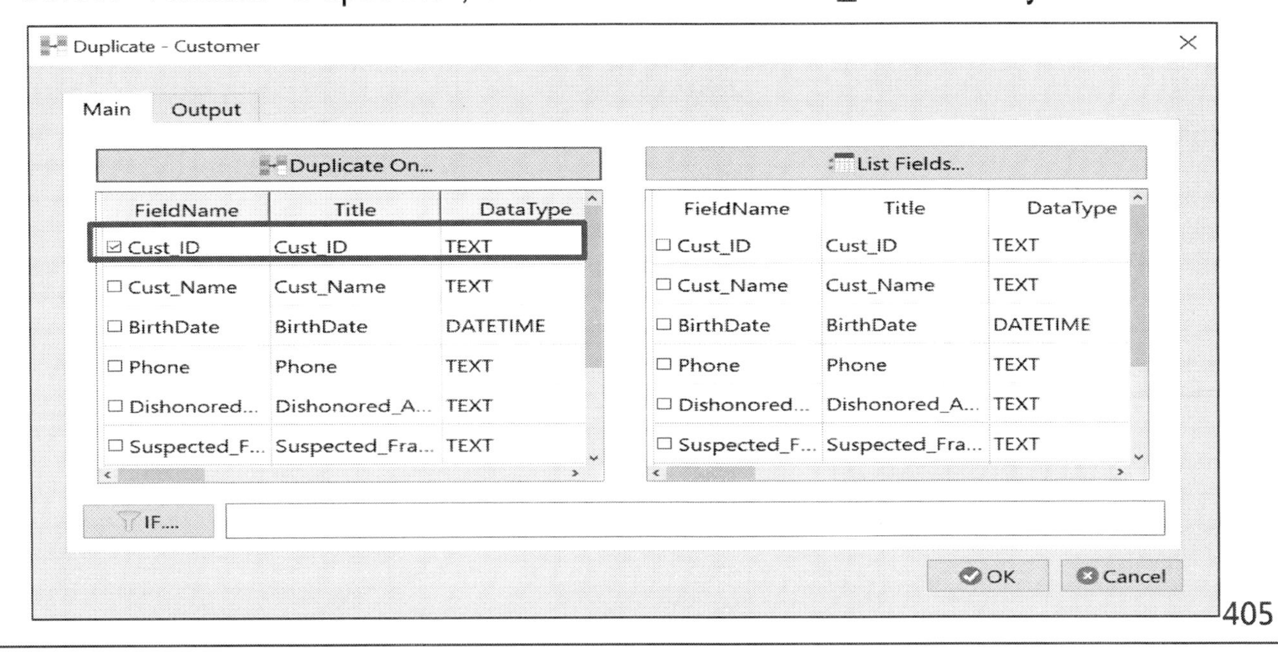

Exercise 7.17

Please check whether there are any duplicated or suspected duplicated customer master files.

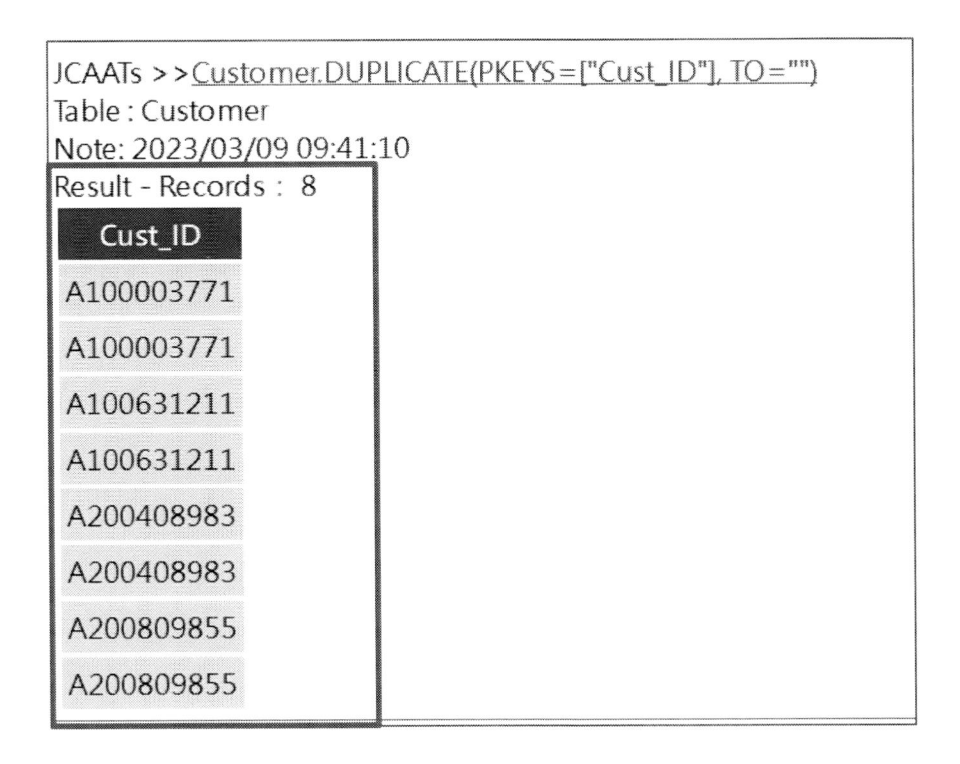

JCAATs >>Customer.DUPLICATE(PKEYS=["Cust_ID"], TO="")
Table : Customer
Note: 2023/03/09 09:41:10
Result - Records : 8

Cust_ID
A100003771
A100003771
A100631211
A100631211
A200408983
A200408983
A200809855
A200809855

Exercise 7.17

Analyze data: Please divide the loan amount of this year (2016) into 6 tiers. Which branch has the highest total loan amount in the tier with the largest amount?

- Select "Analyze>Stratify", choose Interval Size as Equal, subtotal the Amount field.
- Add a Filter Condition: Loan_Date.between(date(2016-01-01), date(2016-12-31))

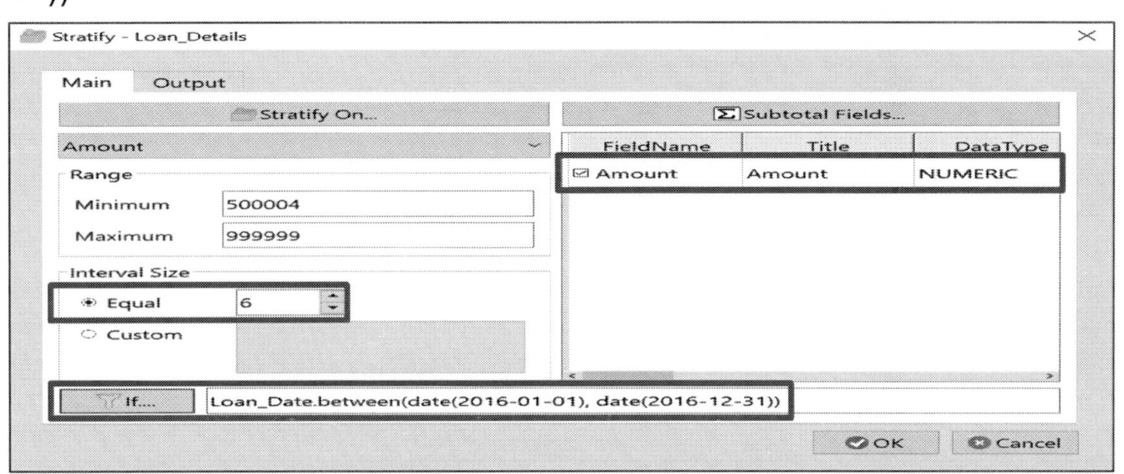

JCAATs-AI Audit Software Copyright © 2023 JACKSOFT.

Exercise 7.17

Analyze data: Please divide the loan amount of this year (2016) into 6 tiers. Which branch has the highest total loan amount in the tier with the largest amount?

- Drill down into the Amount_interval to access more information.

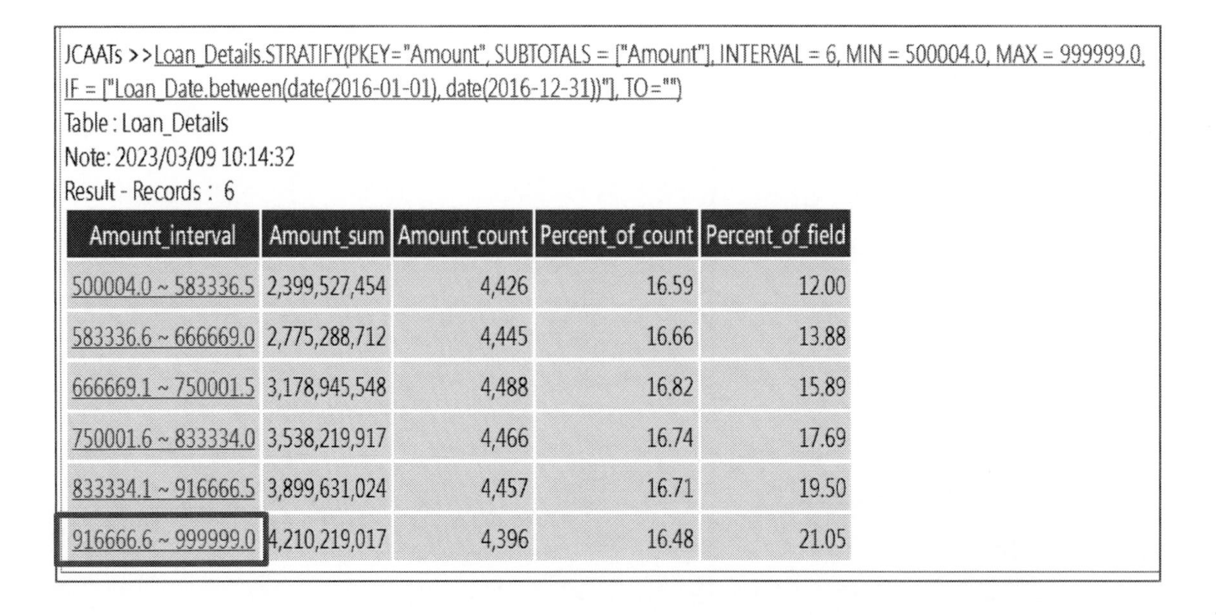

JCAATs >>Loan_Details.STRATIFY(PKEY="Amount", SUBTOTALS = ["Amount"], INTERVAL = 6, MIN = 500004.0, MAX = 999999.0, IF = ["Loan_Date.between(date(2016-01-01), date(2016-12-31))"], TO="")
Table : Loan_Details
Note: 2023/03/09 10:14:32
Result - Records : 6

Amount_interval	Amount_sum	Amount_count	Percent_of_count	Percent_of_field
500004.0 ~ 583336.5	2,399,527,454	4,426	16.59	12.00
583336.6 ~ 666669.0	2,775,288,712	4,445	16.66	13.88
666669.1 ~ 750001.5	3,178,945,548	4,488	16.82	15.89
750001.6 ~ 833334.0	3,538,219,917	4,466	16.74	17.69
833334.1 ~ 916666.5	3,899,631,024	4,457	16.71	19.50
916666.6 ~ 999999.0	4,210,219,017	4,396	16.48	21.05

Exercise 7.17

Analyze data: Please divide the loan amount of this year (2016) into 6 tiers. Which branch has the highest total loan amount in the tier with the largest amount?

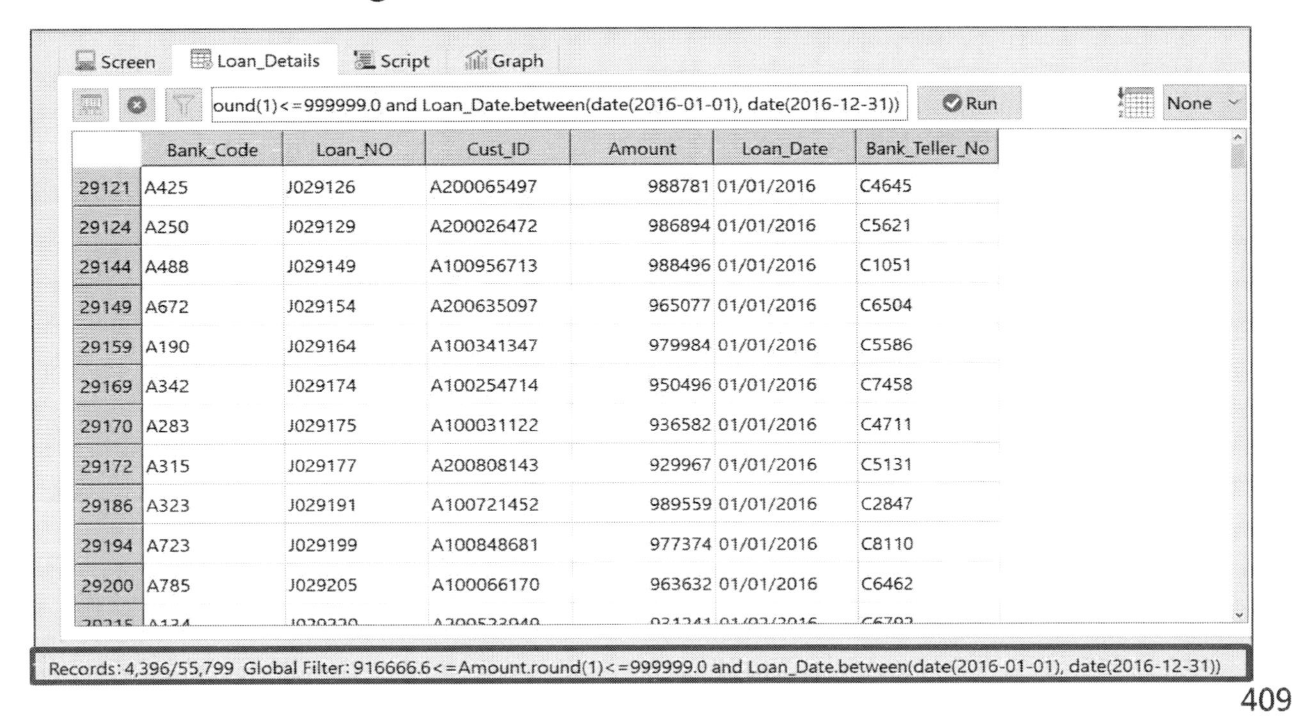

	Bank_Code	Loan_NO	Cust_ID	Amount	Loan_Date	Bank_Teller_No
29121	A425	J029126	A200065497	988781	01/01/2016	C4645
29124	A250	J029129	A200026472	986894	01/01/2016	C5621
29144	A488	J029149	A100956713	988496	01/01/2016	C1051
29149	A672	J029154	A200635097	965077	01/01/2016	C6504
29159	A190	J029164	A100341347	979984	01/01/2016	C5586
29169	A342	J029174	A100254714	950496	01/01/2016	C7458
29170	A283	J029175	A100031122	936582	01/01/2016	C4711
29172	A315	J029177	A200808143	929967	01/01/2016	C5131
29186	A323	J029191	A100721452	989559	01/01/2016	C2847
29194	A723	J029199	A100848681	977374	01/01/2016	C8110
29200	A785	J029205	A100066170	963632	01/01/2016	C6462

JCAATs-AI Audit Software
Copyright © 2023 JACKSOFT.

Exercise 7.17

Analyze data: Please divide the loan amount of this year (2016) into 6 tiers. Which branch has the highest total loan amount in the tier with the largest amount?

- Select "Analyze>Classify", choose the "Bank_Code" field to classify and subtotal by Amount.

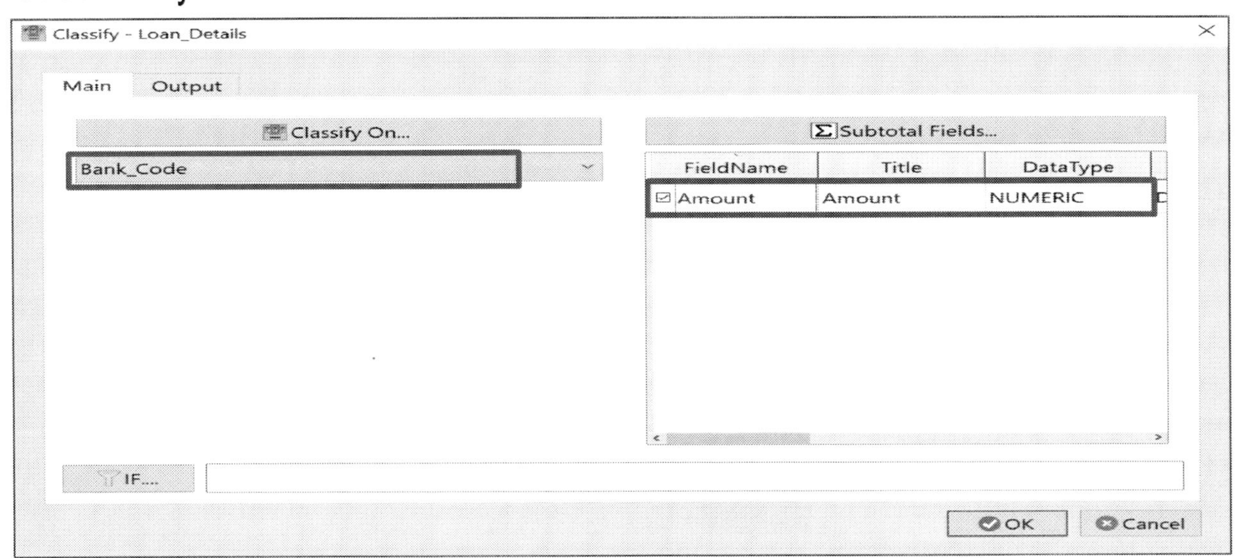

Exercise 7.17

Analyze data: Please divide the loan amount of this year (2016) into 6 tiers. Which branch has the highest total loan amount in the tier with the largest amount?

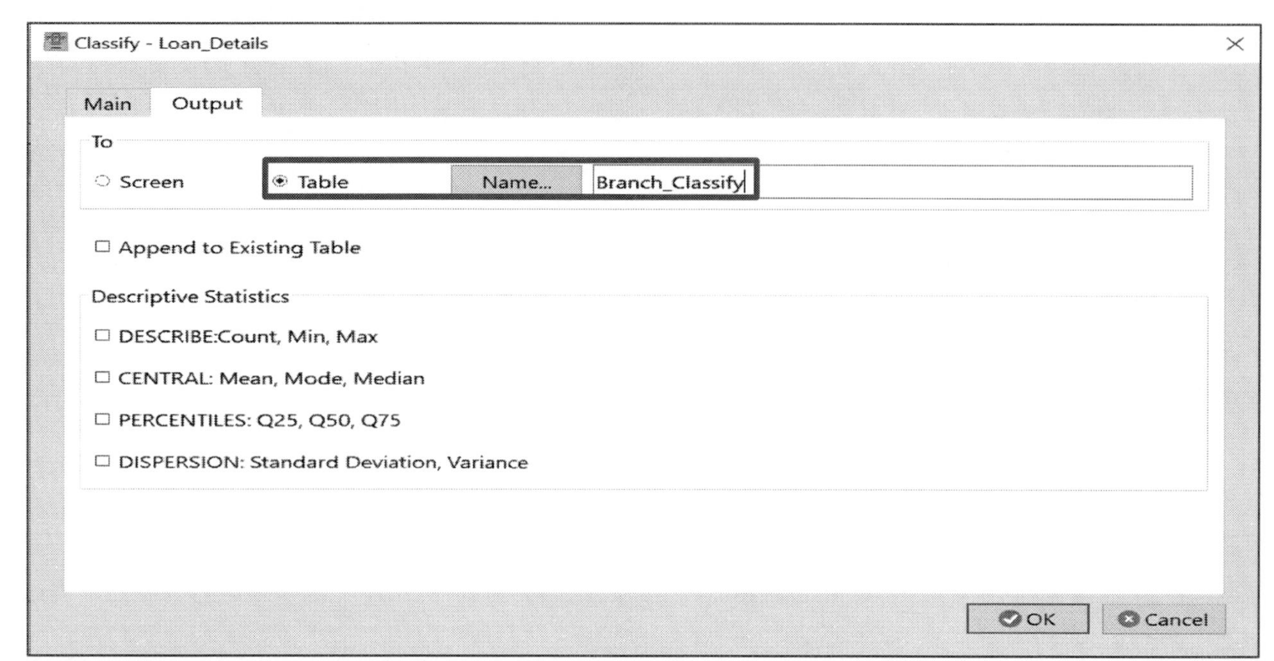

JCAATs-AI Audit Software Copyright © 2023 JACKSOFT.

Exercise 7.17

Analyze data: Please divide the loan amount of this year (2016) into 6 tiers. Which branch has the highest total loan amount in the tier with the largest amount?

- Right-click and select "Quick Sort-Descending"

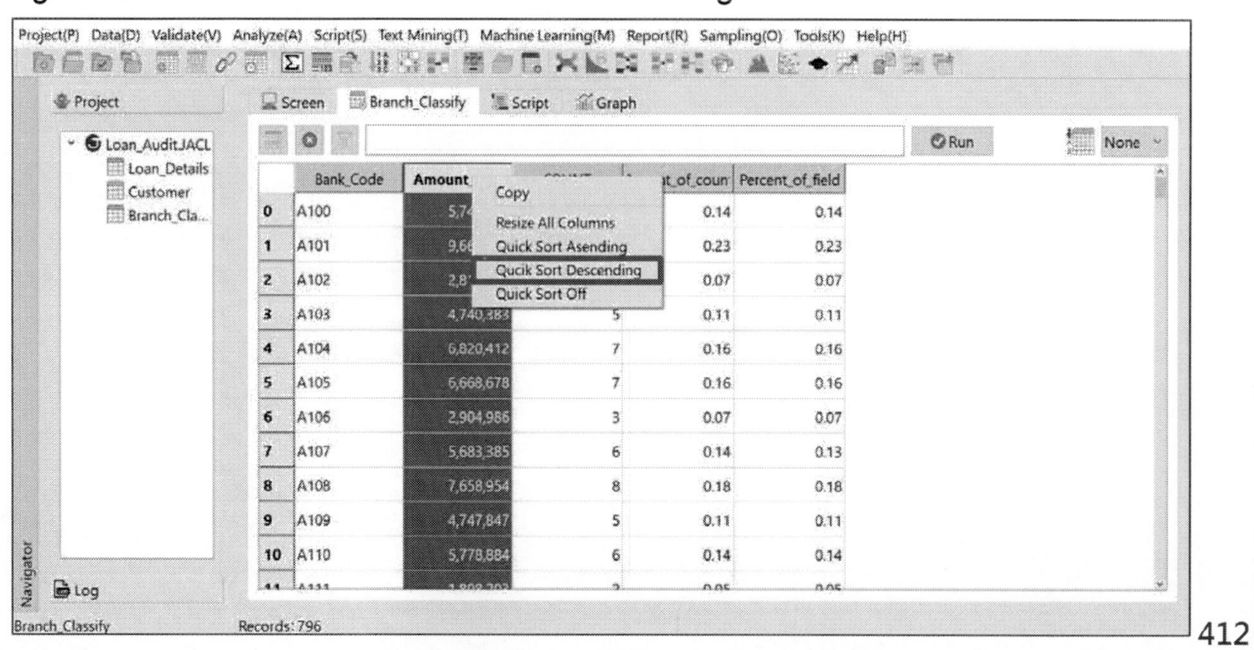

Exercise 7.17

Analyze data: Please divide the loan amount of this year (2016) into 6 tiers. Which branch has the highest total loan amount in the tier with the largest amount?

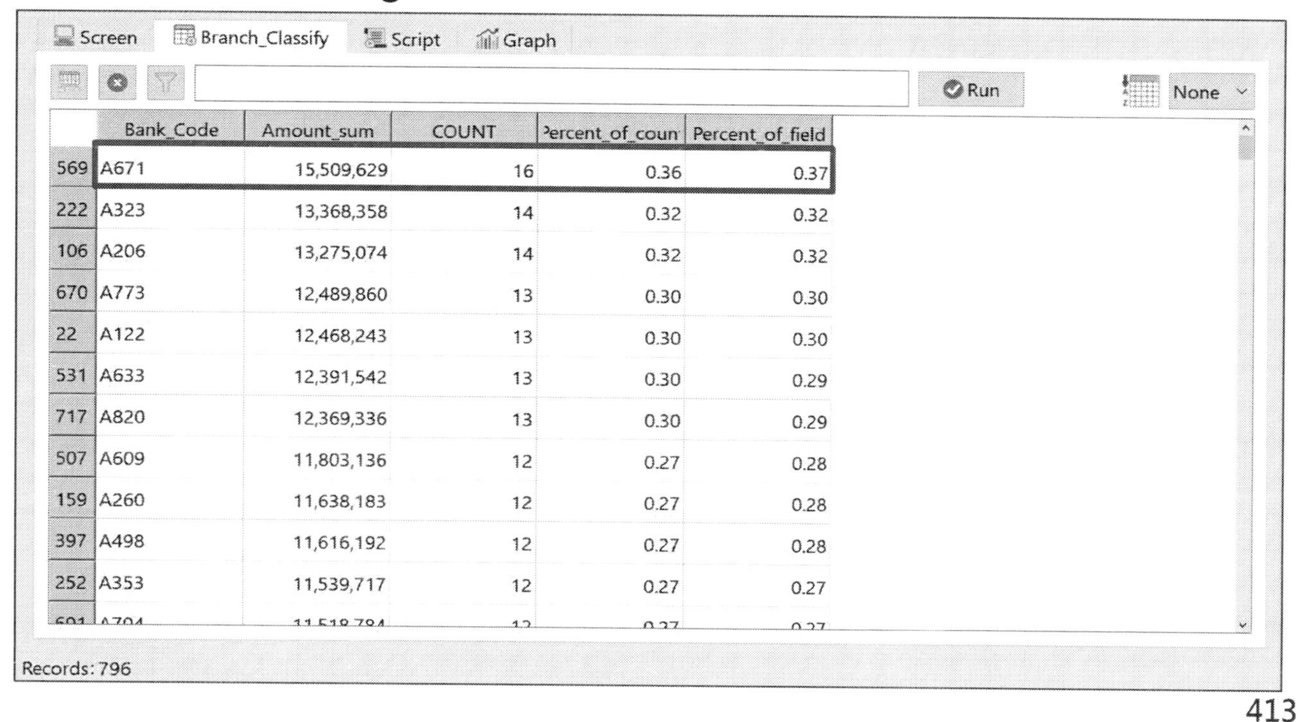

	Bank_Code	Amount_sum	COUNT	Percent_of_coun	Percent_of_field
569	A671	15,509,629	16	0.36	0.37
222	A323	13,368,358	14	0.32	0.32
106	A206	13,275,074	14	0.32	0.32
670	A773	12,489,860	13	0.30	0.30
22	A122	12,468,243	13	0.30	0.30
531	A633	12,391,542	13	0.30	0.29
717	A820	12,369,336	13	0.30	0.29
507	A609	11,803,136	12	0.27	0.28
159	A260	11,638,183	12	0.27	0.28
397	A498	11,616,192	12	0.27	0.28
252	A353	11,539,717	12	0.27	0.27

JCAATs-AI Audit Software

Exercise 7.17

Analyze data: Please divide the loan amount of this year (2016) into tiers of 0, 650,000, 800,000, 900,000, and 1,500,000. Which branch has the highest concentration of loans in the tier with the lowest total amount?

- Select "Analyze>Stratify", choose Interval Size as Custom, subtotal the Amount field.
- Add a Filter Condition: Loan_Date.between(date(2016-01-01), date(2016-12-31))

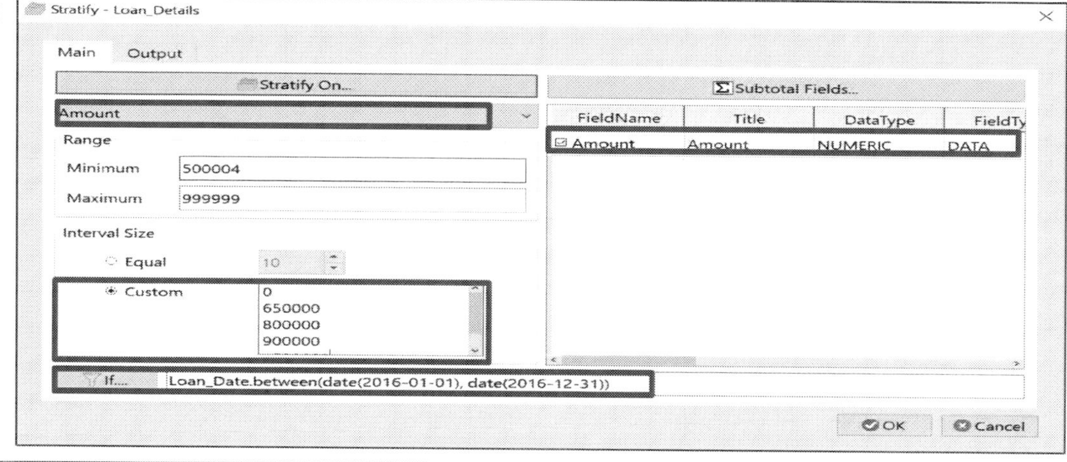

Exercise 7.17

Analyze data: Please divide the loan amount of this year (2016) into tiers of 0, 650,000, 800,000, 900,000, and 1,500,000. Which branch has the highest concentration of loans in the tier with the lowest total amount?

- Drill down into the Amount_interval to access more information.

JCAATs >>Loan_Details.STRATIFY(PKEY="Amount", SUBTOTALS = ["Amount"], FREES = ["0.0","650000.0","800000.0","900000.0","1500000.0"], MIN = 500004.0, MAX = 999999.0, IF = ["Loan_Date.between(date(2016-01-01), date(2016-12-31))"], TO="")
Table : Loan_Details
Note: 2023/03/09 10:22:00
Result - Records : 4

Amount_interval	Amount_sum	Amount_count	Percent_of_count	Percent_of_field
0.0 ~ 650000.0	4,623,542,950	8,034	30.11	23.12
650000.1 ~ 800000.0	5,787,420,369	7,978	29.90	28.93
800000.1 ~ 900000.0	4,563,096,309	5,370	20.13	22.81
900000.1 ~ 1500000.0	5,027,772,044	5,296	19.85	25.14

JCAATs-AI Audit Software
Copyright © 2023 JACKSOFT.

Exercise 7.17

Analyze data: Please divide the loan amount of this year (2016) into tiers of 0, 650,000, 800,000, 900,000, and 1,500,000. Which branch has the highest concentration of loans in the tier with the lowest total amount?

Screen | Loan_Details | Script | Graph

ound(1)<=900000.0 and Loan_Date.between(date(2016-01-01), date(2016-12-31)) Run None

	Bank_Code	Loan_NO	Cust_ID	Amount	Loan_Date	Bank_Teller_No
29135	A106	J029140	A100365696	805675	01/01/2016	C7381
29139	A253	J029144	A200788002	807307	01/01/2016	C6288
29141	A550	J029146	A200781692	801476	01/01/2016	C5681
29143	A866	J029148	A200089139	807086	01/01/2016	C1817
29151	A199	J029156	A200559056	828453	01/01/2016	C1879
29166	A389	J029171	A100995685	806226	01/01/2016	C4535
29168	A849	J029173	A100786631	847383	01/01/2016	C3960
29175	A822	J029180	A200085355	881907	01/01/2016	C8874
29179	A357	J029184	A101024425	806565	01/01/2016	C7672
29180	A485	J029185	A100925763	857765	01/01/2016	C8591
29181	A139	J029186	A100160557	823995	01/01/2016	C5218

Records: 5,370/55,799 Global Filter: 800000.1<=Amount.round(1)<=900000.0 and Loan_Date.between(date(2016-01-01), date(2016-12-31))

Exercise 7.17

Analyze data: Please divide the loan amount of this year (2016) into tiers of 0, 650,000, 800,000, 900,000, and 1,500,000. Which branch has the highest concentration of loans in the tier with the lowest total amount?

- Select "Analyze>Classify", choose the "Bank_Code" field to classify

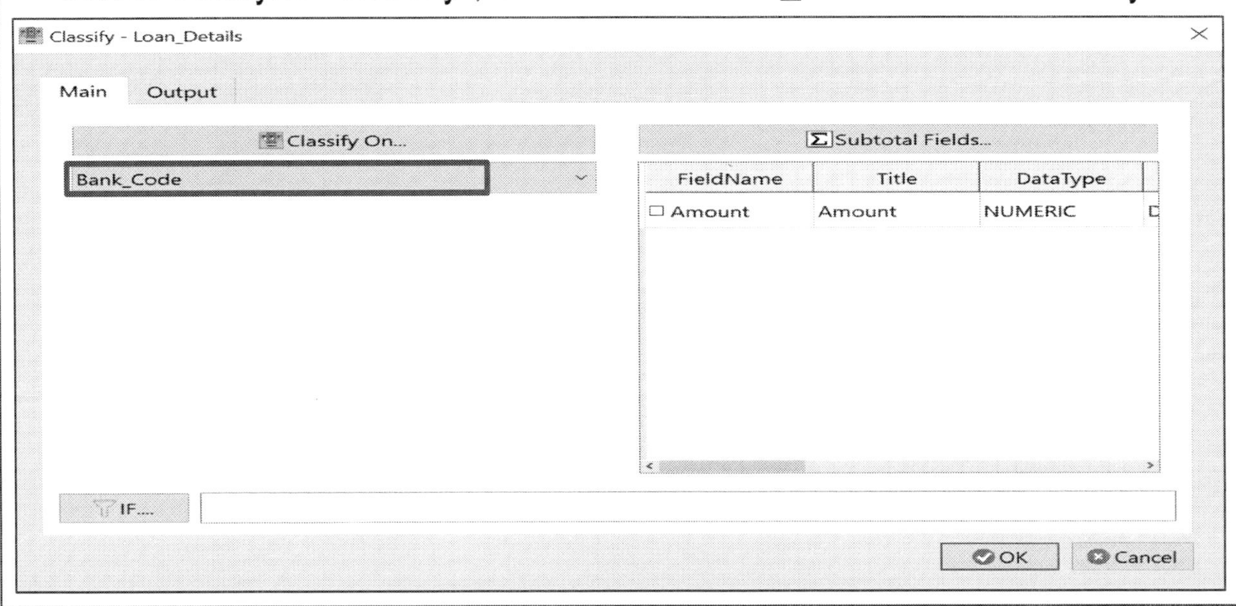

Exercise 7.17

Analyze data: Please divide the loan amount of this year (2016) into tiers of 0, 650,000, 800,000, 900,000, and 1,500,000. Which branch has the highest concentration of loans in the tier with the lowest total amount?

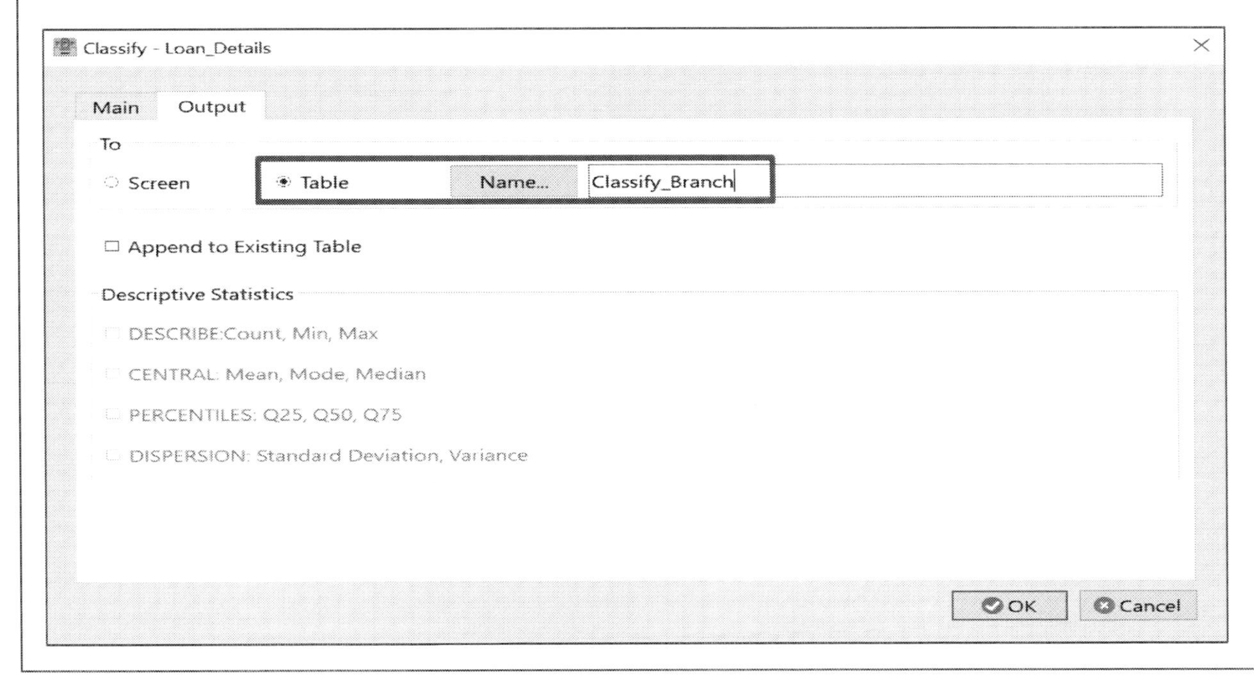

Exercise 7.17

Analyze data: Please divide the loan amount of this year (2016) into tiers of 0, 650,000, 800,000, 900,000, and 1,500,000. Which branch has the highest concentration of loans in the tier with the lowest total amount?

- Right-click and select "Quick Sort-Ascending"

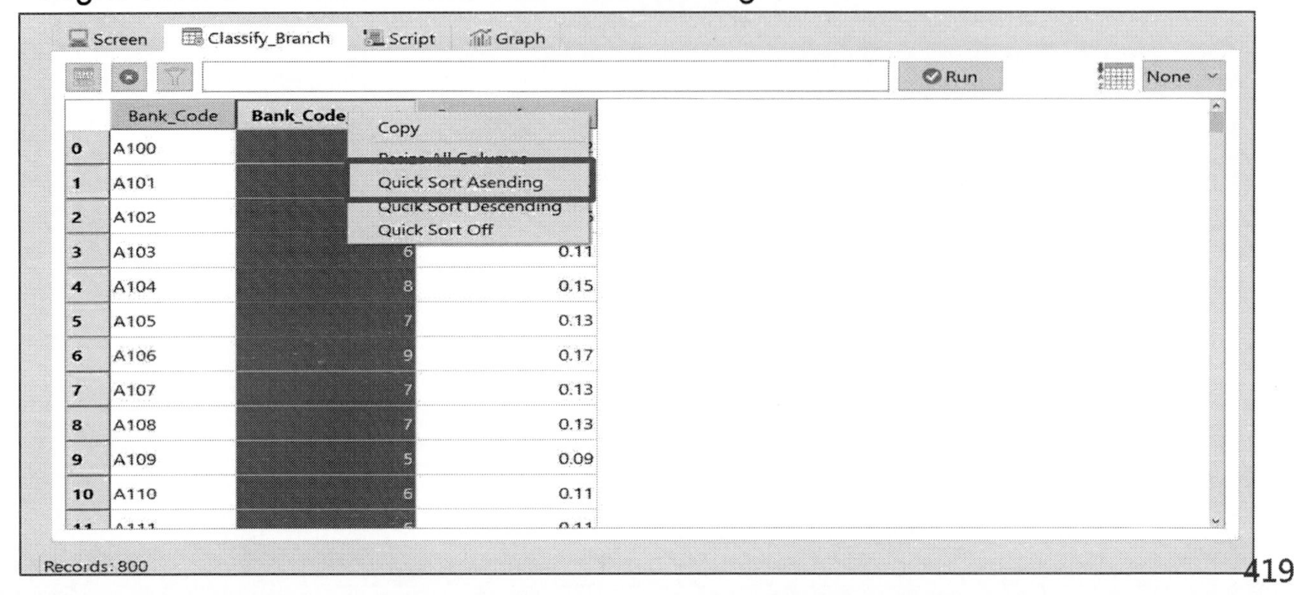

Exercise 7.17

Analyze data: Please divide the loan amount of this year (2016) into tiers of 0, 650,000, 800,000, 900,000, and 1,500,000. Which branch has the highest concentration of loans in the tier with the lowest total amount?

Screen | Classify_Branch | Script | Graph

Run | None

	Bank_Code	Bank_Code_count	Percent_of_count
454	A554	19	0.35
552	A652	15	0.28
247	A347	14	0.26
759	A859	14	0.26
79	A179	14	0.26
377	A477	14	0.26
199	A299	13	0.24
320	A420	13	0.24
750	A850	13	0.24
556	A656	13	0.24
736	A836	13	0.24

Records: 800

Exercise 7.17

Analyze data: As of the end of November 2016, the total amount of loans that have been overdue for more than six months (183 days) is .

- Select "Analyze>Age", define the Cutoff Date as 11/30/2016, set the Aging Periods as 0 and 184, subtotal the Amount field.

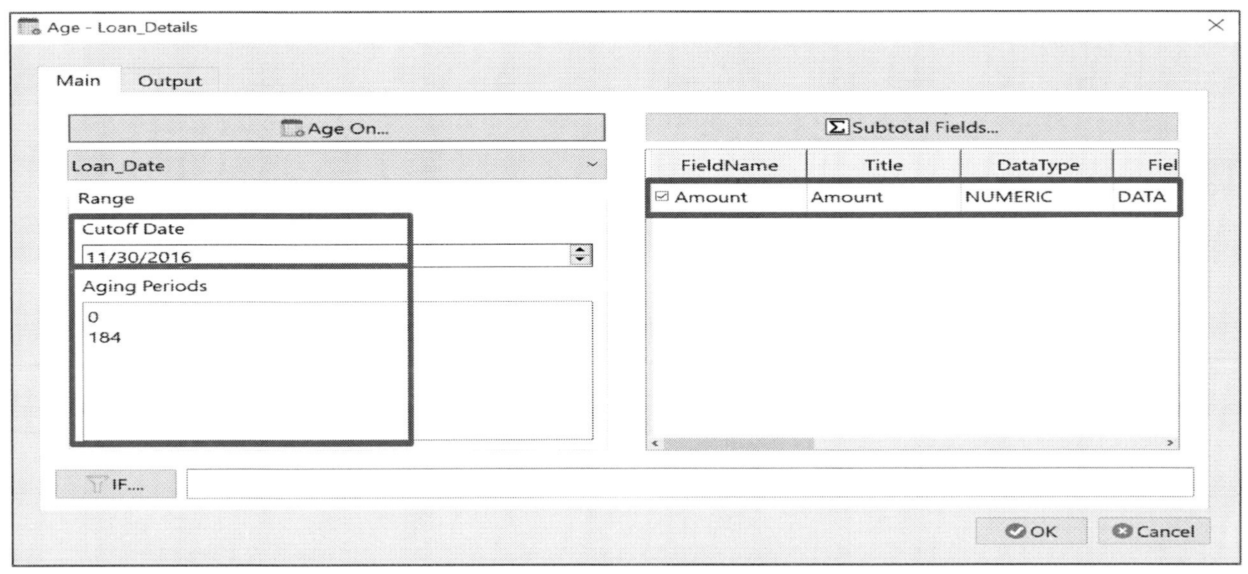

JCAATs-AI Audit Software
Copyright © 2023 JACKSOFT.

Exercise 7.17

Analyze data: As of the end of November 2016, the total amount of loans that have been overdue for more than six months (183 days) is .

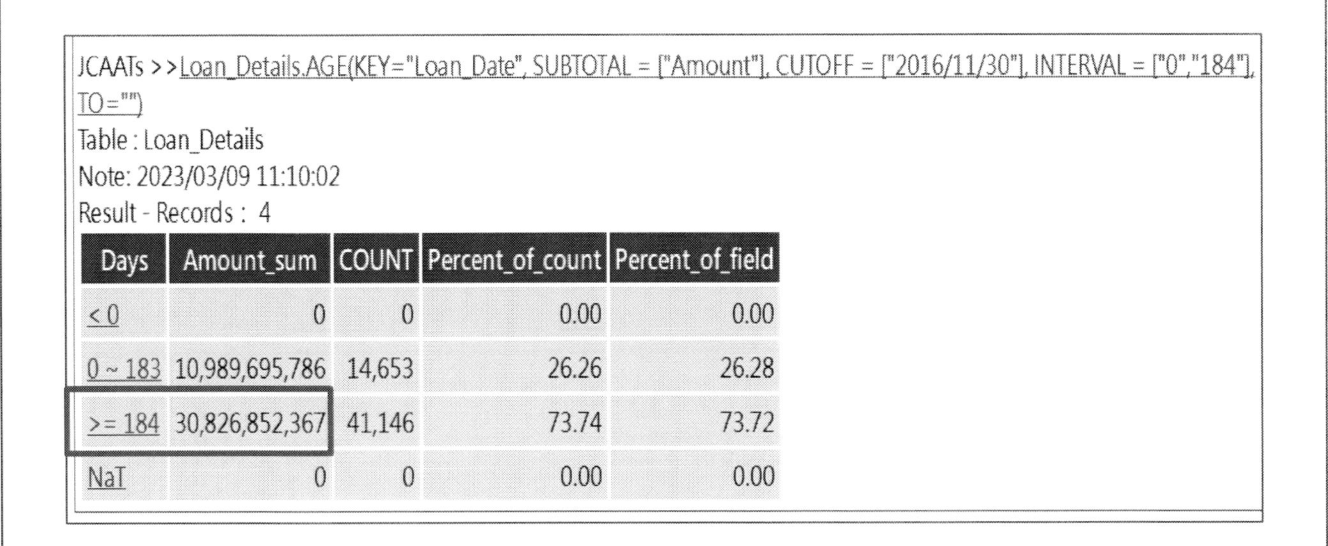

JCAATs >>Loan_Details.AGE(KEY="Loan_Date", SUBTOTAL = ["Amount"], CUTOFF = ["2016/11/30"], INTERVAL = ["0","184"], TO="")
Table : Loan_Details
Note: 2023/03/09 11:10:02
Result - Records : 4

Days	Amount_sum	COUNT	Percent_of_count	Percent_of_field
< 0	0	0	0.00	0.00
0 ~ 183	10,989,695,786	14,653	26.26	26.28
>= 184	30,826,852,367	41,146	73.74	73.72
NaT	0	0	0.00	0.00

Exercise 7.17

Analyze data: The employee with the highest total amount of loans disbursed is .

- Open data set we need (Loan_Details)
- Select "Analyze>Classify", choose the Bank_Teller_No field to classify, and subtotal the Amount field

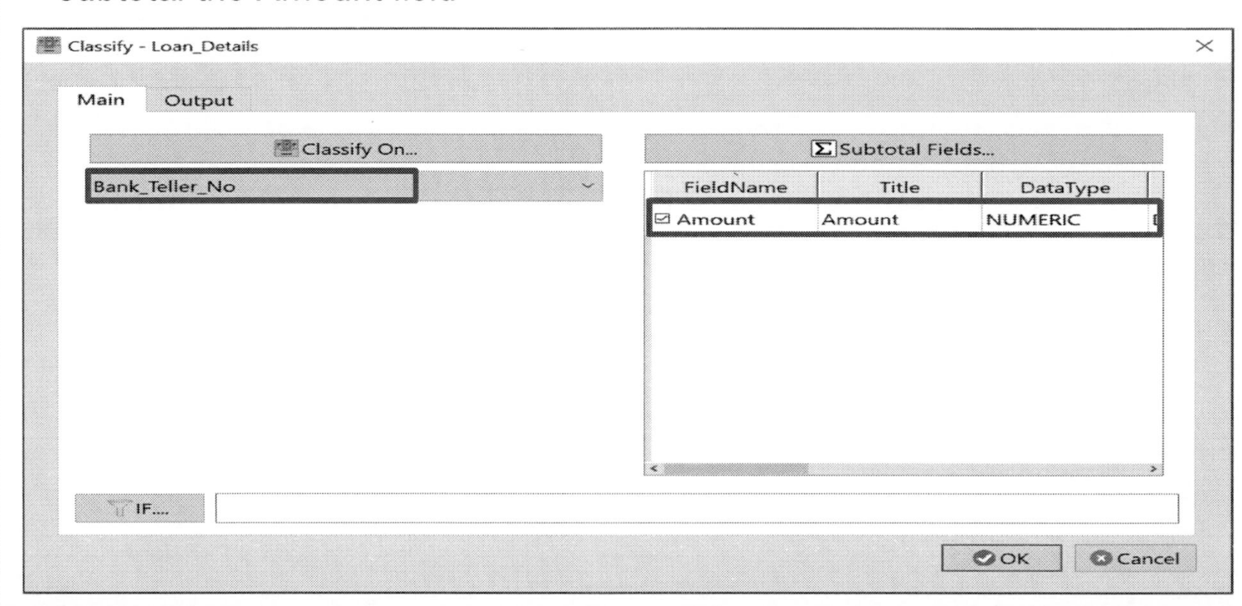

JCAATs-AI Audit Software Copyright © 2023 JACKSOFT.

Exercise 7.17

Analyze data: The employee with the highest total amount of loans disbursed is .

- Right-click and select "Quick Sort-Descending"

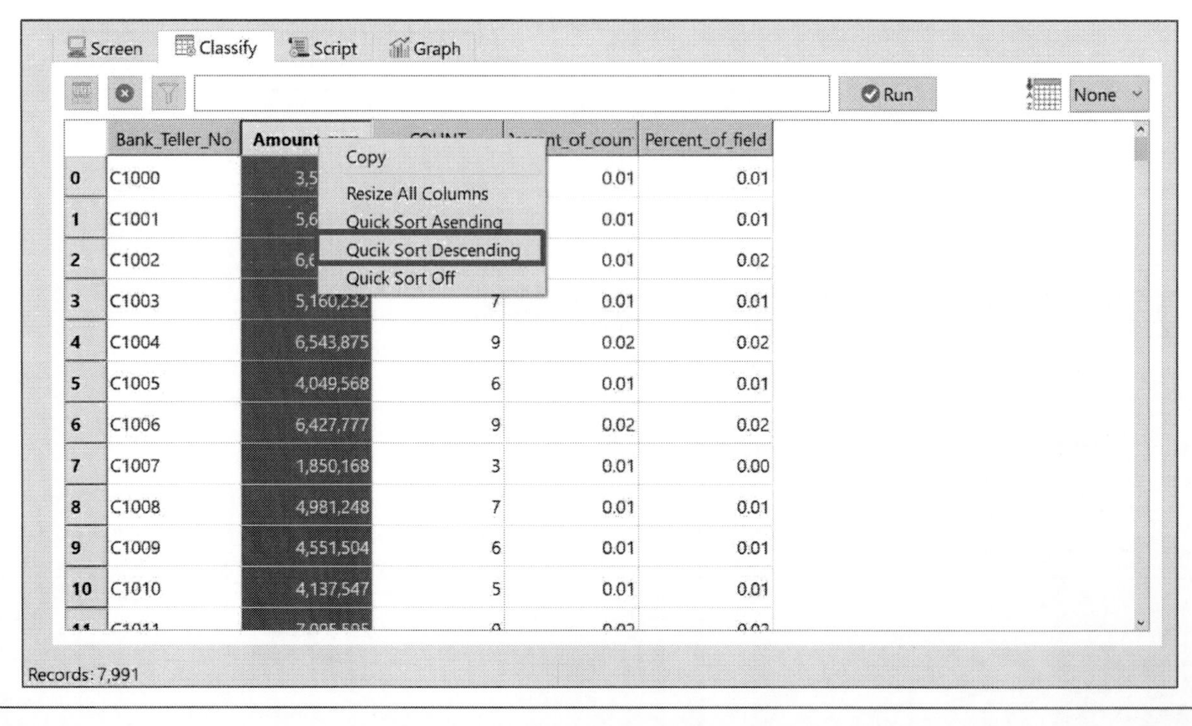

Exercise 7.17

Analyze data: The employee with the highest total amount of loans disbursed is .

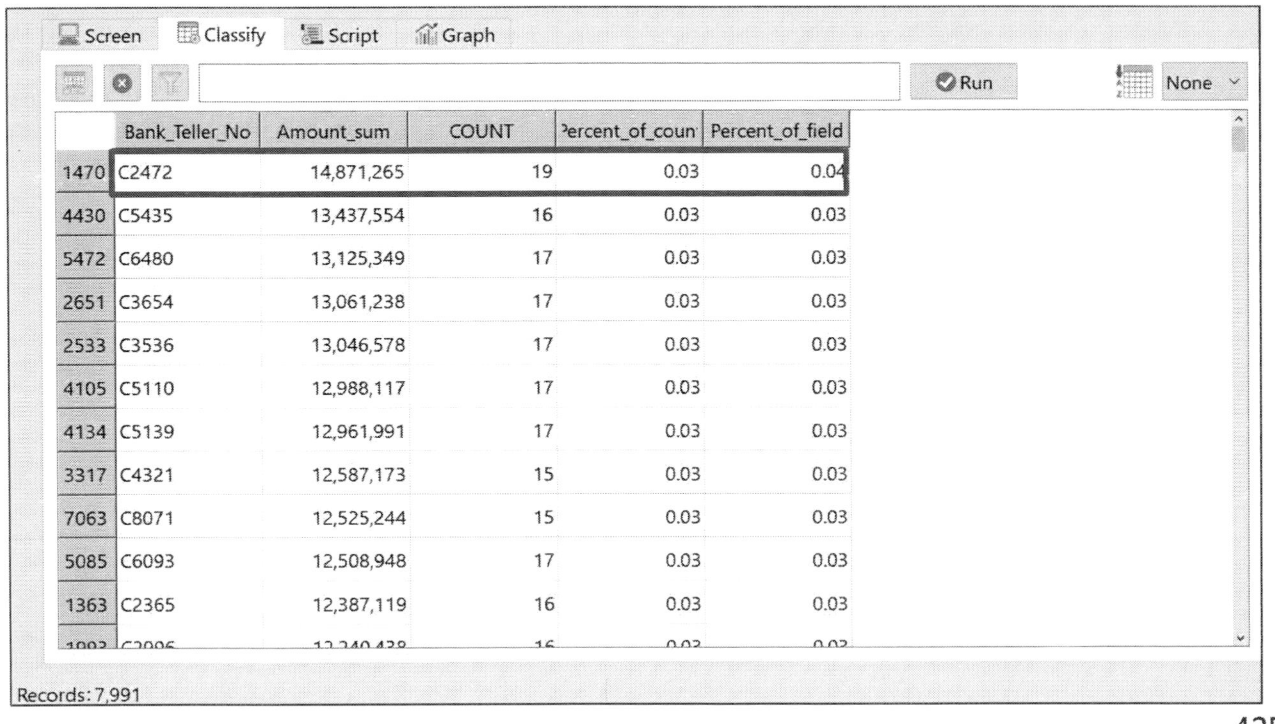

	Bank_Teller_No	Amount_sum	COUNT	Percent_of_coun	Percent_of_field
1470	C2472	14,871,265	19	0.03	0.04
4430	C5435	13,437,554	16	0.03	0.03
5472	C6480	13,125,349	17	0.03	0.03
2651	C3654	13,061,238	17	0.03	0.03
2533	C3536	13,046,578	17	0.03	0.03
4105	C5110	12,988,117	17	0.03	0.03
4134	C5139	12,961,991	17	0.03	0.03
3317	C4321	12,587,173	15	0.03	0.03
7063	C8071	12,525,244	15	0.03	0.03
5085	C6093	12,508,948	17	0.03	0.03
1363	C2365	12,387,119	16	0.03	0.03

Exercise 7.17

Analyze data: The employee with the highest total amount of daily loans is .

- Open data set we need (Loan_Details)
- Select "Analyze>Summarize", choose the Bank_Teller_No and Loan_Date fields to summarize, and subtotal the Amount field

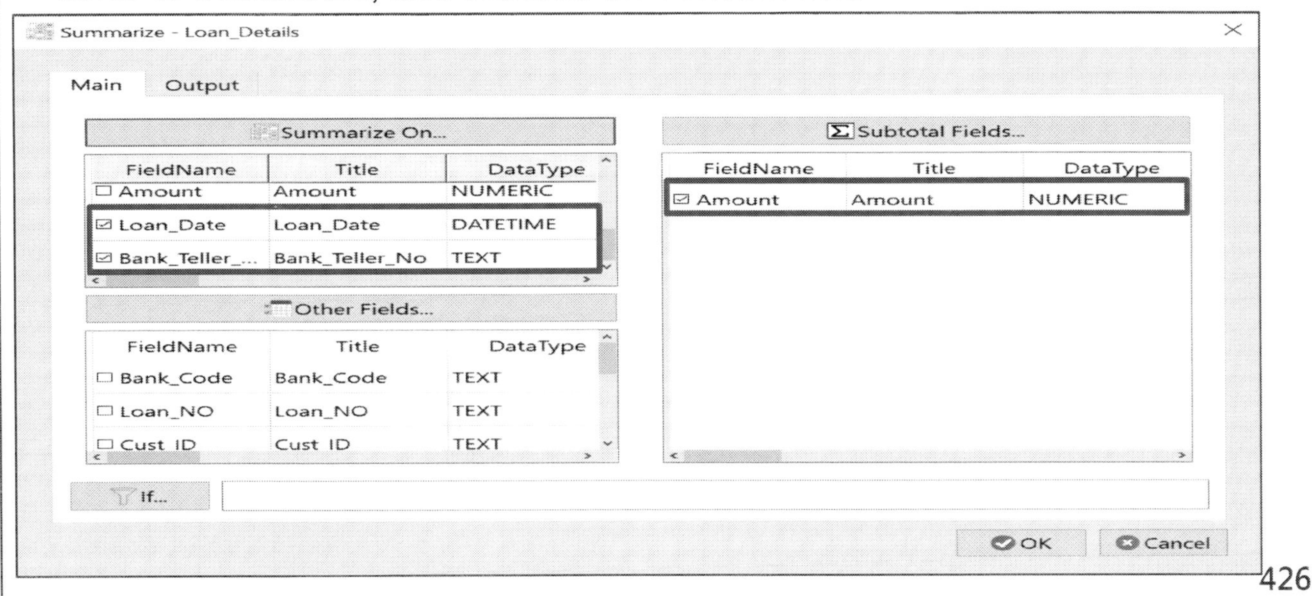

Exercise 7.17

Analyze data: The employee with the highest total amount of daily loans is .

- Right-click and select "Quick Sort-Descending"

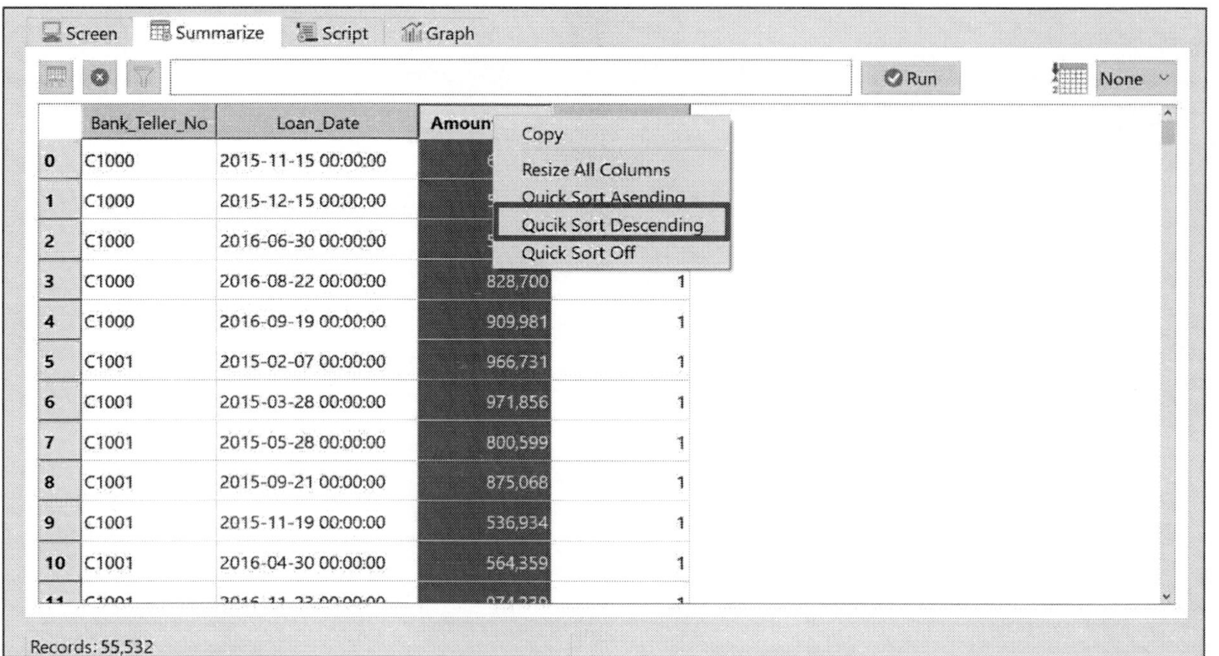

Exercise 7.17

Analyze data: The employee with the highest total amount of daily loans is .

Screen | Summarize | Script | Graph | Run | None

	Bank_Teller_No	Loan_Date	Amount_sum	COUNT
49907	C8207	2015-01-23 00:00:00	1,984,885	2
3140	C1458	2016-04-10 00:00:00	1,966,187	2
713	C1115	2016-01-20 00:00:00	1,945,070	2
36980	C6340	2015-12-14 00:00:00	1,944,172	2
38511	C6553	2015-12-13 00:00:00	1,919,599	2
28421	C5103	2016-02-13 00:00:00	1,906,312	2
9467	C2365	2016-10-02 00:00:00	1,897,114	2
30869	C5457	2016-06-30 00:00:00	1,885,159	2
11563	C2668	2015-11-22 00:00:00	1,874,791	2
44275	C7392	2015-01-22 00:00:00	1,871,930	2
54687	C8881	2016-07-09 00:00:00	1,866,118	2

Records: 55,532

Exercise 7.17
Analyze data: Have loans been granted to non-bank customers?

- Select "Analyze>Join", choose "Loan_Details" as the primary table, "Customer" as secondary, choose "Cust_ID" as key, and list out the "Loan_NO" and "Amount" fields.

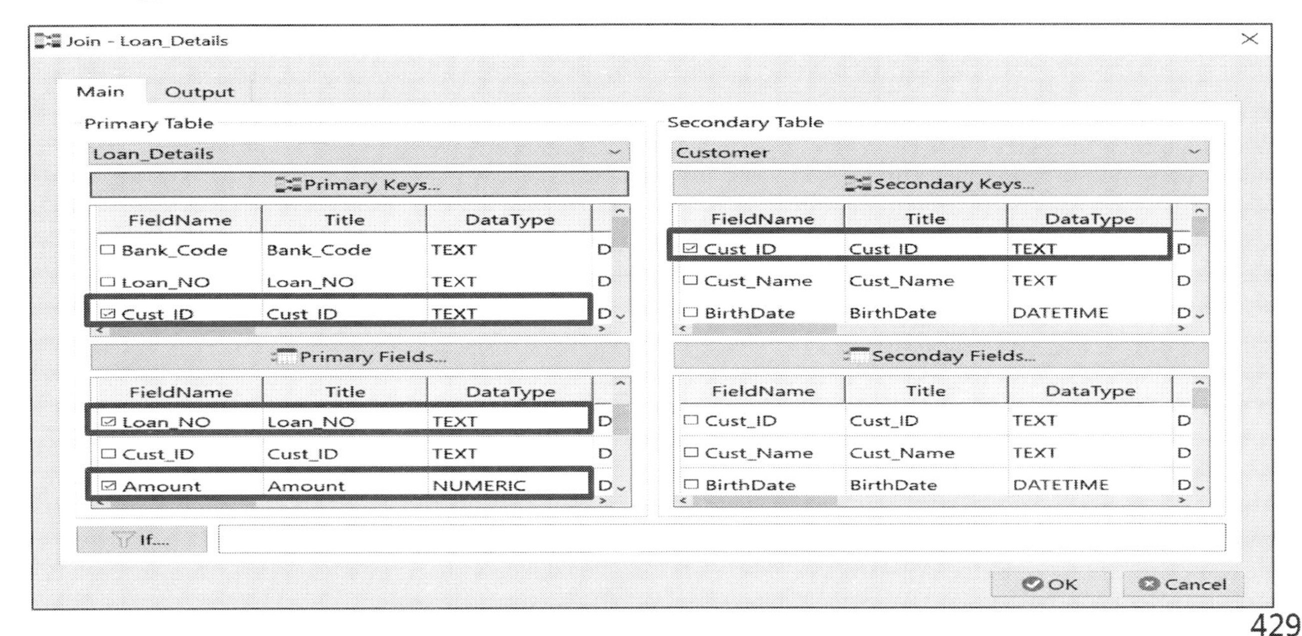

JCAATs-AI Audit Software Copyright © 2023 JACKSOFT.

Exercise 7.17
Analyze data: Have loans been granted to non-bank customers?

- Select Join_Types as Unmatch Primary

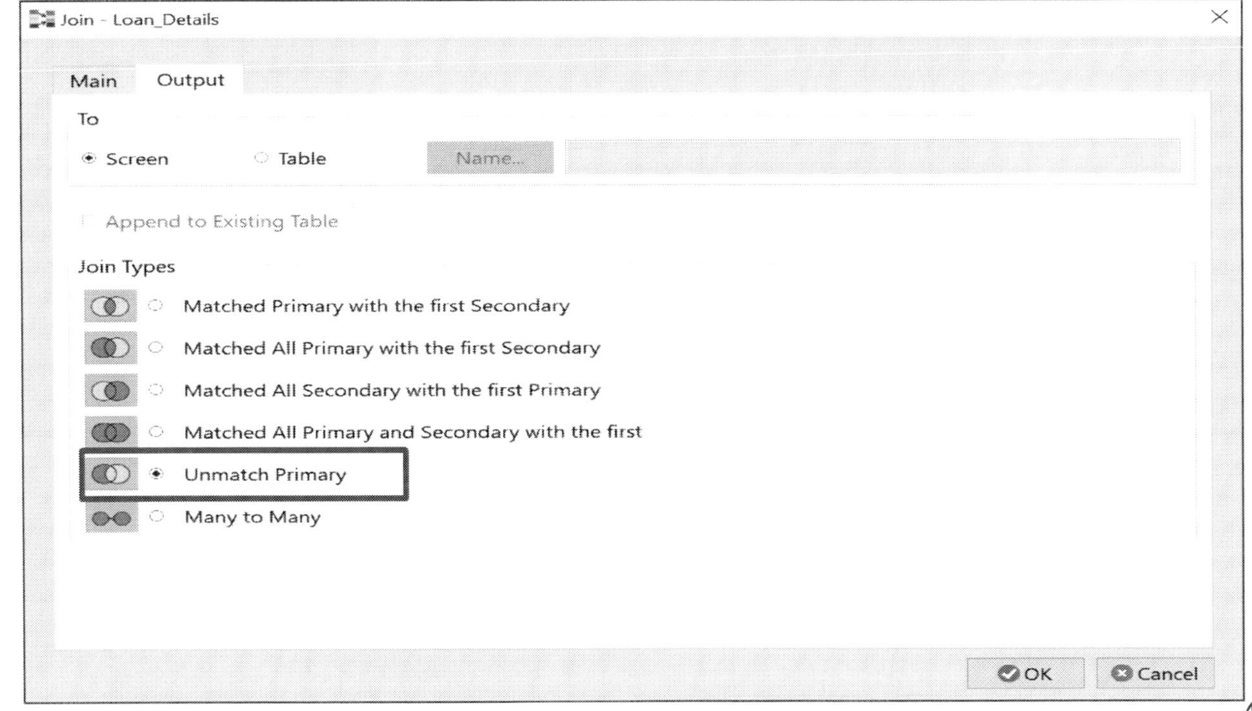

Exercise 7.17

Analyze data: Have loans been granted to non-bank customers?

JCAATs >>Loan_Details.JOIN(SECONDARY_TABLE = Customer, PKEYS=['Cust_ID'], SKEYS=['Cust_ID'], TYPE = "UNMATCHED", PFIELDS = ["Loan_NO","Amount"], TO="")
Table : Loan_Details
Note: 2023/03/09 11:20:02
Result - Records : 3

Cust_ID	Loan_NO	Amount
A100542782	J000070	794,019
A100936597	J053586	713,672
A101031368	J000087	608,864

Exercise 7.17

Analyze data: Have loans been granted to rejected or suspected fraudulent customers?

- Select "Customer" as the primary table, "Loan_Details" as secondary, choose "Cust_ID" as the key, list out the "Cust_Name" in the primary field, "Loan_NO" and "Amount" in the secondary fields.
- Add a Filter Condition: Dishonored_Account == "Y" or Suspected_Fraud == "Y"

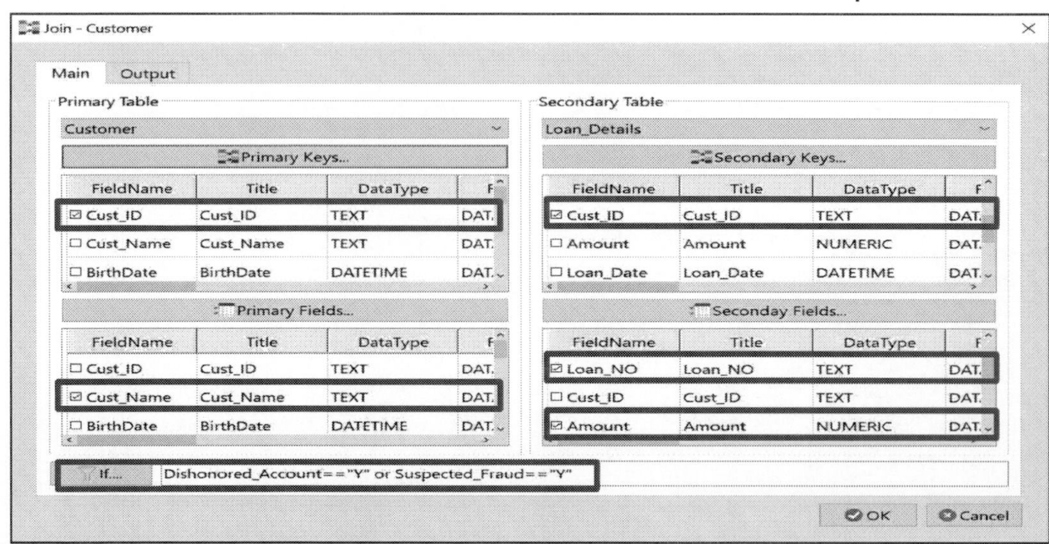

Exercise 7.17
Analyze data: Have loans been granted to rejected or suspected fraudulent customers?

- Select Join_Types as Matched Primary with the first Secondary

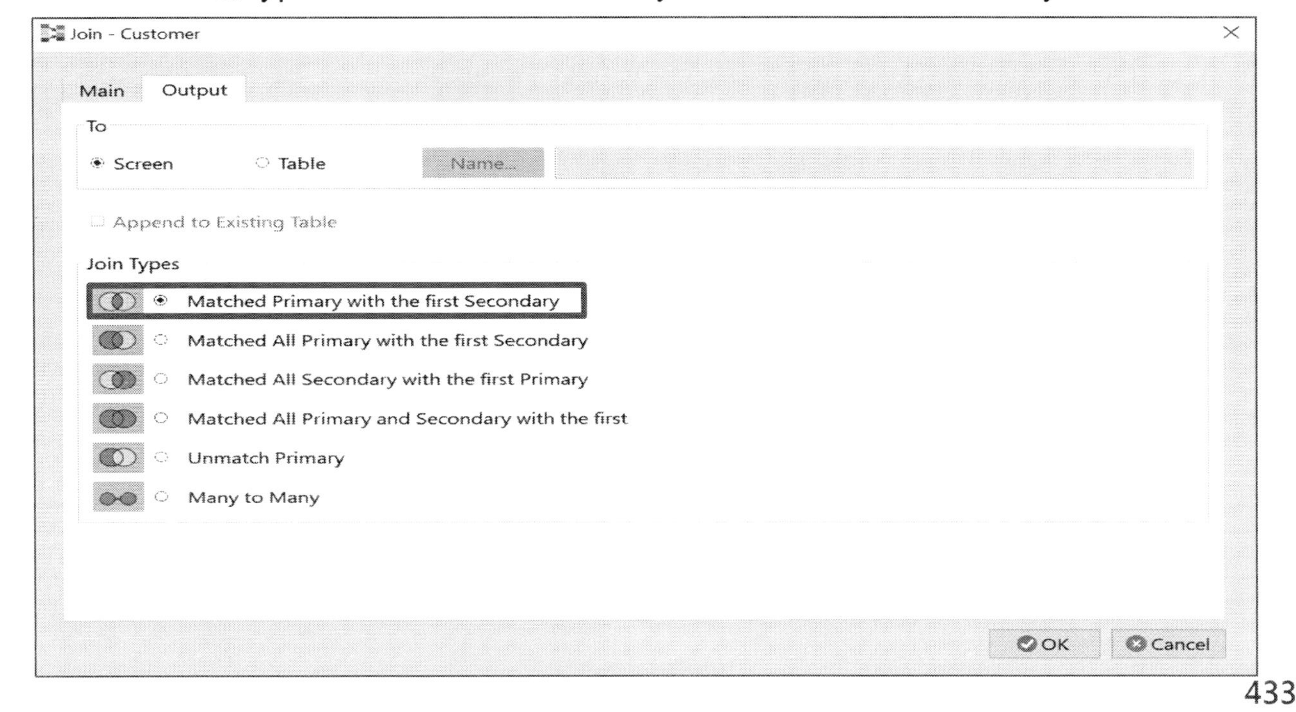

Exercise 7.17
Analyze data: Have loans been granted to rejected or suspected fraudulent customers?

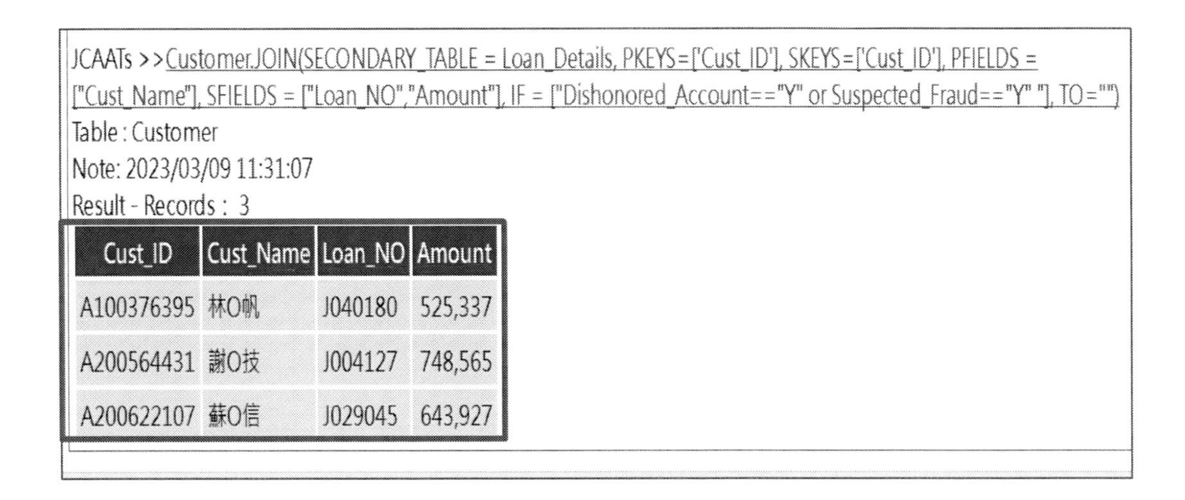

JCAATs >>Customer.JOIN(SECONDARY_TABLE = Loan_Details, PKEYS=['Cust_ID'], SKEYS=['Cust_ID'], PFIELDS = ["Cust_Name"], SFIELDS = ["Loan_NO","Amount"], IF = ["Dishonored_Account=="Y" or Suspected_Fraud=="Y" "], TO="")
Table : Customer
Note: 2023/03/09 11:31:07
Result - Records : 3

Cust_ID	Cust_Name	Loan_NO	Amount
A100376395	林O帆	J040180	525,337
A200564431	謝O玟	J004127	748,565
A200622107	蘇O信	J029045	643,927

Exercise 7.17

Analyze data: Have loans been granted to customers with annual income (calculated based on two months' income) below 300,000?

- Select "Customer" as the primary table, "Loan_Details" as secondary, choose "Cust_ID" as the key, list out the field "Bank_Teller_No"
- Add a Filter Condition: Year_Income < 300000

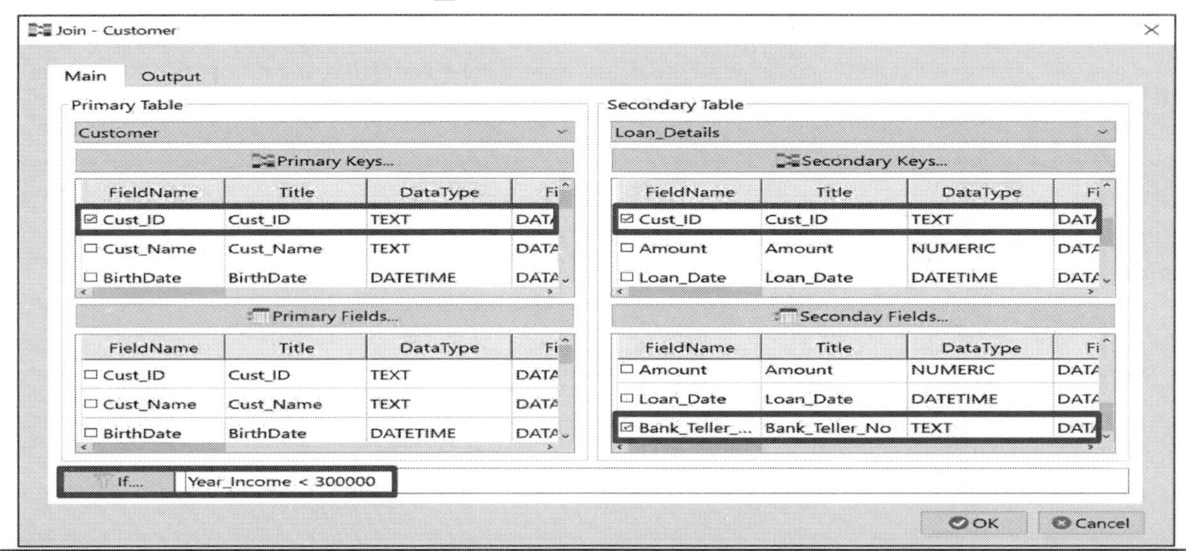

Exercise 7.17

Analyze data: Have loans been granted to customers with annual income (calculated based on two months' income) below 300,000?

- Select Join_Types as Matched Primary with the first Secondary

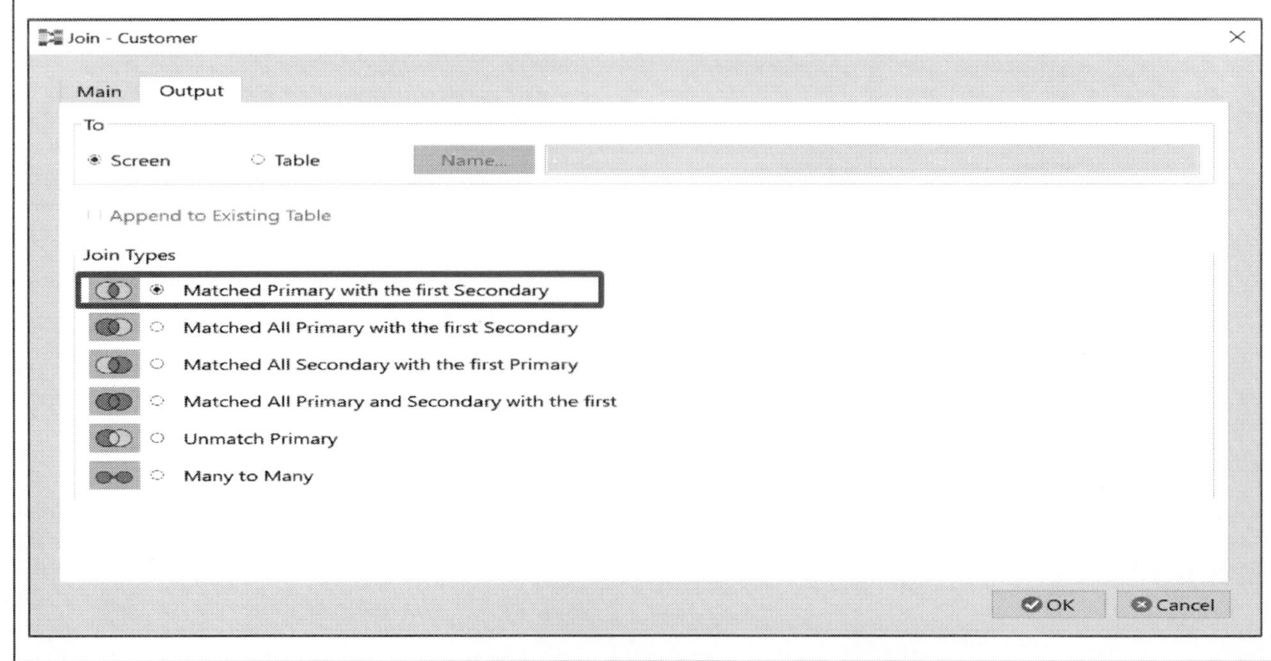

Exercise 7.17

Analyze data: Have loans been granted to customers with annual income (calculated based on two months' income) below 300,000?

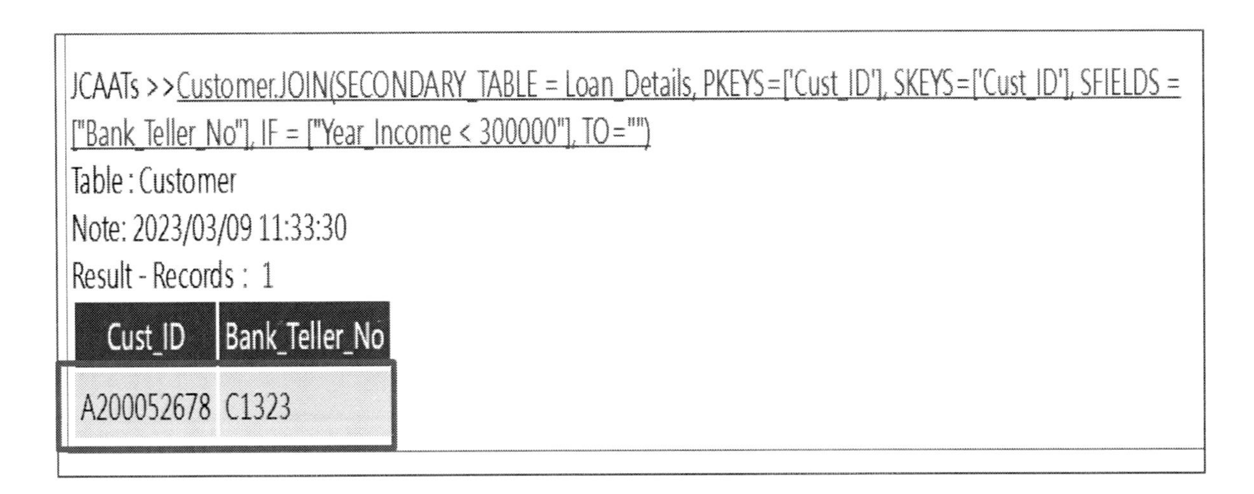
JCAATs >>Customer.JOIN(SECONDARY_TABLE = Loan_Details, PKEYS=['Cust_ID'], SKEYS=['Cust_ID'], SFIELDS = ["Bank_Teller_No"], IF = ["Year_Income < 300000"], TO="")
Table : Customer
Note: 2023/03/09 11:33:30
Result - Records : 1

Cust_ID	Bank_Teller_No
A200052678	C1323

Copyright © 2023 JACKSOFT.

Exercise 7.17

Analyze data: Whether there is a situation of lending exceeding the customer's credit limit.

- Select "Customer" as the primary table, "Loan_Details" as secondary, choose "Cust_ID" as the key, list out the "Credit_Limit" in the primary field, "Loan_NO" and "Amount" in the secondary fields.

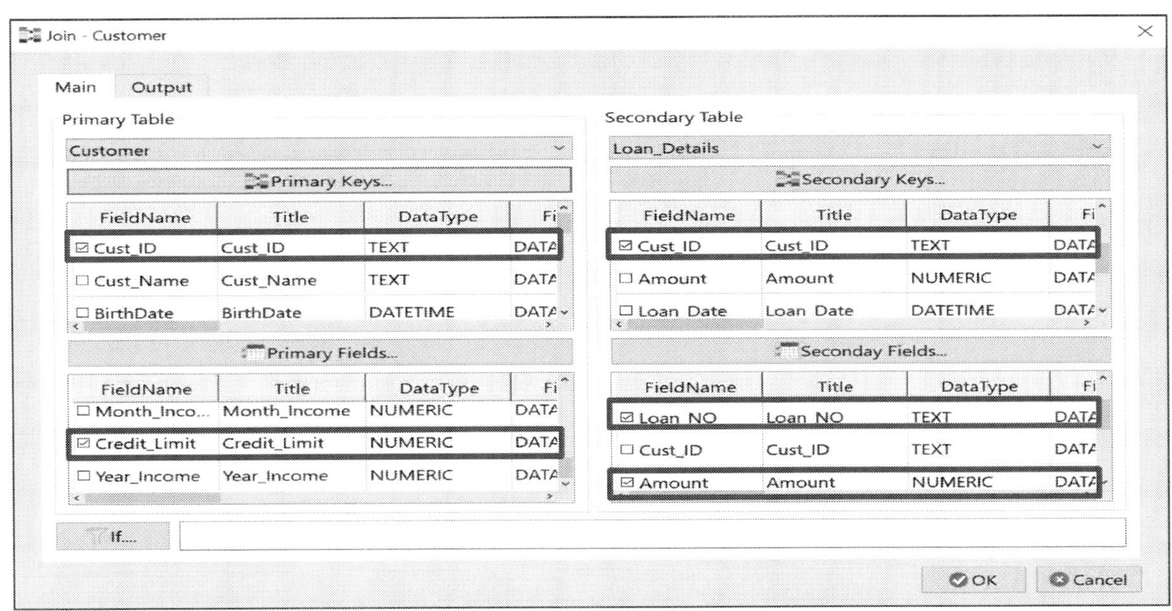

Exercise 7.17
Analyze data: Whether there is a situation of lending exceeding the customer's credit limit.

- Output to a new table
- Select Join_Types as Matched Primary with the first Secondary

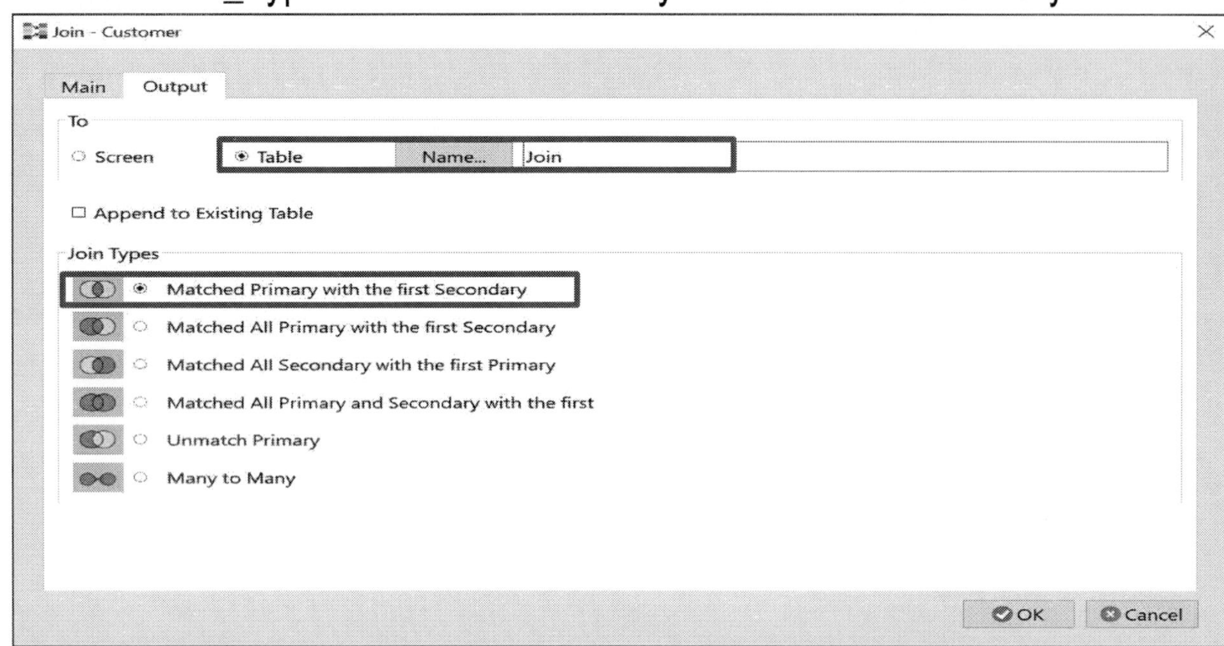

Exercise 7.17
Analyze data: Whether there is a situation of lending exceeding the customer's credit limit.

- Set Filter: Amount > Credit_Limit

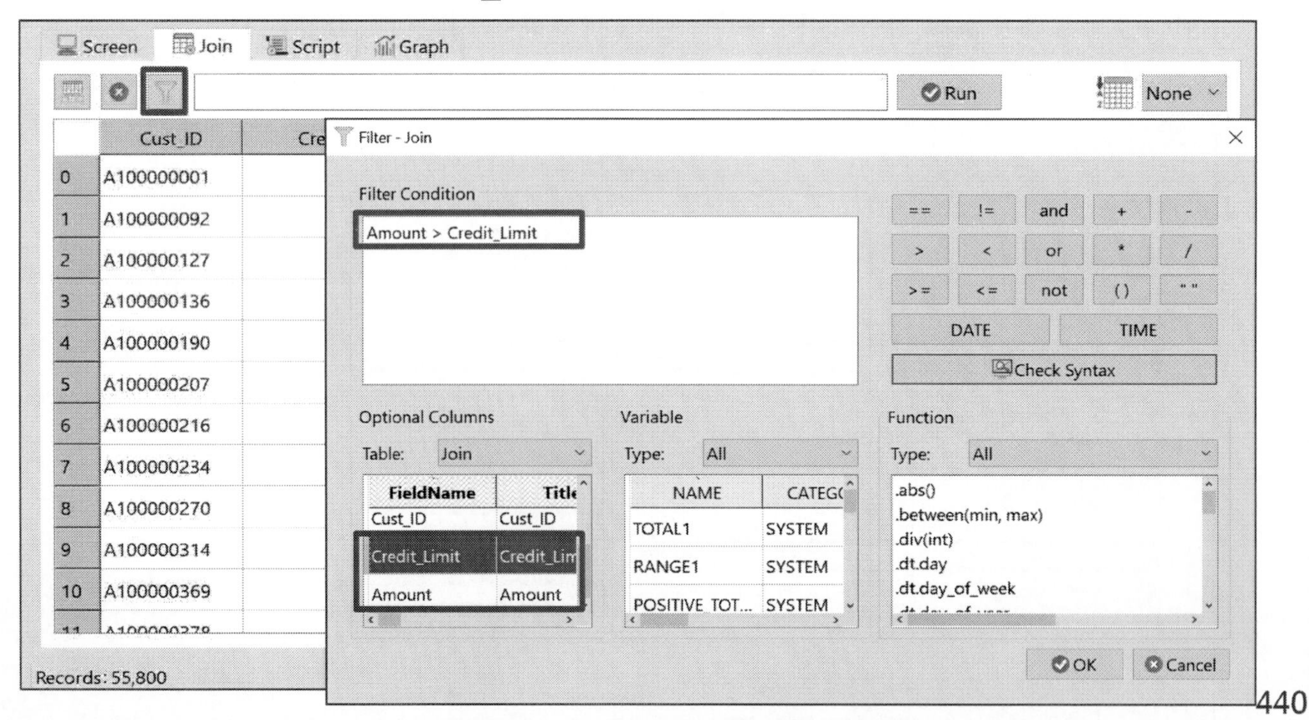

Exercise 7.17
Analyze data: Whether there is a situation of lending exceeding the customer's credit limit.

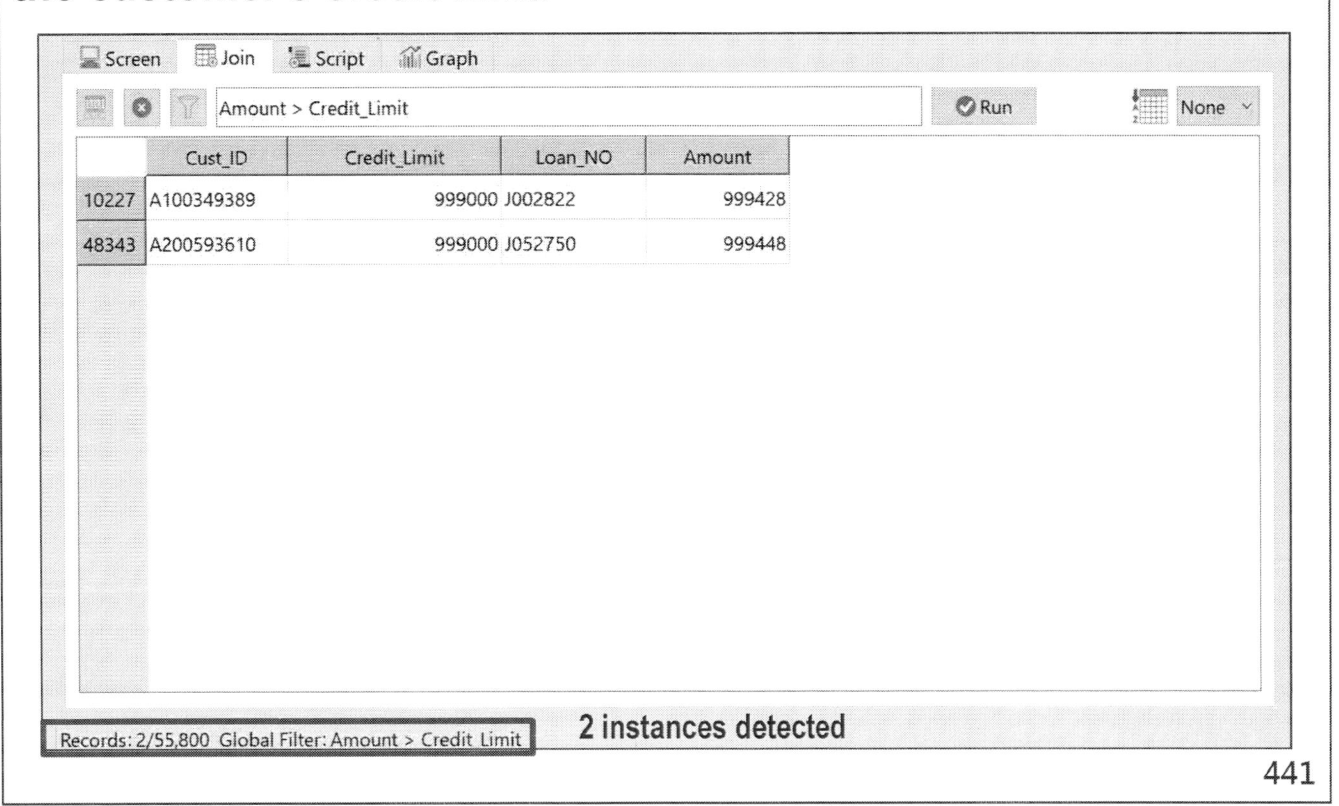

Exercise 7.18: A Case Study

- Please create a project file named "AR_Audit" , and import relevant data tables according to the following data format.

1. Accounts Receivable Detail File(AR.txt):

Field Name	Type	Decimal	Length	Note
No	CHAR		6	Customer Code
Date	DATE		10	Invoice Date
Due	DATE		10	Due Date
Ref	CHAR		6	Reference Number
Type	CHAR		2	Transaction Type
Amount	NUMERIC	2	22	Transaction Amount

Exercise 7.18: A Case Study

2 Customer Data(Customer.xlsx):

Field Name	Type	Decimal	Length	Note
No	CHAR		6	Customer Code
Name	CHAR		25	Name
Address	CHAR		25	Address
City	CHAR		15	City
State	CHAR		2	State
Zip	CHAR		5	Zip Code
Limit	NUMERIC	0	5	Credit Limit
Sales_Rep_No	CHAR		5	Sales_Rep_No

Exercise 7.18: A Case Study

3. Inventory data (Inventory.del):

Field Name	Type	Decimal	Length	Note
ProdNo	CHAR		10	Product Number
ProdCls	CHAR		2	Product Type
ProdDesc	CHAR		25	Product Description
ProdStat	CHAR		1	Product Stat
Location	CHAR		2	Location
SalePr	NUMERIC	2	10	Sales Price
MinQty	NUMERIC	0	5	Minimum Inv Level
QtyOH	NUMERIC	0	5	Inventory Quantity
UnCst	NUMERIC	2	5	Unit Cost
Value	NUMERIC	2	5	Value
MktVal	NUMERIC	2	10	Market Value
CstDte	DATE		10	Cost Date
PrcDte	DATE		10	Price Date

Exercise 7.18: A Case Study

1. To perform the audit for bonus distribution, please use the Accounts Receivable Detail File (AR) for calculation. The company policy is that the bonus distribution is calculated at 8% of the sales amount. When goods are sold, the system will record the transaction in both sales revenue and accounts receivable. How many sales transactions with a bonus exceeding USD 250 are there, and what is the total bonus amount accumulated?

____

____

2. What is the total number of bonus transactions and the accumulated bonus amount issued throughout the year?

____

____

Exercise 7.18: A Case Study

3. Your manager plans to analyze the cost and selling price of inventory. Please use the Inventory data table to calculate the inventory write-down provision for the current period. What is the amount of the inventory write-down provision that should be made for this period?

____

4. Please calculate the inventory information for each category in the right table according to the following inventory classification policy.

Gross Profit Margin	Category
Below 20%	Low
20%-50%	Middle
Above 50%	High

Category	Inventory Quantity	Inventory Cost	Market Value
Low			
Middle			
High			

____

____

Exercise 7.18: A Case Study

5. You needs to perform an audit on the sales and collection cycle. Please use the Accounts Receivable Detail File (AR) and the Customer data table to perform the check. Is there any case of ghost customers?
   (Note: "Ghost customers" refers to non-existent or fraudulent customers that are created for the purpose of generating fictitious sales transactions and inflating revenue.)
   – Please list the customer numbers.

   ____

   ____

6. Are there any customers who have not conducted transactions in the current period? Please list their names.

   ____

   ____

Exercise 7.18
Please create a project file named "AR_Audit"

1. Create a new folder.
2. Click JCAATs-AI audit software.
3. Click "Project" > "Select New Project."
4. Define a project name.
5. Save.

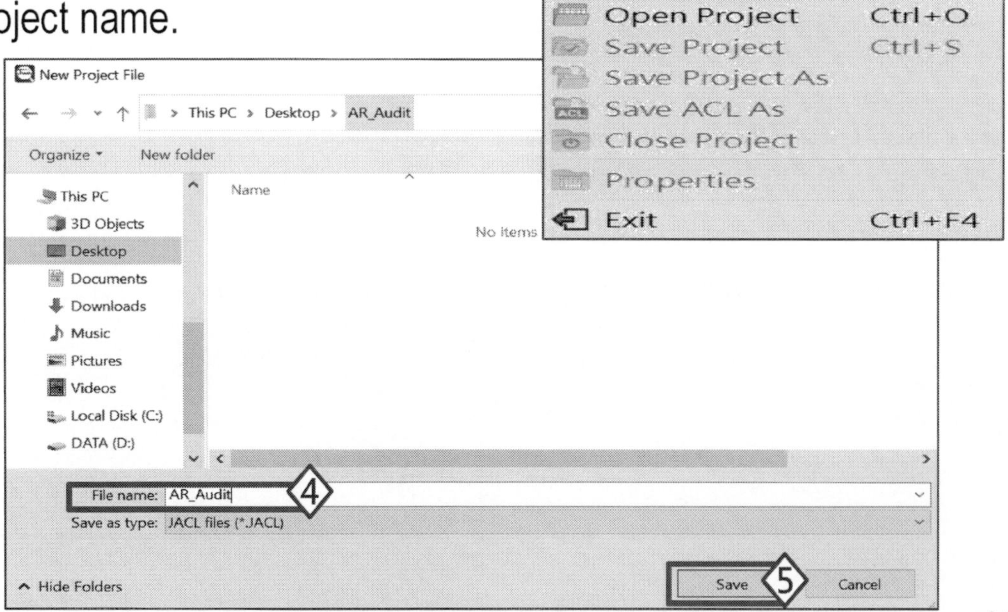

Exercise 7.18

Please import the Accounts Receivable Detail File according to the file format (Schema) provided below.

- Select Delimited Text File as the File Format
- Click on "Next" to proceed

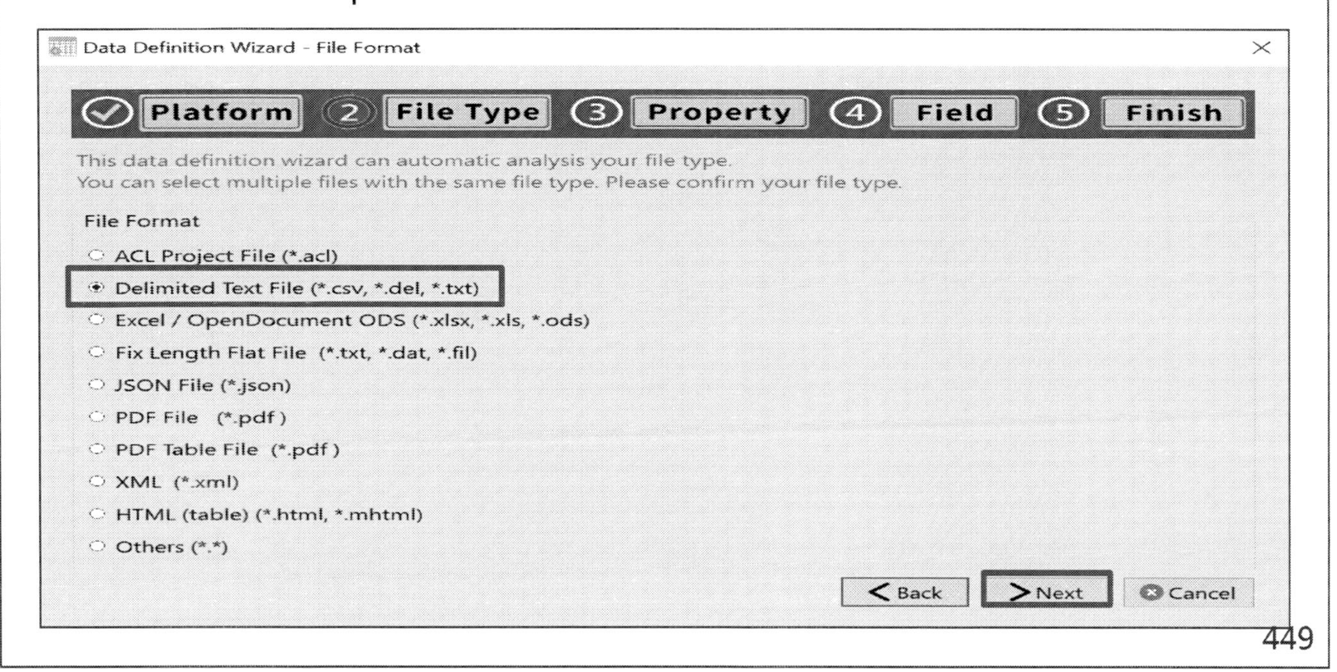

JCAATs-AI Audit Software
Copyright © 2023 JACKSOFT.

Exercise 7.18

Please import the Accounts Receivable Detail File according to the file format (Schema) provided below.

- Set the data file separator as "Space"

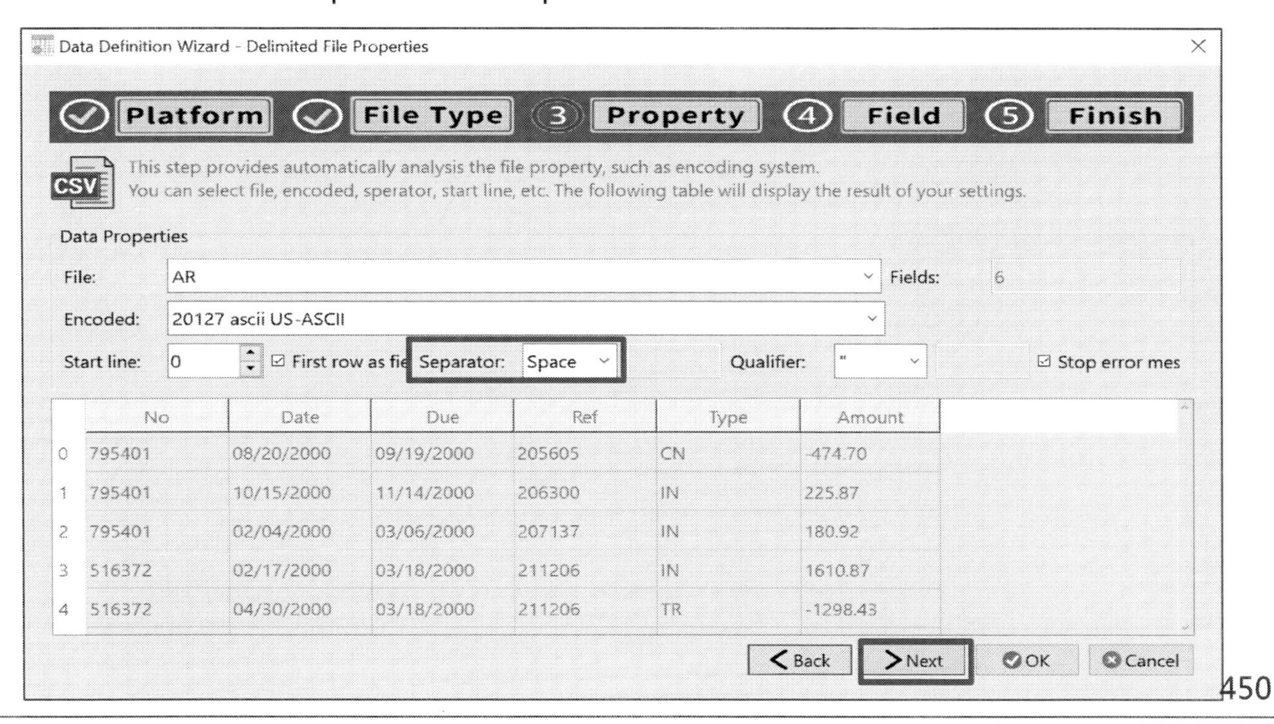

	No	Date	Due	Ref	Type	Amount
0	795401	08/20/2000	09/19/2000	205605	CN	-474.70
1	795401	10/15/2000	11/14/2000	206300	IN	225.87
2	795401	02/04/2000	03/06/2000	207137	IN	180.92
3	516372	02/17/2000	03/18/2000	211206	IN	1610.87
4	516372	04/30/2000	03/18/2000	211206	TR	-1298.43

Exercise 7.18

Please import the Accounts Receivable Detail File according to the file format (Schema) provided below.

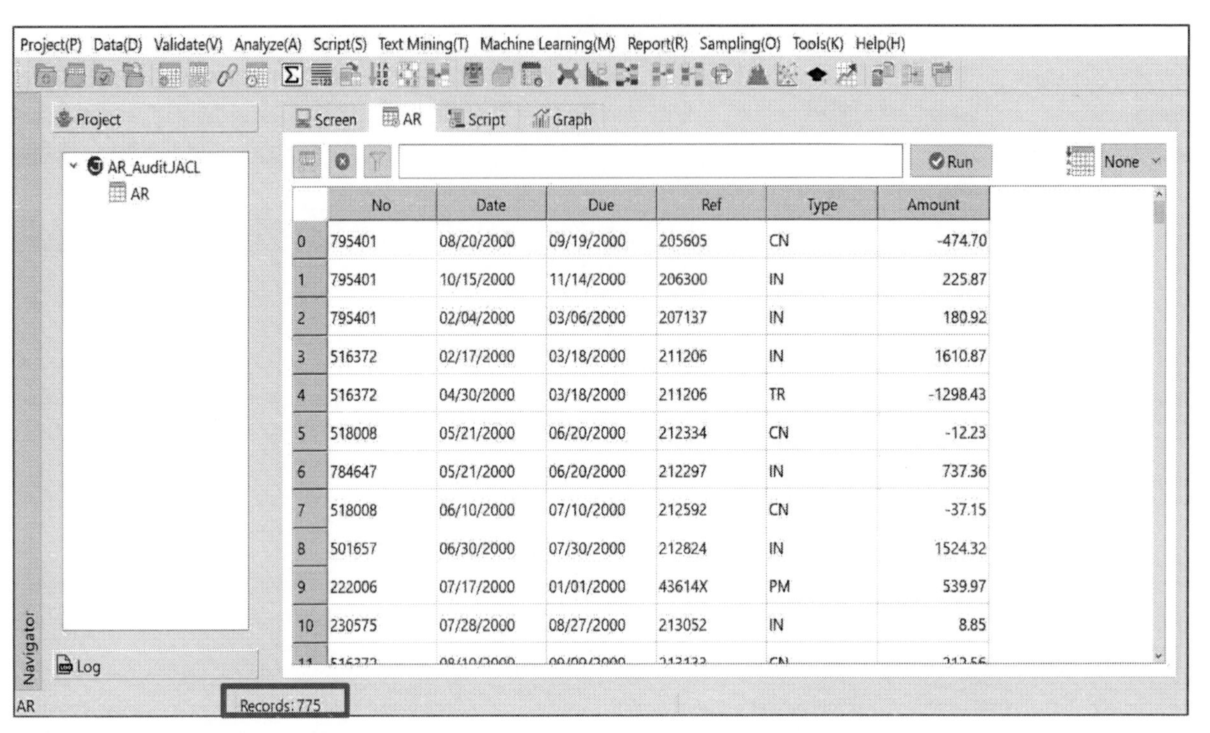

	No	Date	Due	Ref	Type	Amount
0	795401	08/20/2000	09/19/2000	205605	CN	-474.70
1	795401	10/15/2000	11/14/2000	206300	IN	225.87
2	795401	02/04/2000	03/06/2000	207137	IN	180.92
3	516372	02/17/2000	03/18/2000	211206	IN	1610.87
4	516372	04/30/2000	03/18/2000	211206	TR	-1298.43
5	518008	05/21/2000	06/20/2000	212334	CN	-12.23
6	784647	05/21/2000	06/20/2000	212297	IN	737.36
7	518008	06/10/2000	07/10/2000	212592	CN	-37.15
8	501657	06/30/2000	07/30/2000	212824	IN	1524.32
9	222006	07/17/2000	01/01/2000	43614X	PM	539.97
10	230575	07/28/2000	08/27/2000	213052	IN	8.85

Exercise 7.18

Please import the Customer Data according to the file format (Schema) provided below.

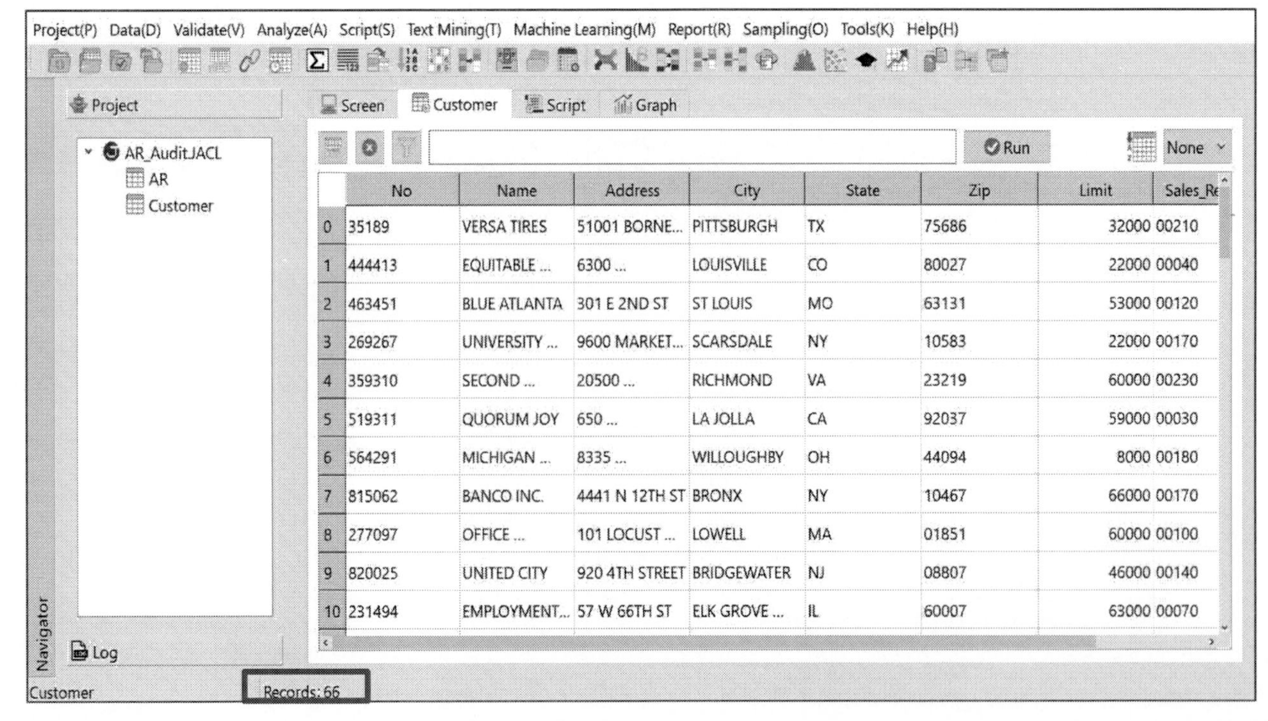

	No	Name	Address	City	State	Zip	Limit	Sales_Re
0	35189	VERSA TIRES	51001 BORNE...	PITTSBURGH	TX	75686	32000	00210
1	444413	EQUITABLE ...	6300 ...	LOUISVILLE	CO	80027	22000	00040
2	463451	BLUE ATLANTA	301 E 2ND ST	ST LOUIS	MO	63131	53000	00120
3	269267	UNIVERSITY ...	9600 MARKET...	SCARSDALE	NY	10583	22000	00170
4	359310	SECOND ...	20500 ...	RICHMOND	VA	23219	60000	00230
5	519311	QUORUM JOY	650 ...	LA JOLLA	CA	92037	59000	00030
6	564291	MICHIGAN ...	8335 ...	WILLOUGHBY	OH	44094	8000	00180
7	815062	BANCO INC.	4441 N 12TH ST	BRONX	NY	10467	66000	00170
8	277097	OFFICE ...	101 LOCUST ...	LOWELL	MA	01851	60000	00100
9	820025	UNITED CITY	920 4TH STREET	BRIDGEWATER	NJ	08807	46000	00140
10	231494	EMPLOYMENT...	57 W 66TH ST	ELK GROVE ...	IL	60007	63000	00070

Exercise 7.18

Please import the Inventory Data according to the file format (Schema) provided below.

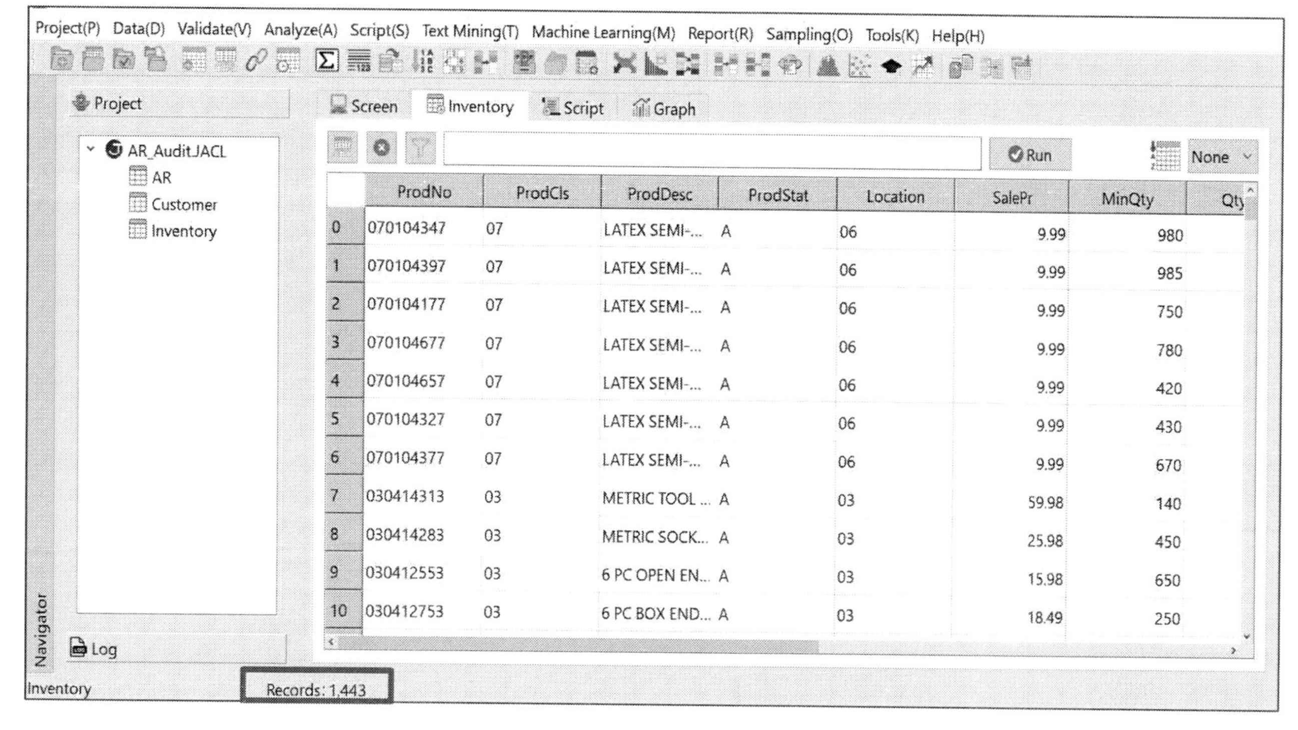

Exercise 7.18

How many sales transactions with a bonus exceeding USD 250 are there, and what is the total bonus amount accumulated?

- Open data set we need (AR)
- Select "Data>Table Layout"

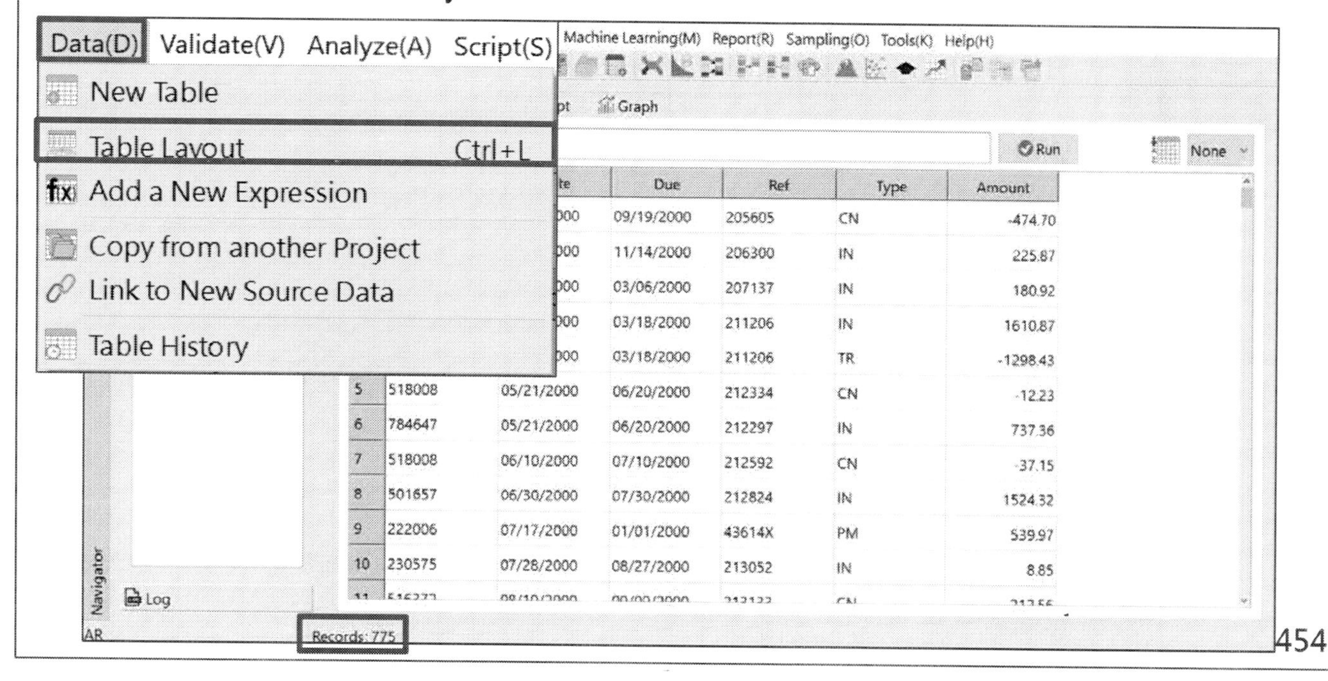

Exercise 7.18

How many sales transactions with a bonus exceeding USD 250 are there, and what is the total bonus amount accumulated?

- Select F(x) to add a computed field

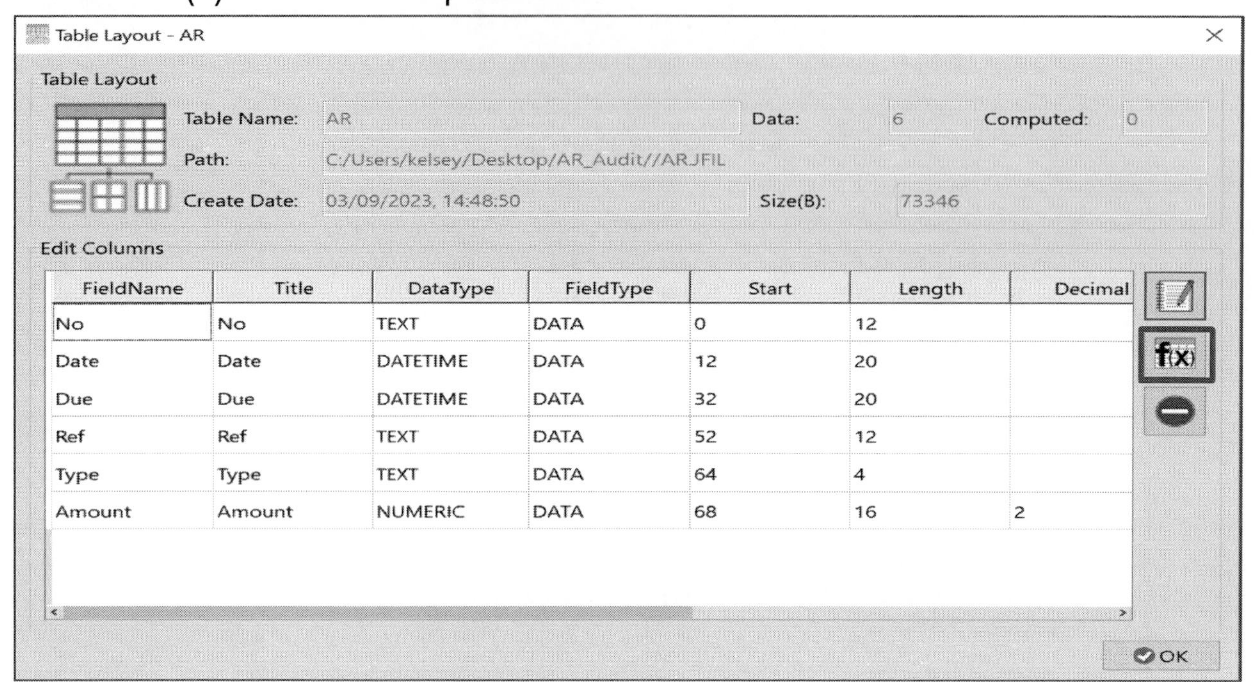

FieldName	Title	DataType	FieldType	Start	Length	Decimal
No	No	TEXT	DATA	0	12	
Date	Date	DATETIME	DATA	12	20	
Due	Due	DATETIME	DATA	32	20	
Ref	Ref	TEXT	DATA	52	12	
Type	Type	TEXT	DATA	64	4	
Amount	Amount	NUMERIC	DATA	68	16	2

Exercise 7.18

How many sales transactions with a bonus exceeding USD 250 are there, and what is the total bonus amount accumulated?

- Add a filter condition to compute the bonus as "Amount * 0.08"

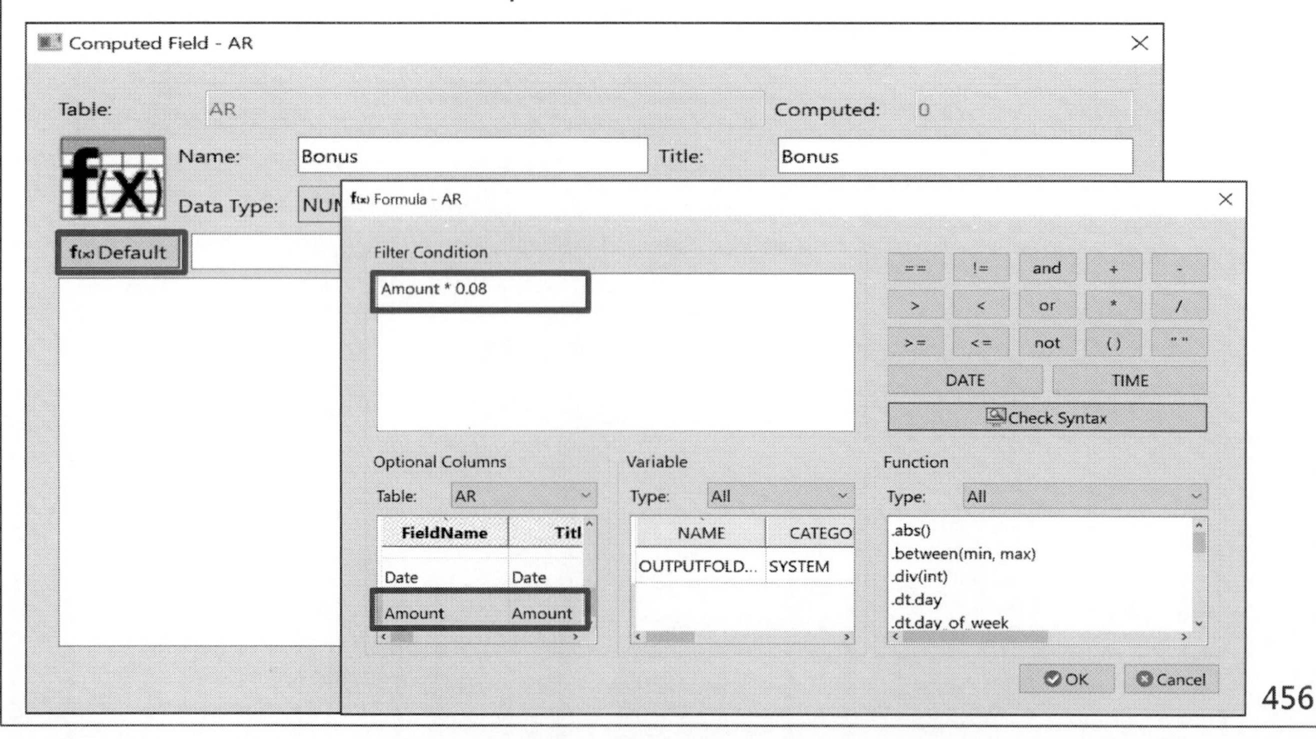

Exercise 7.18

How many sales transactions with a bonus exceeding USD 250 are there, and what is the total bonus amount accumulated?

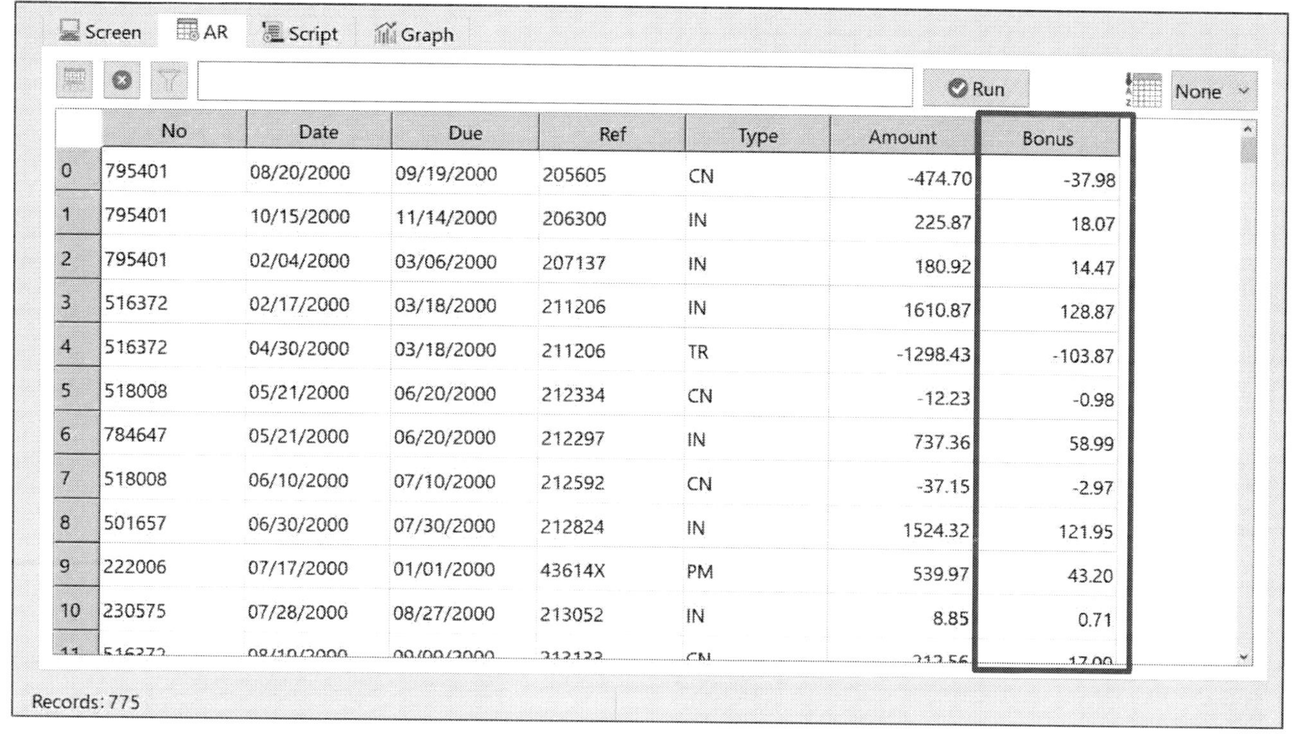

	No	Date	Due	Ref	Type	Amount	Bonus
0	795401	08/20/2000	09/19/2000	205605	CN	-474.70	-37.98
1	795401	10/15/2000	11/14/2000	206300	IN	225.87	18.07
2	795401	02/04/2000	03/06/2000	207137	IN	180.92	14.47
3	516372	02/17/2000	03/18/2000	211206	IN	1610.87	128.87
4	516372	04/30/2000	03/18/2000	211206	TR	-1298.43	-103.87
5	518008	05/21/2000	06/20/2000	212334	CN	-12.23	-0.98
6	784647	05/21/2000	06/20/2000	212297	IN	737.36	58.99
7	518008	06/10/2000	07/10/2000	212592	CN	-37.15	-2.97
8	501657	06/30/2000	07/30/2000	212824	IN	1524.32	121.95
9	222006	07/17/2000	01/01/2000	43614X	PM	539.97	43.20
10	230575	07/28/2000	08/27/2000	213052	IN	8.85	0.71

JCAATs-AI Audit Software
Copyright © 2023 JACKSOFT.

Exercise 7.18

How many sales transactions with a bonus exceeding USD 250 are there, and what is the total bonus amount accumulated?

- Select "Analyze>Statistics" to obtain the statistics for the bonus
- Add a Filter condition: Bonus > 250

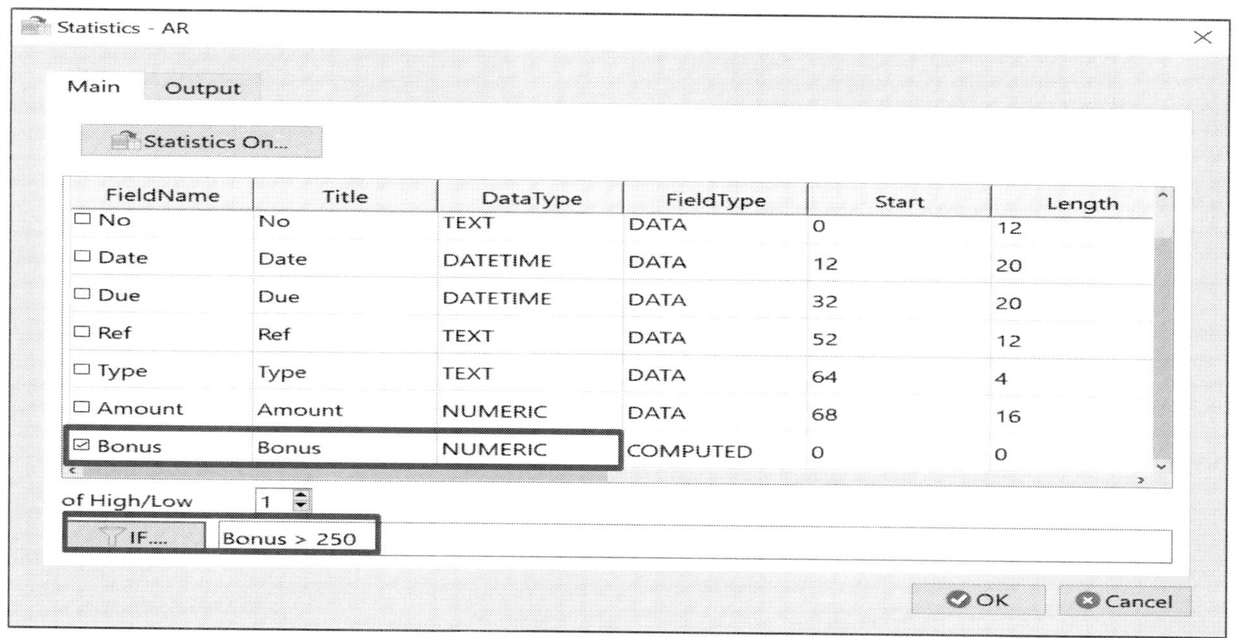

FieldName	Title	DataType	FieldType	Start	Length
☐ No	No	TEXT	DATA	0	12
☐ Date	Date	DATETIME	DATA	12	20
☐ Due	Due	DATETIME	DATA	32	20
☐ Ref	Ref	TEXT	DATA	52	12
☐ Type	Type	TEXT	DATA	64	4
☐ Amount	Amount	NUMERIC	DATA	68	16
☑ Bonus	Bonus	NUMERIC	COMPUTED	0	0

Exercise 7.18

How many sales transactions with a bonus exceeding USD 250 are there, and what is the total bonus amount accumulated?

JCAATs >>AR.STATISTICS(PKEYS=["Bonus"], MAX=1, IF = ["Bonus > 250"], TO="")
Table : AR
Note: 2023/03/09 15:16:03
Result - Records : 13

Table_Name	Field_Name	Data_Type	Factor	Value
AR	Bonus	NUMERIC	Count	15.00
AR	Bonus	NUMERIC	Total	7,294.36
AR	Bonus	NUMERIC	Abs. Value	7,294.36
AR	Bonus	NUMERIC	Minimum	252.20
AR	Bonus	NUMERIC	Maximum	1,707.52
AR	Bonus	NUMERIC	Range	1,455.32
AR	Bonus	NUMERIC	Positive Count	15.00
AR	Bonus	NUMERIC	Positive Total	7,294.36
AR	Bonus	NUMERIC	Negative Count	0.00

Exercise 7.18

What is the total number of bonus transactions and the accumulated bonus amount issued throughout the year?

- Select "Analyze>Statistics" to obtain the statistics for the bonus
- Add a Filter condition: Date.between(date(2000-01-01), date(2000-03-31))

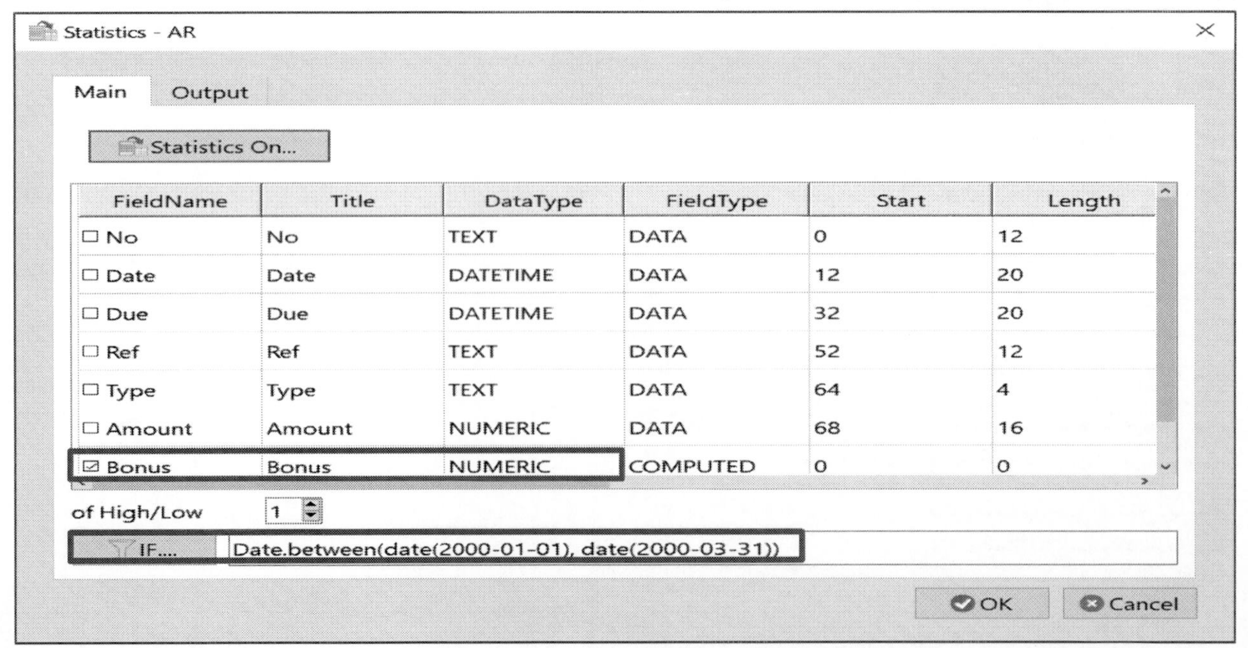

Exercise 7.18
What is the total number of bonus transactions and the accumulated bonus amount issued throughout the year?

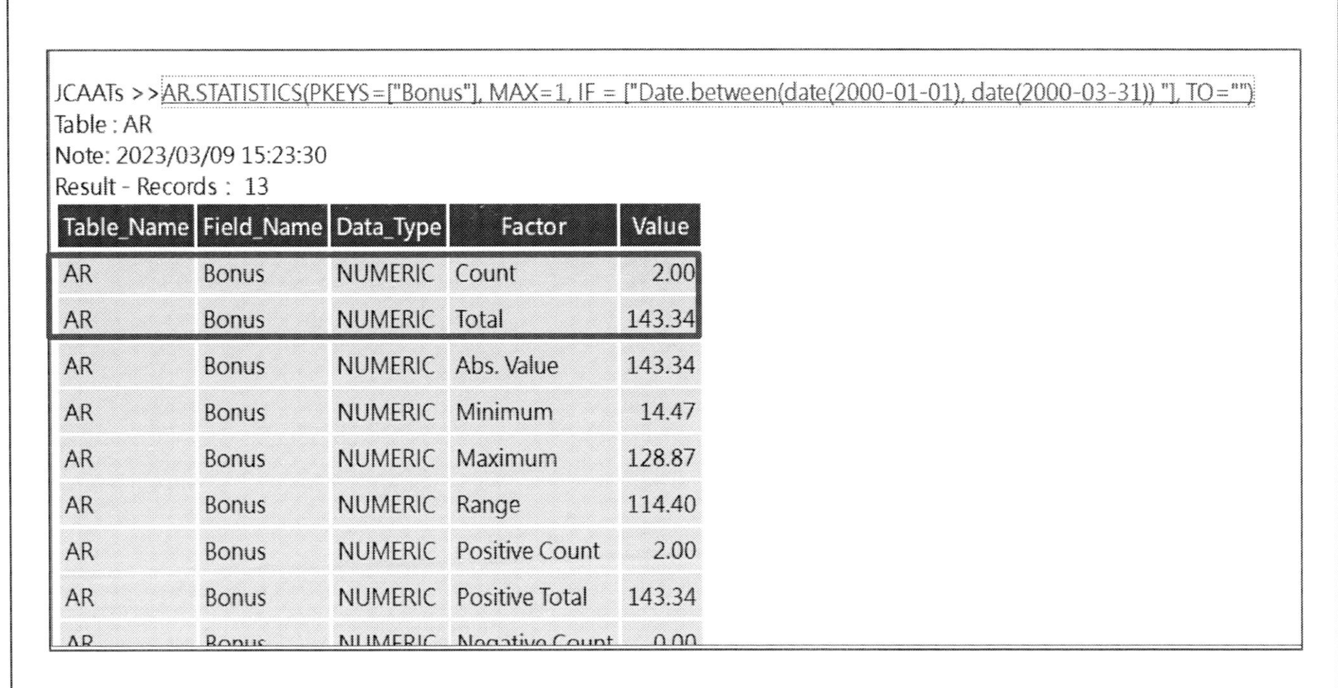
JCAATs >>AR.STATISTICS(PKEYS=["Bonus"], MAX=1, IF = ["Date.between(date(2000-01-01), date(2000-03-31)) "], TO="")
Table : AR
Note: 2023/03/09 15:23:30
Result - Records : 13

Table_Name	Field_Name	Data_Type	Factor	Value
AR	Bonus	NUMERIC	Count	2.00
AR	Bonus	NUMERIC	Total	143.34
AR	Bonus	NUMERIC	Abs. Value	143.34
AR	Bonus	NUMERIC	Minimum	14.47
AR	Bonus	NUMERIC	Maximum	128.87
AR	Bonus	NUMERIC	Range	114.40
AR	Bonus	NUMERIC	Positive Count	2.00
AR	Bonus	NUMERIC	Positive Total	143.34
AR	Bonus	NUMERIC	Negative Count	0.00

Exercise 7.18
What is the total number of bonus transactions and the accumulated bonus amount issued throughout the year?

- Select "Analyze>Statistics" to obtain the statistics for the bonus
- Add a Filter condition: Date.between(date(2000-04-01), date(2000-06-30))

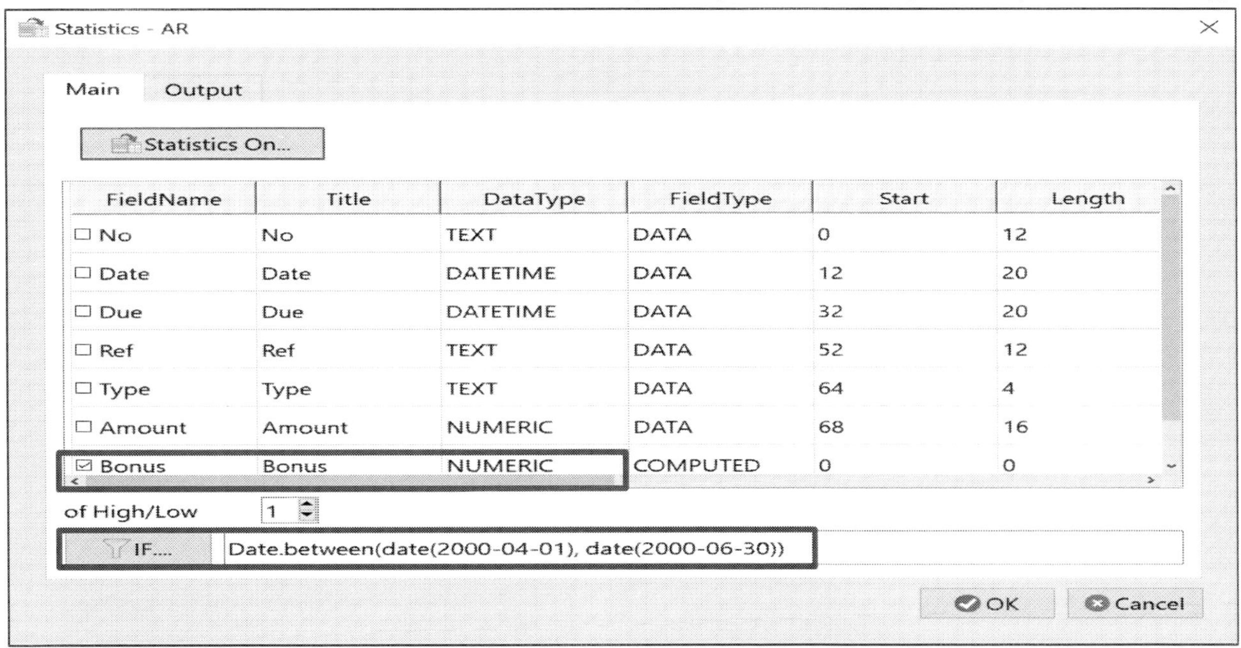

FieldName	Title	DataType	FieldType	Start	Length
☐ No	No	TEXT	DATA	0	12
☐ Date	Date	DATETIME	DATA	12	20
☐ Due	Due	DATETIME	DATA	32	20
☐ Ref	Ref	TEXT	DATA	52	12
☐ Type	Type	TEXT	DATA	64	4
☐ Amount	Amount	NUMERIC	DATA	68	16
☑ Bonus	Bonus	NUMERIC	COMPUTED	0	0

Exercise 7.18
What is the total number of bonus transactions and the accumulated bonus amount issued throughout the year?

JCAATs >>AR.STATISTICS(PKEYS=["Bonus"], MAX=1, IF = ["Date.between(date(2000-04-01), date(2000-06-30)) "], TO="")
Table : AR
Note: 2023/03/09 15:22:58
Result - Records : 13

Table_Name	Field_Name	Data_Type	Factor	Value
AR	Bonus	NUMERIC	Count	5.00
AR	Bonus	NUMERIC	Total	73.11
AR	Bonus	NUMERIC	Abs. Value	288.76
AR	Bonus	NUMERIC	Minimum	-103.87
AR	Bonus	NUMERIC	Maximum	121.95
AR	Bonus	NUMERIC	Range	225.82
AR	Bonus	NUMERIC	Positive Count	2.00
AR	Bonus	NUMERIC	Positive Total	180.93
AR	Bonus	NUMERIC	Negative Count	3.00

Exercise 7.18
What is the total number of bonus transactions and the accumulated bonus amount issued throughout the year?

- Select "Analyze>Statistics" to obtain the statistics for the bonus
- Add a Filter condition: Date.between(date(2000-07-01), date(2000-09-30))

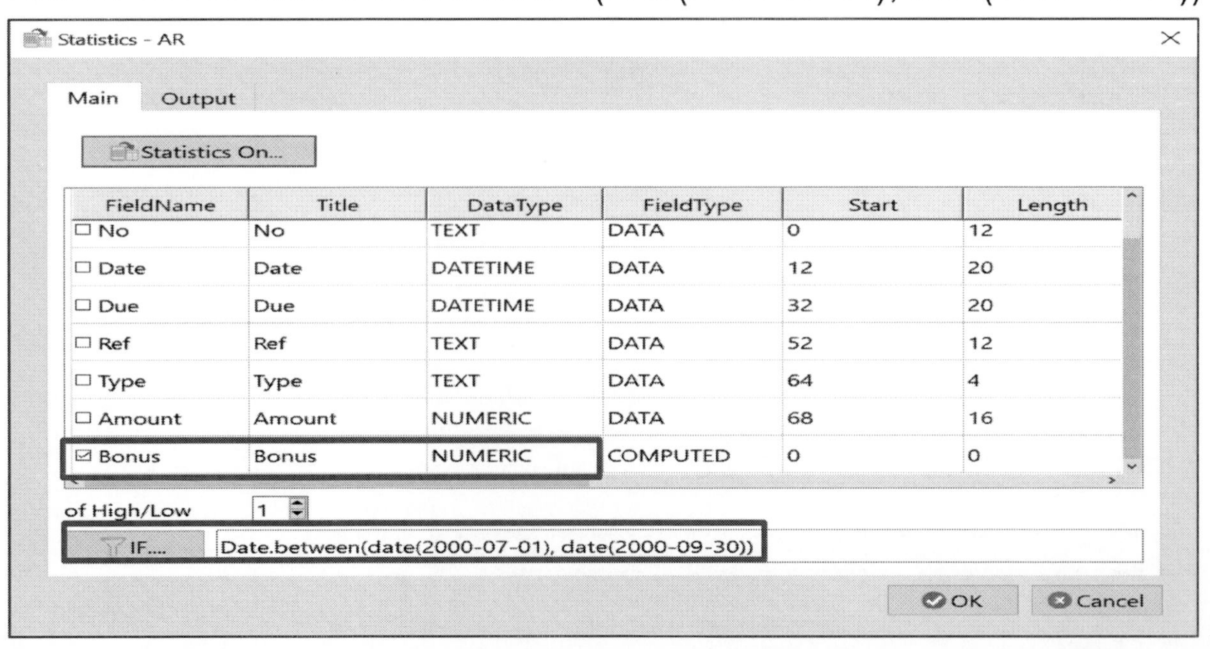

Exercise 7.18
What is the total number of bonus transactions and the accumulated bonus amount issued throughout the year?

JCAATs >>AR.STATISTICS(PKEYS=["Bonus"], MAX=1, IF = ["Date.between(date(2000-07-01), date(2000-09-3
Table : AR
Note: 2023/03/09 15:24:33
Result - Records : 13

Table_Name	Field_Name	Data_Type	Factor	Value
AR	Bonus	NUMERIC	Count	131.00
AR	Bonus	NUMERIC	Total	10,884.30
AR	Bonus	NUMERIC	Abs. Value	11,330.11
AR	Bonus	NUMERIC	Minimum	-42.69
AR	Bonus	NUMERIC	Maximum	443.94
AR	Bonus	NUMERIC	Range	486.63
AR	Bonus	NUMERIC	Positive Count	108.00
AR	Bonus	NUMERIC	Positive Total	11,107.21
AR	Bonus	NUMERIC	Negative Count	23.00

Exercise 7.18
What is the total number of bonus transactions and the accumulated bonus amount issued throughout the year?

- Select "Analyze>Statistics" to obtain the statistics for the bonus
- Add a Filter condition: Date.between(date(2000-10-01), date(2000-12-31))

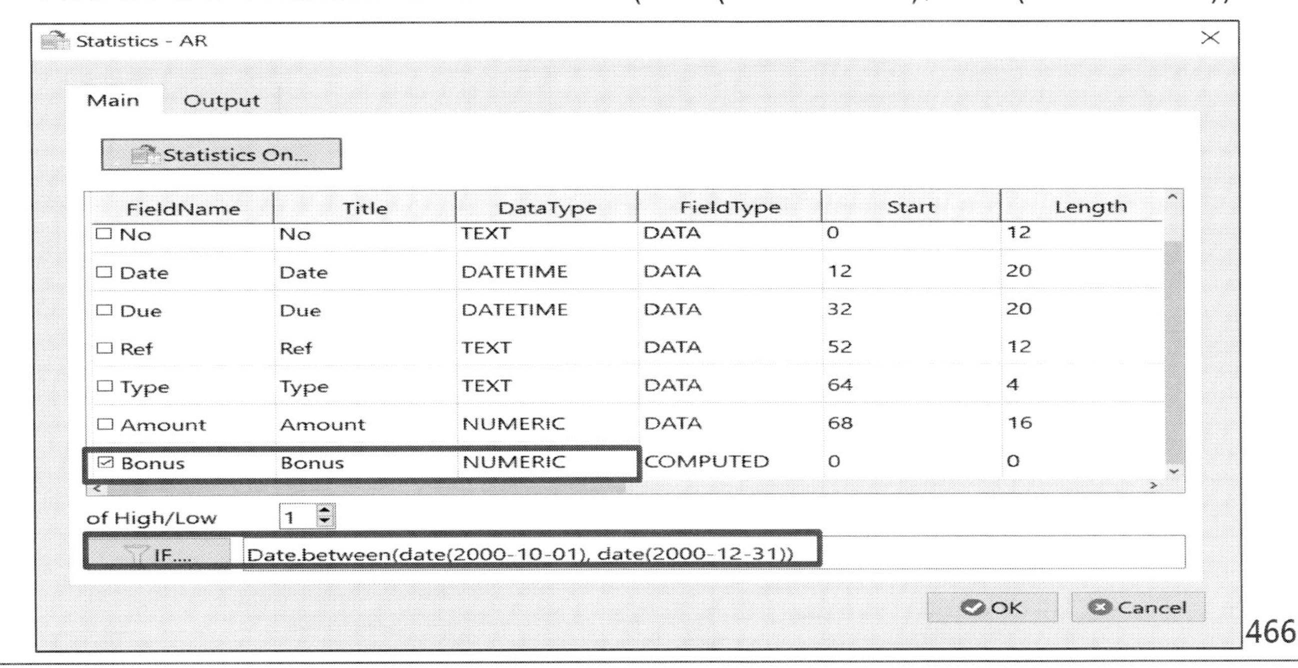

Exercise 7.18

What is the total number of bonus transactions and the accumulated bonus amount issued throughout the year?

JCAATs >>AR.STATISTICS(PKEYS=["Bonus"], MAX=1, IF = ["Date.between(date(2000-10-01), date(2000-12-31)) "], TO="")
Table : AR
Note: 2023/03/09 15:25:29
Result - Records : 13

Table_Name	Field_Name	Data_Type	Factor	Value
AR	Bonus	NUMERIC	Count	637.00
AR	Bonus	NUMERIC	Total	30,013.43
AR	Bonus	NUMERIC	Abs. Value	38,695.47
AR	Bonus	NUMERIC	Minimum	-286.64
AR	Bonus	NUMERIC	Maximum	1,707.52
AR	Bonus	NUMERIC	Range	1,994.16
AR	Bonus	NUMERIC	Positive Count	500.00
AR	Bonus	NUMERIC	Positive Total	34,354.45
AR	Bonus	NUMERIC	Negative Count	135.00

Exercise 7.18

What is the amount of the inventory write-down provision that should be made for this period?

- Select F(x) to add a computed field
- Add a formula "Value - MktVal" to compute the Loss

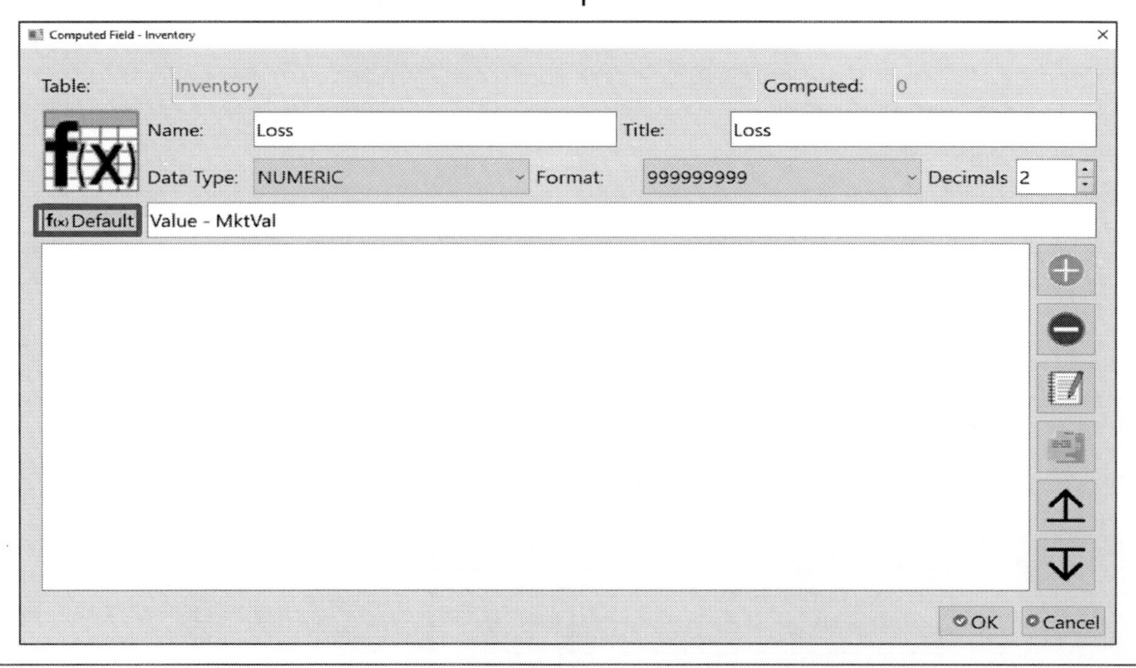

Exercise 7.18

What is the amount of the inventory write-down provision that should be made for this period?

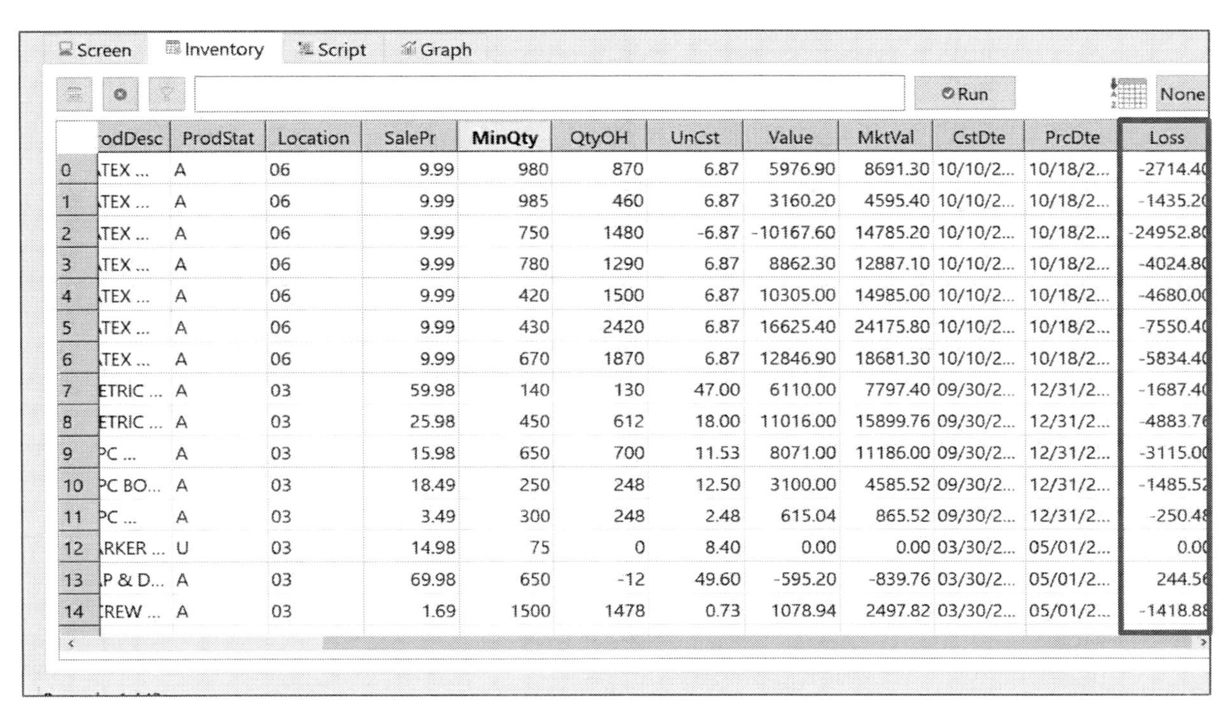

	odDesc	ProdStat	Location	SalePr	MinQty	QtyOH	UnCst	Value	MktVal	CstDte	PrcDte	Loss
0	TEX ...	A	06	9.99	980	870	6.87	5976.90	8691.30	10/10/2...	10/18/2...	-2714.40
1	TEX ...	A	06	9.99	985	460	6.87	3160.20	4595.40	10/10/2...	10/18/2...	-1435.20
2	TEX ...	A	06	9.99	750	1480	-6.87	-10167.60	14785.20	10/10/2...	10/18/2...	-24952.80
3	TEX ...	A	06	9.99	780	1290	6.87	8862.30	12887.10	10/10/2...	10/18/2...	-4024.80
4	TEX ...	A	06	9.99	420	1500	6.87	10305.00	14985.00	10/10/2...	10/18/2...	-4680.00
5	TEX ...	A	06	9.99	430	2420	6.87	16625.40	24175.80	10/10/2...	10/18/2...	-7550.40
6	TEX ...	A	06	9.99	670	1870	6.87	12846.90	18681.30	10/10/2...	10/18/2...	-5834.40
7	ETRIC ...	A	03	59.98	140	130	47.00	6110.00	7797.40	09/30/2...	12/31/2...	-1687.40
8	ETRIC ...	A	03	25.98	450	612	18.00	11016.00	15899.76	09/30/2...	12/31/2...	-4883.76
9	PC ...	A	03	15.98	650	700	11.53	8071.00	11186.00	09/30/2...	12/31/2...	-3115.00
10	PC BO...	A	03	18.49	250	248	12.50	3100.00	4585.52	09/30/2...	12/31/2...	-1485.52
11	PC ...	A	03	3.49	300	248	2.48	615.04	865.52	09/30/2...	12/31/2...	-250.48
12	RKER ...	U	03	14.98	75	0	8.40	0.00	0.00	03/30/2...	05/01/2...	0.00
13	P & D...	A	03	69.98	650	-12	49.60	-595.20	-839.76	03/30/2...	05/01/2...	244.56
14	REW ...	A	03	1.69	1500	1478	0.73	1078.94	2497.82	03/30/2...	05/01/2...	-1418.88

Exercise 7.18

What is the amount of the inventory write-down provision that should be made for this period?

- Select Validate>Total to calculate the total loss.
- Add a filter condition: Loss > 0

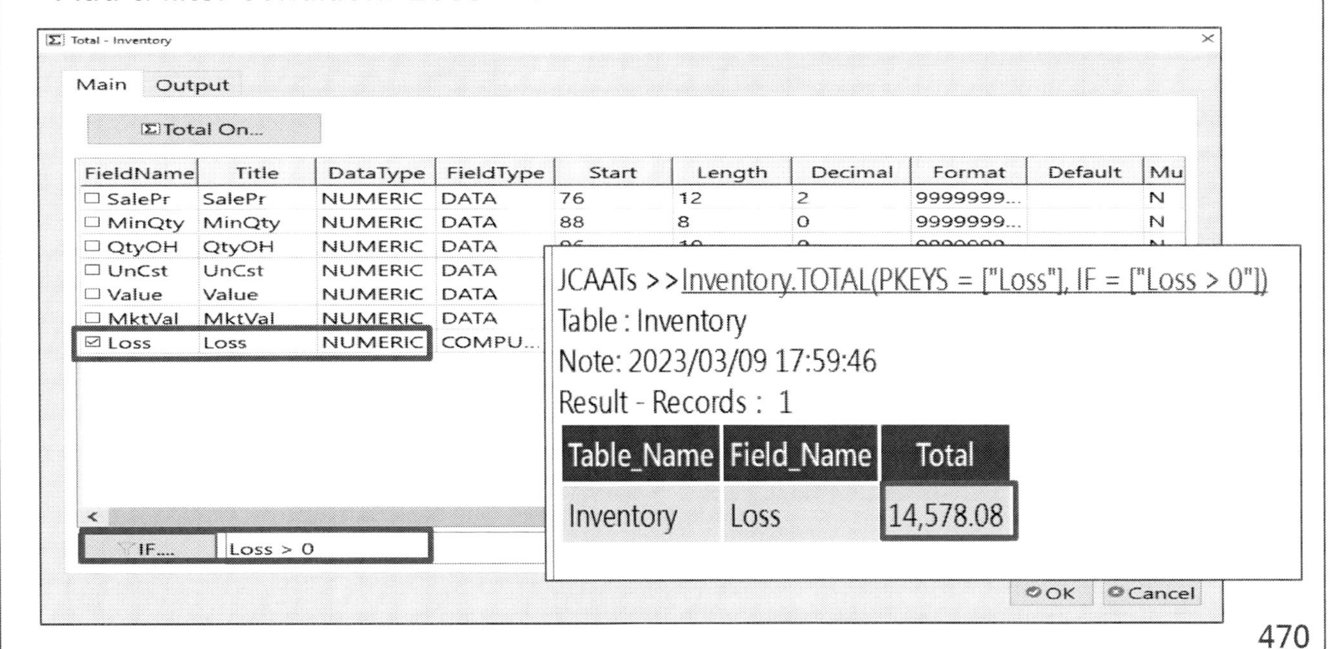

Exercise 7.18

Please calculate the inventory information for each category in the right table according to the following inventory classification policy.

- Select F(x) to add a computed field
- Add a formula "High", click on "Add" button to Apply filter condition.

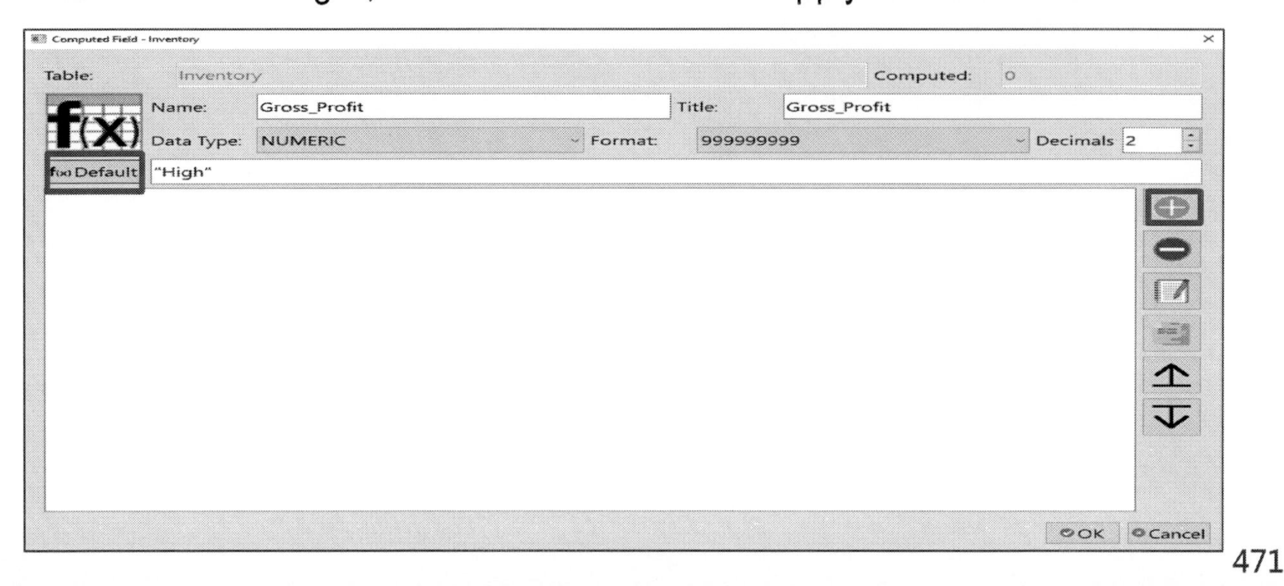

JCAATs-AI Audit Software

Exercise 7.18

Please calculate the inventory information for each category in the right table according to the following inventory classification policy.

- Open the Filter Editor and set the filter condition as "(SalePr - UnCst) / SalePr < 0.20".
- Open the Formula Editor and define it as "Low".

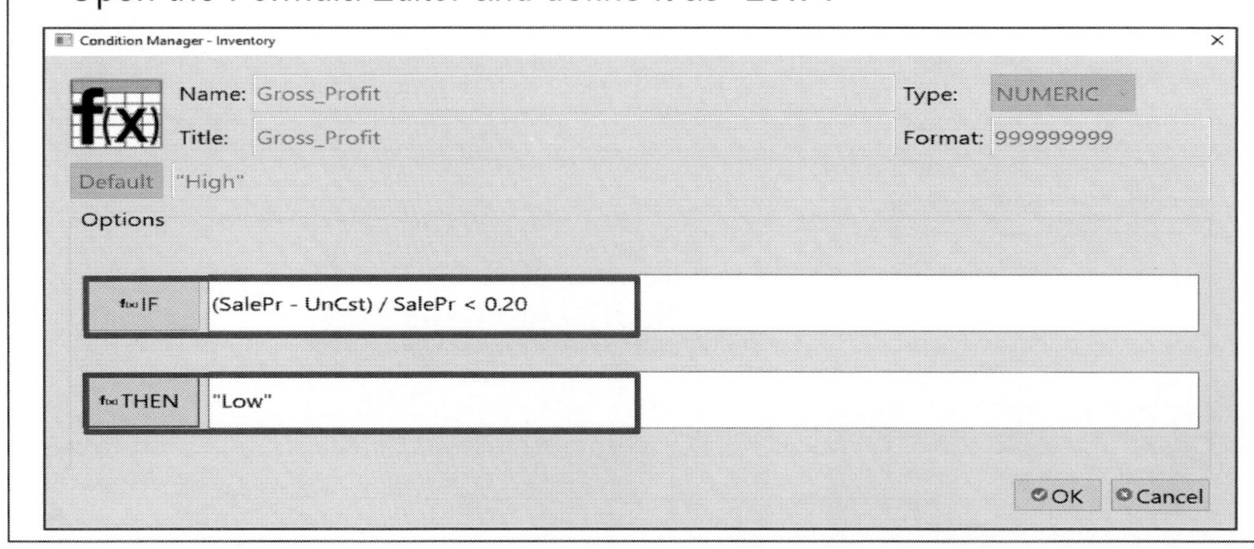

Exercise 7.18

Please calculate the inventory information for each category in the right table according to the following inventory classification policy.

- Select "Copy" and edit the copied condition.
- Modify the filter to "0.20 <= (SalePr - UnCst) / SalePr <= 0.50" and the formula to "Mid"

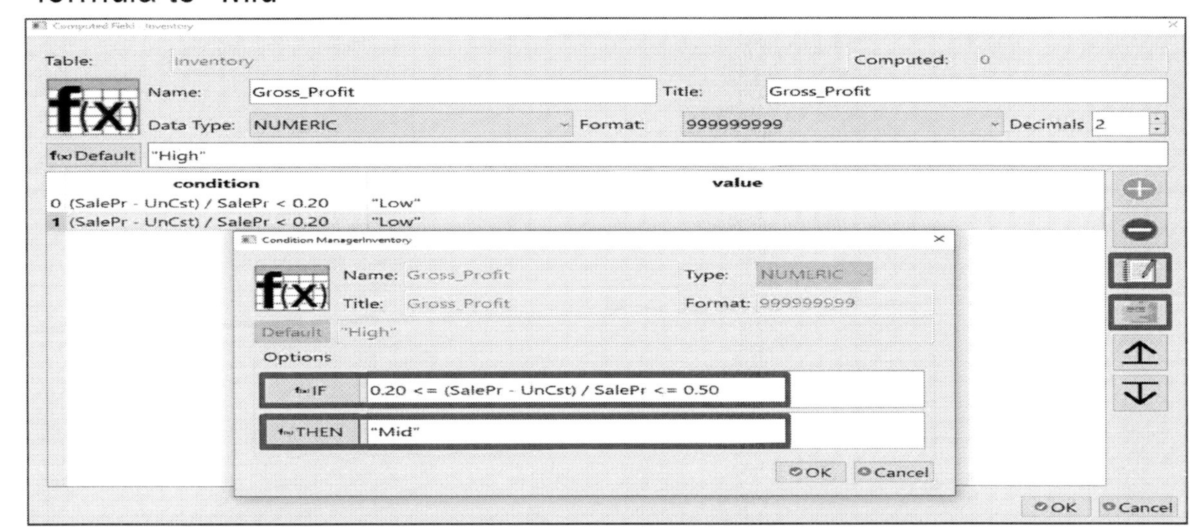

Exercise 7.18

Please calculate the inventory information for each category in the right table according to the following inventory classification policy.

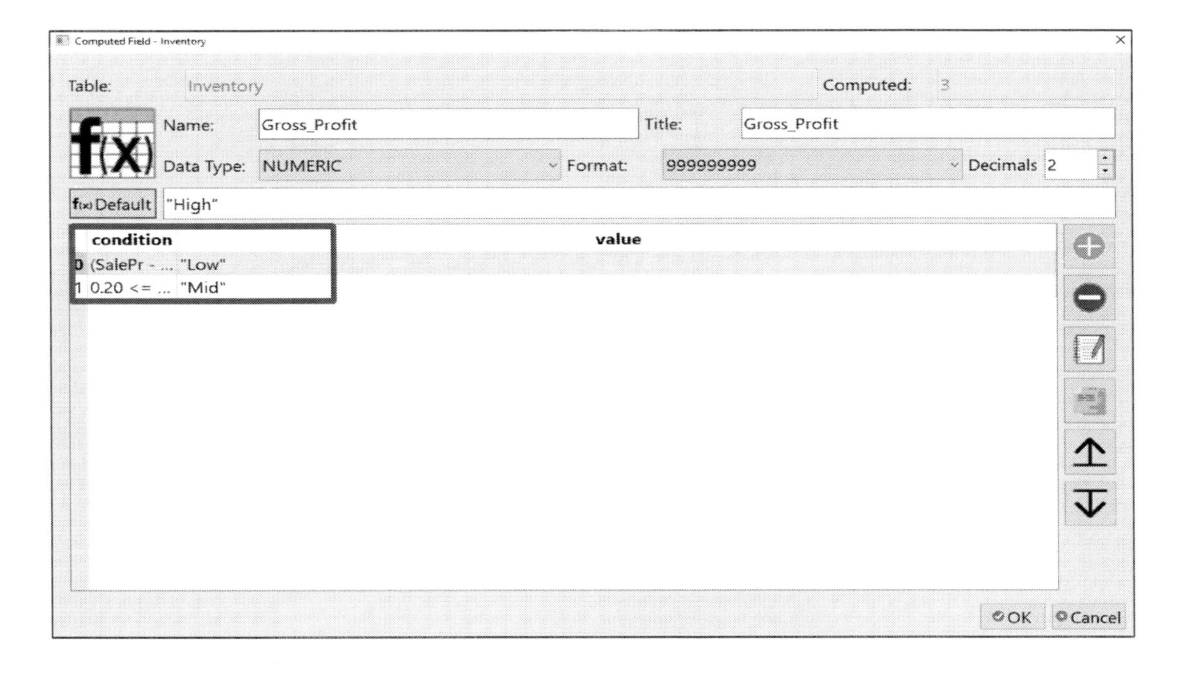

Exercise 7.18

Please calculate the inventory information for each category in the right table according to the following inventory classification policy.

- Select "Analyze>Classify" to classify the gross profit
- Subtotal the "QtyOH", "Value" and "MktVal" fields

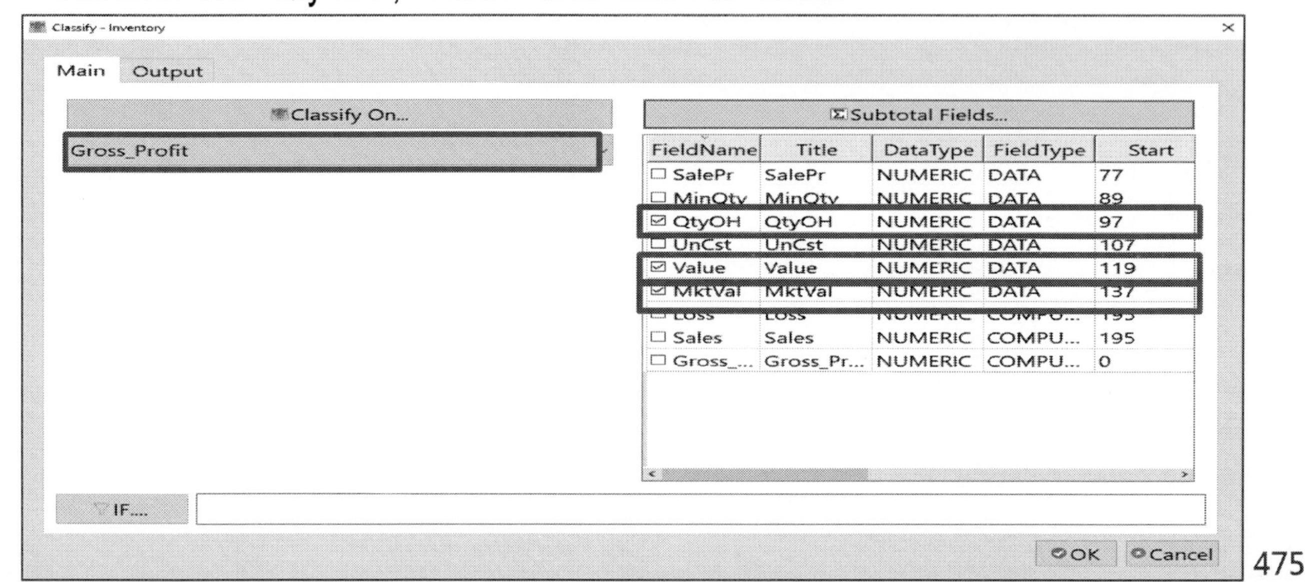

JCAATs-AI Audit Software

Exercise 7.18

Please calculate the inventory information for each category in the right table according to the following inventory classification policy.

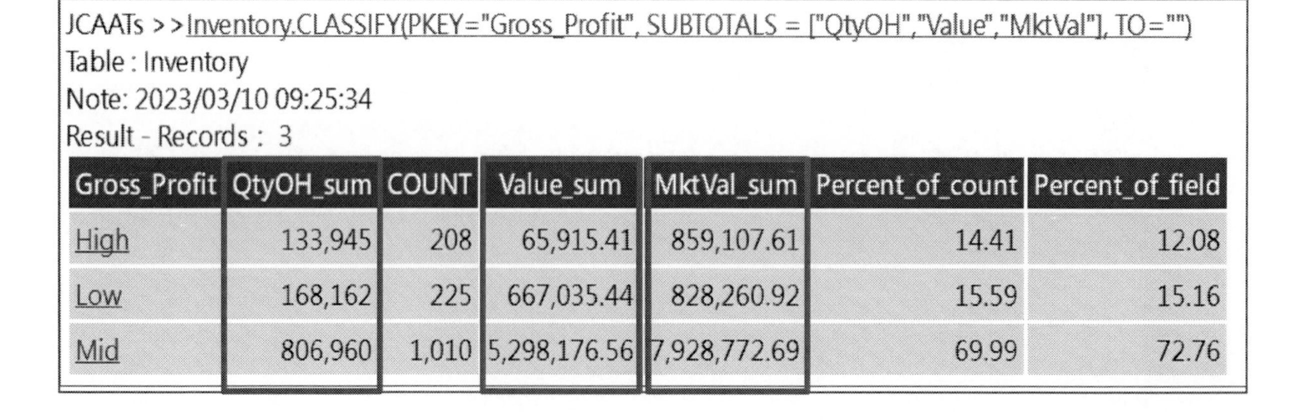

JCAATs >>Inventory.CLASSIFY(PKEY="Gross_Profit", SUBTOTALS = ["QtyOH","Value","MktVal"], TO="")
Table : Inventory
Note: 2023/03/10 09:25:34
Result - Records : 3

Gross_Profit	QtyOH_sum	COUNT	Value_sum	MktVal_sum	Percent_of_count	Percent_of_field
High	133,945	208	65,915.41	859,107.61	14.41	12.08
Low	168,162	225	667,035.44	828,260.92	15.59	15.16
Mid	806,960	1,010	5,298,176.56	7,928,772.69	69.99	72.76

Exercise 7.18

Please use the Accounts Receivable Detail File (AR) and the Customer data table to perform the check. Is there any case of ghost customers?

- Select "AR" as the primary table, "Customer" as secondary, choose "No" as the key

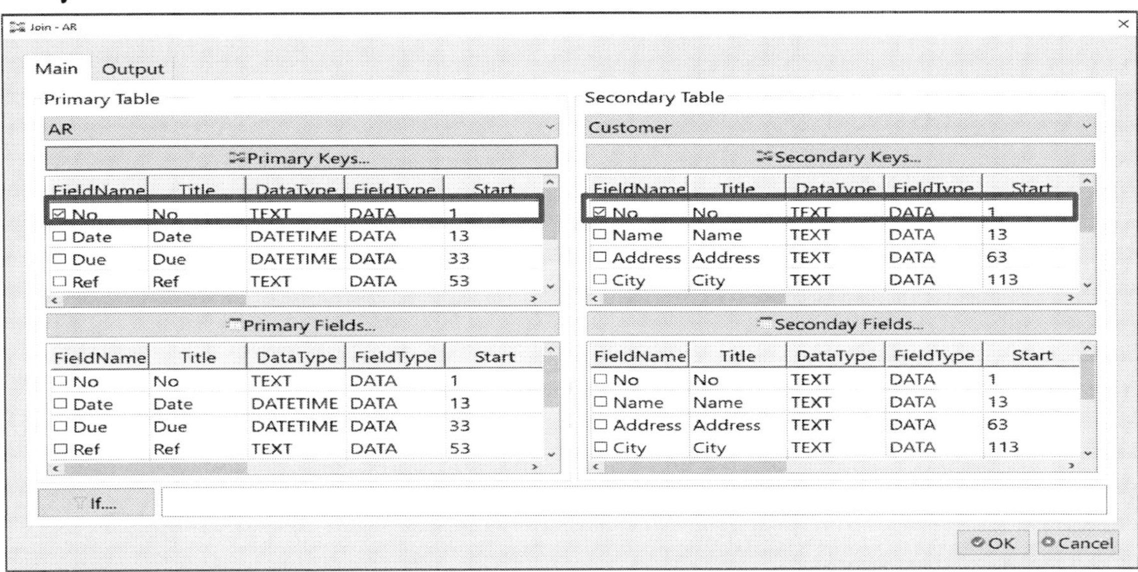

Exercise 7.18

Please use the Accounts Receivable Detail File (AR) and the Customer data table to perform the check. Is there any case of ghost customers?

- Select Join_Types as Unmatch Primary

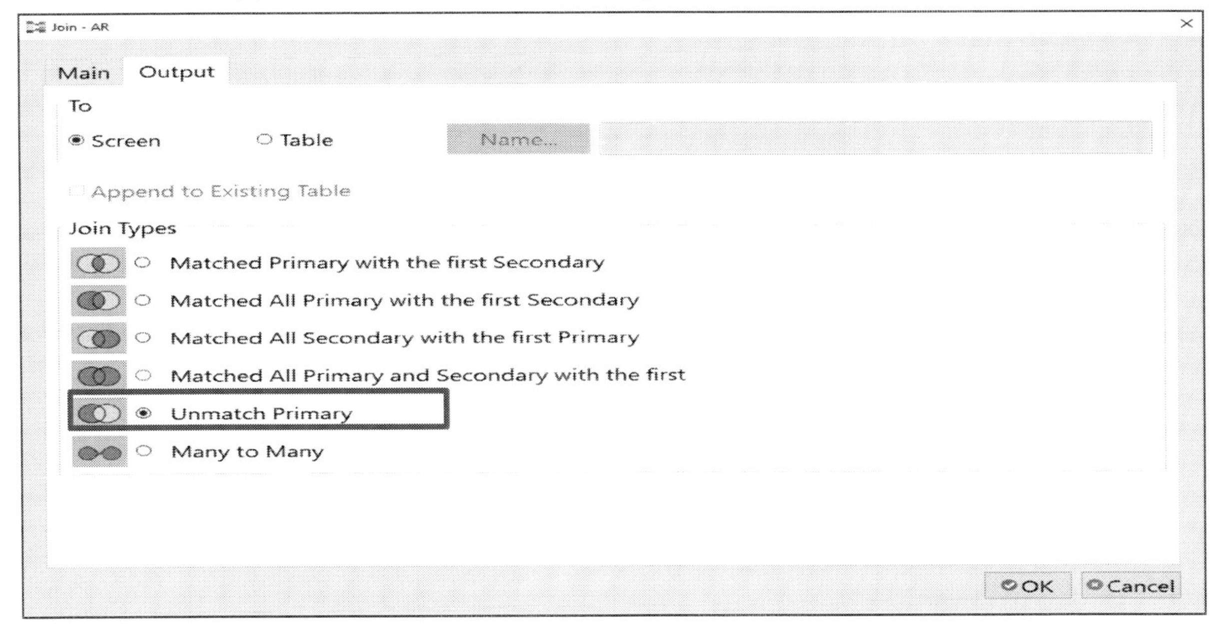

Exercise 7.18

Please use the Accounts Receivable Detail File (AR) and the Customer data table to perform the check. Is there any case of ghost customers?

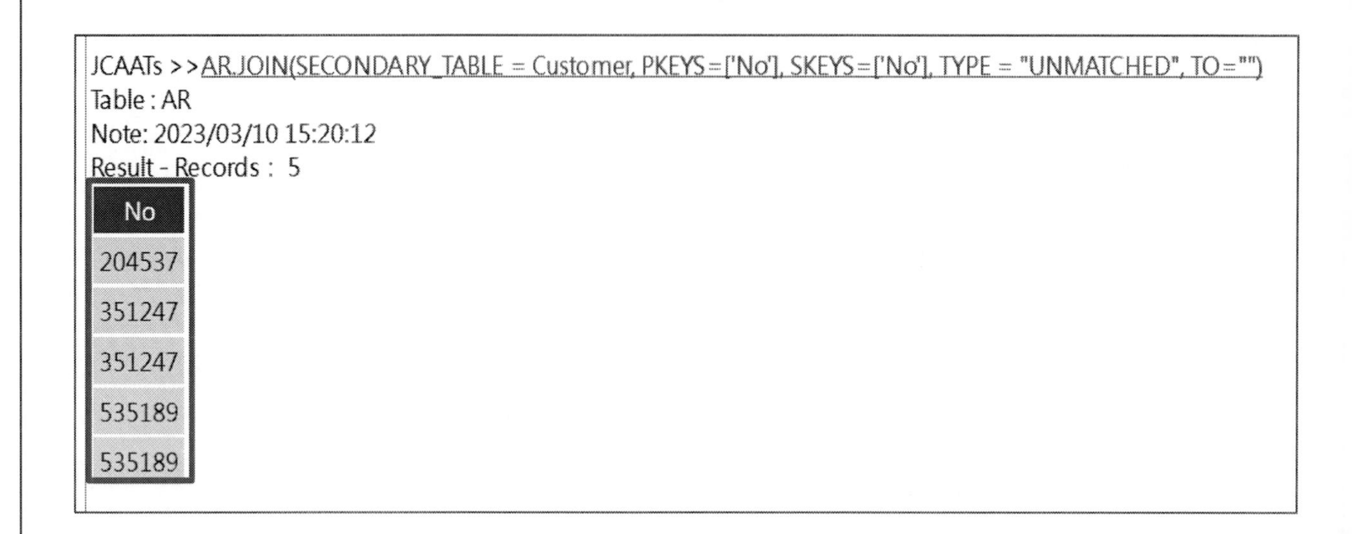

Exercise 7.18

Are there any customers who have not conducted transactions in the current period? Please list their names.

- Select "Customer" as the primary table, “AR" as secondary, choose “No" as the key, list out the “Name" in the primary field

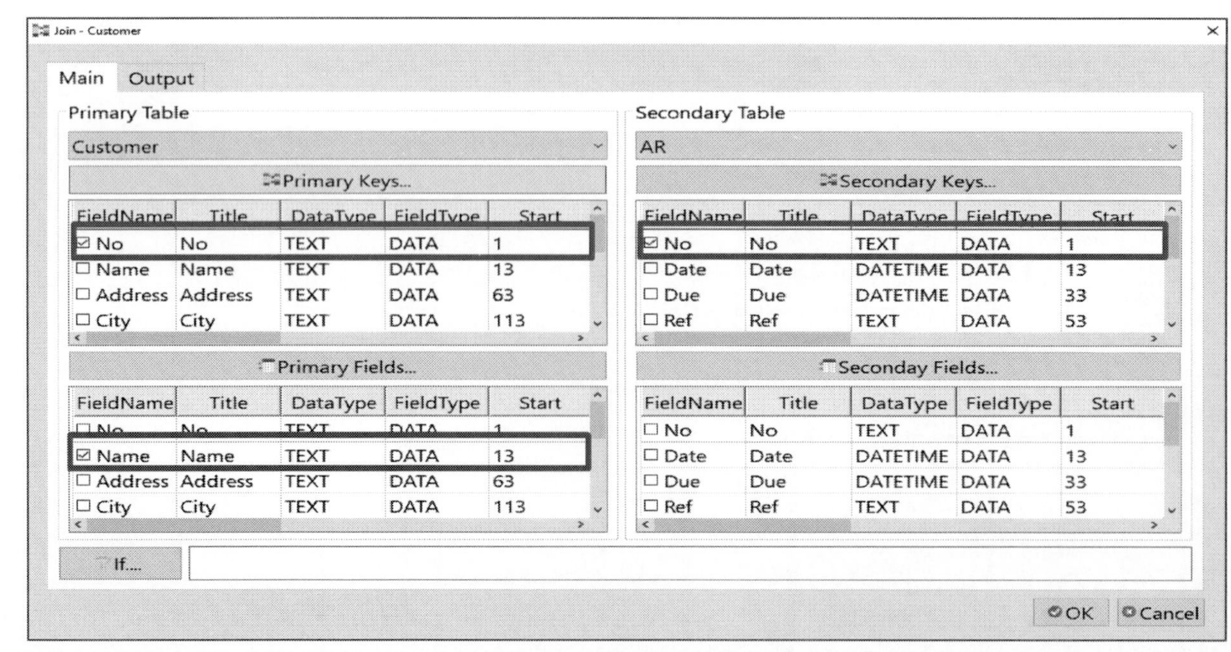

Exercise 7.18
Are there any customers who have not conducted transactions in the current period? Please list their names.

- Select Join_Types as Unmatch Primary

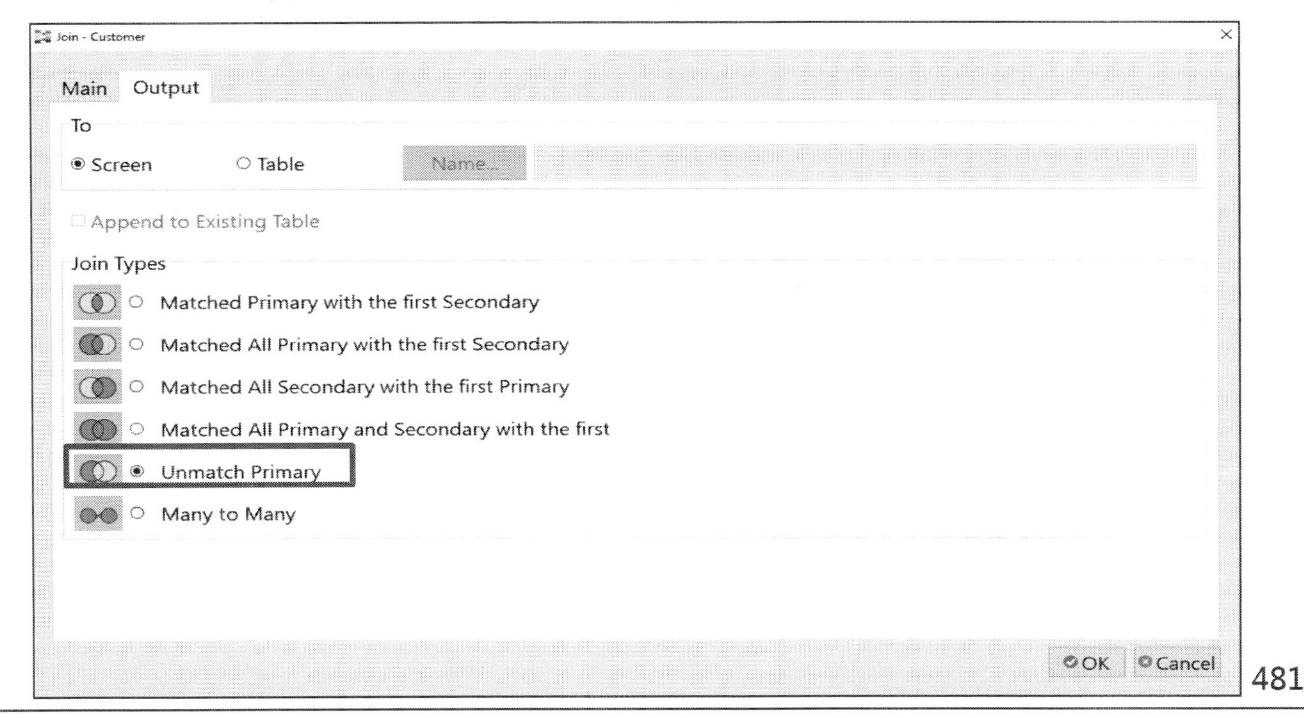

Exercise 7.18
Are there any customers who have not conducted transactions in the current period? Please list their names.

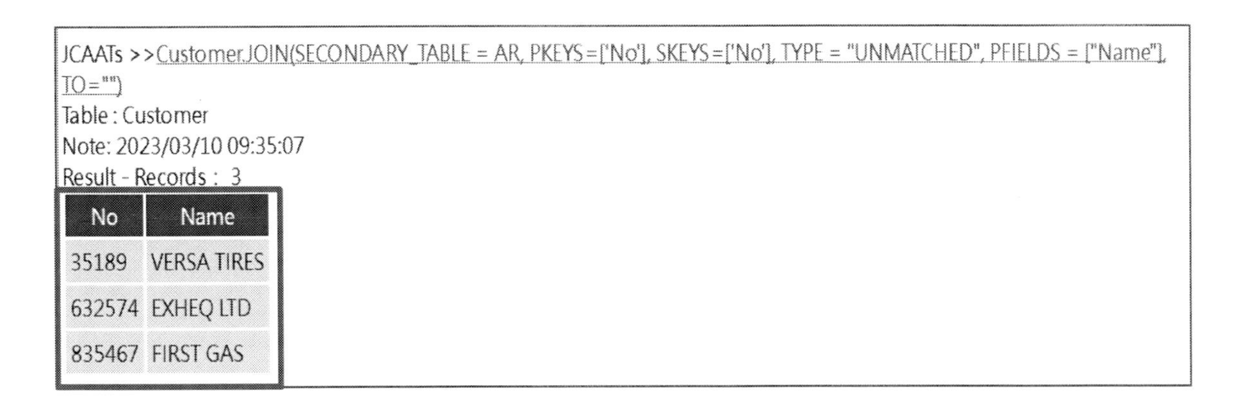
JCAATs >>Customer.JOIN(SECONDARY_TABLE = AR, PKEYS=['No'], SKEYS=['No'], TYPE = "UNMATCHED", PFIELDS = ["Name"], TO="")
Table : Customer
Note: 2023/03/10 09:35:07
Result - Records : 3

No	Name
35189	VERSA TIRES
632574	EXHEQ LTD
835467	FIRST GAS

Exercise 7.19: A Case Study

- To comply with the latest international anti-money laundering regulations and strengthen the audit process, it is necessary to obtain the customer data file (CUSTOMER) and loan data file (LOANS), and download the latest blacklist of sanctioned entities from the Office of Foreign Asset Control (OFAC) of the United States Treasury Department, in order to begin the audit process.

Exercise 7.19: A Case Study

- Please create a project file, and import relevant data tables according to the following data format.

1. Customer Data (Customer.txt):

Field Name	Type	Decimal	Length	Note
CustNum	C	0	7	Customer Number
CustName	C	0	25	Customer Name
Address	C	0	25	Address
City	C	0	15	City
State	C	0	5	State
Zip	C	0	5	Zip Code
CustLimit	N	2	22	Credit Limit

Exercise 7.19: A Case Study

2. Loan data (LOAN.xls):

Field Name	Type	Decimal	Length	Note
CustNum	C	0	7	Customer Number
LoanNumber	C	0	10	Loan Number
Date	D	0	10	Loan Date
Amount	N	2	9	Loan Amount
Limit	N	0	7	Credit Limit

Exercise 7.19: A Case Study

3. Sdn List (Sdn.xml):

Field Name	Type	Decimal	Length	Note
uid	N	0	5	Uid
lastName	C	0	113	Last Name
sdnType	C	0	10	Sdn Type
firstName	C	0	52	First Name
title	C	0	198	Title
remarks	C	0	1060	Remarks

Exercise 7.19: A Case Study

1. Verify data
   i. Verify whether there are errors in the customer data file.
      □ No, the data is correct.
      □ Yes, there is an error in the row and column and the reason: ________________.
   ii. Verify whether there are errors in the loan data file.
      □ No, the data is correct.
      □ Yes, there is an error in the row and column and the reason: ________________.
   iii. Verify whether there are errors in the sdn data file.
      □ No, the data is correct.
      □ Yes, there is an error in the row and column and the reason: ________________.

JCAATs-AI Audit Software
Copyright © 2023 JACKSOFT.

Exercise 7.19: A Case Study

2. Validate the data
   1) The record number in customer data is _______ .
   2) The record number in sdn data is _______ .
   3) The record number in loan data is _________ .
   4) The total amount of loans is _________ .
3. Validate the data
   1) Please check whether there are any duplicated customer numbers.
      □ No, there is no duplicated information.
      □ Yes, the duplicated information is:
   2) Please check whether there are any duplicated Loan numbers.
      □ No, there is no duplicated information.
      □ Yes, the duplicated information is:

Exercise 7.19: A Case Study

4. Please check if there are any anomalies where loan customers are not recorded in the customer file.

   □ No, there are no instances where loan customers are not recorded in the customer file

   □ Yes, please provide the CUSTNUM, LoanNumber, and Amount fields for customers who are not recorded in the file.

5. Please check whether the customer file contains the OFAC regulatory blacklist (SDN)

   □ No, there are no customer that are on the SDN list.

   □ Yes, please provide the CUSTNUM, and CustName

Exercise 7.19: A Case Study

6. Please check whether any loans have been granted that exceed the customer's total credit limit.

   □ No, there are no loans that have been granted that exceed the customer's total credit limit.

   □ Yes, please provide the CUSTNUM, and CustName

Exercise 7.19

Please import the customer data according to the file format (Schema) provided below.

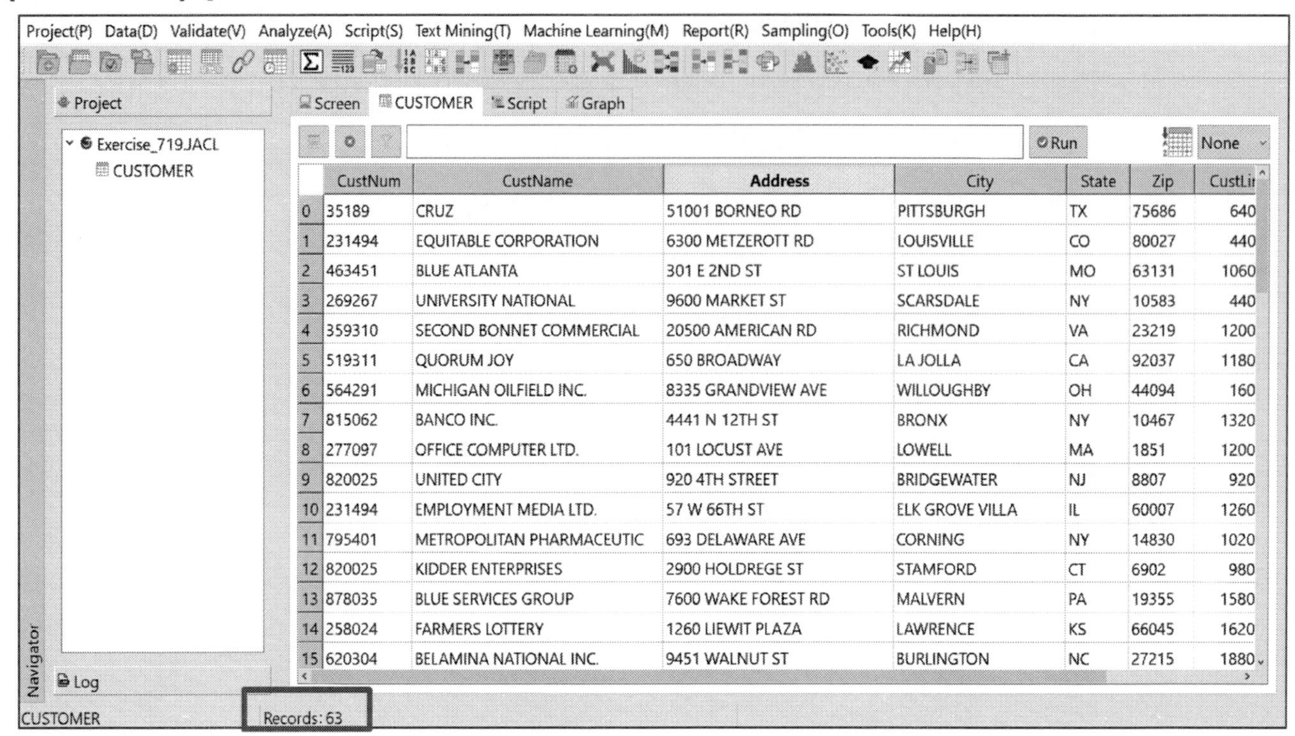

	CustNum	CustName	Address	City	State	Zip	CustLi
0	35189	CRUZ	51001 BORNEO RD	PITTSBURGH	TX	75686	640
1	231494	EQUITABLE CORPORATION	6300 METZEROTT RD	LOUISVILLE	CO	80027	440
2	463451	BLUE ATLANTA	301 E 2ND ST	ST LOUIS	MO	63131	1060
3	269267	UNIVERSITY NATIONAL	9600 MARKET ST	SCARSDALE	NY	10583	440
4	359310	SECOND BONNET COMMERCIAL	20500 AMERICAN RD	RICHMOND	VA	23219	1200
5	519311	QUORUM JOY	650 BROADWAY	LA JOLLA	CA	92037	1180
6	564291	MICHIGAN OILFIELD INC.	8335 GRANDVIEW AVE	WILLOUGHBY	OH	44094	160
7	815062	BANCO INC.	4441 N 12TH ST	BRONX	NY	10467	1320
8	277097	OFFICE COMPUTER LTD.	101 LOCUST AVE	LOWELL	MA	1851	1200
9	820025	UNITED CITY	920 4TH STREET	BRIDGEWATER	NJ	8807	920
10	231494	EMPLOYMENT MEDIA LTD.	57 W 66TH ST	ELK GROVE VILLA	IL	60007	1260
11	795401	METROPOLITAN PHARMACEUTIC	693 DELAWARE AVE	CORNING	NY	14830	1020
12	820025	KIDDER ENTERPRISES	2900 HOLDREGE ST	STAMFORD	CT	6902	980
13	878035	BLUE SERVICES GROUP	7600 WAKE FOREST RD	MALVERN	PA	19355	1580
14	258024	FARMERS LOTTERY	1260 LIEWIT PLAZA	LAWRENCE	KS	66045	1620
15	620304	BELAMINA NATIONAL INC.	9451 WALNUT ST	BURLINGTON	NC	27215	1880

Exercise 7.19

Please import the loan data according to the file format (Schema) provided below.

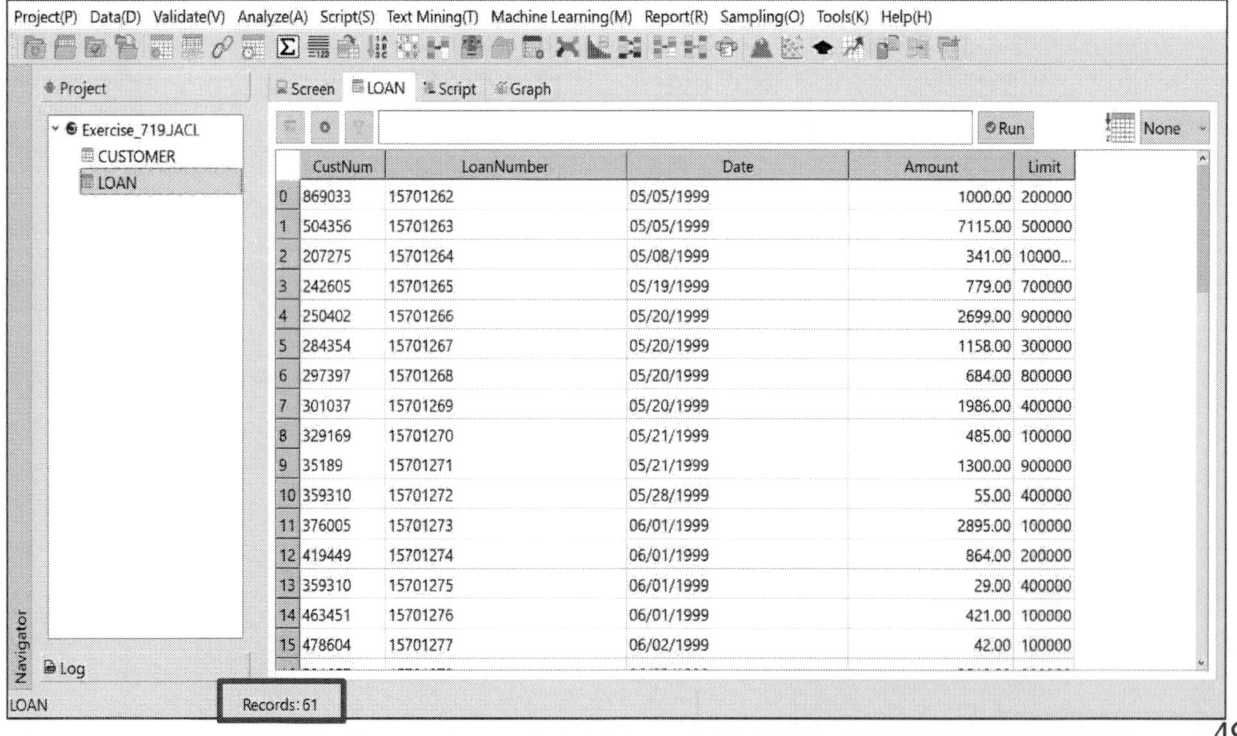

	CustNum	LoanNumber	Date	Amount	Limit
0	869033	15701262	05/05/1999	1000.00	200000
1	504356	15701263	05/05/1999	7115.00	500000
2	207275	15701264	05/08/1999	341.00	10000...
3	242605	15701265	05/19/1999	779.00	700000
4	250402	15701266	05/20/1999	2699.00	900000
5	284354	15701267	05/20/1999	1158.00	300000
6	297397	15701268	05/20/1999	684.00	800000
7	301037	15701269	05/20/1999	1986.00	400000
8	329169	15701270	05/21/1999	485.00	100000
9	35189	15701271	05/21/1999	1300.00	900000
10	359310	15701272	05/28/1999	55.00	400000
11	376005	15701273	06/01/1999	2895.00	100000
12	419449	15701274	06/01/1999	864.00	200000
13	359310	15701275	06/01/1999	29.00	400000
14	463451	15701276	06/01/1999	421.00	100000
15	478604	15701277	06/02/1999	42.00	100000

Exercise 7.19
Please import the OFAC sdn list according to the file format (Schema) provided below.

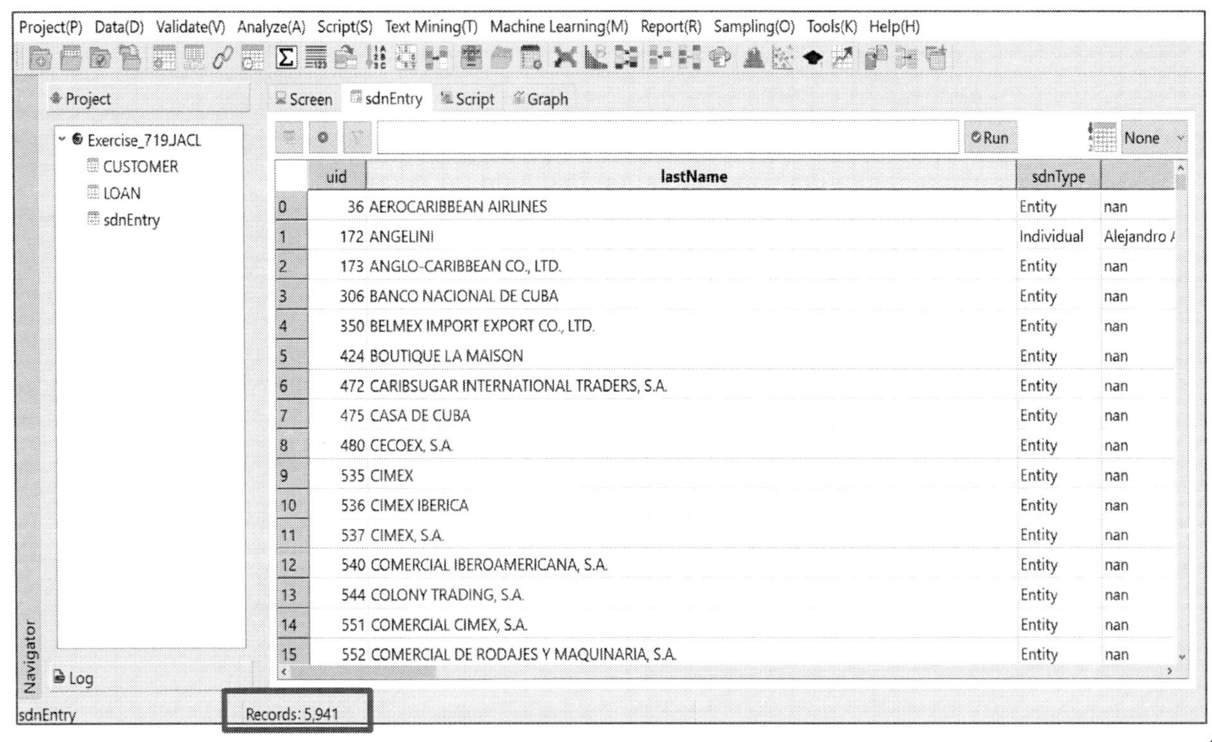

Exercise 7.19
Verify whether there are errors in the customer data file.

- Open data set we need (Customer)
- Select "Validate>Verify"

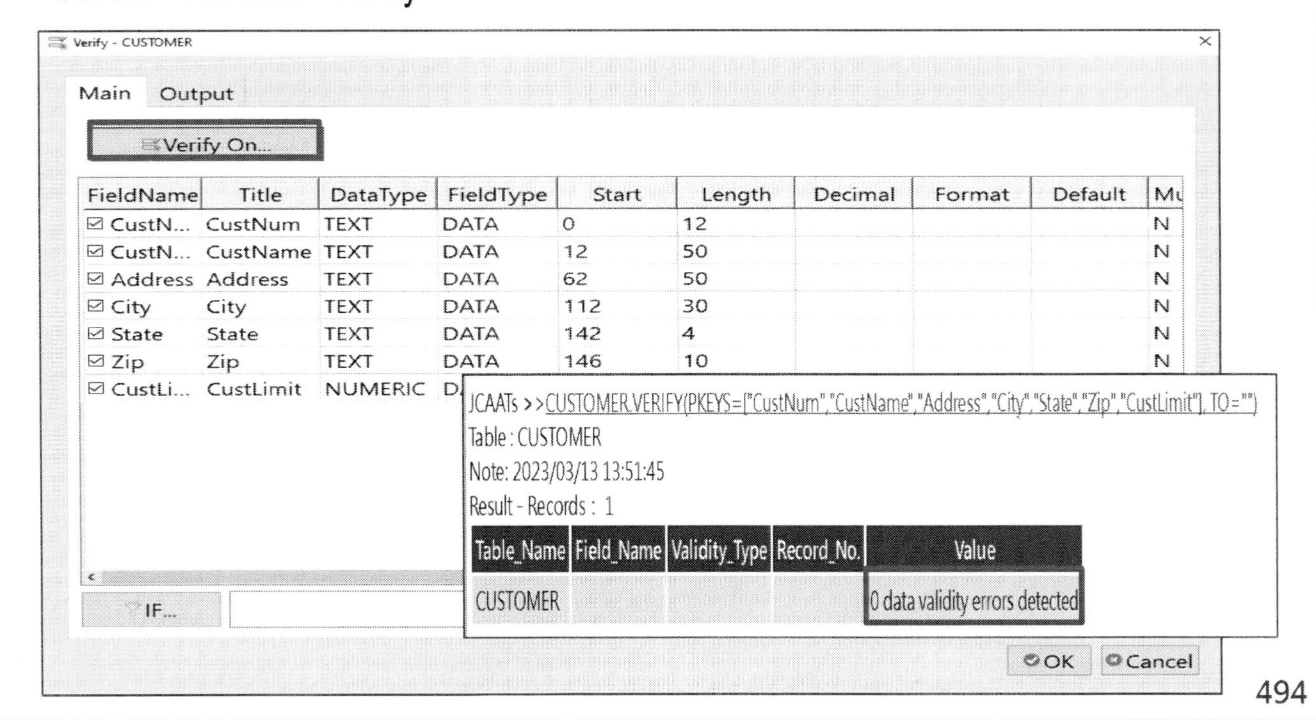

Exercise 7.19

Verify whether there are errors in the loan data file.

- Open data set we need (Loan)
- Select "Validate>Verify"

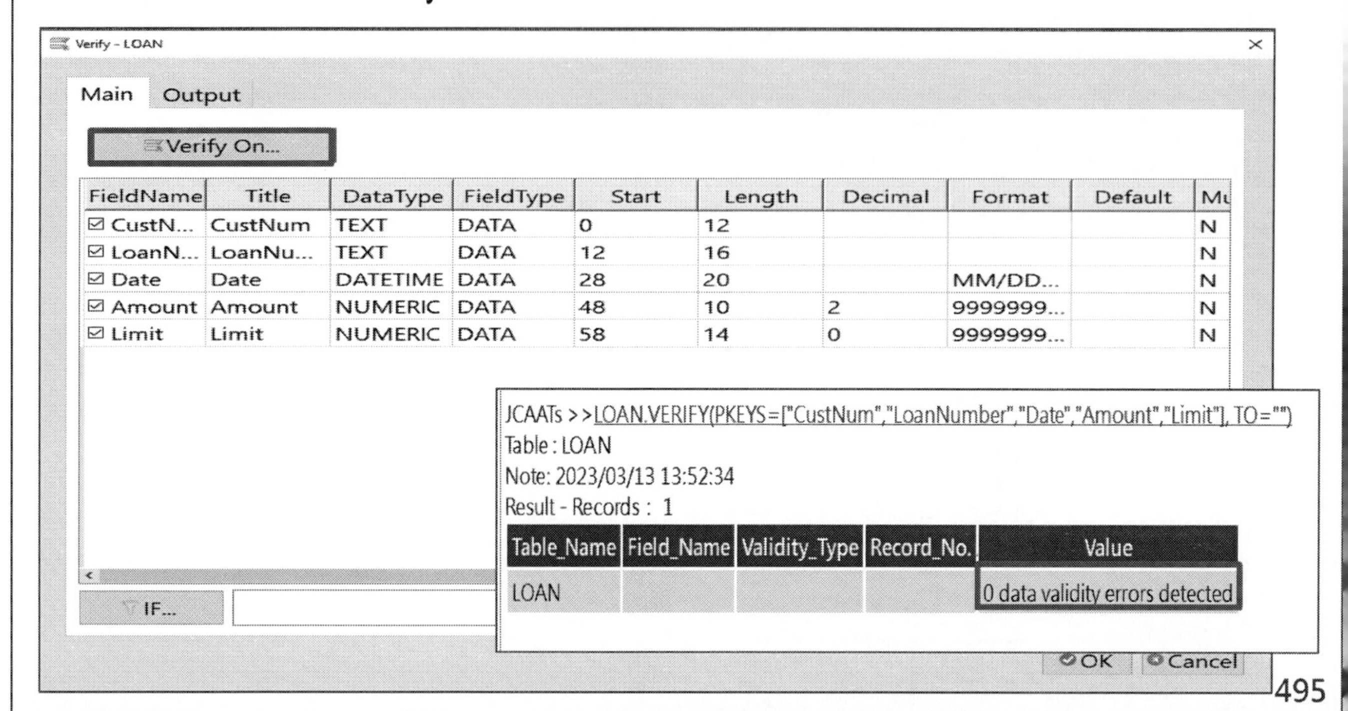

Exercise 7.19

Verify whether there are errors in the sdn data file.

- Open data set we need (Loan)
- Select "Validate>Verify"

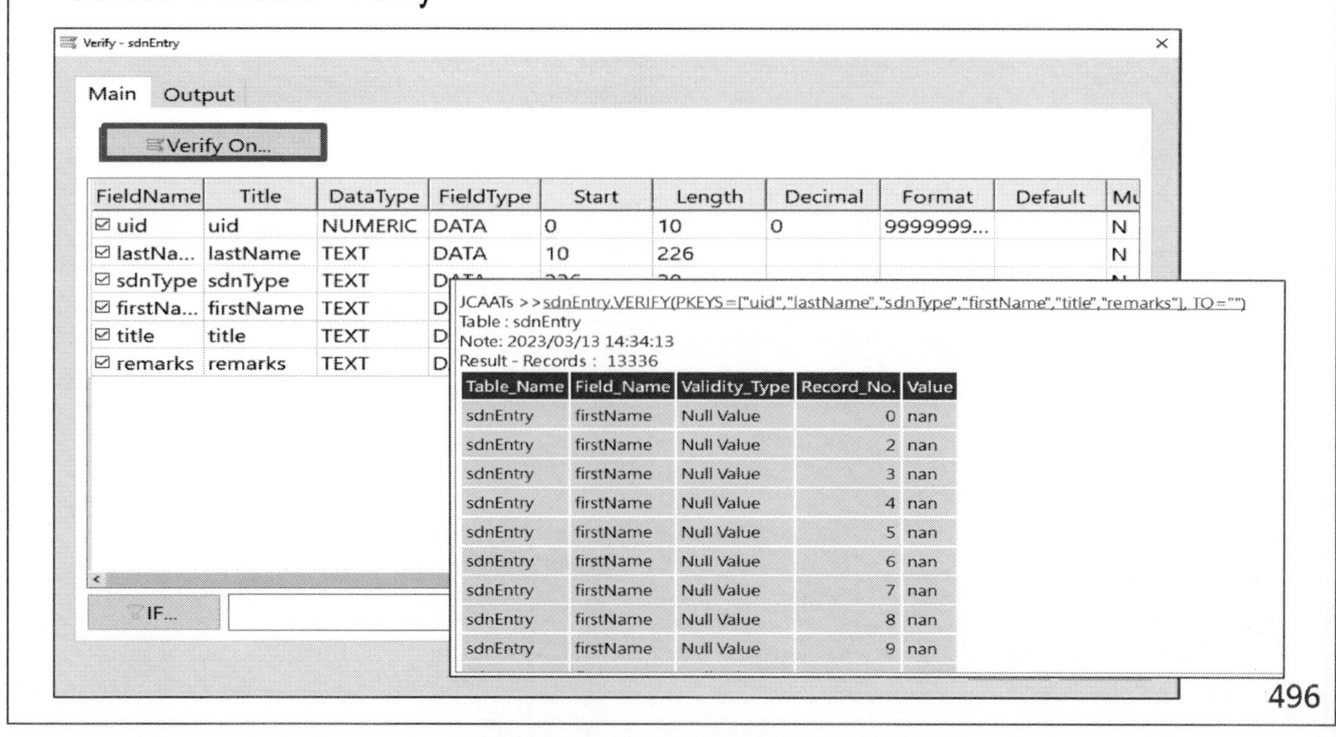

Exercise 7.19
Validate the data: The record number in customer data

- Select "Validate>Count"

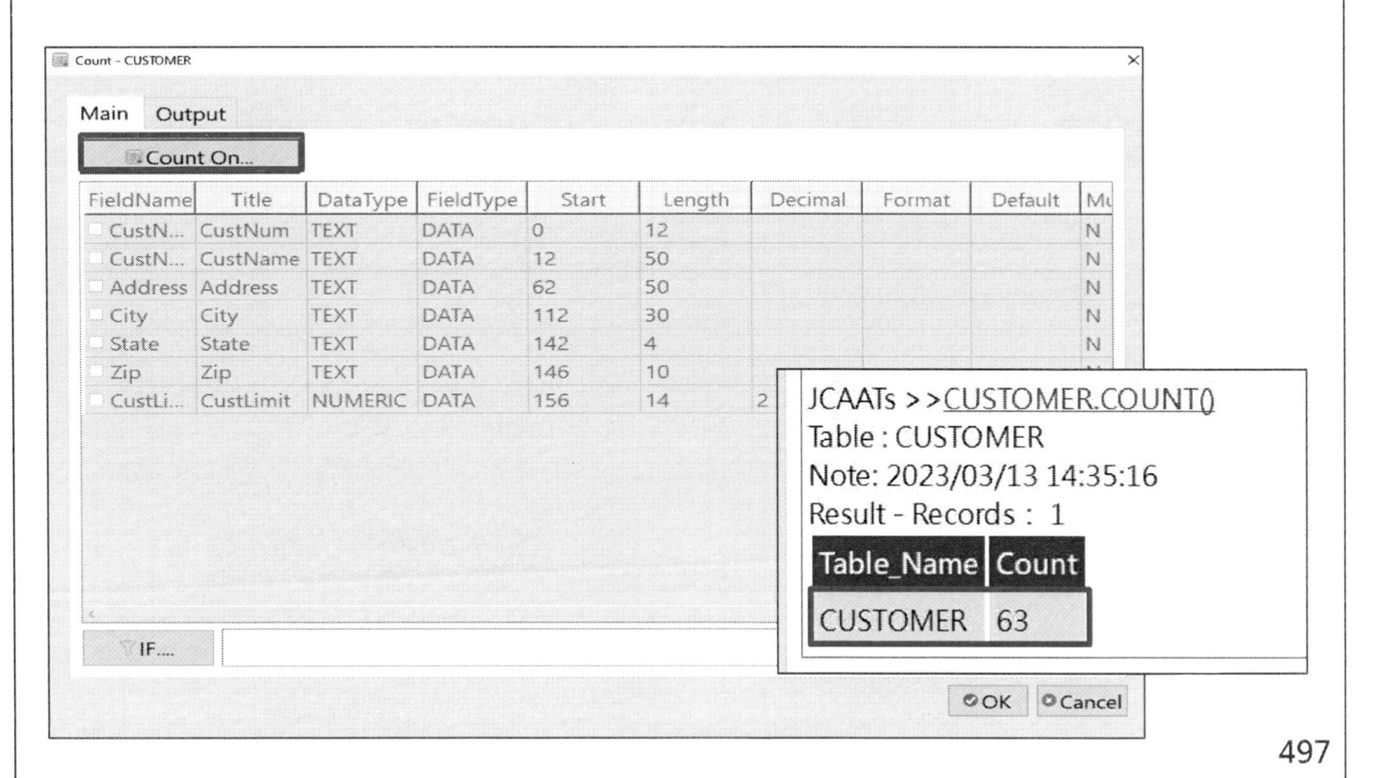

Exercise 7.19
Validate the data: The record number in sdn data

- Select "Validate>Count"

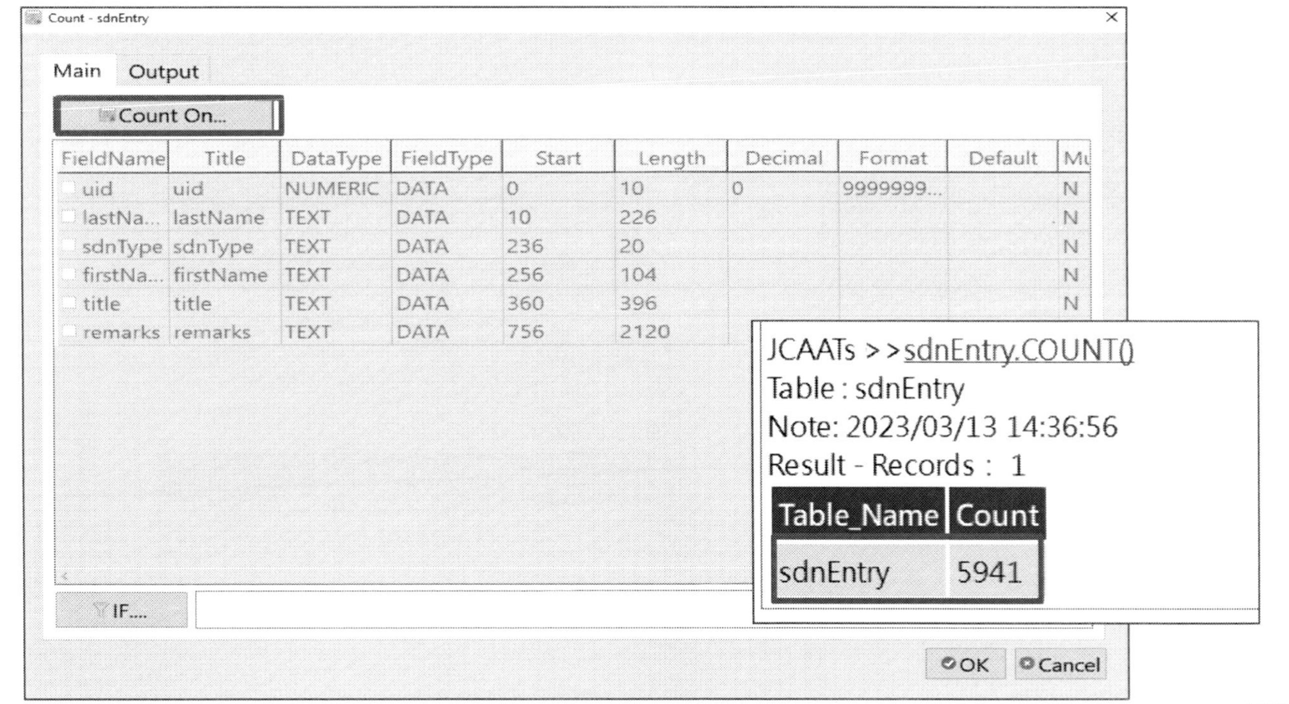

Exercise 7.19
Validate the data: The record number in loan data

- Select "Validate>Count"

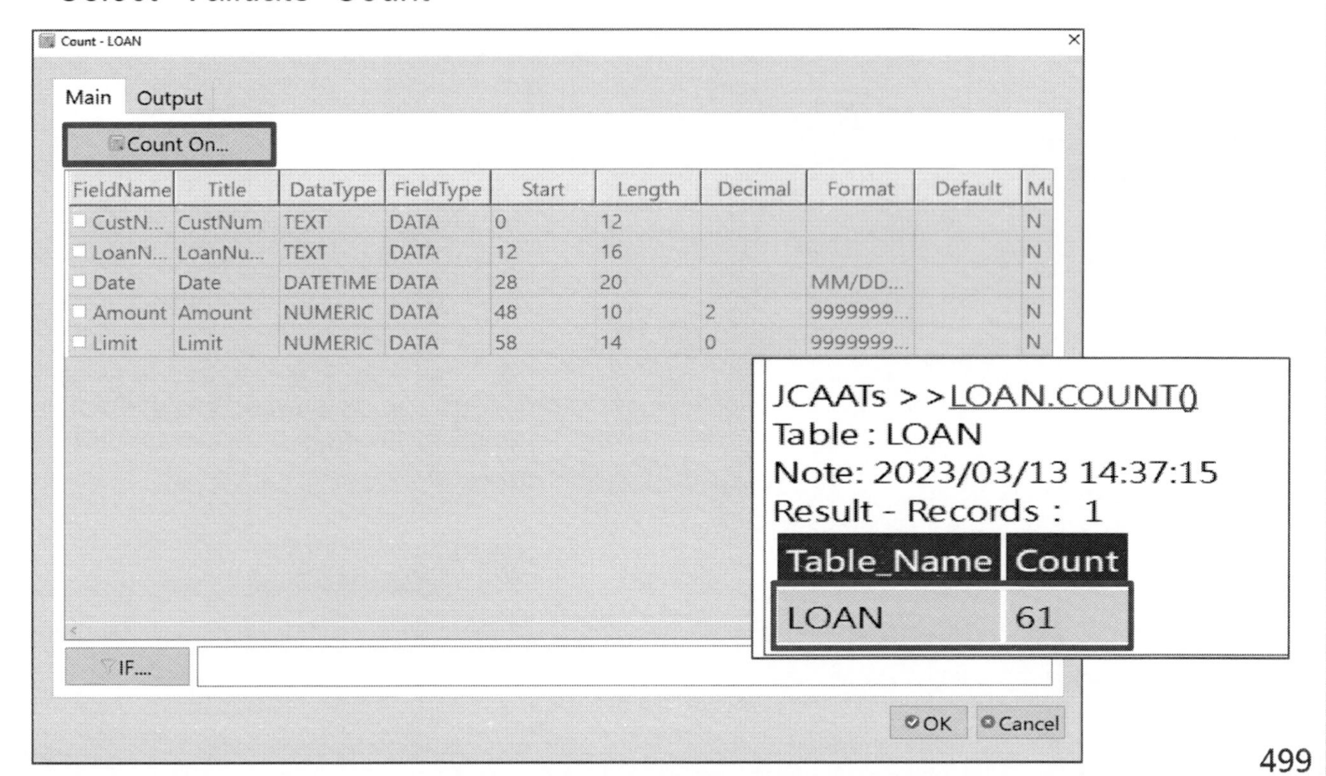

Exercise 7.19
Validate the data: The total amount of loans

- Select "Validate>Total"

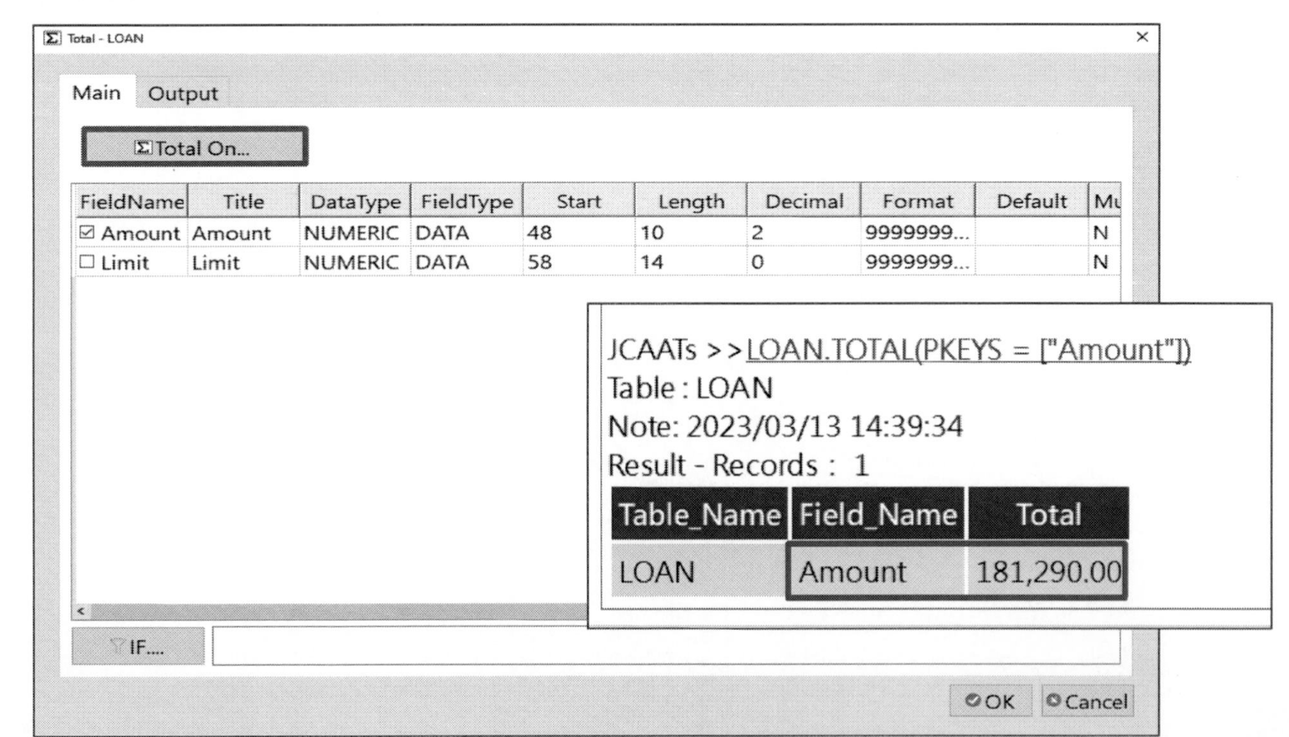

Exercise 7.19
Please check whether there are any duplicated customer numbers.

- Select "Validate > Duplicate", and choose the "CustNum" field to analyze the duplicate data

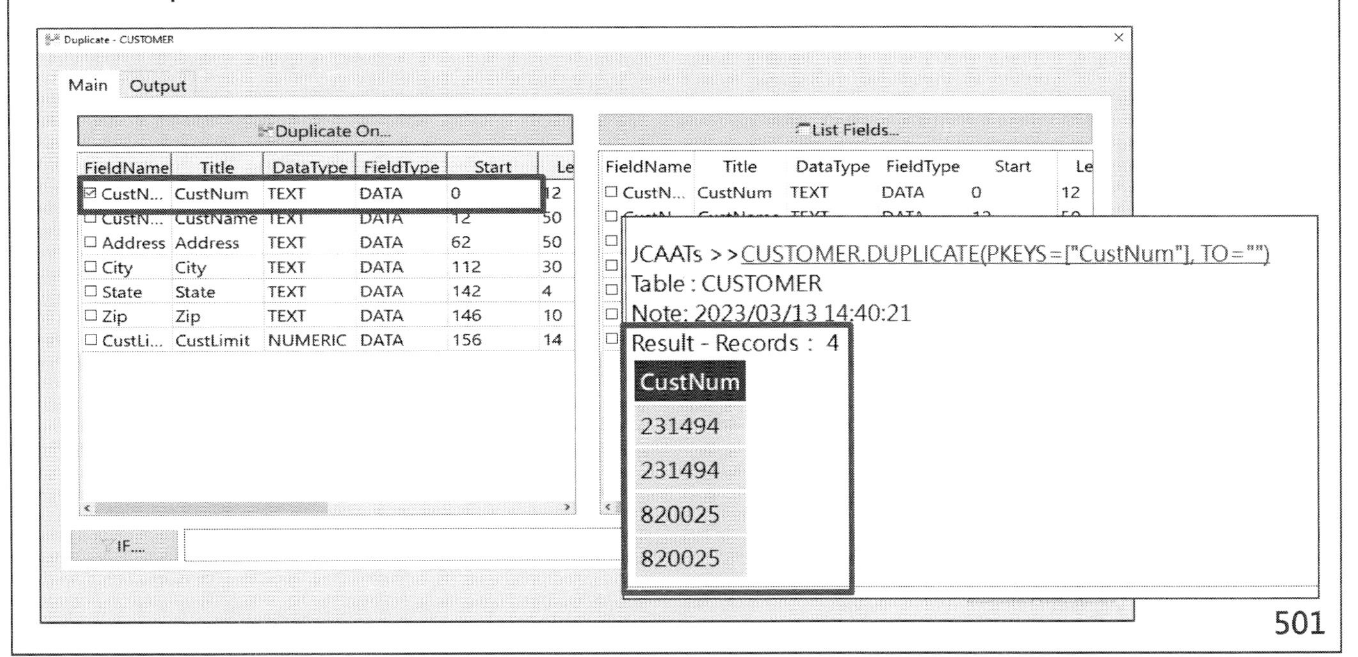

Exercise 7.19
Please check whether there are any duplicated Loan numbers.

- Select "Validate > Duplicate", and choose the "LoanNumber" field to analyze the duplicate data

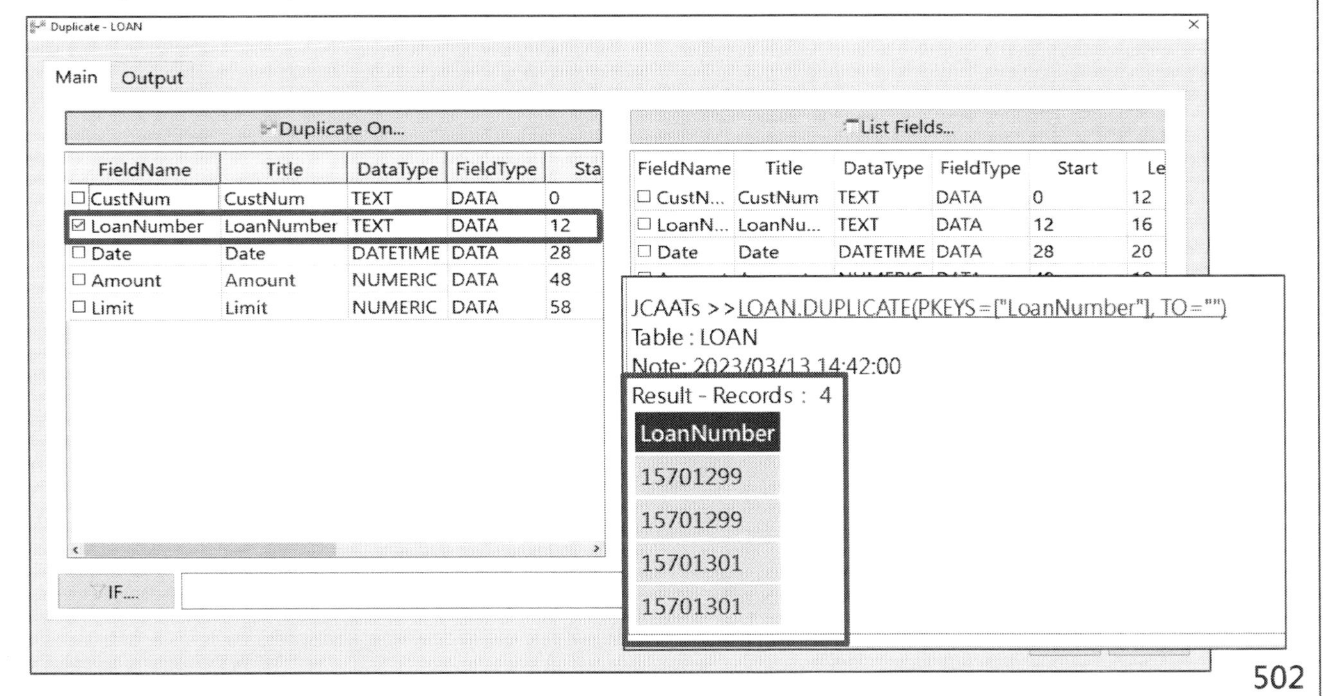

Exercise 7.19

Please check if there are any anomalies where loan customers are not recorded in the customer file.

- Select "Analyze > Join", choose "Loan" as the primary table and "Customer" as secondary. Choose "CustNum" as the key and list out the "Date" and "Amount" fields in the primary table.

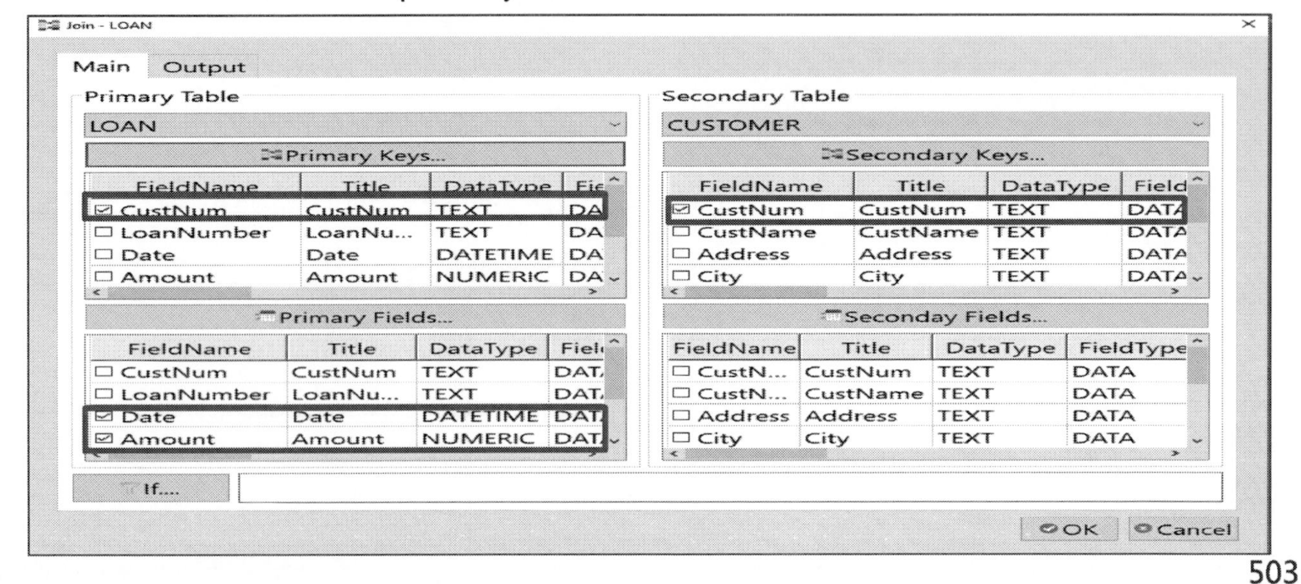

Exercise 7.19

Please check if there are any anomalies where loan customers are not recorded in the customer file.

- Select Join_Types as Unmatch Primary

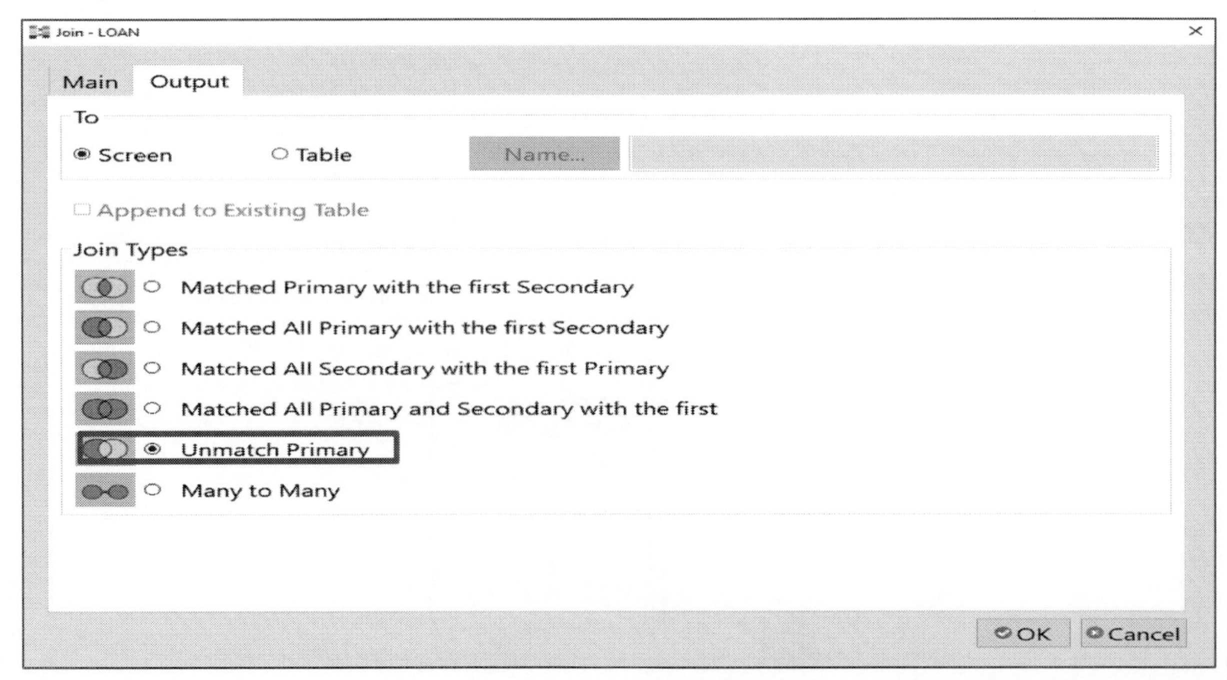

Exercise 7.19
Please check if there are any anomalies where loan customers are not recorded in the customer file.

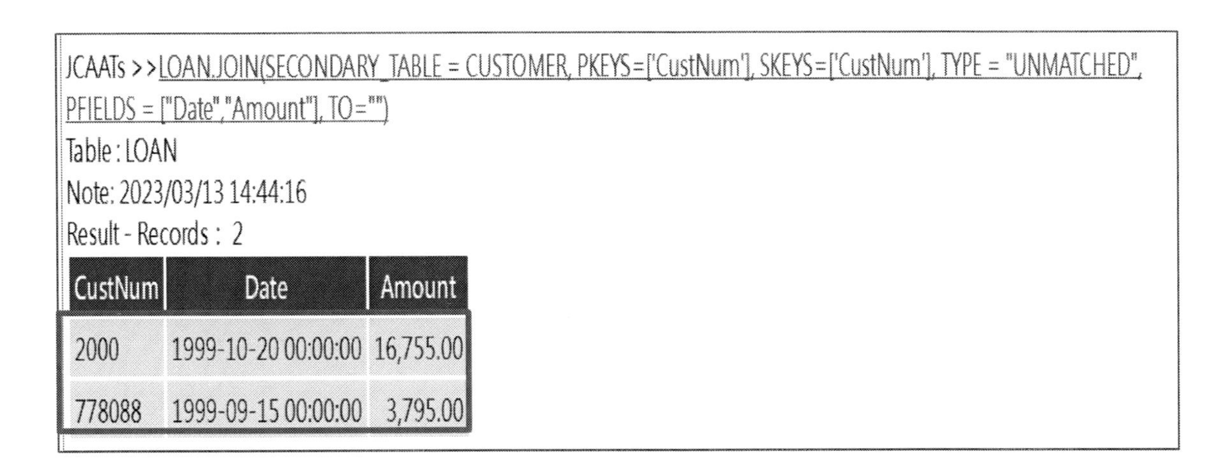
JCAATs >>LOAN.JOIN(SECONDARY_TABLE = CUSTOMER, PKEYS=['CustNum'], SKEYS=['CustNum'], TYPE = "UNMATCHED", PFIELDS = ["Date","Amount"], TO="")
Table : LOAN
Note: 2023/03/13 14:44:16
Result - Records : 2

CustNum	Date	Amount
2000	1999-10-20 00:00:00	16,755.00
778088	1999-09-15 00:00:00	3,795.00

Exercise 7.19
Please check whether the customer file contains the OFAC regulatory blacklist (SDN).

- Select "Analyze > Join", choose "Customer" as the primary table and "sdnEntry" as secondary. Choose "CustNum" as the key and list out the field "CustNum".

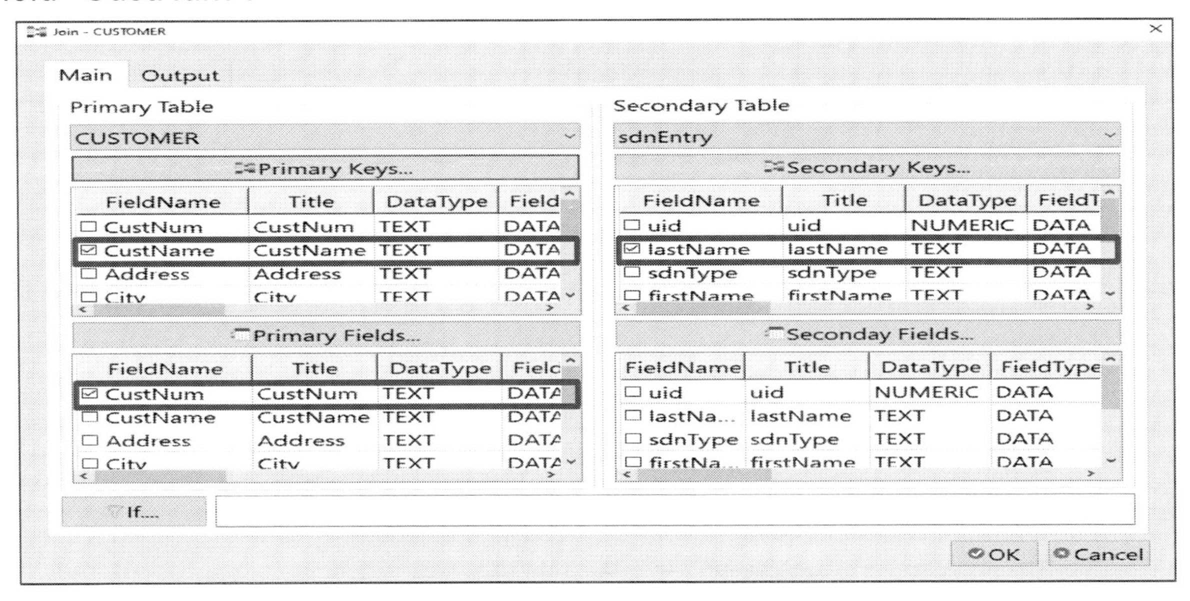

Exercise 7.19
Please check whether the customer file contains the OFAC regulatory blacklist (SDN).

- Select Join_Types as Matched Primary with the first Secondary

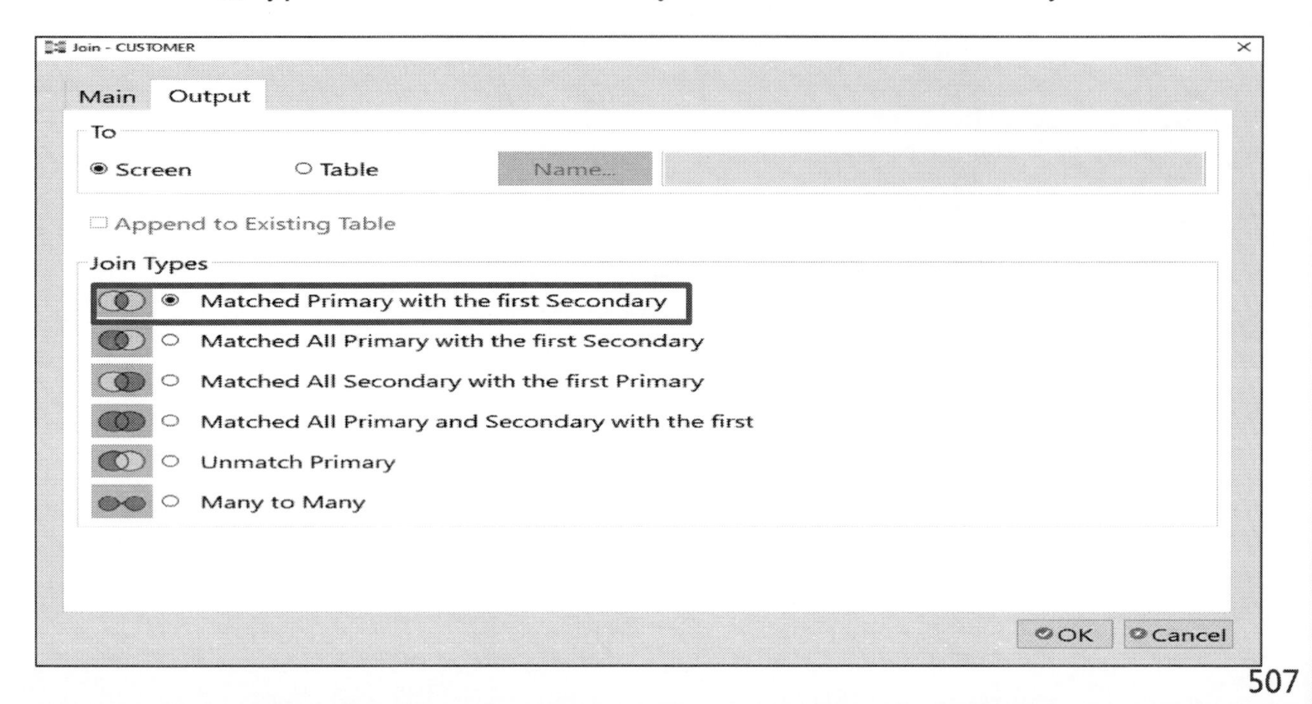

JCAATs-AI Audit Software
Copyright © 2023 JACKSOFT.

Exercise 7.19
Please check whether the customer file contains the OFAC regulatory blacklist (SDN).

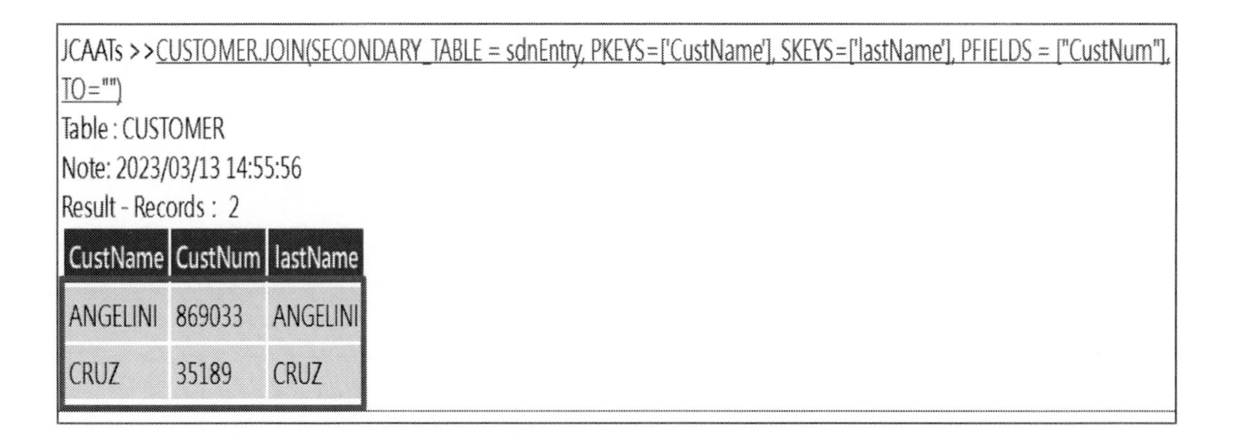
JCAATs >>CUSTOMER.JOIN(SECONDARY_TABLE = sdnEntry, PKEYS=['CustName'], SKEYS=['lastName'], PFIELDS = ["CustNum"], TO="")

Table : CUSTOMER

Note: 2023/03/13 14:55:56

Result - Records : 2

CustName	CustNum	lastName
ANGELINI	869033	ANGELINI
CRUZ	35189	CRUZ

Exercise 7.19

Please check whether any loans have been granted that exceed the customer's total credit limit.

- Select “Analyze>Classify” to classify the CustNum field
- Subtotal the "Amount" field

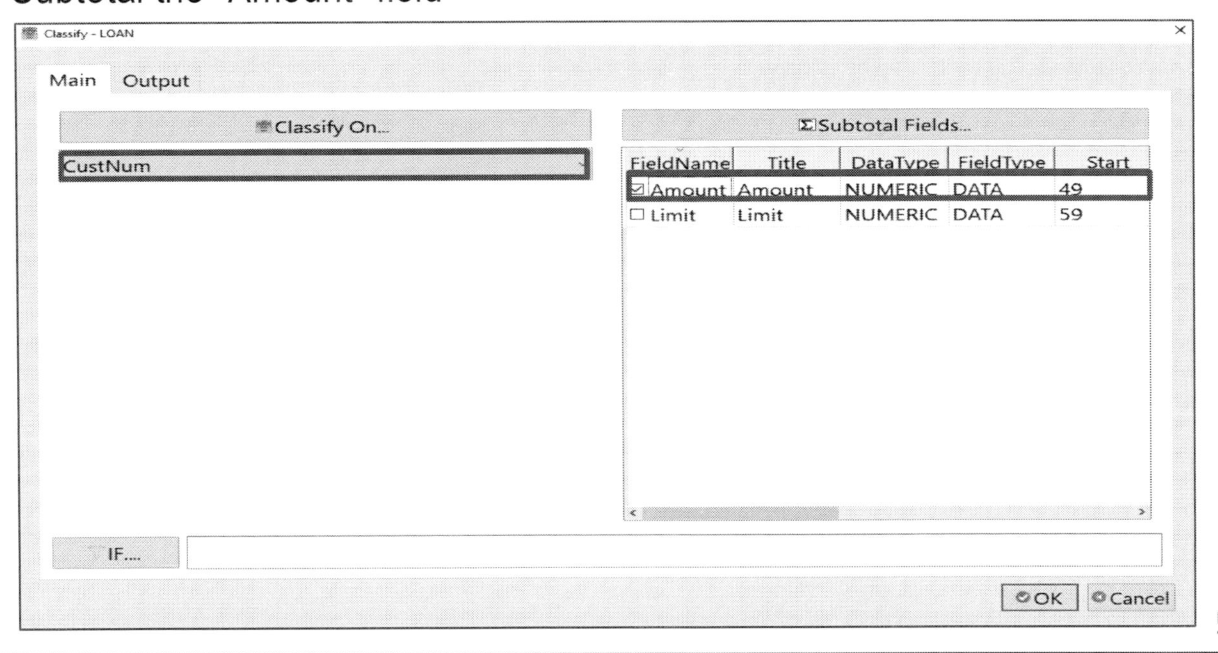

Exercise 7.19

Please check whether any loans have been granted that exceed the customer's total credit limit.

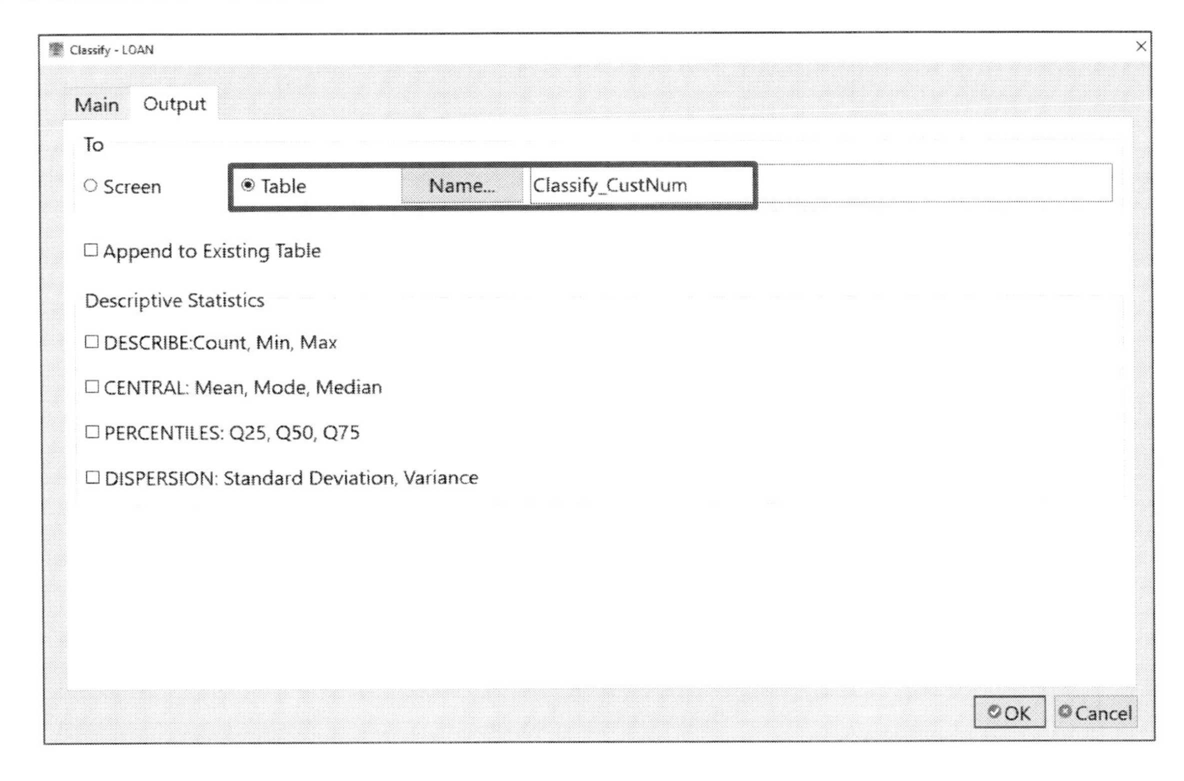

Exercise 7.19

Please check whether any loans have been granted that exceed the customer's total credit limit.

- Select "Analyze > Join", choose "Classify_CustNum" as the primary table and "Customer" as secondary. Choose "CustNum" as the key and list out the field "Amount_sum" in the primary field and "CustLimit" in secondary.

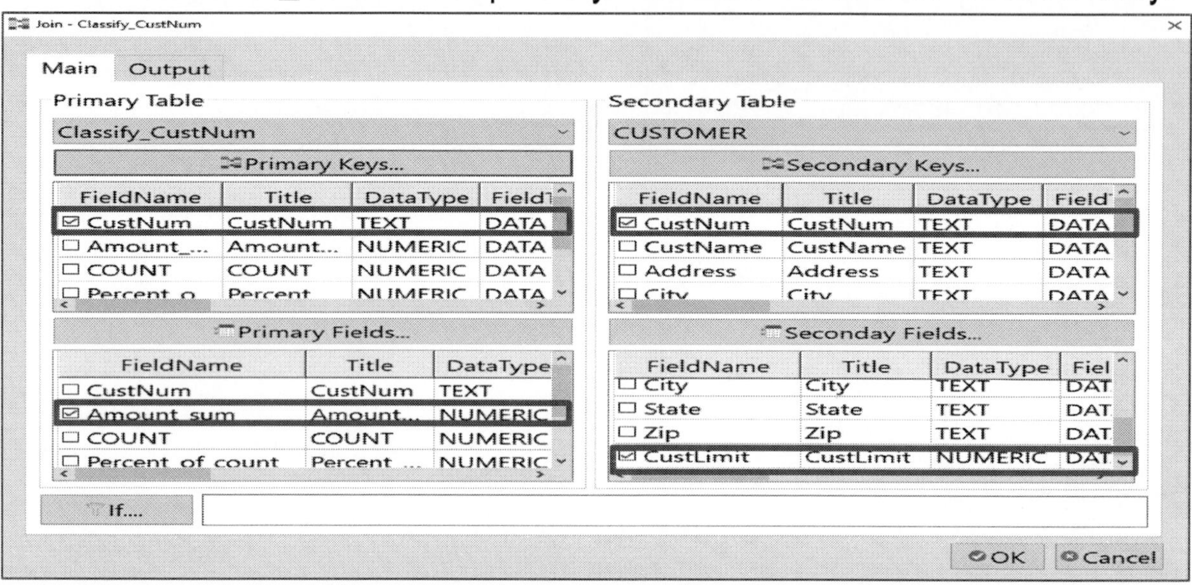

JCAATs-AI Audit Software
Copyright © 2023 JACKSOFT.

Exercise 7.19

Please check whether any loans have been granted that exceed the customer's total credit limit.

- Select Join_Types as Matched All Primary with the first Secondary

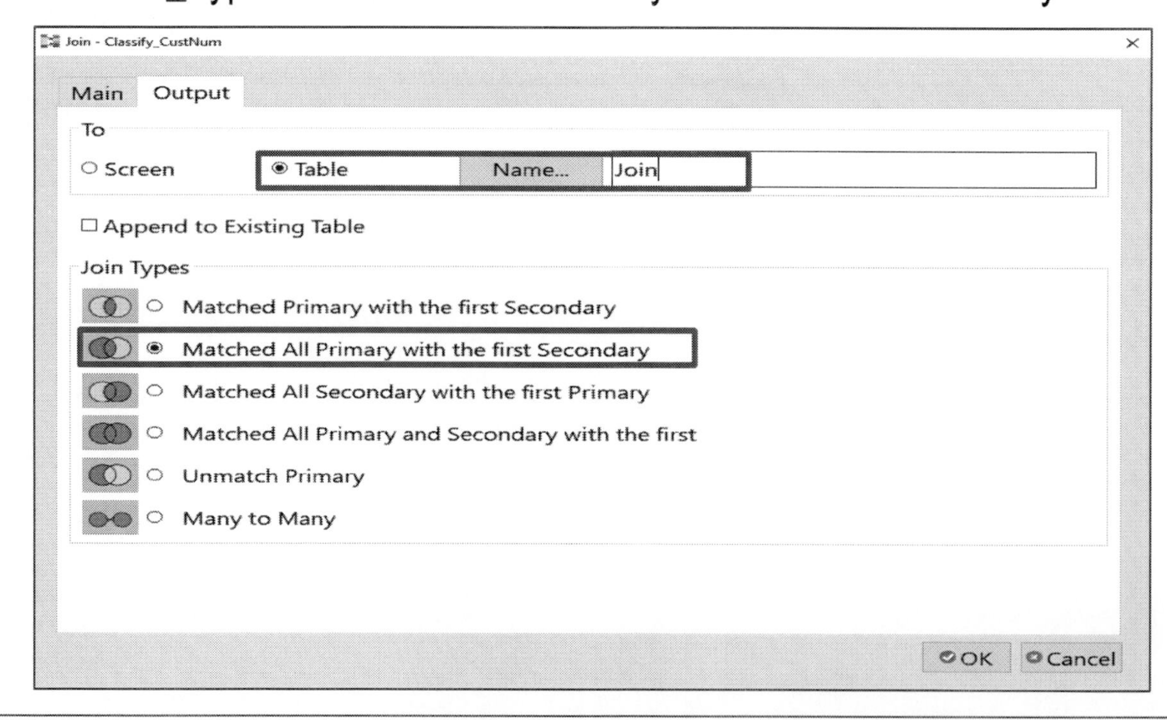

Exercise 7.19
Please check whether any loans have been granted that exceed the customer's total credit limit.

- Select "Validate > Clean" to obtain a refined dataset

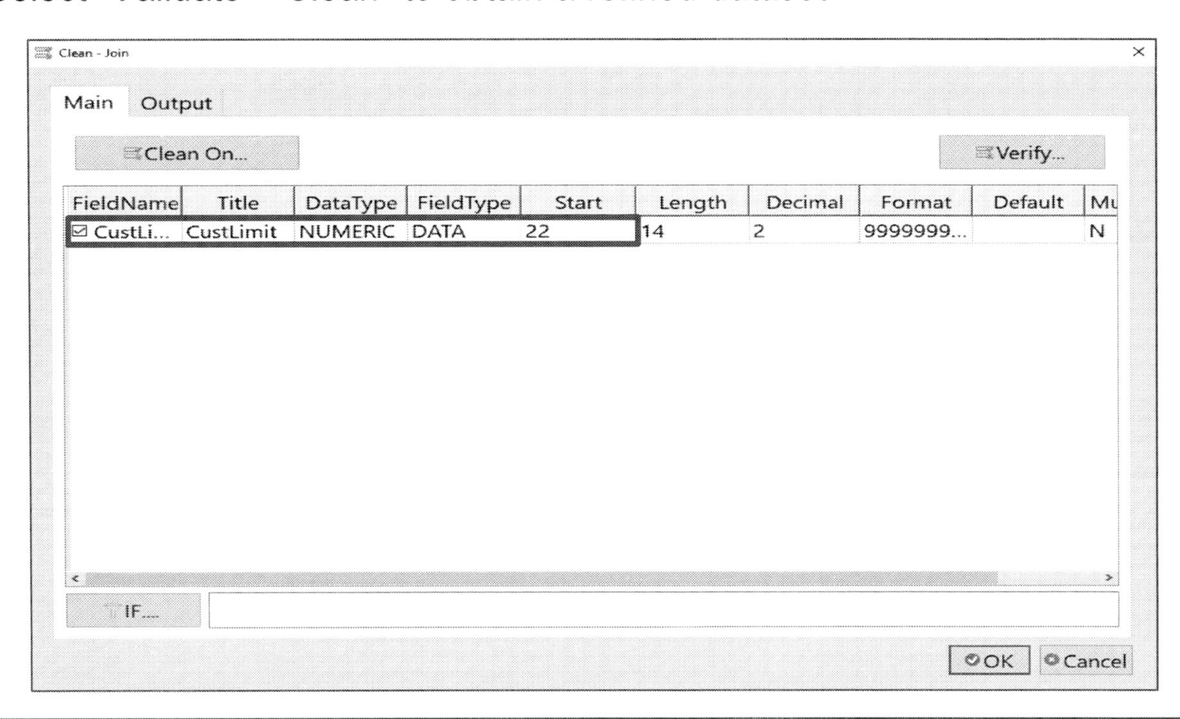

JCAATs-AI Audit Software Copyright © 2023 JACKSOFT.

Exercise 7.19
Please check whether any loans have been granted that exceed the customer's total credit limit.

- Select the "Numerical Imputation" processing approach for missing values and set it as 0.

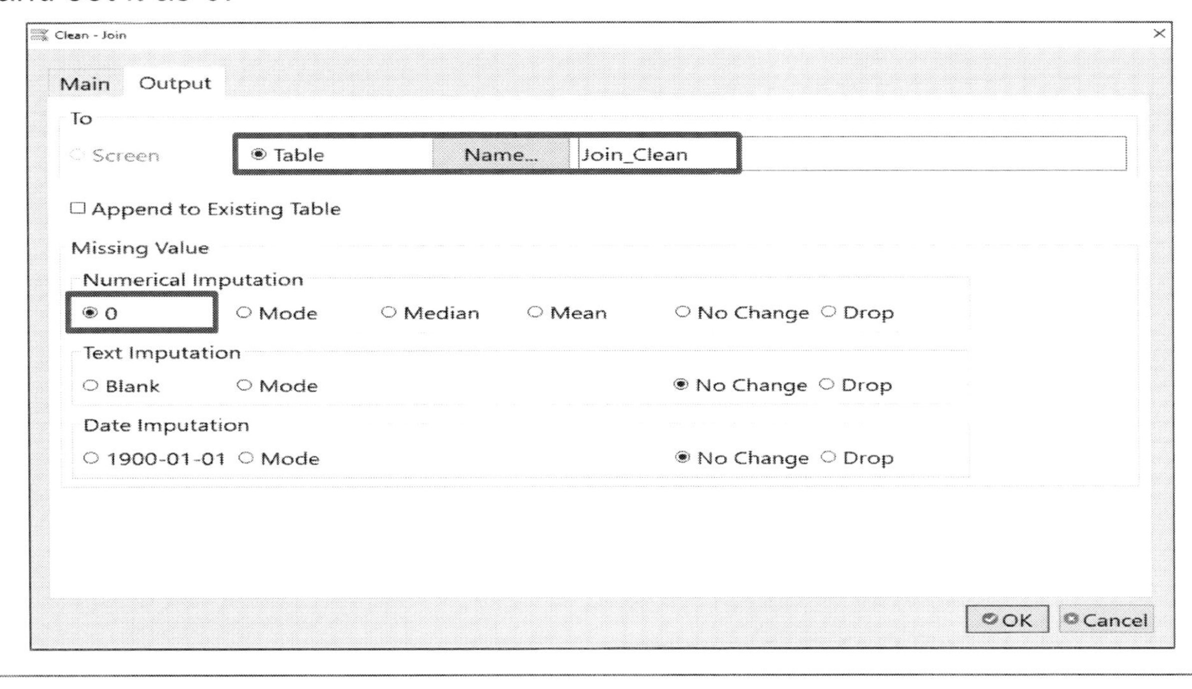

Exercise 7.19

Please check whether any loans have been granted that exceed the customer's total credit limit.

- Set a filter condition: Amount_sum > CustLimit

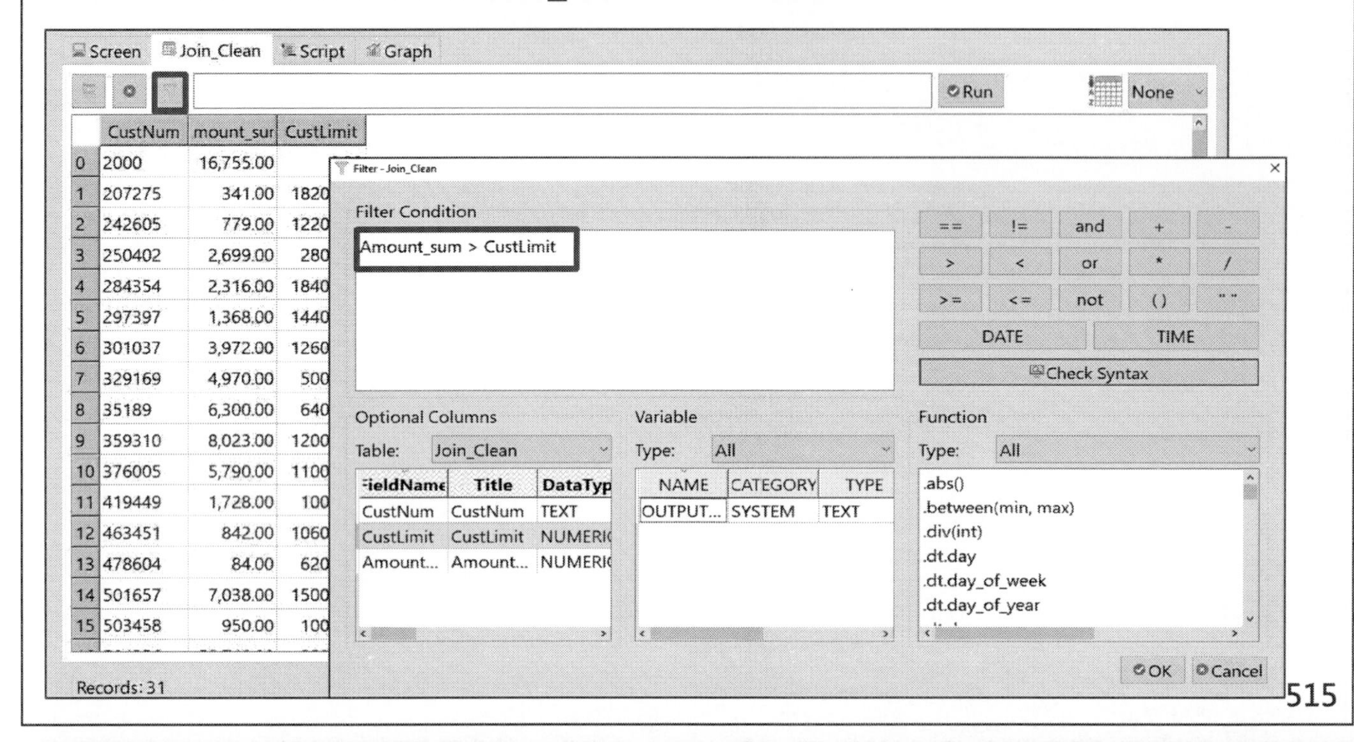

JCAATs-AI Audit Software
Copyright © 2023 JACKSOFT.

Exercise 7.19

Please check whether any loans have been granted that exceed the customer's total credit limit.

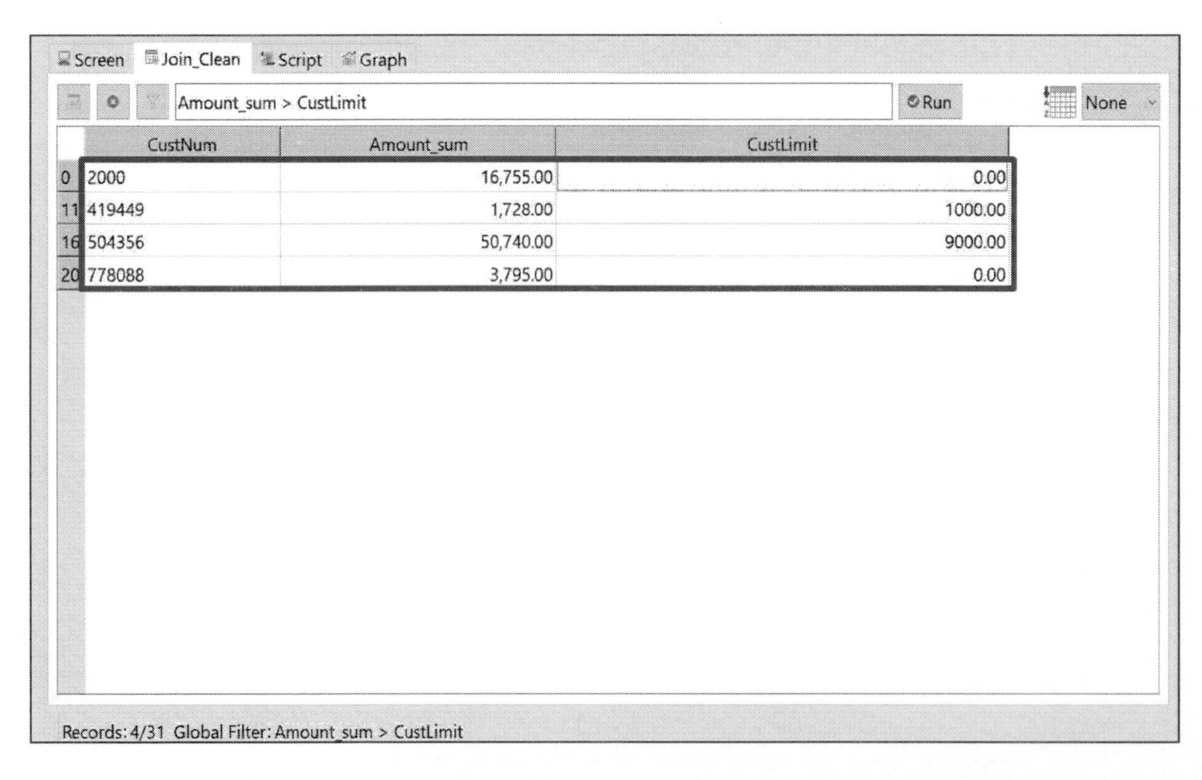

	CustNum	Amount_sum	CustLimit
0	2000	16,755.00	0.00
11	419449	1,728.00	1000.00
16	504356	50,740.00	9000.00
20	778088	3,795.00	0.00

JCAATs - AI Audit Software

Python Based Computer-Assisted Audit Techniques (CAATs)

Data Analysis and Smart Audit

Chapter 8 . Script

Outline:

1. New Script
2. Edit Script
3. Check Script Grammar
4. Folder Setting
5. Product Execute

New Script

- The audit program of JCAATs is written in Python syntax and has the file extension ".py".
- Script on the menu>>New Script to add a script for editing
- Log data can be saved as a script by converting the selected data.
- JCAATs allows you to copy from another JCAATs' projects and auto saved them as scripts in the current project.
- JCAATs allows you to open a Python program in the script editor. You can save it as a new script name in the project. You may need to learn advanced skills to run the python program within JCAATs.

Script Editor

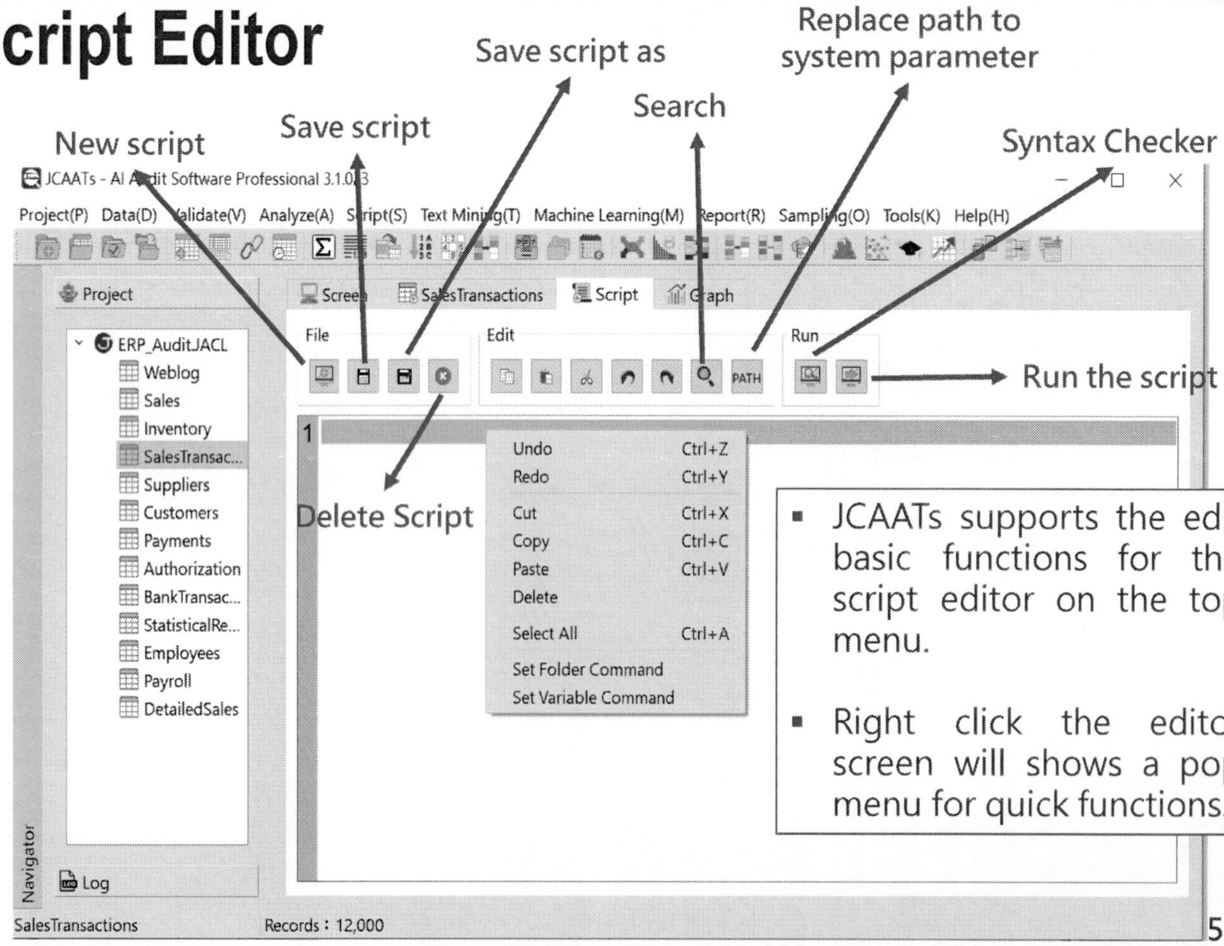

- JCAATs supports the edit basic functions for the script editor on the top menu.
- Right click the editor screen will shows a pop menu for quick functions.

New Script

- Script>>New script, enter a name for the script, and it will be added to the tree diagram. You can then start editing the script in Python syntax.

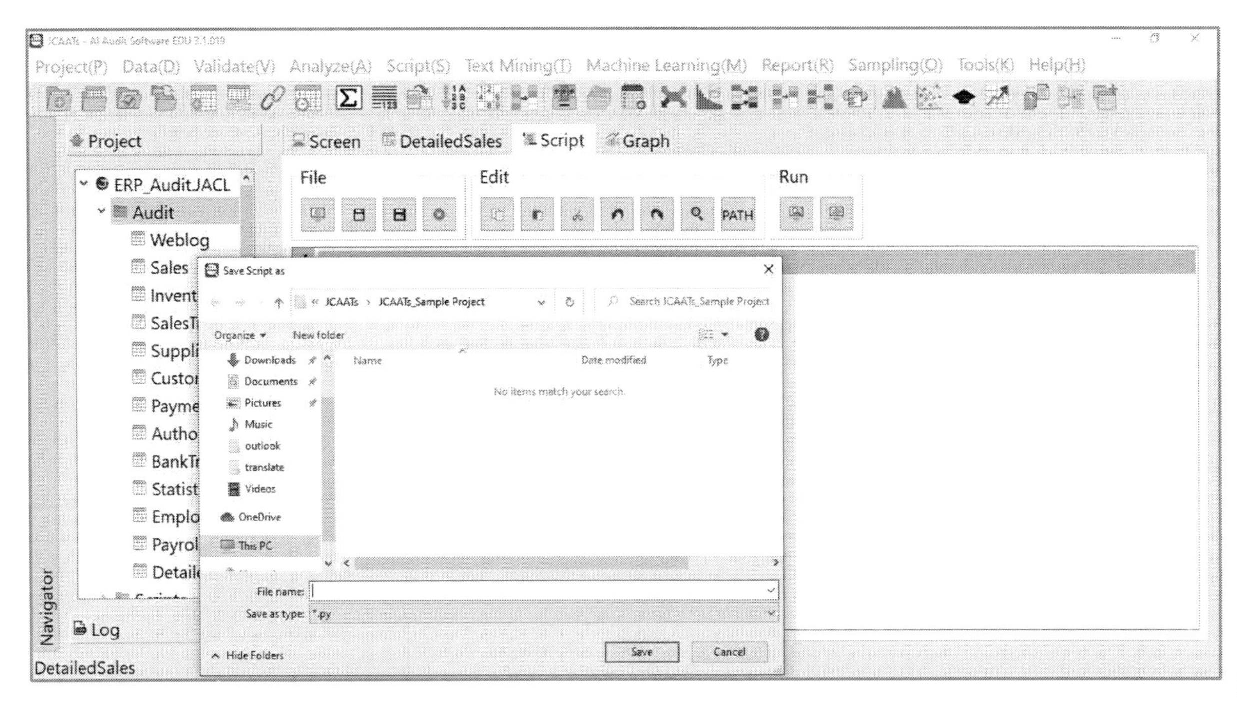

Save logs as a script

- The log tab is a user interface element that displays this log information in a formatted and readable manner. To operate the log, you can select specific log entries by clicking on them, and then right-clicking to show a menu of available functions. One common function is "Save Script As," which allows you to save the log information as a script file that can be edited or modified as needed.

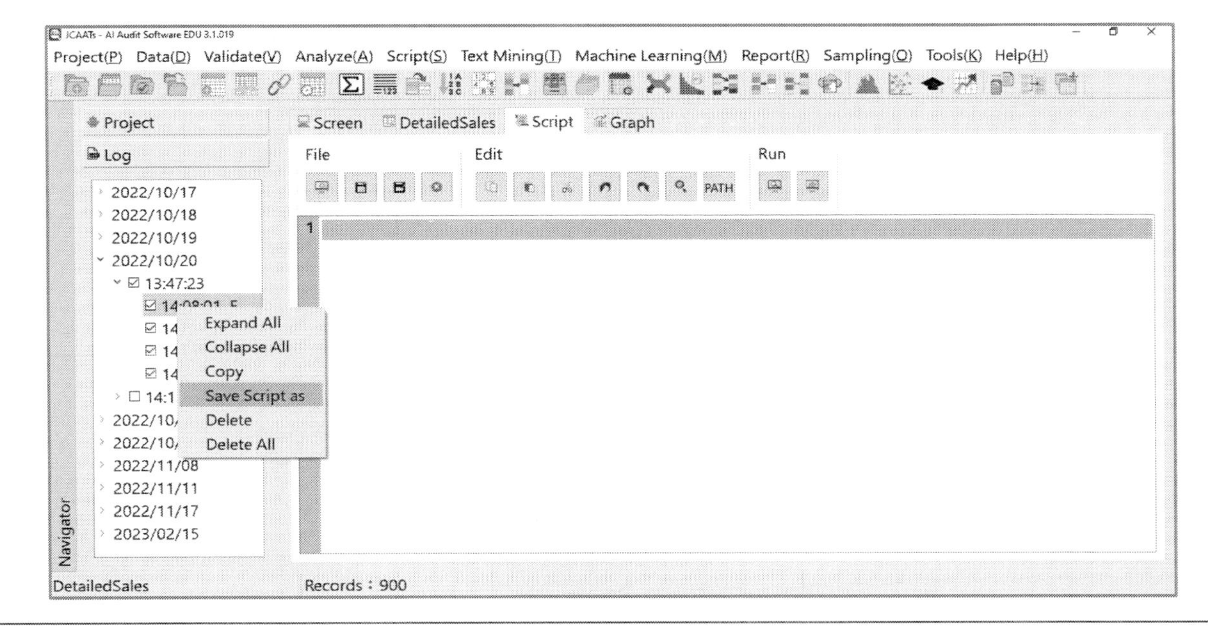

Open Python Script

- Select Open Python Script and save a new file to save the script into this project.

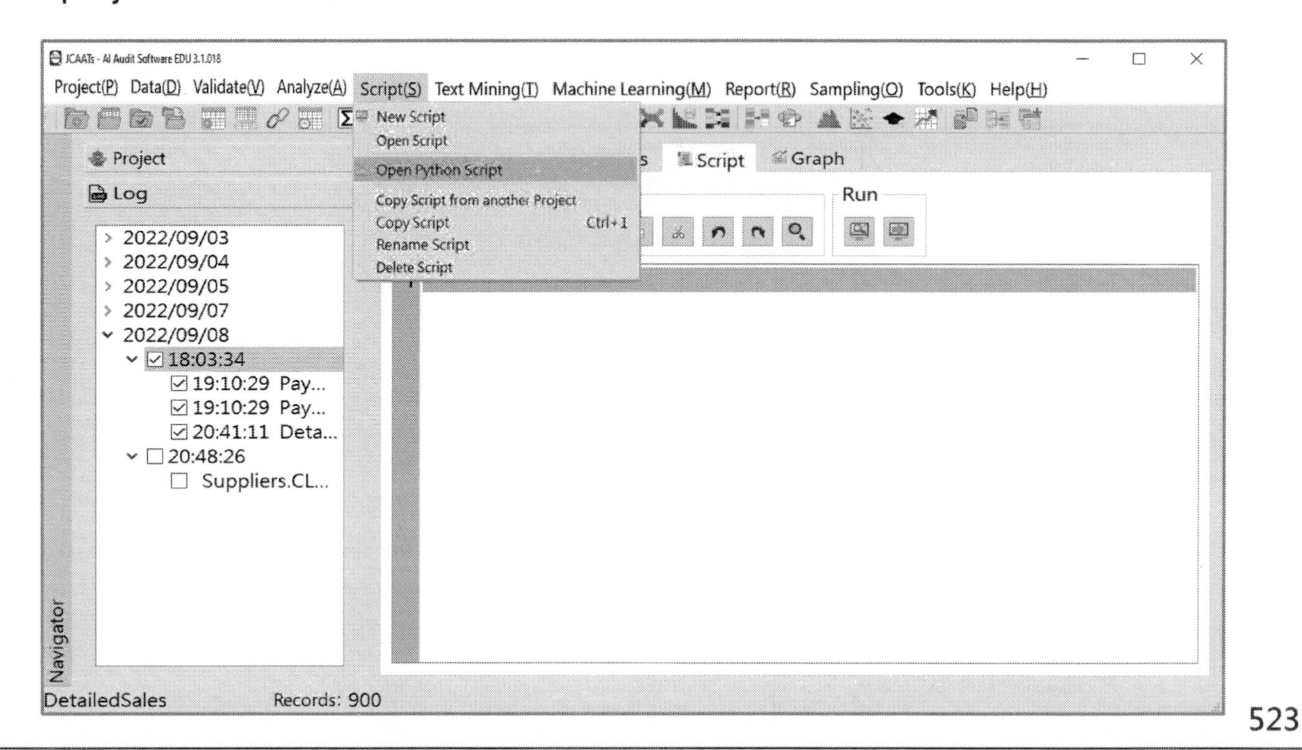

Edit Script

- Go to the navigator and click on the script to be opened. The script will then be opened in the script area of the main screen where the user can edit or execute it.

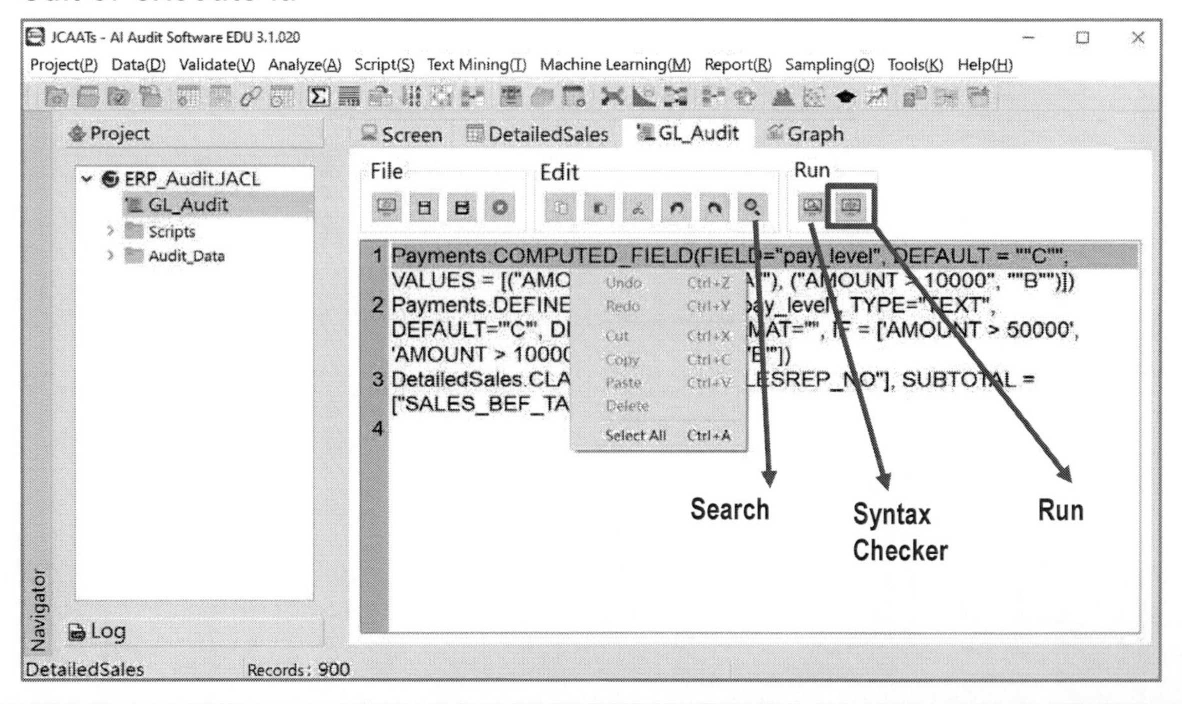

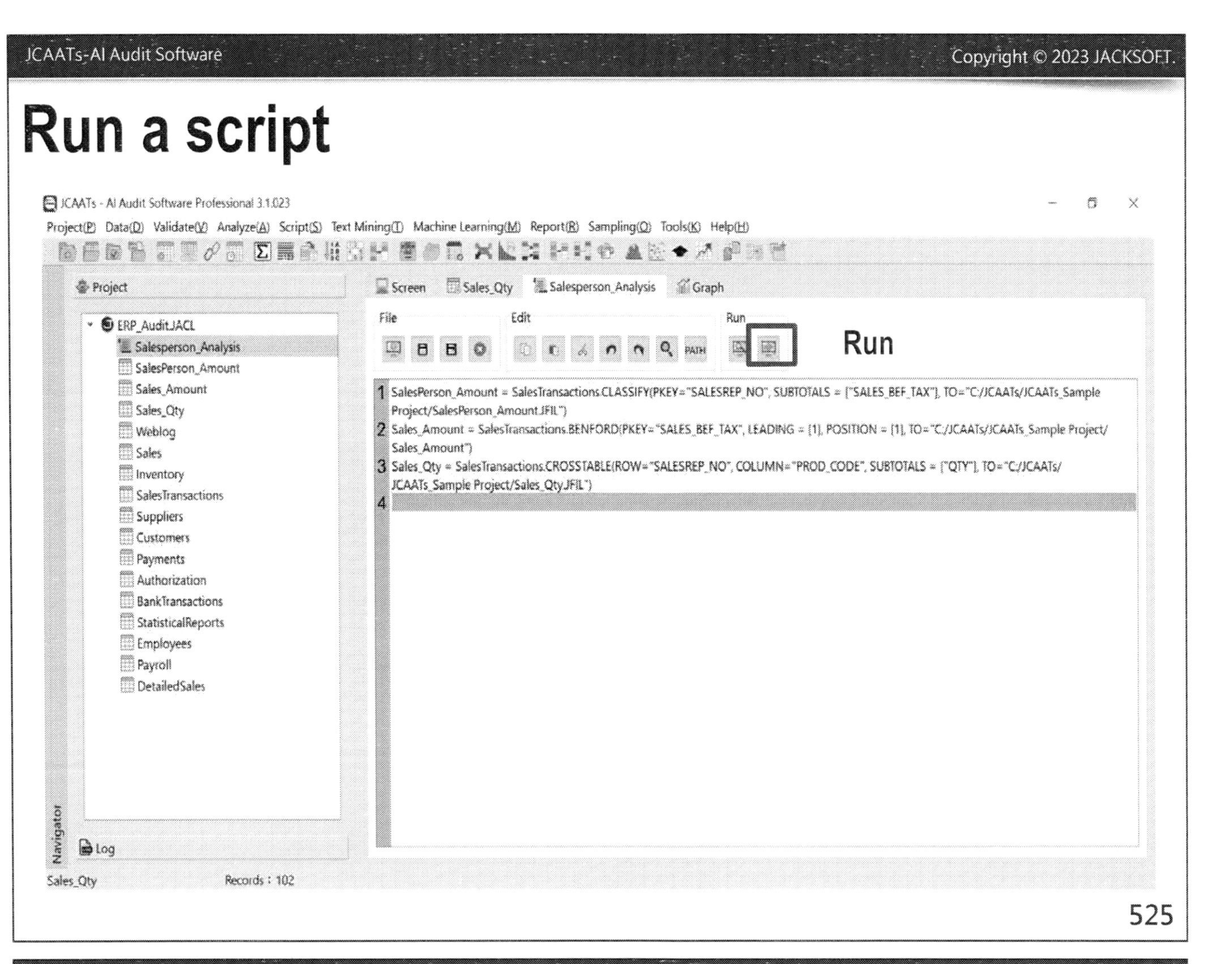

Syntax Checker

- Click the syntax checker to check each scripts line by line.
- An error message window will be displayed if there is an error.
- Use "**Search**" to quick find and replace the error position.

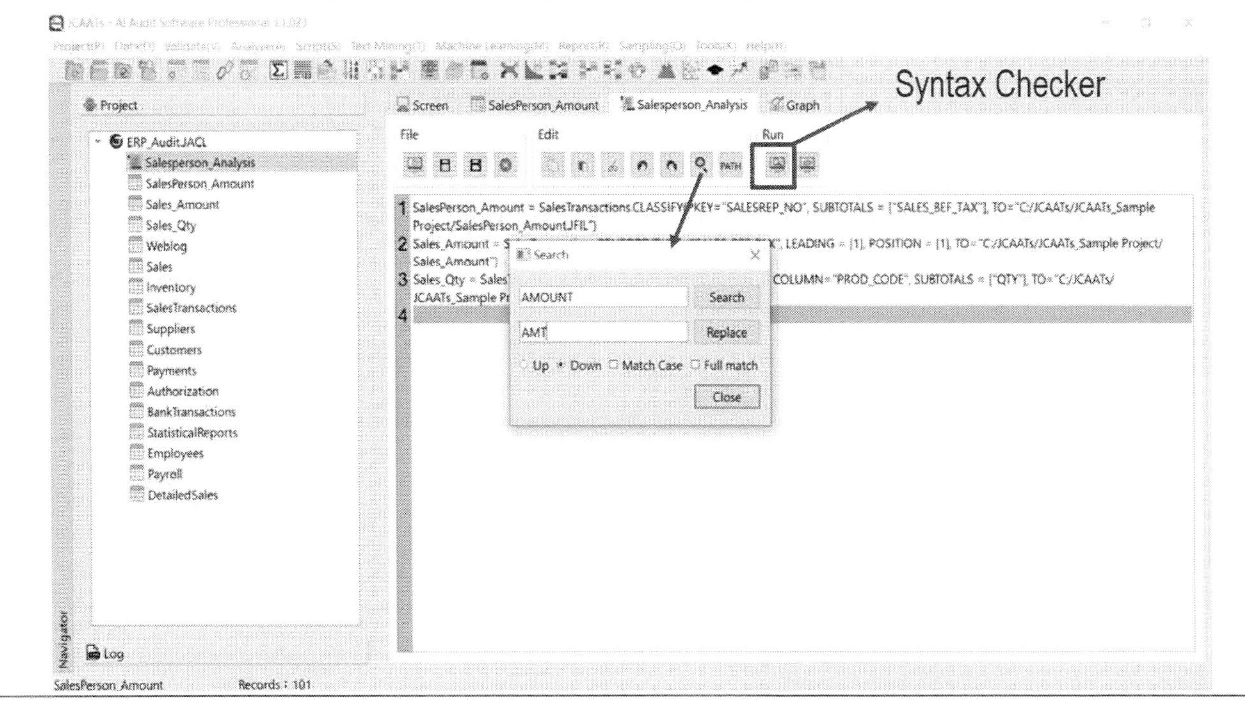

System Parameter- self.v_path_project

- JCAATs Log is a fixed path file used for recording. To ensure smooth component execution on other computers, the path should be changed to a variable method.
- The "system parameter-project path" parameter can be used. Otherwise, the path on the other computer may be different from the path on the development machine, resulting in an error that the path does not exist.

- JCAATs system parameter-project path:

  self.v_path_project

PATH
- Change to system path parameter

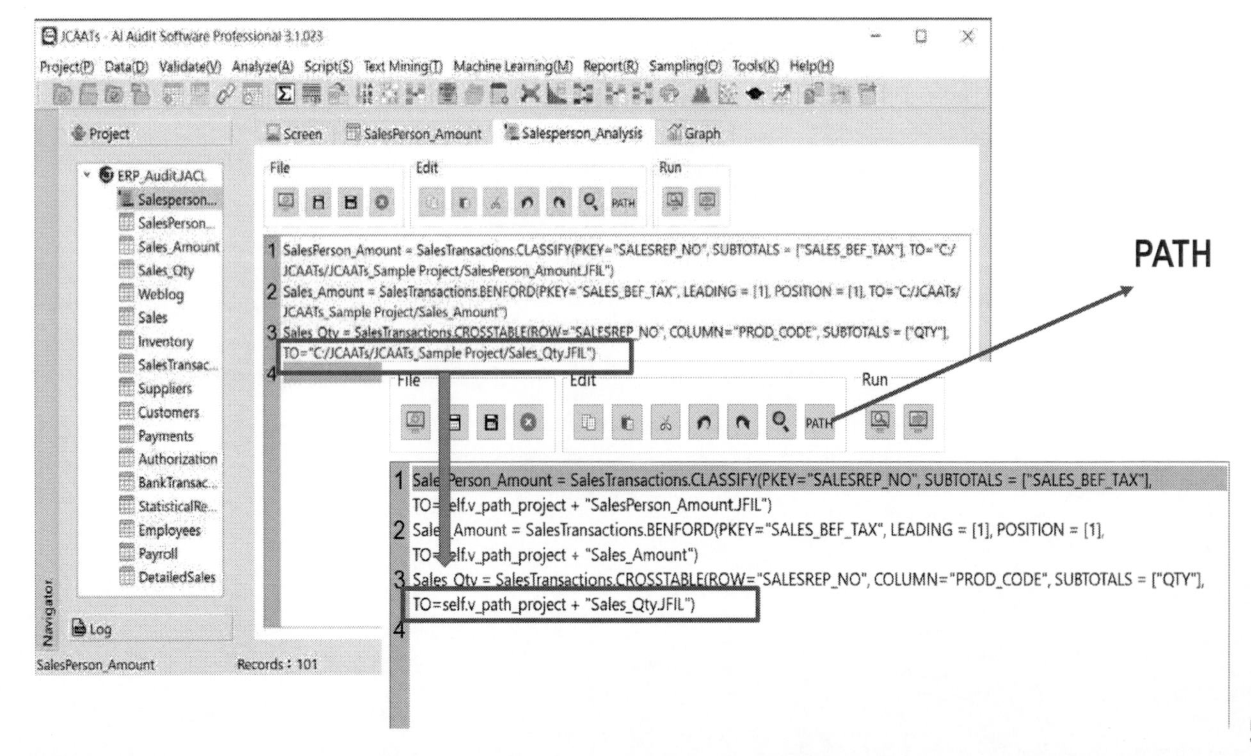

Do Script: self.DO_SCRIPT

- JCAATs provides the following commands that allow you to execute **scripts** and execute multiple **scripts**:

self.DO_SCRIPT ("XXX")

For example: self.DO_SCRIPT ("A_1")

- If you want to execute a script using self.DO_SCRIPT, you must ensure that the script has been created and saved in the project beforehand.

- JCAATs uses the same comment command as Python, where a comment **line** starts with the # symbol.

Variable: self.SET_VAR

- JCAATs provides the following commands that allow you to store a constant into a variable:

self.SET_VAR (variable, value, data type**)**

For example: self.SET_VAR("X", ""1"", "TEXT")

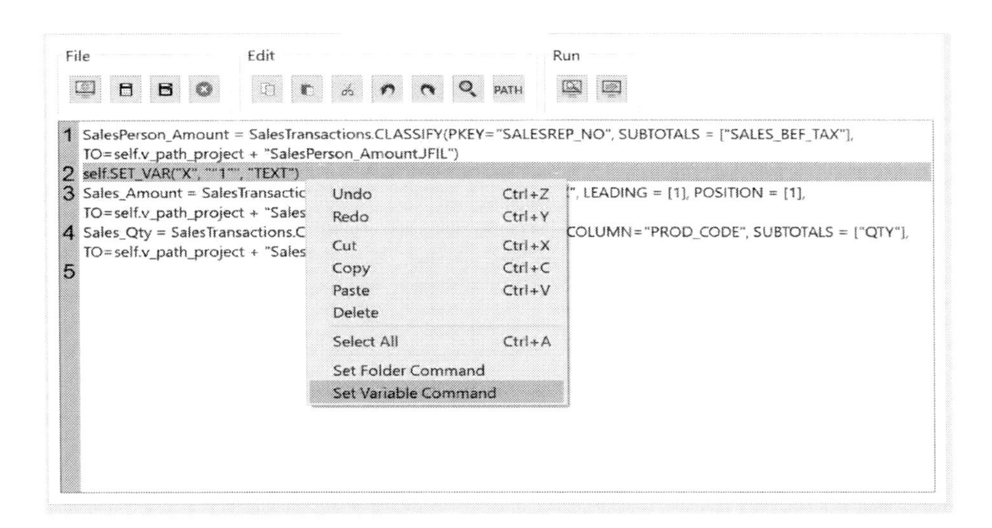

Example: Main Script

- The main script includes three different subscripts and two different folders.

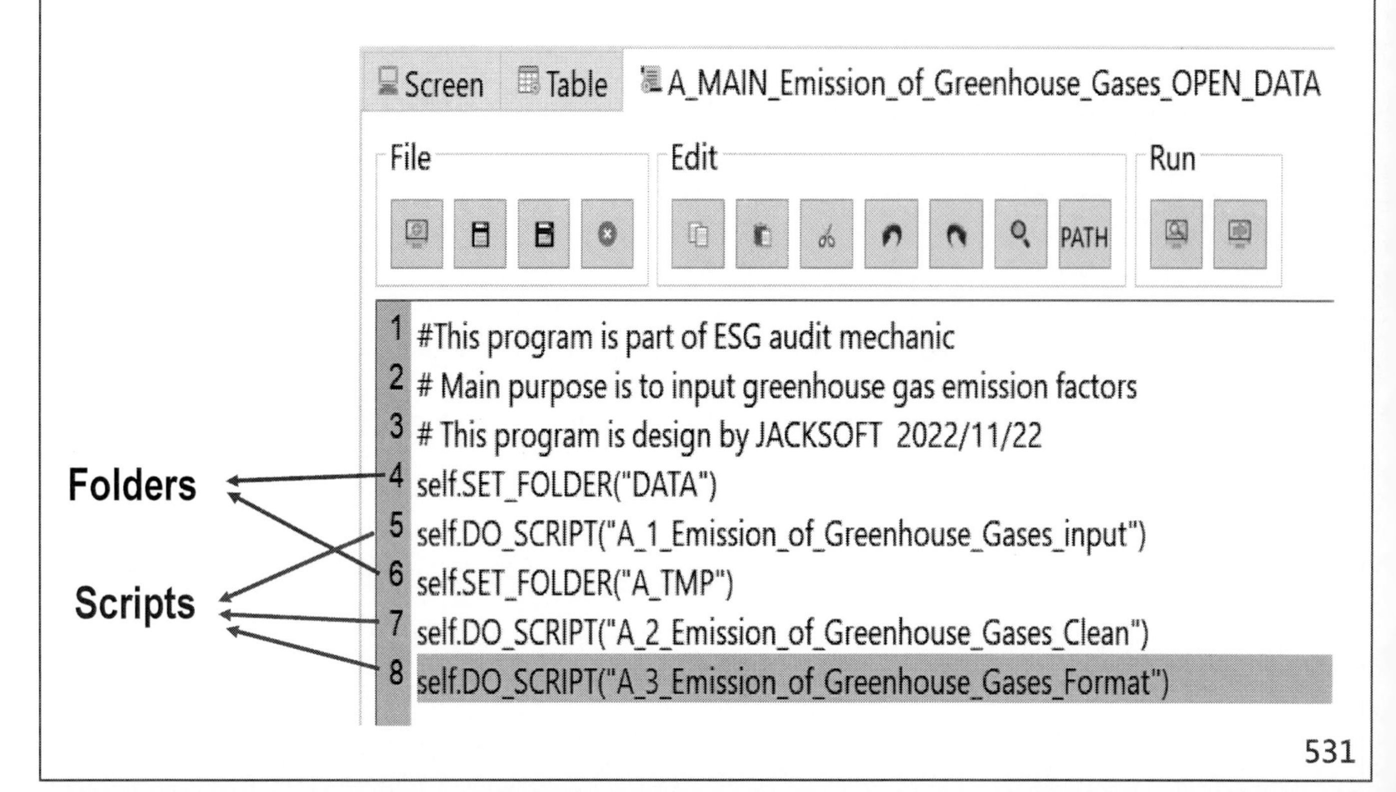

JCAATs Learning notes：

JCAATs - AI Audit Software

Python Based Computer-Assisted Audit Techniques (CAATs)

Data Analysis and Smart Audit

Chapter 9 . Text Mining

Outline:

1. Natural Language Processing
2. Fuzzy Analysis for Text
   - Fuzzy Duplicate Command
   - Fuzzy Join Command
3. Token Analysis for Text
   - Keyword Command
   - Text Cloud Command
4. Sentimental Analysis for Text
   - Sentiment Command

JCAATs - AI Audit Software

1. Nature Language Processing

Nature Language Processing

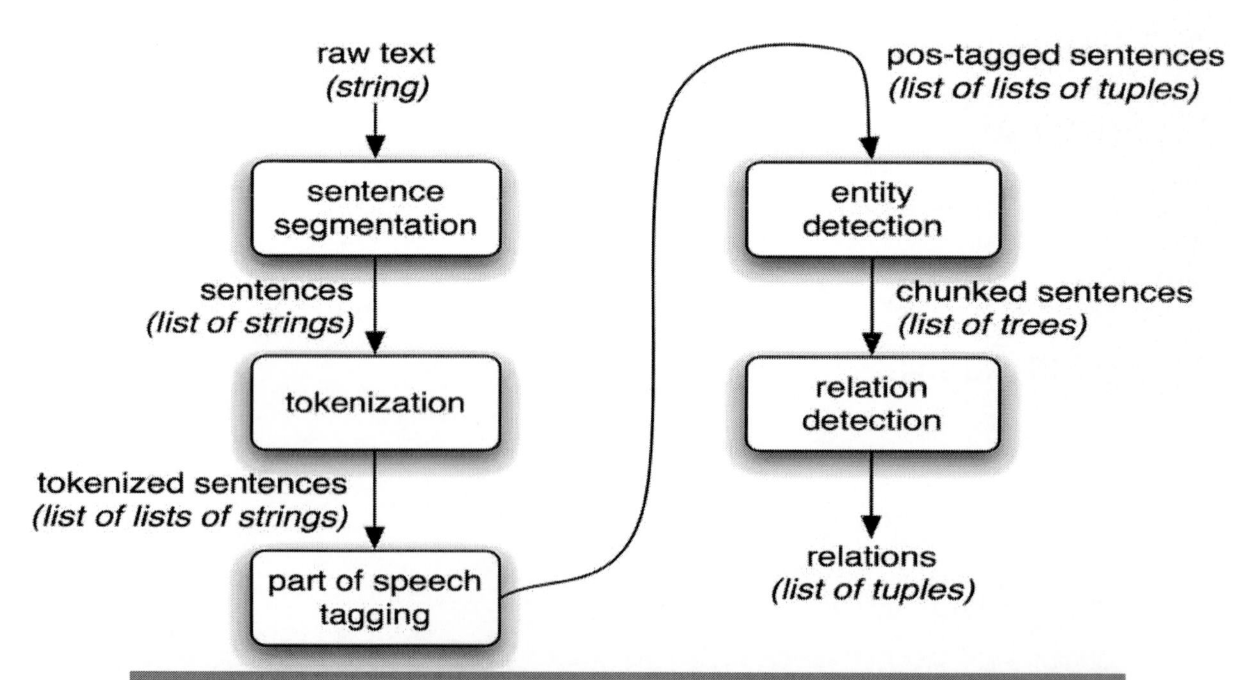

Source: Liu, 2011.

What can you do with Text Mining?

- JCAATs based on NLTK python library.
- It can process 31 languages, such as English, Chinese, German, French, Spanish, Chinese, Japanese, Korean, Arabic, India, etc.
- What can you do with text mining within JCAATs 3.1?
  - **Fuzzy Duplicate** - Clustering similar documents
  - **Fuzzy Join** - Topic detection
  - **Keyword** - Named entity recognition
  - **Text Cloud** - Trend detection
  - **Sentiment** - Sentiment analysis for positive or negative of the sentence.

JCAATs - AI Audit Software

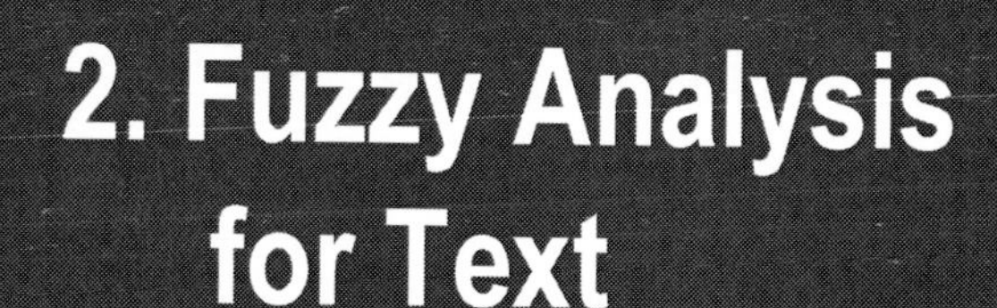

Fuzzy analysis for text

- Fuzzy analysis for text is mainly a mechanism to deal with the similarity of sentences. JCAATs has two commands:
  - **Fuzzy Duplicate** - Clustering similar documents
  - **Fuzzy Join** - Topic detection
- JCAATs use the **Levenshtein Distance** as the fuzzy mechanism to process the text.
- **Levenshtein Distance**
  - It also known as **Edit Distance**, is a metric used to measure the difference between two sequences of characters.
  - The Levenshtein Distance between two strings is defined as the minimum number of insertions, deletions, or substitutions required to transform one string into the other. In other words, it is the minimum number of operations needed to make two strings equal.

Source: Wikipedia, 2023

An example of the Levenshtein Distance

- What is the Levenshtein Distance between the words “dread" and “fraud" ?
- Solution:
  - This is because we can transform “dread" into “fraud".
    - » Step 1: by replacing the “d" with an “f",
    - » Step 2: by deleting the "e",
    - » Step 3: by inserting the “u",
  - The distance is 3:

Thresholds of Fuzzy Analysis for Text

- **Fuzzy analysis for text** computation using Levenshtein distance can be costly, worst-case complete calculation has time complexity and memory space complexity. There are several optimization techniques to improve amortized complexity to avoid complete Levenshtein distance calculation as a pre-selected threshold.

- JCAATs involve two thresholds for these analysis:
  - **Edit Distance:** controls how much two evaluate sentences can differ within a threshold of edit distance.
  - **Difference Percentage:** controls the proportion of an individual value that can be different.

$$\text{Difference Percentage} = \frac{\text{Levenshtein Distance}}{\text{Length of the shorter string}}$$

Example

- If a user sets an **edit distance threshold** of 4 and a **difference percentage** of 40% for the following entries in a database of addresses field. For example, the entries for "12 Main St" , "12 Man St." , "12 Main Street" , and "No 123 Main Street".

String 1	String 2	Length difference	Levenshtein distance	Difference percentage	Fuzzy Group
12 Main St	12 Man St.	0	2	20%	Yes
12 Main St	12 Main Street	4	4	40%	Yes
12 Main St	12 Man Street	3	5	50%	No
12 Main St	No.12 Main Street	7	7	70%	No
12 Man St.	12 Main Street	4	4	40%	Yes
12 Man St.	12 Man Street	3	4	40%	Yes
12 Man St.	No.12 Main Street	7	8	80%	No

Main tab - Fuzzy Duplicates

- Fuzzy duplicate refers to a type of duplicate record or data entry in a dataset that may not be an exact match, but shares similarities with another record in the dataset.

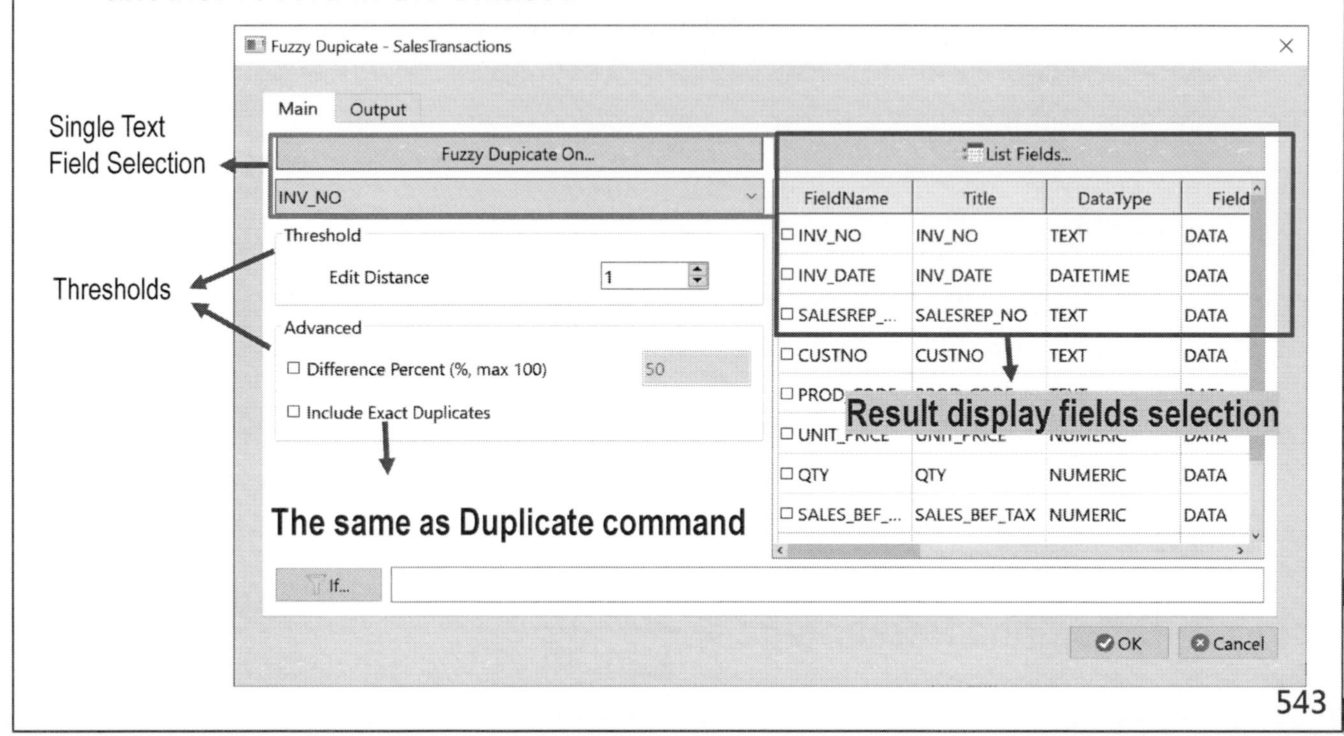

Main tab - Fuzzy Join

- Fuzzy join is a technique used in data analytics to combine data from different sources based on approximate matches between data records. Fuzzy join algorithms allow for matching records that may contain slight differences or errors.

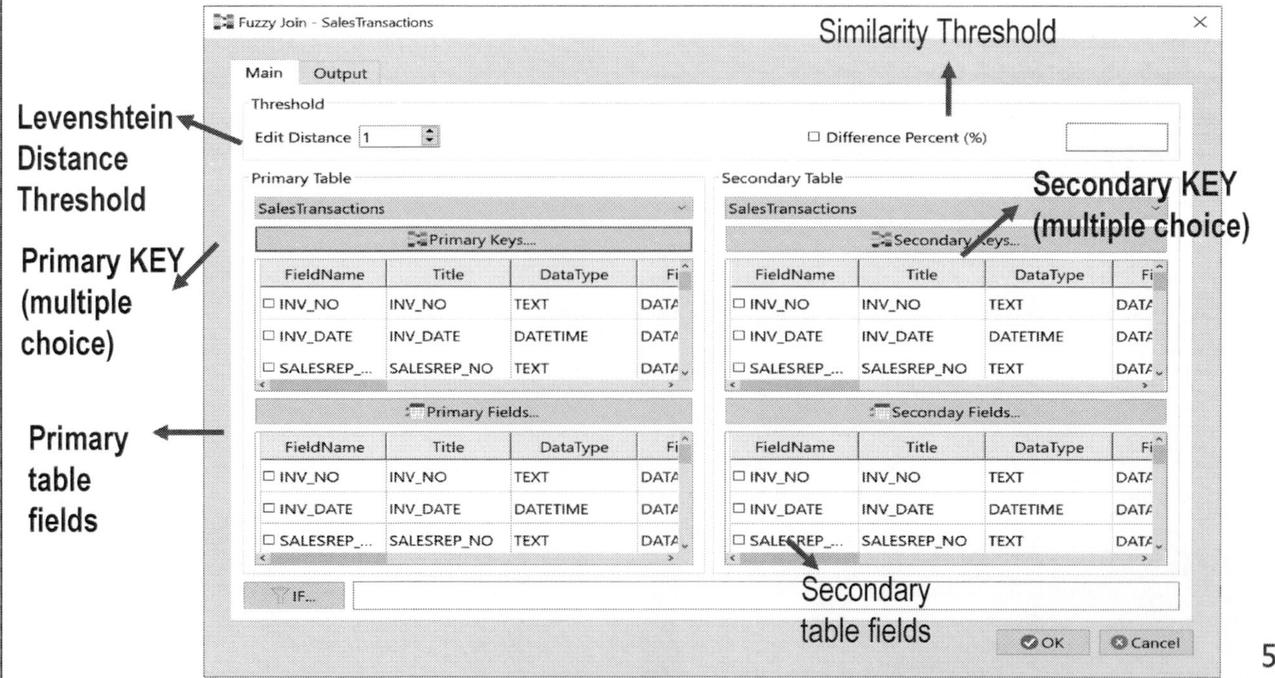

Output tab - Fuzzy Join Type

- Select Join type in the output tab.

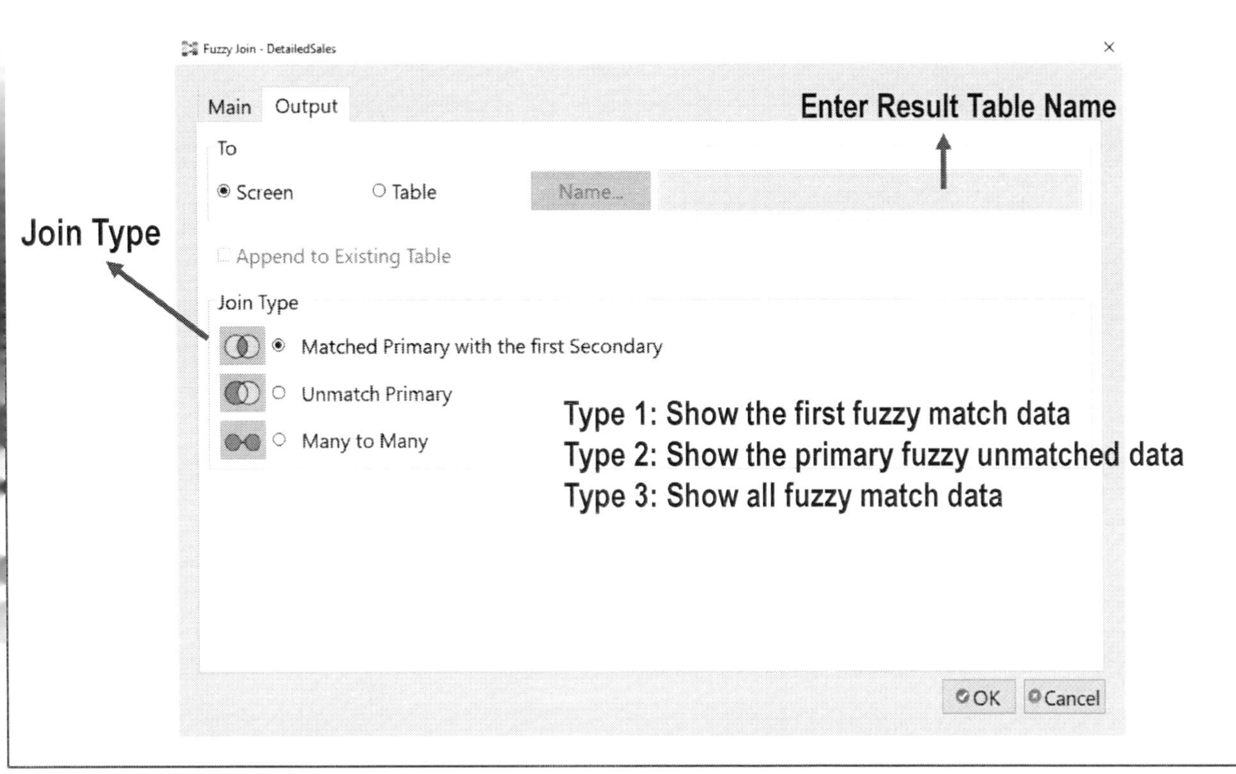

Example: Fuzzy Join

EX

Payroll(Primary)	
EMP_NO	EMP_Street
01	1296 Longman Crescent
02	Number 1 South Shamian Street
03	1220 Southwest 3rd Avenue
04	110 Ninth Avenue, South
05	215 N. Main Street

Employee(Secondary)	
EMP_NO	Street
01	1296 Shortman Crescent
02	Nunber 1 South Shamian Street
03	1220 Southwest 3rd Avenue
04	110 Ninth Avenue, South
05	215 W. Main Street

Result

Fuzzy Match

Join Match

EMP_NO	EMP_Street	Street
01	1296 Longman Crescent	1296 Shortman Crescent
02	Number 1 South Shamian Street	Nunber 1 South Shamian Street
03	1220 Southwest 3rd Avenue	1220 Southwest 3rd Avenue
04	110 Ninth Avenue, South	110 Ninth Avenue, South
05	215 N. Main Street	215 W. Main Street

JCAATs - AI Audit Software

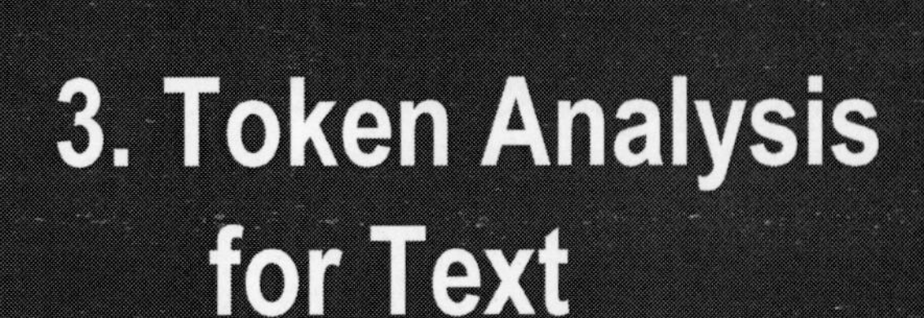

3. Token Analysis for Text

Token Analysis for Text

- Token analysis is a technique used in natural language processing (NLP) to break down a piece of text into individual tokens, which are the smallest units of meaning in the text.

- JCAATs supports **word-level tokenization** techniques which involves breaking the text down into individual words or combined words.

- In JCAATs, there are two commands for token analysis:
  - **Keyword**: it is the process of researching, identifying and analyzing the keywords and phrases that used in a text field of the dataset.
  - **TextCloud:** it, also called a word cloud or tag cloud, is a graphical representation of text data in which the size of each word corresponds to its frequency or importance within the text.

Dictionaries

- Dictionaries commonly used in token analysis to improve the quality and accuracy of token analysis.

- JCAATs supports three dictionaries for text mining in JCAATs, i.e.
  - **Stop word dictionaries**: These are dictionaries that contain a list of common words that are considered to be of little value for text analysis.
  - **Named Entity Recognition (NER) dictionaries**: These are dictionaries that contain lists of named entities such as people, organizations, locations and domain-specific name.
  - **Sentiment dictionary**: a collection of words or phrases that have been annotated with sentiment scores or labels indicating their emotional polarity, such as positive, negative, or neutral.

- JCAATs allows users to build their own dictionaries for text analysis.

Term frequency (TF)

- **Term frequency (TF)** is a metric used in natural language processing and text analysis to quantify the number of times a particular word or term appears in a given text file of the dataset. It is a simple yet powerful approach to understanding the importance of a word in a piece of text.
- **TF** is typically calculated by dividing the number of times a term appears in a document by the total number of terms in the document. The resulting value is a ratio that represents the frequency of occurrence of the term in the document.
- For example, suppose we have a document that contains 100 words, and the term "data" appears 5 times in that document. The TF for the term "data" in that document would be 5/100 = 0.05.

TF-IDF: Term Frequency - Inverse Document Frequency

TF–IDF is a numerical statistic that is intended to reflect how important a word is to a document in a collection or a text field of table.

It is one of the most popular term-weighting schemes for text mining today.

TF-IDF

$$Score_{t,d} = tf_{t,d} \times idf_t$$

Source: https://en.wikipedia.org/wiki/Tf-idf

The **TF–IDF** value increases proportionally to the number of times a word appears in the document and is offset by the number of documents in the text field that contain the word, which helps to adjust for the fact that some words appear more frequently in general.

Main tab - Keyword

- The purpose of keyword analysis is to understand what keywords and phrases are relevant to a particular topic, and how frequently they are used in the dataset.

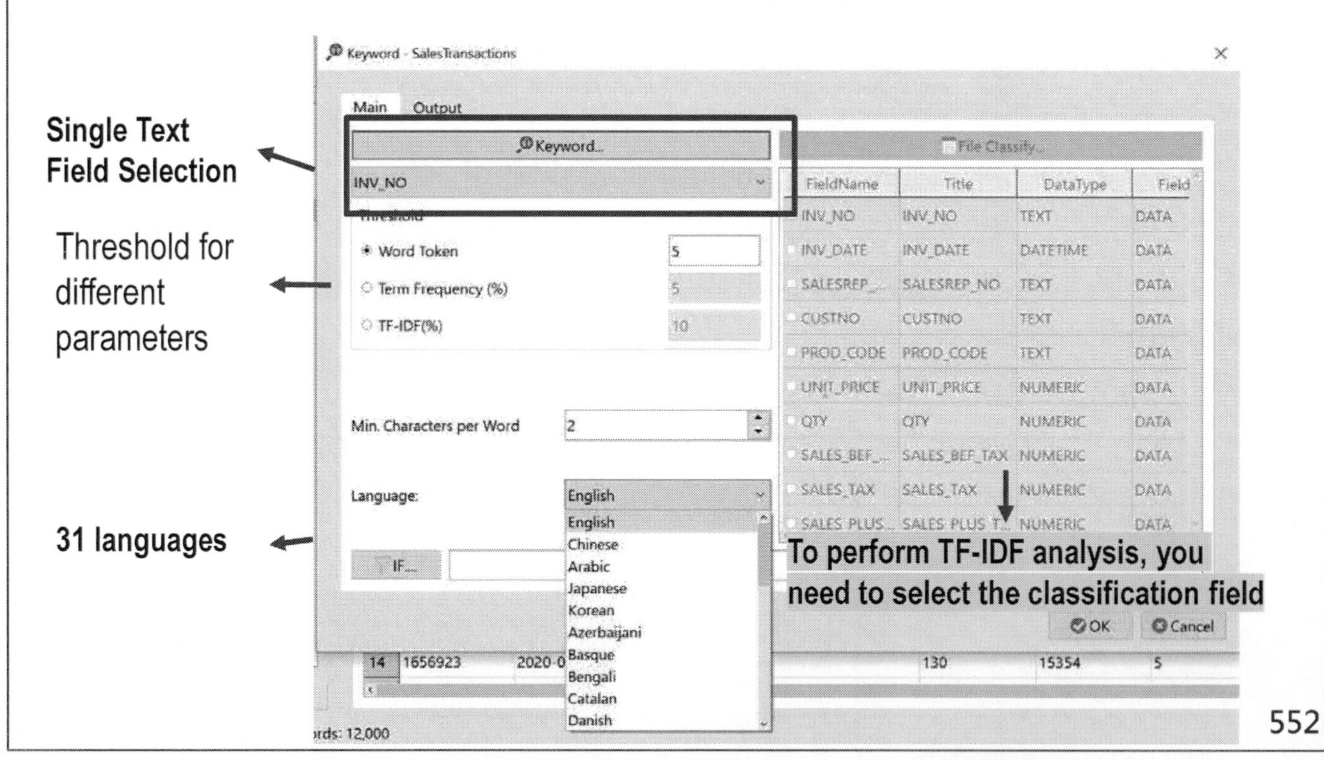

Keyword - Dictionary

- Text Mining can use the NLTK standard dictionary or customize dictionary. Customize Dictionary: Tools>>Dictionary Tool, and upload a dictionary file.

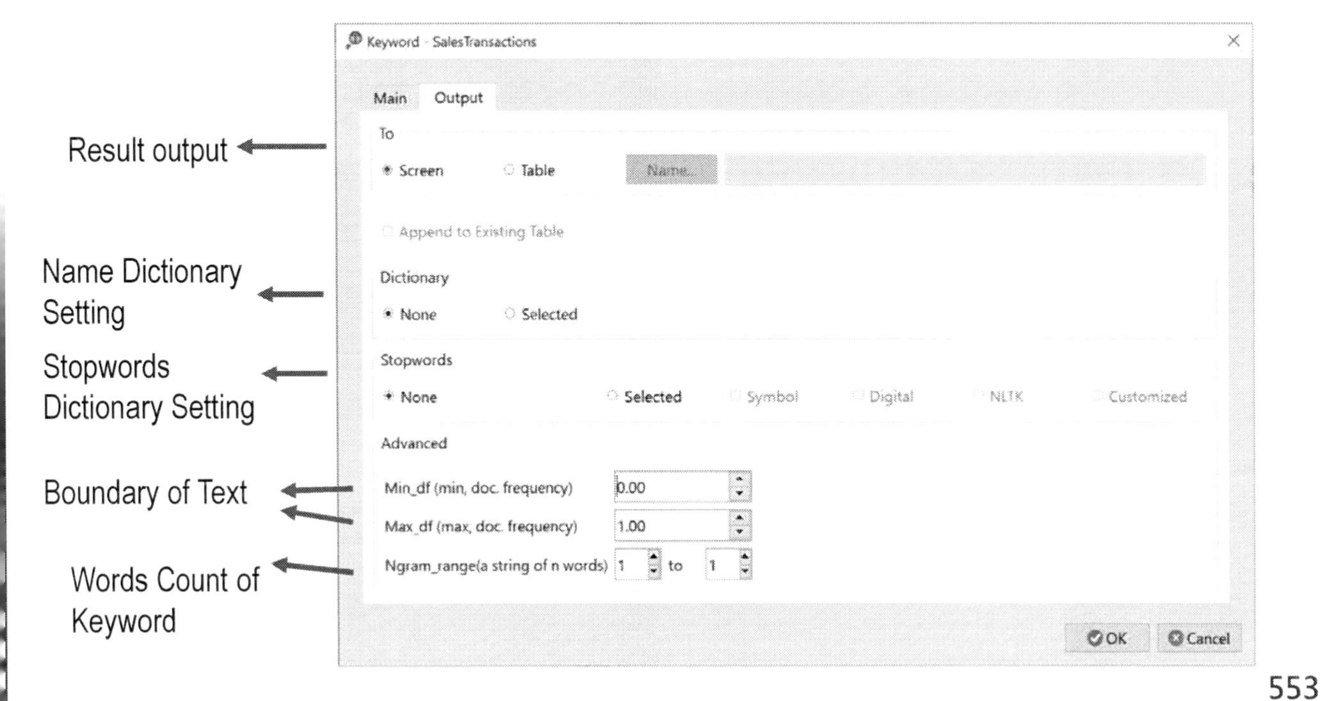

Result Table of Keyword Command

- The result table includes two columns, i.e. Word and TF-IDF.

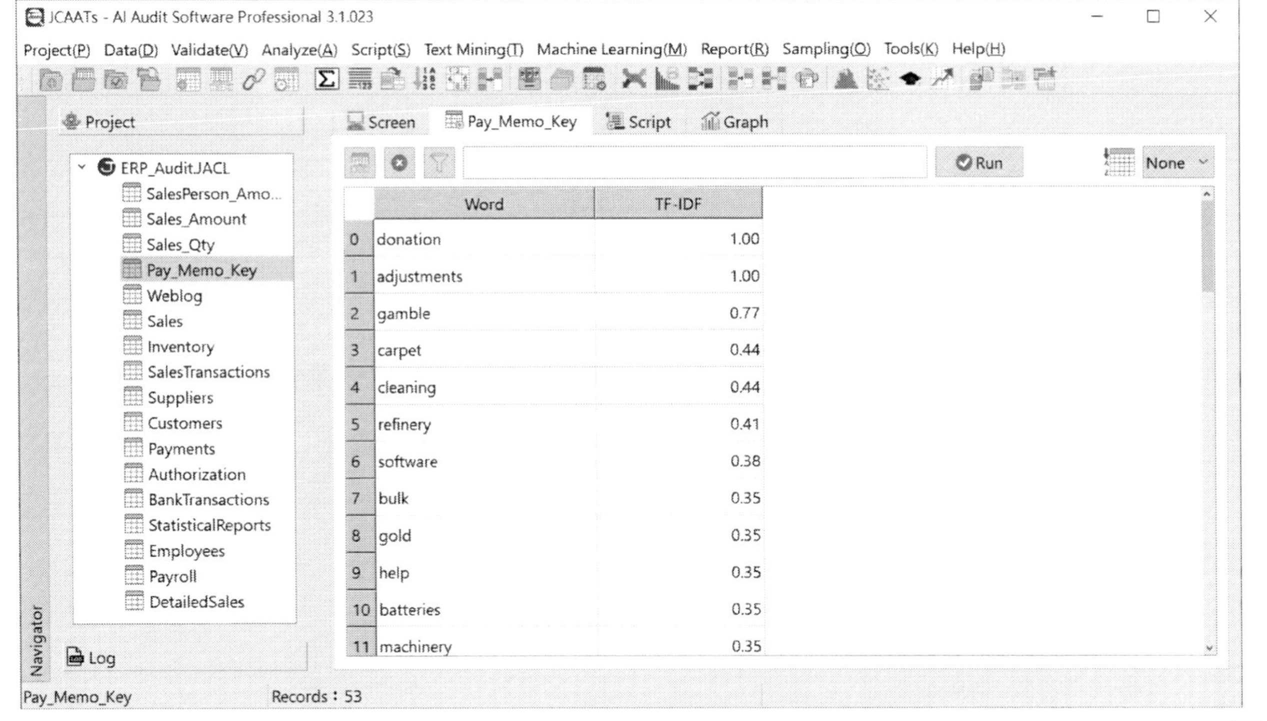

	Word	TF-IDF
0	donation	1.00
1	adjustments	1.00
2	gamble	0.77
3	carpet	0.44
4	cleaning	0.44
5	refinery	0.41
6	software	0.38
7	bulk	0.35
8	gold	0.35
9	help	0.35
10	batteries	0.35
11	machinery	0.35

Main tab - Text Cloud

- A **text cloud** or **word cloud** is a visualization of word frequency in a given text as a weighted list. The technique has been popularly used to visualize the topical content in text mining.

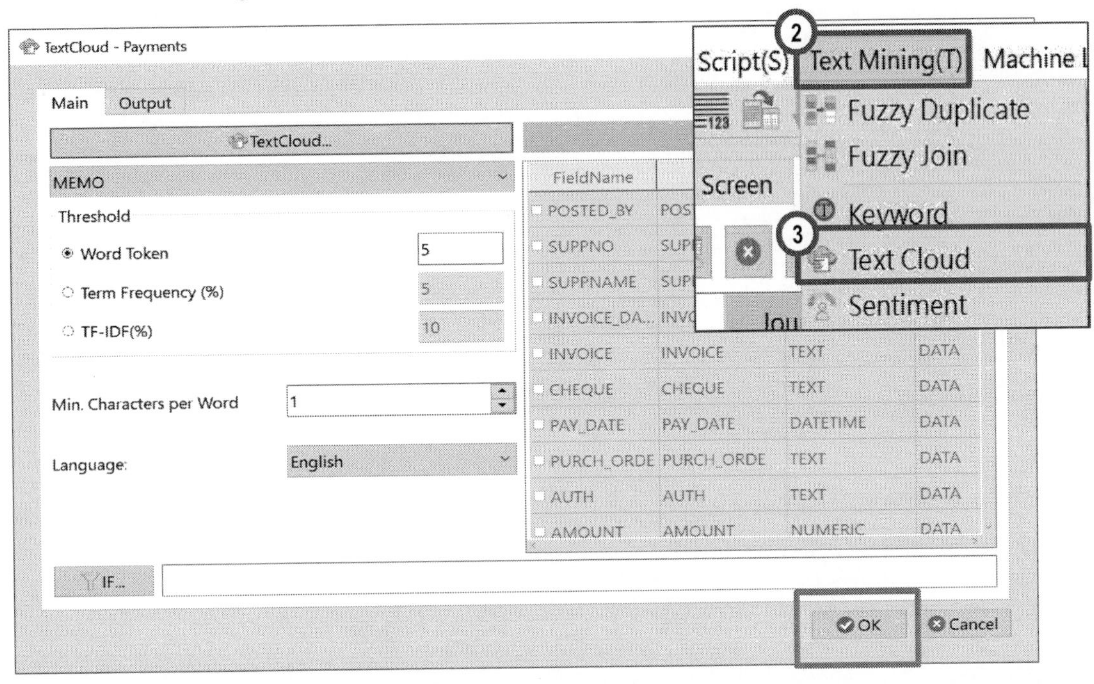

Result Graph of Textcloud Command

- The most frequently occurring words are displayed in larger font sizes and the less frequently occurring words are displayed in smaller font sizes.

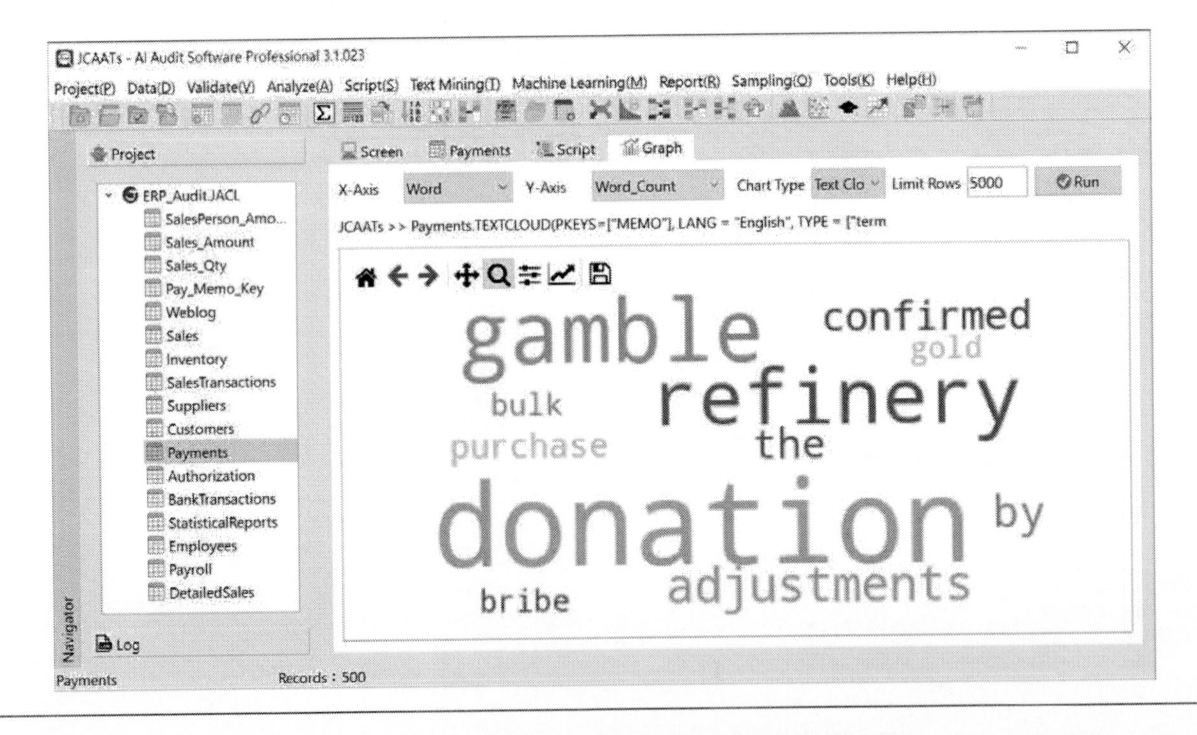

JCAATs - AI Audit Software

3. Sentimental Analysis

Sentimental Analysis

- The goal of sentimental analysis is to identify and extract subjective information from the text, such as opinions, emotions, attitudes, and intentions, and to classify them into **positive, negative, or neutral** categories.

- It can be used in internal audit to analyze and interpret large volumes of text data from various sources, such as **financial reports**, **internal memos**, **employee feedback**, and **customer complaints**, to identify potential risks, issues, and opportunities for improvement.

- However, it is important to note that sentimental analysis is not foolproof and may still require human interpretation and judgment to make accurate assessments and decisions.

Sentimental Analysis Cases

- Whether the text should be positive but have negative statement, such as review of loan approval opinions.

- Whether the text should be negative but have positive statement, such as not approved, not admitted, evaluated.

- Analyzing positive and negative comments on survey data to justify the correctness of the survey analysis.

- Analyzing any follow up of the negative comments of customer feedback from social media.

JCAATs-AI Audit Software
Copyright © 2023 JACKSOFT.

Main tab - Sentiment

- The use of semantic analysis can help auditors to identify potential risks and issues more quickly and efficiently than traditional manual methods.

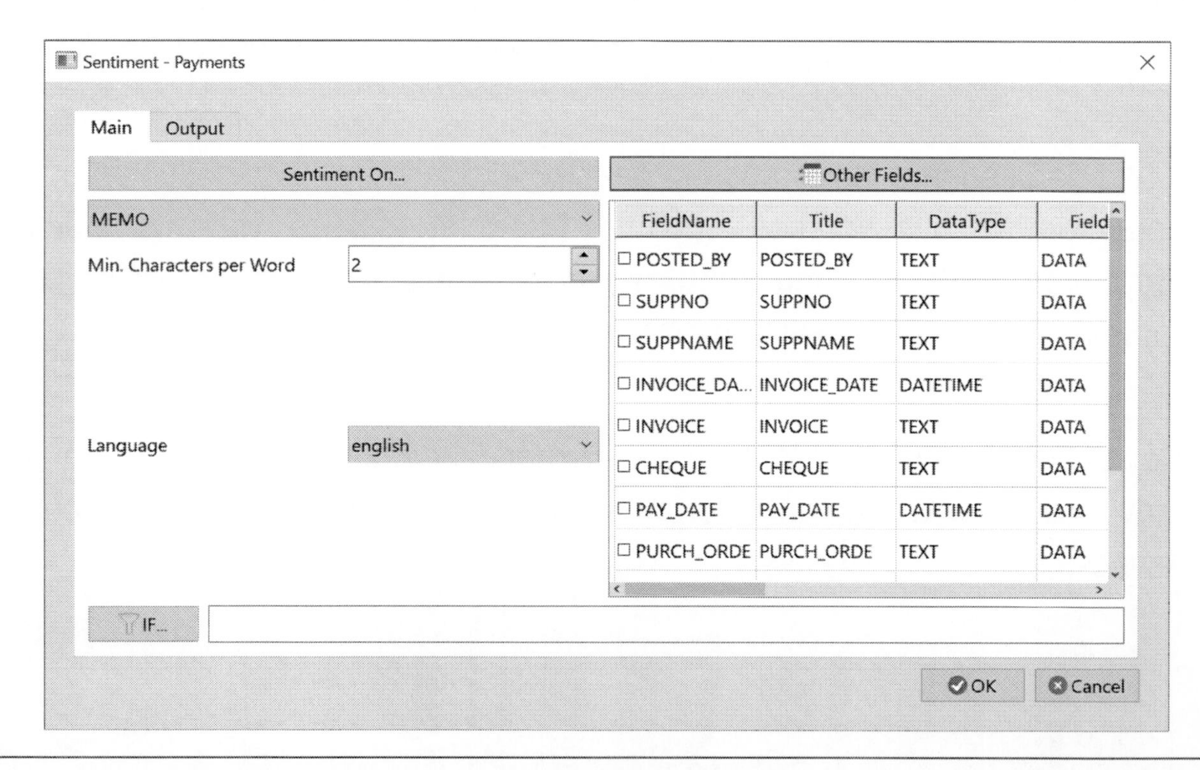

Result Table of Sentiment Command

- The result includes 5 new fields, i.e. negative word rate, positive word rate, neutral word rate, compound rate, and sentiment result.

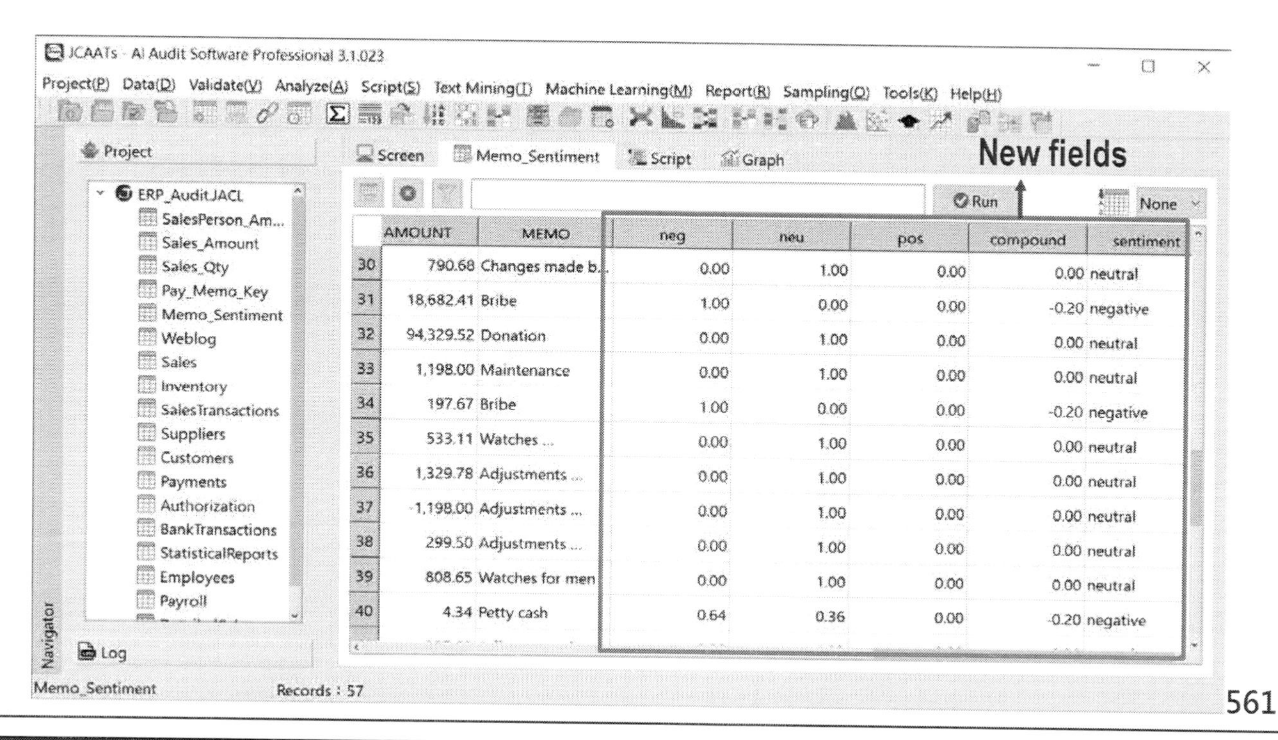

	AMOUNT	MEMO	neg	neu	pos	compound	sentiment
30	790.68	Changes made b...	0.00	1.00	0.00	0.00	neutral
31	18,682.41	Bribe	1.00	0.00	0.00	-0.20	negative
32	94,329.52	Donation	0.00	1.00	0.00	0.00	neutral
33	1,198.00	Maintenance	0.00	1.00	0.00	0.00	neutral
34	197.67	Bribe	1.00	0.00	0.00	-0.20	negative
35	533.11	Watches ...	0.00	1.00	0.00	0.00	neutral
36	1,329.78	Adjustments ...	0.00	1.00	0.00	0.00	neutral
37	-1,198.00	Adjustments ...	0.00	1.00	0.00	0.00	neutral
38	299.50	Adjustments ...	0.00	1.00	0.00	0.00	neutral
39	808.65	Watches for men	0.00	1.00	0.00	0.00	neutral
40	4.34	Petty cash	0.64	0.36	0.00	-0.20	negative

JCAATs Learning notes：

JCAATs - AI Audit Software

Python Based Computer-Assisted Audit Techniques (CAATs)

Data Analysis and Smart Audit

Chapter 10. Machine Learning

Outline:

1. Machine Learning Procedure
2. Supervised Machine Learning
3. Unsupervised Machine Learning

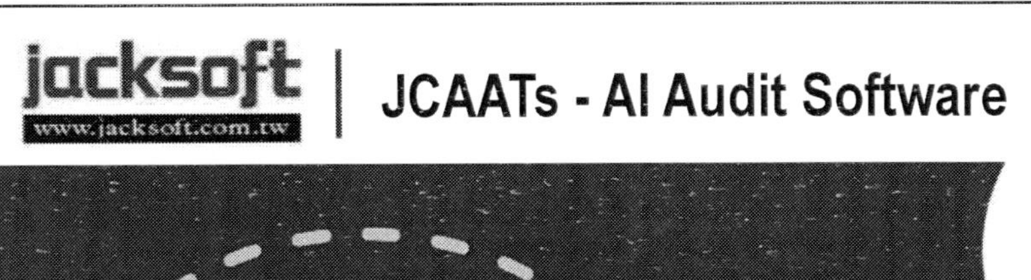

1. Machine Learning

Types of Machine Learning

- JCAATs support several types of machine learning (ML) , including:
  - **Supervised learning**: In this type of ML, the algorithm is trained on a labeled dataset where both the input and output variables are provided. The goal is to learn a mapping between the input and output variables so that the model can predict the output for new input data.
  - **Unsupervised learning**: In this type of ML, the algorithm is trained on an unlabeled dataset where only the input variables are provided. The goal is to discover patterns and relationships in the data without any prior knowledge of the output.
  - **Semi-supervised learning**: This is a combination of **supervised** and **unsupervised** learning, where the algorithm is trained on a dataset that contains both labeled and unlabeled data. The goal is to use the labeled data to guide the learning process and improve the performance of the model.

Supervised Learning Procedure

1. **Model Selection**: Select an appropriate ML algorithm that best suits the problem at hand.
2. **Model Training**: Train the selected ML model using the prepared data. This involves feeding the model with input data and adjusting the training **pipeline** to minimize the prediction error.
3. **Model Evaluation**: Evaluate the performance of the trained model using appropriate metrics, such as **accuracy**, **precision**, **recall**, **F1-score**, etc.
4. **Model Tuning**: Fine-tune the model by adjusting the hyper-parameters based on the evaluation results. This may involve repeating steps 1-3 until the desired performance is achieved.
5. **Model Deployment**: Deploy the trained and tuned model in a production environment, where it can be used to make predictions on new input data.
6. **Model Maintenance**: Continuously monitor and update the deployed model to ensure that it remains accurate and up-to-date.

JCAATs Machine Learning Commands

- JCAATs import **Scikit-learn** python library with several reproofed machine learning algorithms.

- What can you do with machine learning within JCAATs 3.1?
  - Supervised machine learning
    - **Train**
    - **Predict**
  - Unsupervised machine learning
    - **Outlier**
    - **Cluster**

JCAATs - AI Audit Software

2. Supervised Machine Learning

JCAATs Supervised Machine Learning

- Supervised machine learning commands:
  - **Train**: Carry out machine learning on the training data, and generate knowledge models and effect analysis values.
  - **Predict**: Import the knowledge model of Machine Learning to predict the data and display the probability.
- EX: Supervised ML applications in auditing such as Anomaly Detection , Risk Assessment, Fraud Detection,etc. The algorithm can learn what is normal for the organization and flag any deviations from that pattern for further investigation.
- Overall, supervised ML can help auditors automate certain tasks, reduce the risk of errors, and improve the accuracy of their findings. However, it is important to note that the use of ML in audit should be complemented by professional judgement and human oversight.

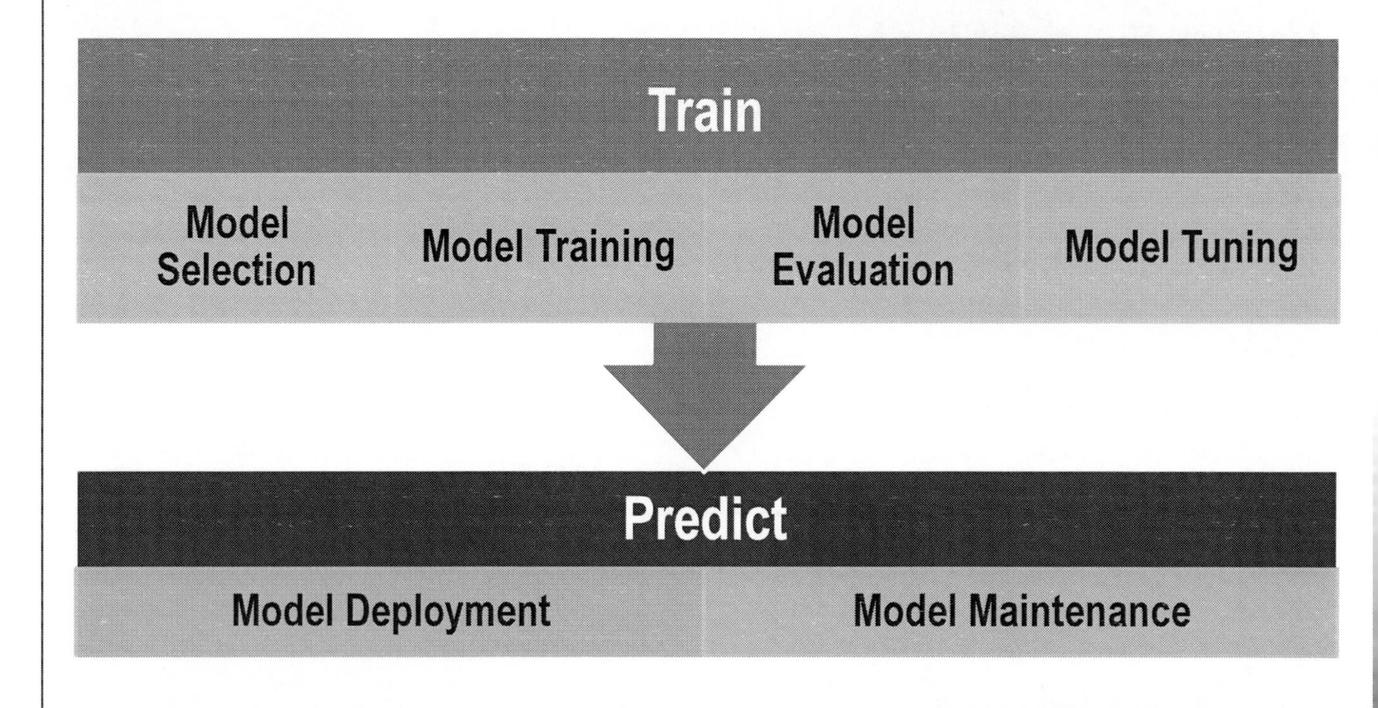
JCAATs-AI Audit Software
Copyright © 2023 JACKSOFT.
Supervised Machine Learning Process
Train
Model Selection
Model Training
Model Evaluation
Model Tuning
Predict
Model Deployment
Model Maintenance
JCAATs through AI technology. Provide a fast and easy way for smart audit.
571

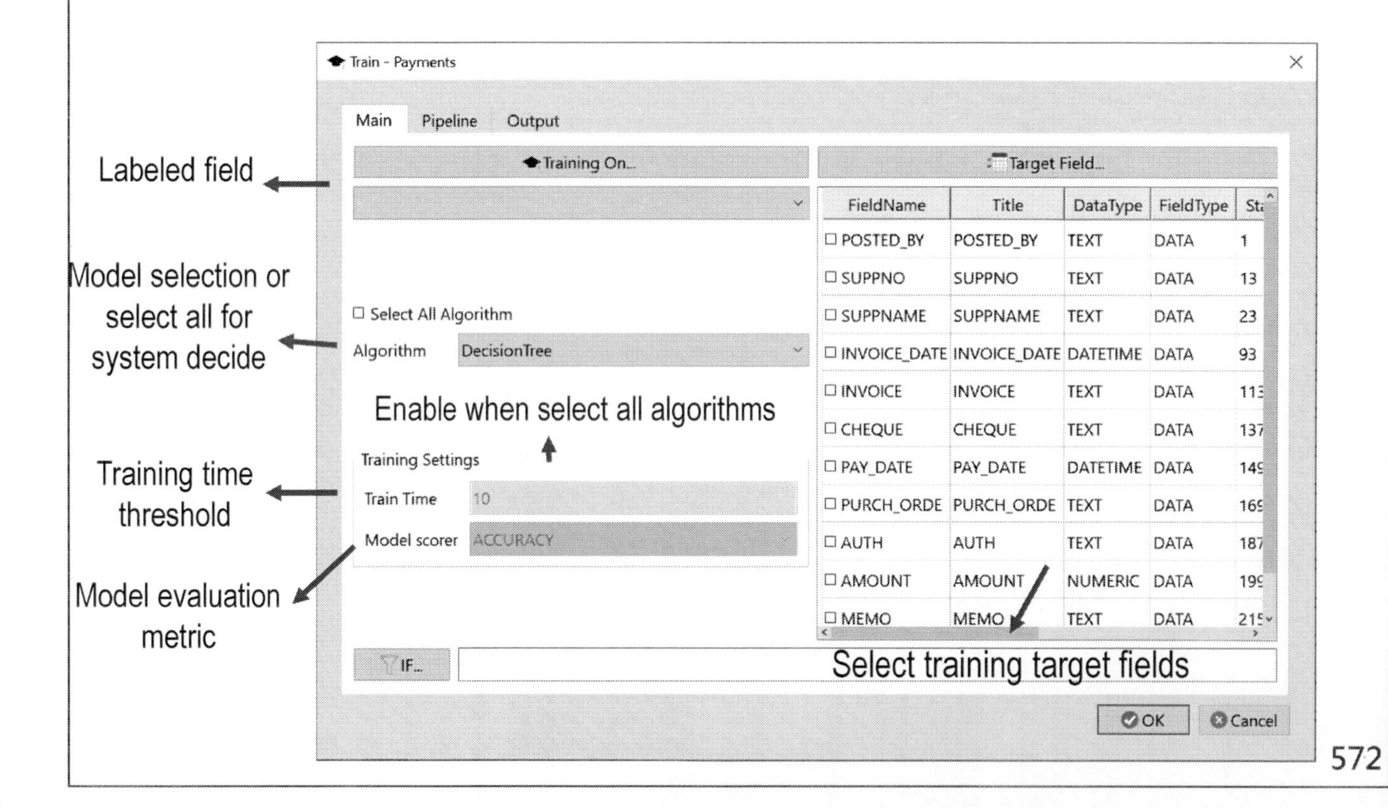
JCAATs-AI Audit Software
Copyright © 2023 JACKSOFT.
Main tab - Train Command
Through flexible interface to start machine learning for classification.
Train - Payments
Main
Pipeline
Output
Training On...
Target Field...
Labeled field
Model selection or select all for system decide
Training time threshold
Model evaluation metric
Select All Algorithm
Algorithm
DecisionTree
Enable when select all algorithms
Training Settings
Train Time
10
Model scorer
ACCURACY
FieldName
Title
DataType
FieldType
POSTED_BY
SUPPNO
SUPPNAME
INVOICE_DATE
INVOICE
CHEQUE
PAY_DATE
PURCH_ORDE
AUTH
AMOUNT
MEMO
TEXT
DATETIME
NUMERIC
DATA
IF...
Select training target fields
OK
Cancel
572

Train - Model Selection

- JCAATs Train command performs both **bi-classification** and **multi-classification**. By choosing the appropriate algorithm for a given classification problem, we can build models that accurately classify instances and provide value in a variety of applications.
- JCAATs provide five machine learning algorithms for your model selection.
  - **Decision Tree**
  - **KNN (k-nearest neighbors)**
  - **Logistic Regression**
  - **Random Forest**
  - **SVM (support vector machines)**
- You also can select all algorithms to let system decided by using your select scorer as the metric for optimization.

Pipeline - Supervised Machine Learning

- A supervised machine learning pipeline is a series of steps that take raw data as input and produce a trained model that can make predictions or classifications on new, unseen data. The steps in JCAATs pipeline include:
  - **Data Preprocessing**: This can involve missing values imputation and normalizing or standardizing features (i.e. label encoder or one-hot encoder) to ensure that they are on the same scale.
  - **Feature Engineering**: Feature engineering involves selecting and transforming the most relevant features for the machine learning model. It include data imbalance process.
  - **Training/Test Data Splitting**: The dataset is split into two parts - a training set and a test set. The training set is used to train the machine learning model, while the test set is used to evaluate the performance of the model.

Customizing Your Pipeline

- By allowing users to customize their machine learning pipeline, JCAATs can provide more transparency and explainability for the learning process and results.
- It is common for an audit dataset, the number of instances in each class (e.g., fraud vs non-fraud) is often highly skewed, with one class being much more prevalent than the other, i.e. **imbalance**. **SMOTE** is a well know technique to improve the performance of machine learning models on imbalanced audit datasets and make more accurate predictions on both the majority and minority classes.
- Overall, JCAATs can also help auditors communicate more effectively with other stakeholders, such as directors, managers, and regulators. It can help to improve the effectiveness and efficiency of the audit process, while also providing more confidence and trust in the audit findings by using machine learning.

Pipeline tab- Train

- JCAATs allows you to customize your their machine learning pipeline.

Data Preprocessing

Feature Engineering

Training/ Test Data Splitting

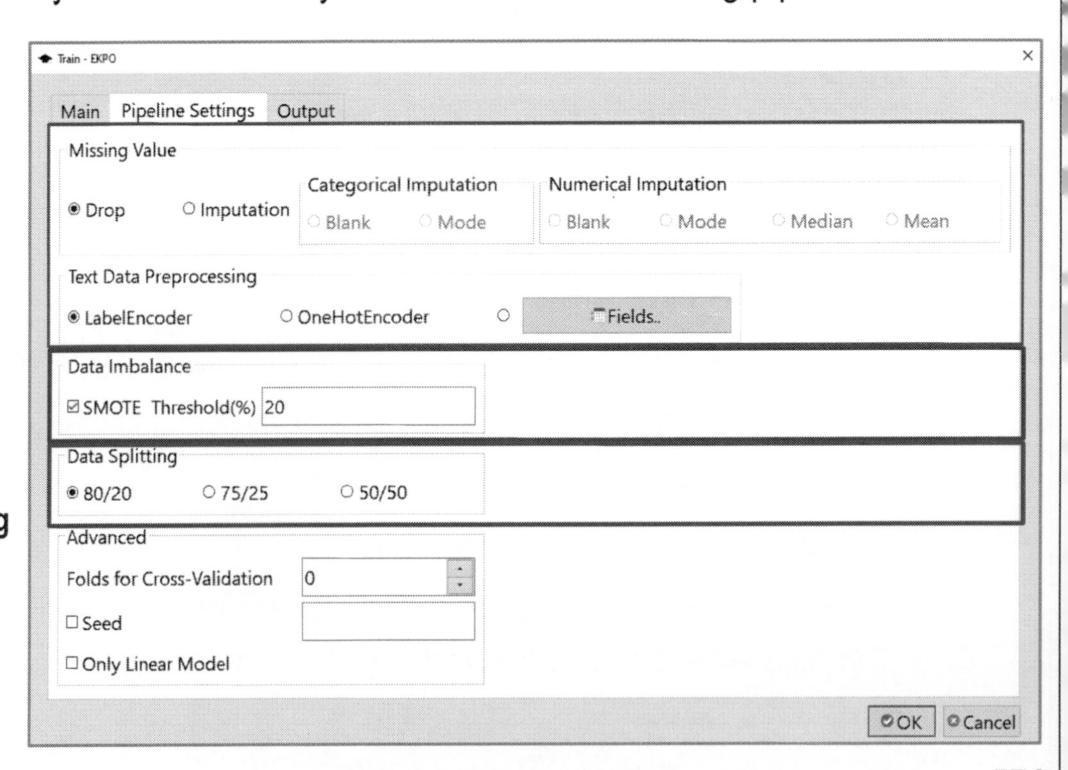

Output tab - Train

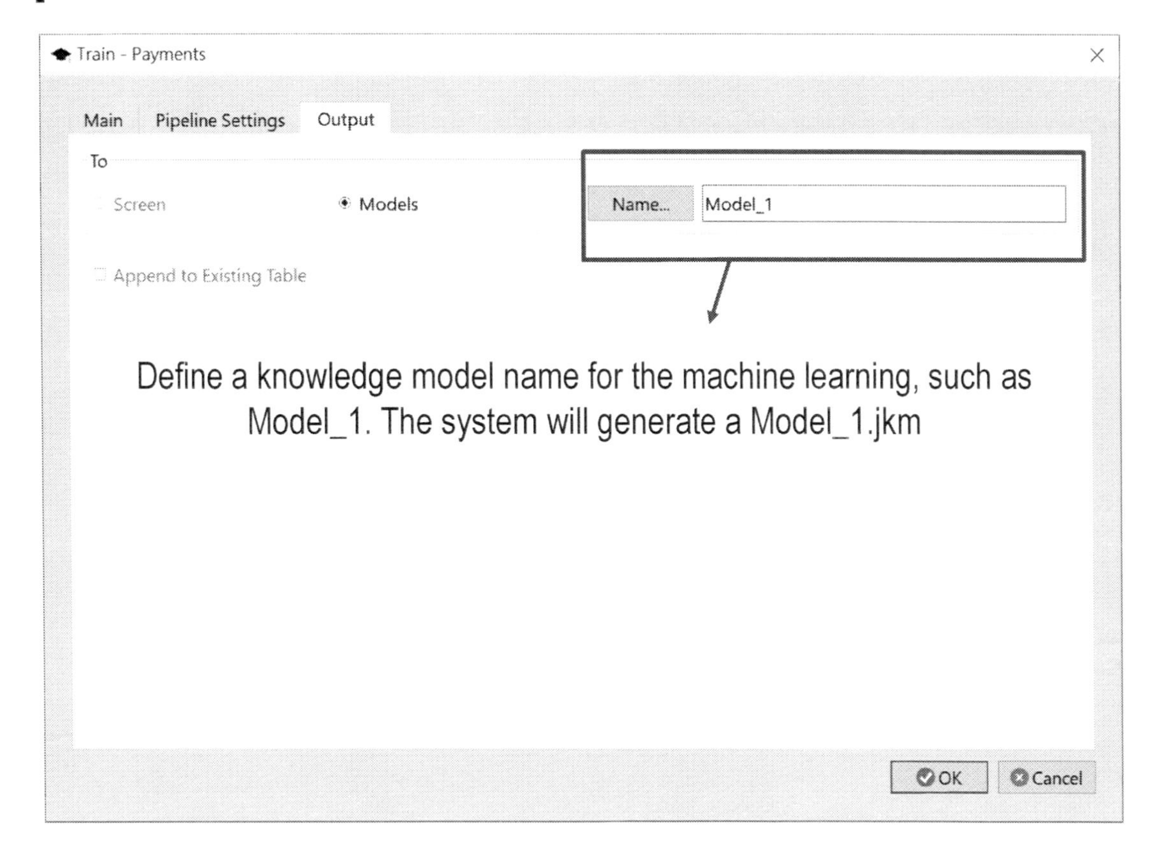

Model Evaluation - Train

- JCAATs supports the following three tables for you to evaluate the supervised machine learning result model.
  - **ConfusionMatrix**
  - **PerformanceMetrics**
  - **SummaryReport**
- JCAATs supports various metrics with **weight** for **multi-classification** as:
  - **Accuracy**: It measures the proportion of correctly predicted outcomes out of the total number of predictions.
  - **Precision**: It measures the proportion of true positive predictions out of the total number of positive predictions.
  - **Recall**: It measures the proportion of true positive predictions out of the total number of actual positive outcomes.
  - **F1 score**: It is the harmonic mean of precision and recall. It combines both metrics to provide an overall measure of the model's performance.

Confusion Matrix

- A confusion matrix is a table that is often used to evaluate the performance of a machine learning classification model. It summarizes the number of correct and incorrect predictions by the model, compared to the actual outcomes in the data.

		Predicted	
		Negative (N)	Positive (P)
Actual	Negative	True Negative (TN)	**False Positive (FP)** **Type I Error**
	Positive	**False Negative (FN)** **Type II Error**	True Positive (TP)

Source: Anuganti Suresh, 2020

Confusion Matrix - Metric Definition

- **For bi-classification:**
  - **Accuracy**: It is defined as (TP + TN) / (TP + TN + FP + FN).
  - **Precision**: It is defined as TP / (TP + FP).
  - **Recall**: It is defined as TP / (TP + FN).
  - **F1 score**: It is defined as 2 * (precision * recall) / (precision + recall).

- **For multi-classification:** This is the weighted average of metric across all classes, with weights based on the number of instances of each class, i.e. count_i of class i.
  - **Weighted precision**: It is defined as Σ(precision_i * count_i) / Σ(count_i), where precision_i is the precision for class i.
  - **Weighted recall**: It is defined as Σ(recall_i * count_i) / Σ(count_i), where recall_i is the recall for class i.
  - **Weighted F1 score**: It is defined as Σ(F1_i * count_i) / Σ(count_i), where F1_i is the F1 score for class i.

Confusion Matrix - Train

- Confusion Matrix Graph enables you to easy explain the learning result.

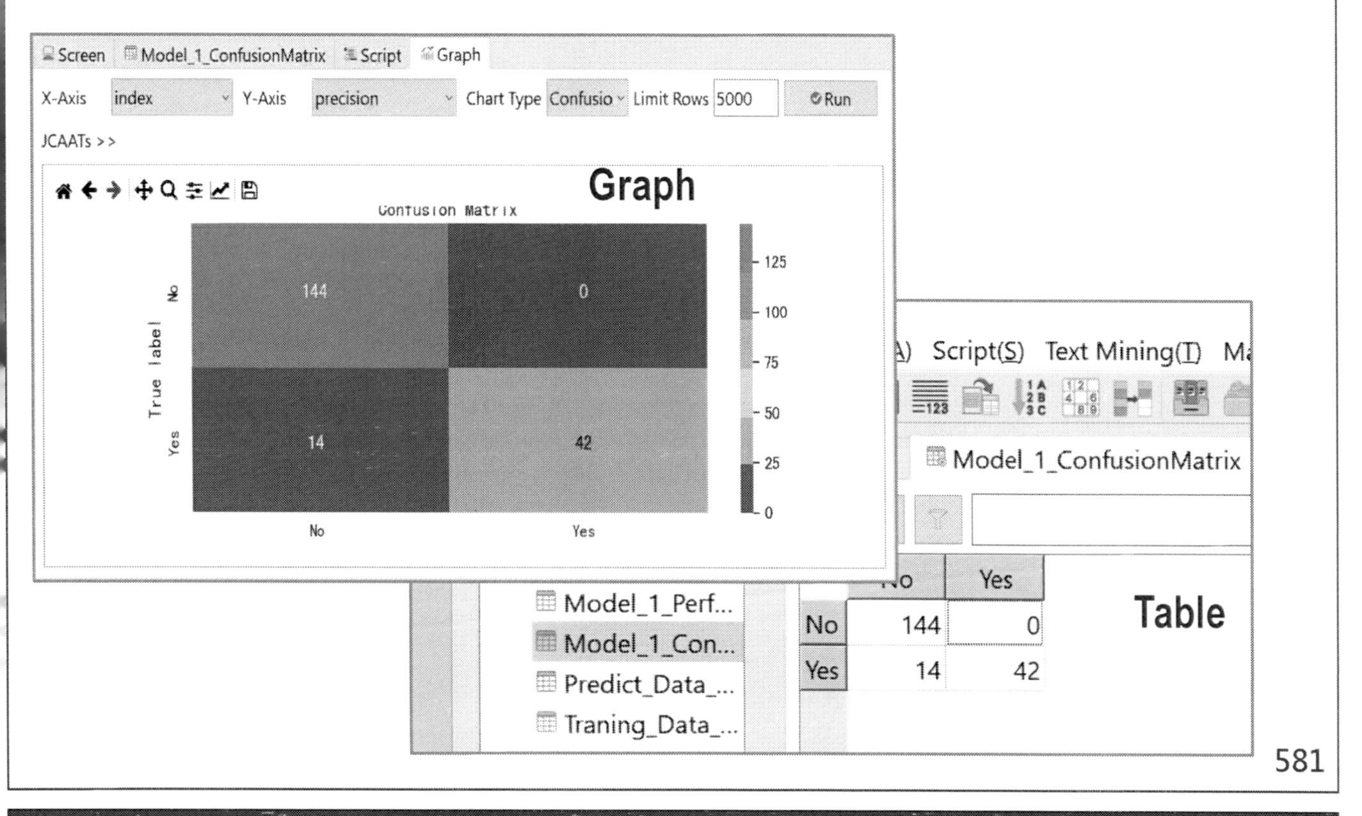

Performance Metrics - Train

- Performance Metrics enables you to see each metrics after training with each your selected learning model.

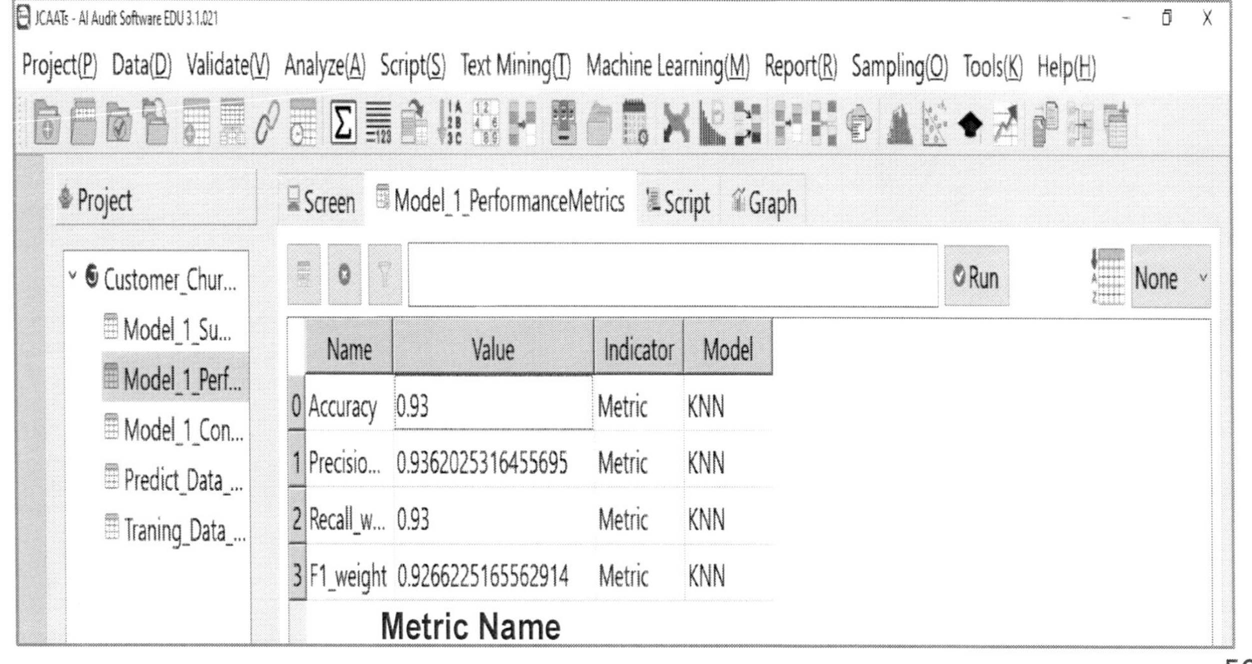

	Name	Value	Indicator	Model
0	Accuracy	0.93	Metric	KNN
1	Precisio...	0.9362025316455695	Metric	KNN
2	Recall_w...	0.93	Metric	KNN
3	F1_weight	0.9266225165562914	Metric	KNN

Summary Report - Train

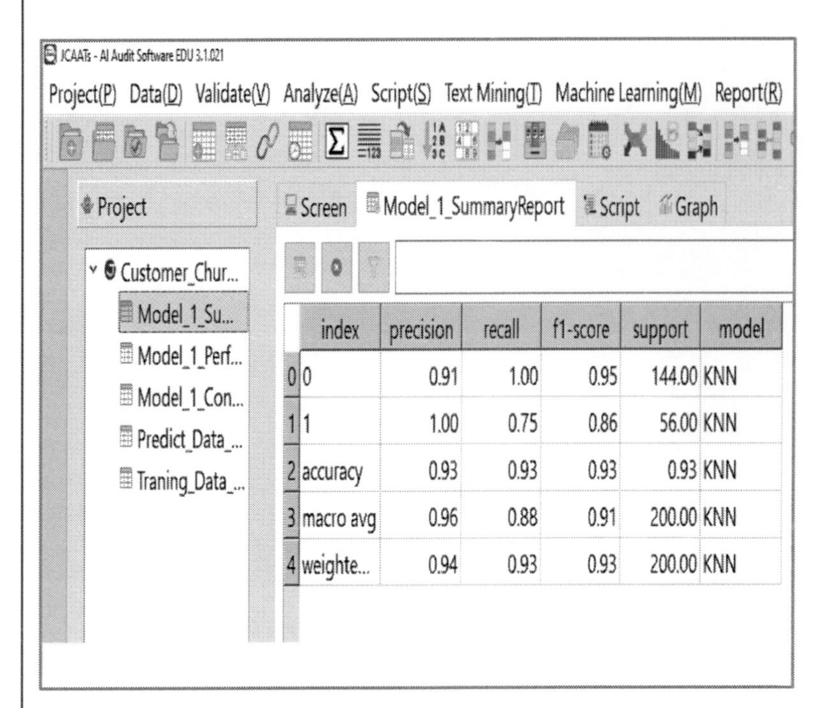

	index	precision	recall	f1-score	support	model
0	0	0.91	1.00	0.95	144.00	KNN
1	1	1.00	0.75	0.86	56.00	KNN
2	accuracy	0.93	0.93	0.93	0.93	KNN
3	macro avg	0.96	0.88	0.91	200.00	KNN
4	weighte...	0.94	0.93	0.93	200.00	KNN

Here's how to interpret the output:

- **Precision**: Out of all the data that the model predicted would get **100%** actually did.
- **Recall**: Out of all the data that the model only predicted this outcome correctly for **75%** of those data.
- **F1 Score**: This value is calculated as 86% by combining Precision and Recall.

- Since this value is very close to 1, it tells us that the model does a good job of predicting.

Main tab - Predict Command

- Start predictive Machine Learning models through flexible interface.

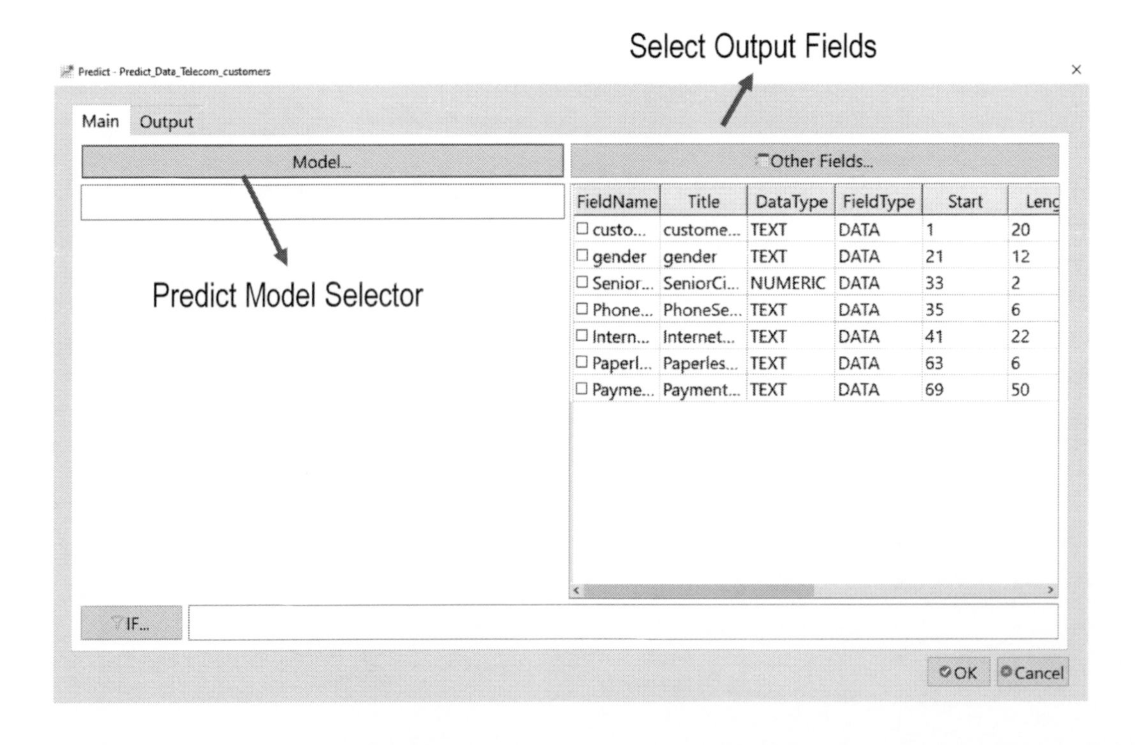

FieldName	Title	DataType	FieldType	Start	Leng
custo...	custome...	TEXT	DATA	1	20
gender	gender	TEXT	DATA	21	12
Senior...	SeniorCi...	NUMERIC	DATA	33	2
Phone...	PhoneSe...	TEXT	DATA	35	6
Intern...	Internet...	TEXT	DATA	41	22
Paperl...	Paperles...	TEXT	DATA	63	6
Payme...	Payment...	TEXT	DATA	69	50

Select Knowledge Model Dialog - Predict

- Click「Model」, select a knowledge model file, *.jkm, from your disk which was generated by Train command.

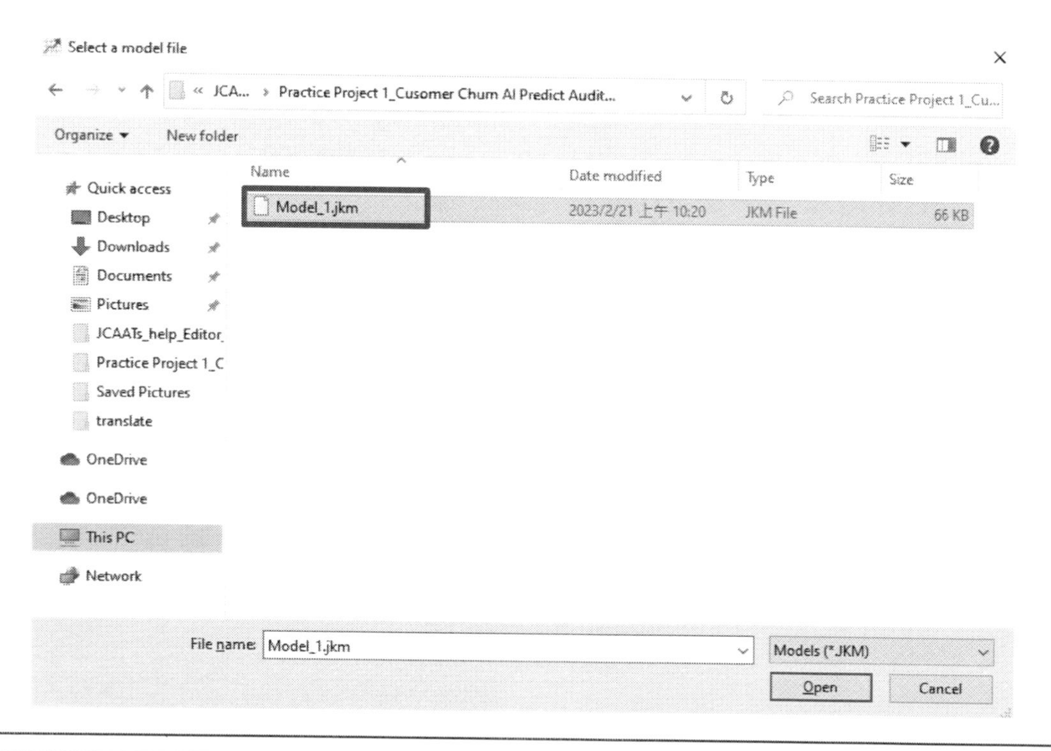

Result Table - Predict Command

- Obtain a table which display the predicted result by your selected knowledge model.

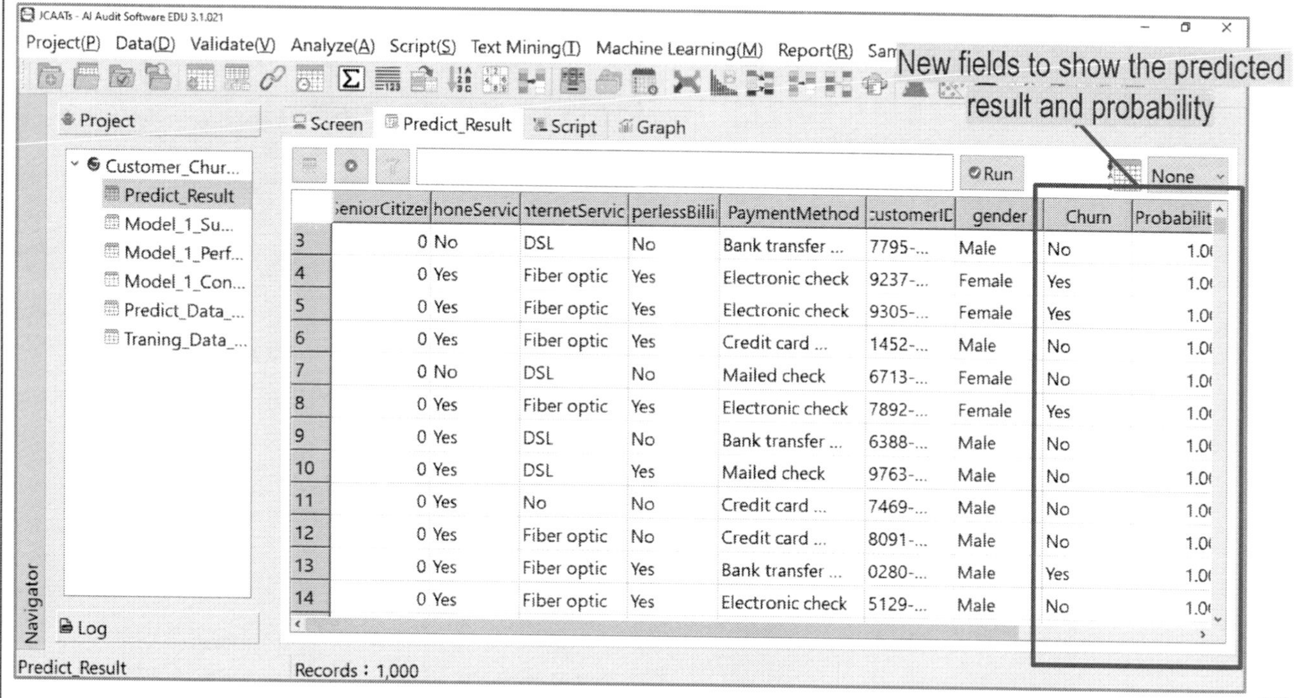

	SeniorCitizer	PhoneService	InternetService	PaperlessBilling	PaymentMethod	customerID	gender	Churn	Probability
3	0	No	DSL	No	Bank transfer ...	7795-...	Male	No	1.0
4	0	Yes	Fiber optic	Yes	Electronic check	9237-...	Female	Yes	1.0
5	0	Yes	Fiber optic	Yes	Electronic check	9305-...	Female	Yes	1.0
6	0	Yes	Fiber optic	Yes	Credit card ...	1452-...	Male	No	1.0
7	0	No	DSL	No	Mailed check	6713-...	Female	No	1.0
8	0	Yes	Fiber optic	Yes	Electronic check	7892-...	Female	Yes	1.0
9	0	Yes	DSL	No	Bank transfer ...	6388-...	Male	No	1.0
10	0	Yes	DSL	Yes	Mailed check	9763-...	Male	No	1.0
11	0	Yes	No	No	Credit card ...	7469-...	Male	No	1.0
12	0	Yes	Fiber optic	No	Credit card ...	8091-...	Male	No	1.0
13	0	Yes	Fiber optic	Yes	Bank transfer ...	0280-...	Male	Yes	1.0
14	0	Yes	Fiber optic	Yes	Electronic check	5129-...	Male	No	1.0

Classify the Predicted Results

- You can perform Classify to analysis the predicted results.

JCAATs >>Predict_Result.CLASSIFY(PKEY="Churn", TO="")
Table : Predict_Result
Note: 2023/02/21 11:12:04
Result - Records : 2

Churn	Churn_count	Percent_of_cour
No	744	74.4
Yes	256	25.6

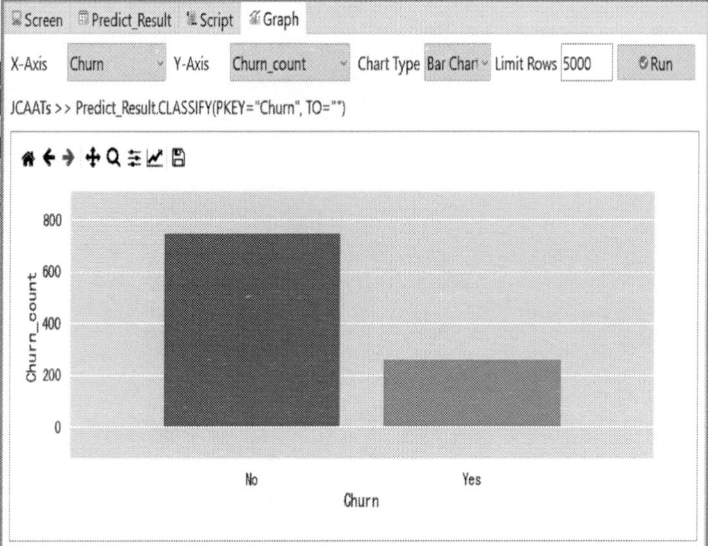

JCAATs - AI Audit Software

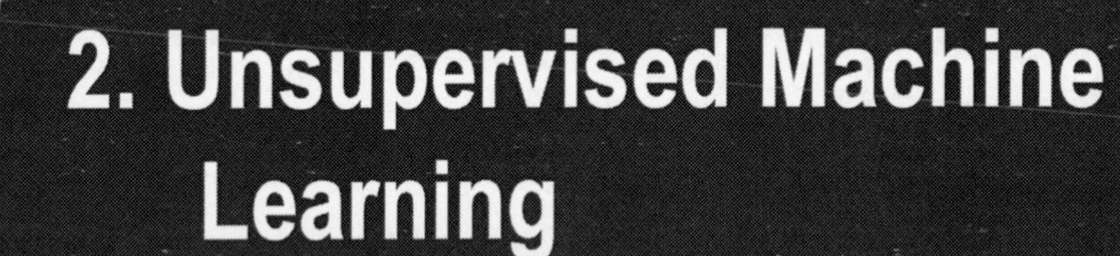

2. Unsupervised Machine Learning

JCAATs Unsupervised Machine Learning

- **Unsupervised Machine Learning:**
  - **Outlier**: It uses statistical **standard deviation** calculations to find data points that are significantly different from other observations.
  - **Cluster**: It used **K-means** algorithm to gather similar samples together to form clusters according to the common attributes between samples. Usually according to distance of classify, distance the more closer the more similar.

- The results after clustering can have the characteristics of the groups with smaller data sets can be tested as a more special group of data.

Outlier

- Outliers can be identified and analyzed using statistical methods, z-score. Z-score indicates how much a given value differs from the standard deviation

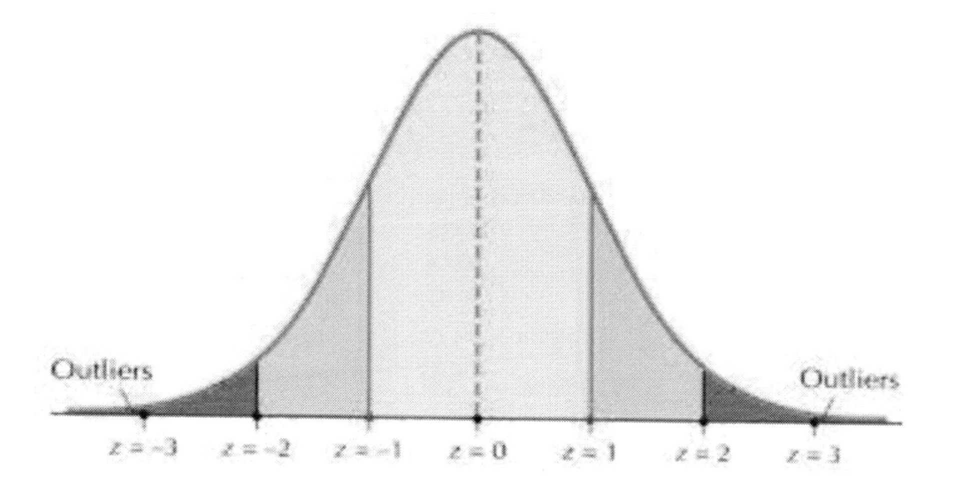

Source: Analytics Vidhya. 2022

Main tab - Outlier Command

- Starting Outlier analysis through flexible interface.

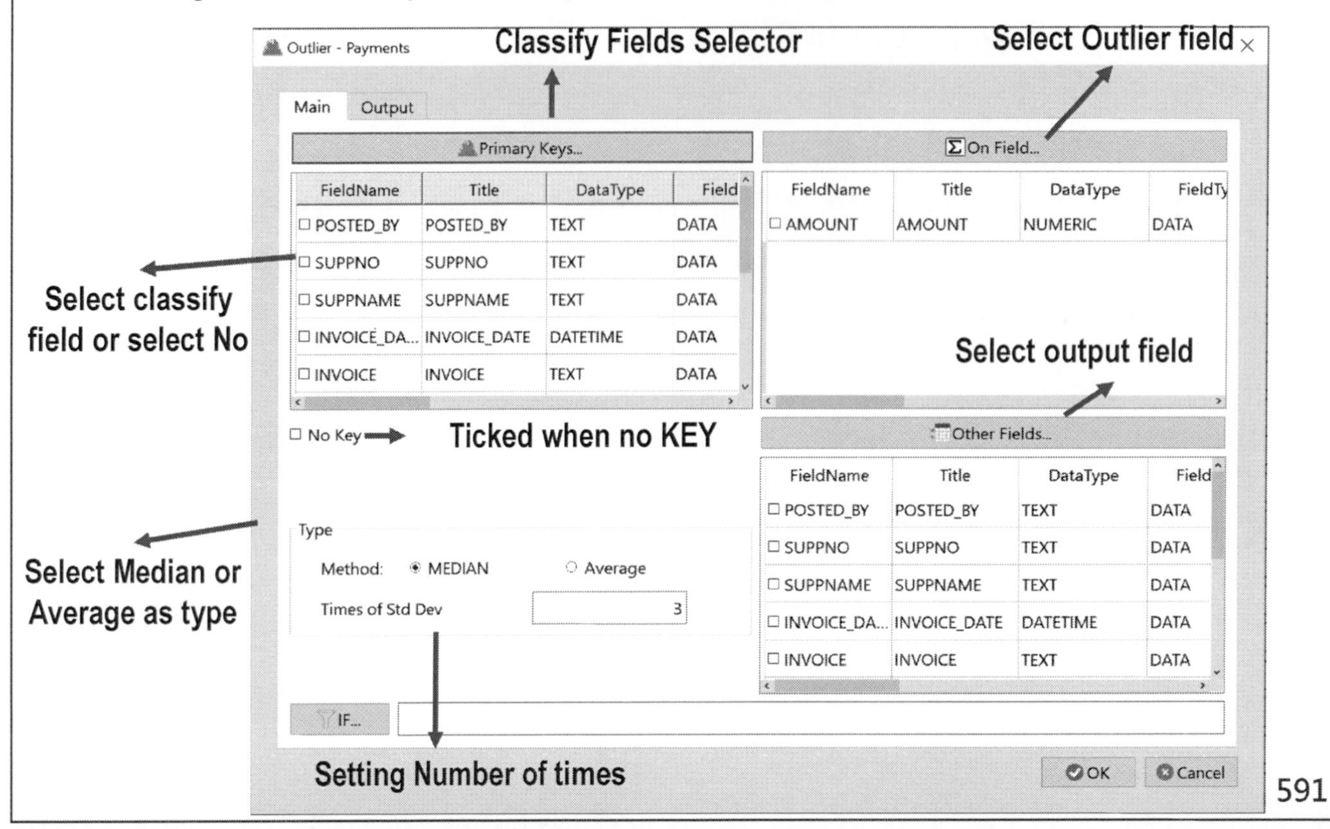

Result Table - Outlier

- After outlier analysis, a table will be brought out that is all outlier which outside the selected standard deviation times.

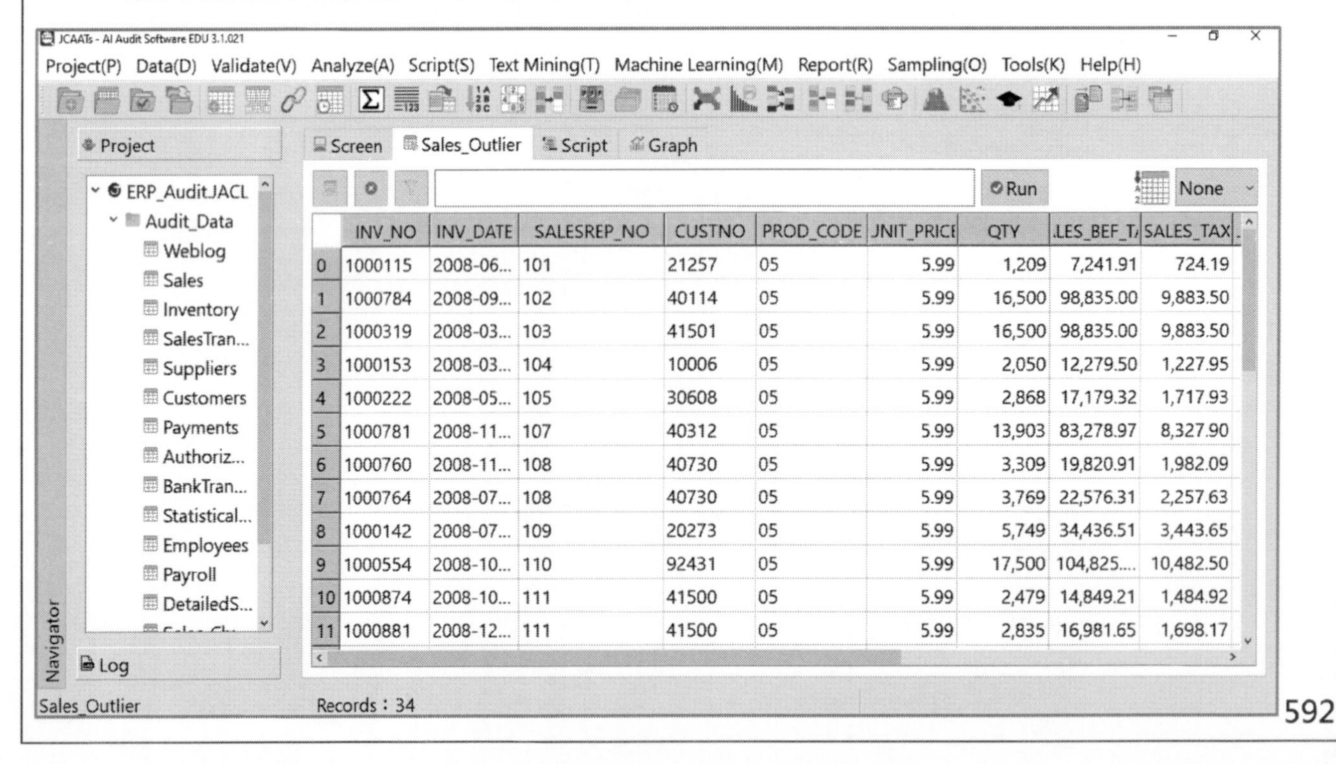

Cluster

- The goal of clustering is to discover the underlying structure or patterns in the data, and to gain insights into the relationships among the data points. JCAATs use **K-means algorithm** to perform clustering.

- Some examples of how clustering could be applied in an audit context:
  - **Fraud detection**: clustering could be used to group transactions with similar amounts, dates, or vendor names, and then investigate any clusters that appear to be outliers or have unusual patterns.
  - **Risk assessment**: clustering could be used to group different types of risks based on their similarity. This could help to identify areas of the business that may be more vulnerable to risk, and prioritize audit resources accordingly.
  - **Compliance monitoring**: clustering could be used to group different types of compliance violations based on their characteristics. This could help to identify patterns or trends in compliance data, and identify areas where additional controls or monitoring may be needed.

K-means Clustering Algorithm

- The algorithm works by iteratively assigning each data point to the closest centroid (a center point of a cluster), and then updating the centroid based on the mean of the data points assigned to it. The algorithm continues to iterate until the centroids no longer move or until a predetermined number of iterations has been reached.

- The questions to use the K-means algorithm:
  - How many seeds do we needs to put into learning?
  - Where is the position for the first grouping seeds?
  - How many times to do needs to learn?
  - How to calculate the distance?

Main tab- Cluster Command

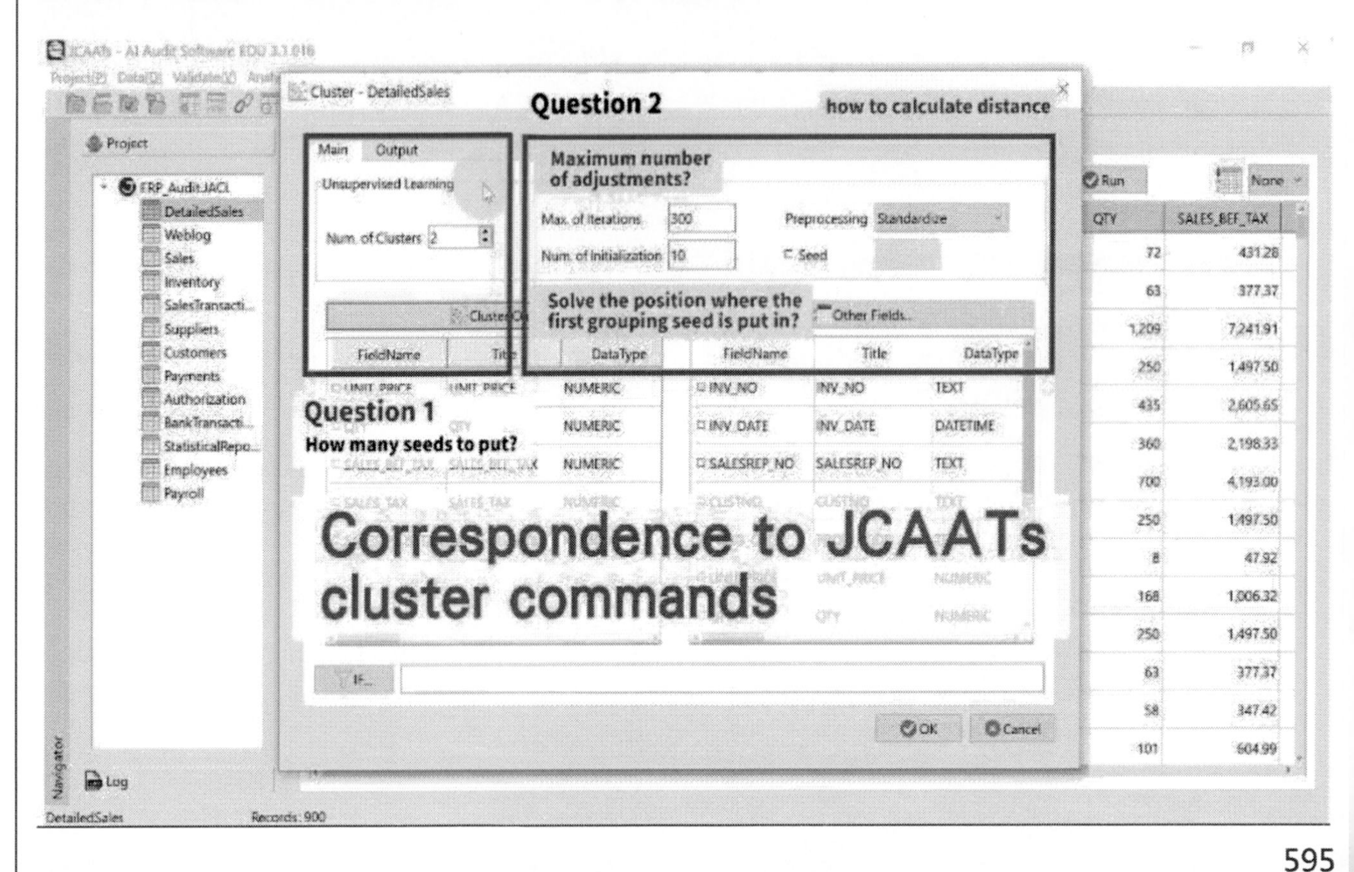

Result Graph- Cluster

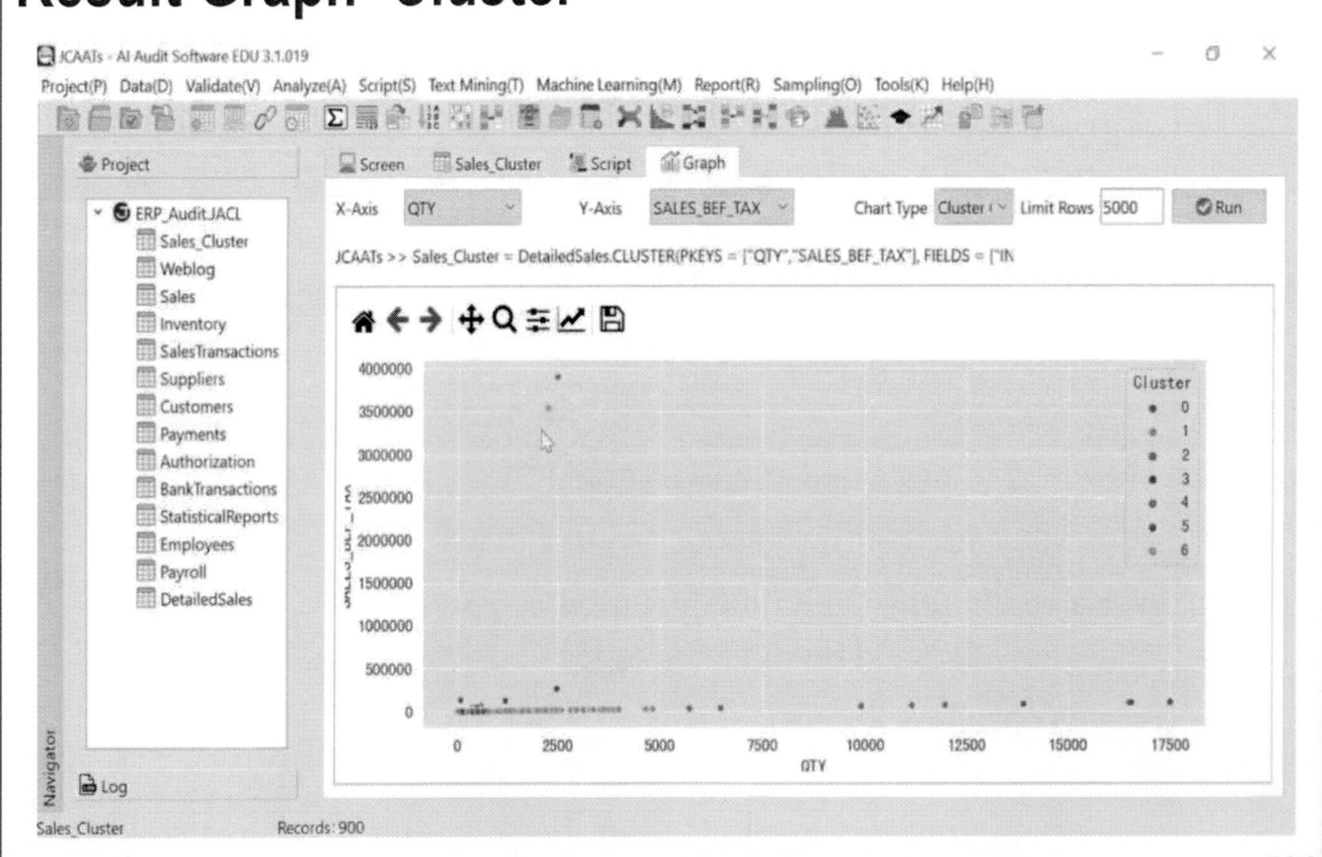

Result Table- Cluster

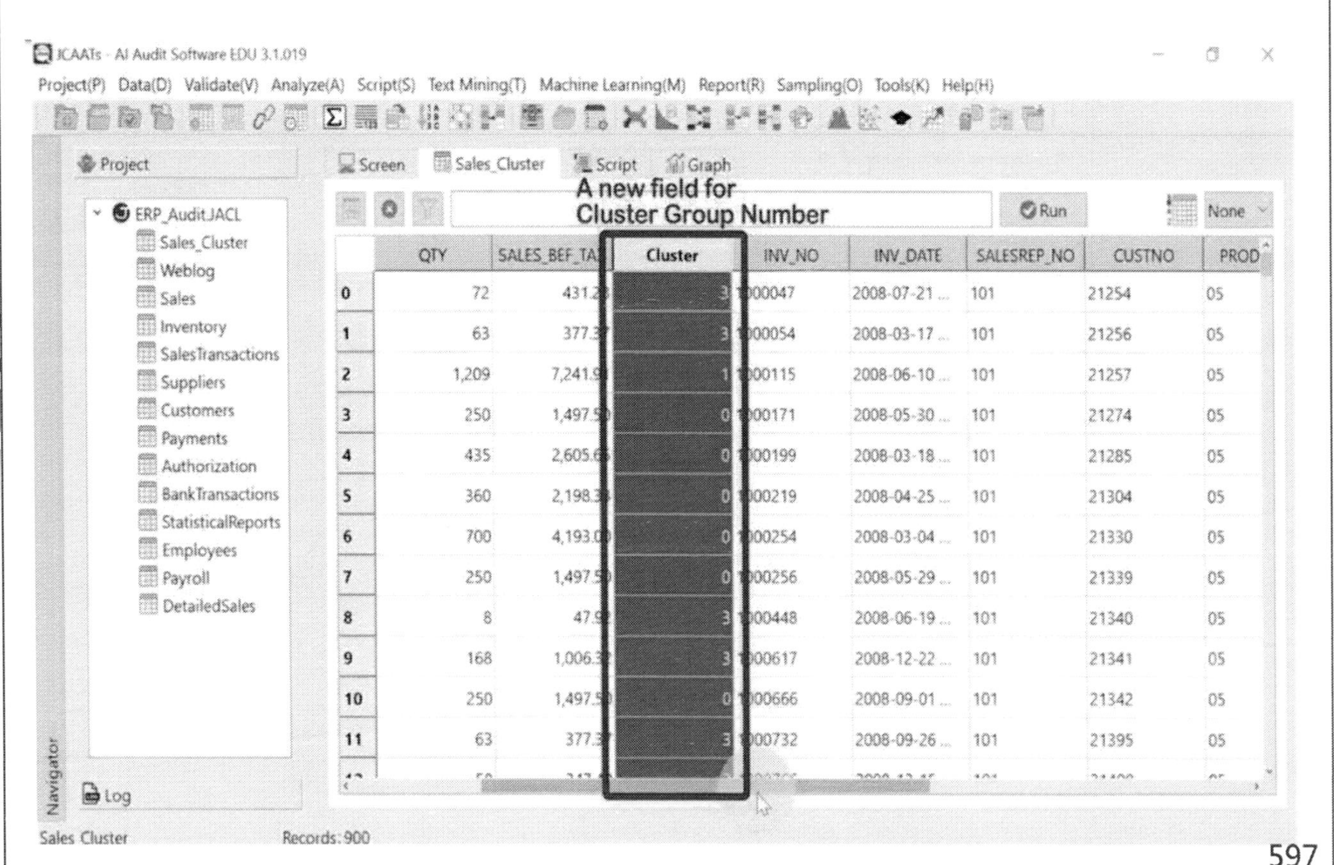

Result Table- Cluster

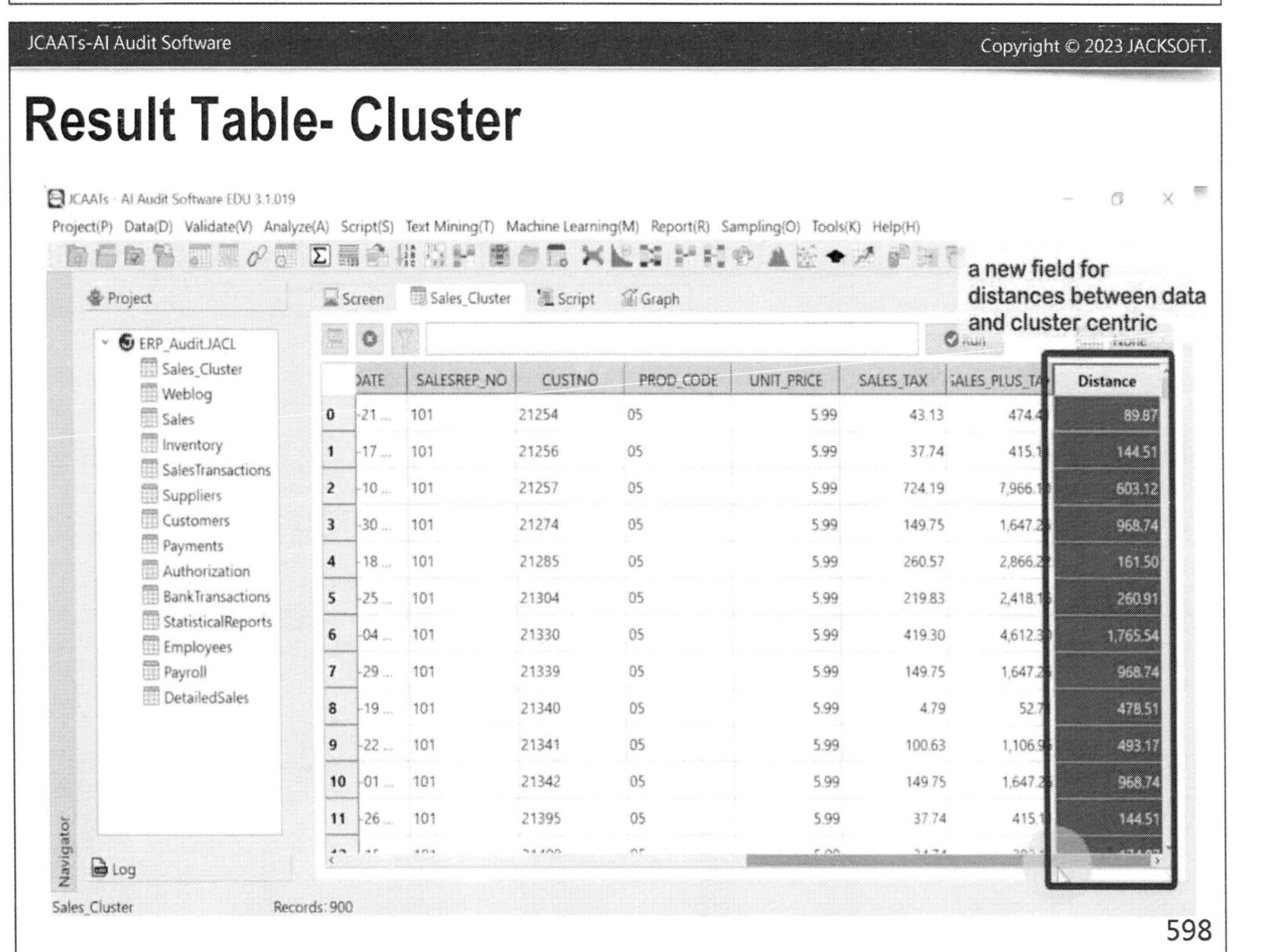

Classify for Deeper Analysis

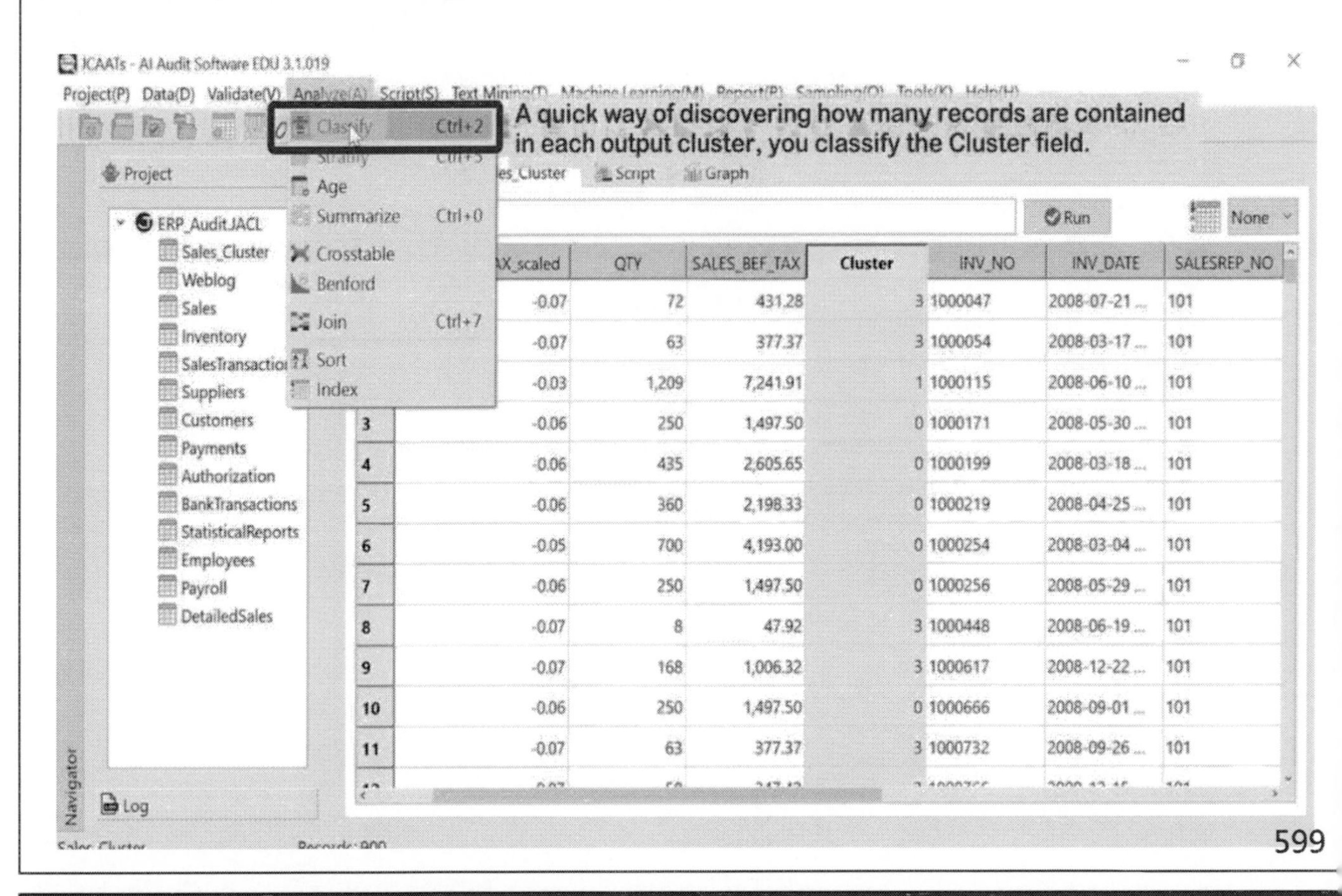

Result Screen - Classify Cluster

- User can understand about each group item through Classify in-depth function.

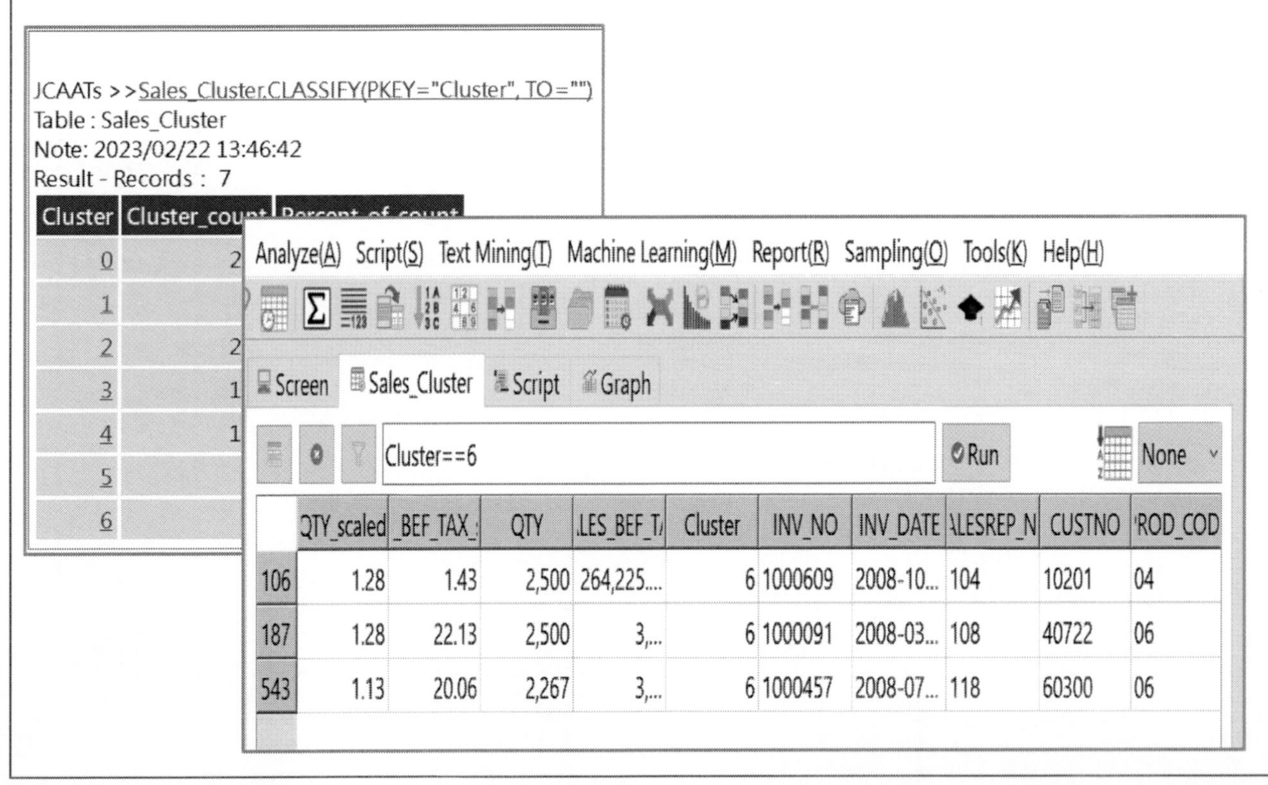

JCAATs Learning notes ：

JCAATs-AI Audit Software
Copyright © 2023 JACKSOFT.

JCAATs Learning notes ：

Python Based Computer-Assisted Audit Techniques (CAATs)

Data Analysis and Smart Audit

Chapter 11. Report

Outline:

1. Report Command List
2. Extract Command
3. Merge Command
4. Export Command
5. Chart Command

JCAATs - AI Audit Software

1. Report Command List

JCAATs Report Commands

- JCAATs focuses on data analysis and smart analysis. All analysis result can be **presented in a new table or appended to an existing table**.
- In JCAATs, there are four report commands:
  - **Extract**: Focus on the content of the table on specific field, display sort of field and correction of missing value on fields. Output the conform to needed table report for user.
  - **Merge**: Save multiple tables into a new table according to the same field name format.
  - **Export**: Export the table into multiple common file format that are easily use by other software.
  - **Chart**: Provide various types of diagram to visualize the data.

JCAATs - AI Audit Software

Extract Command

- The Extract command is used to create a new table from an existing table, by selecting only the needed records and fields. This can help to accelerate future processing by reducing the amount of data that needs to be analyzed or manipulated.
- Here are the general steps to use the Extract command to create a new table:

1. Identify the existing table from which you want to extract data.
2. Determine the specific records and fields that you need to extract, based on the requirements of your analysis or processing.
3. Select either "**Values**" or "**Formulas**", depending on which type of data you want to extract.
4. Save the selected data into a **new table name or append to an existing table**.

Main tab - Extract Command

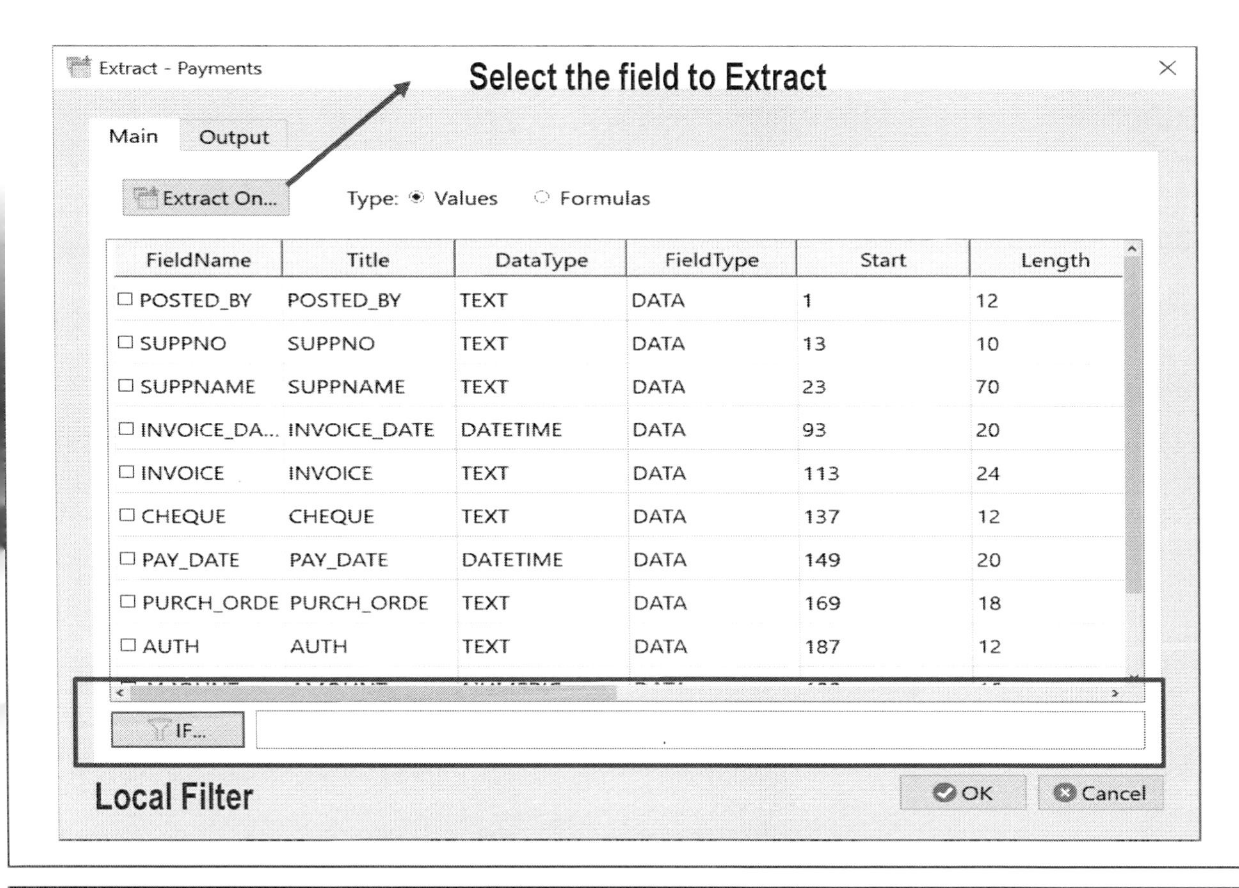

Output tab - Extract Command

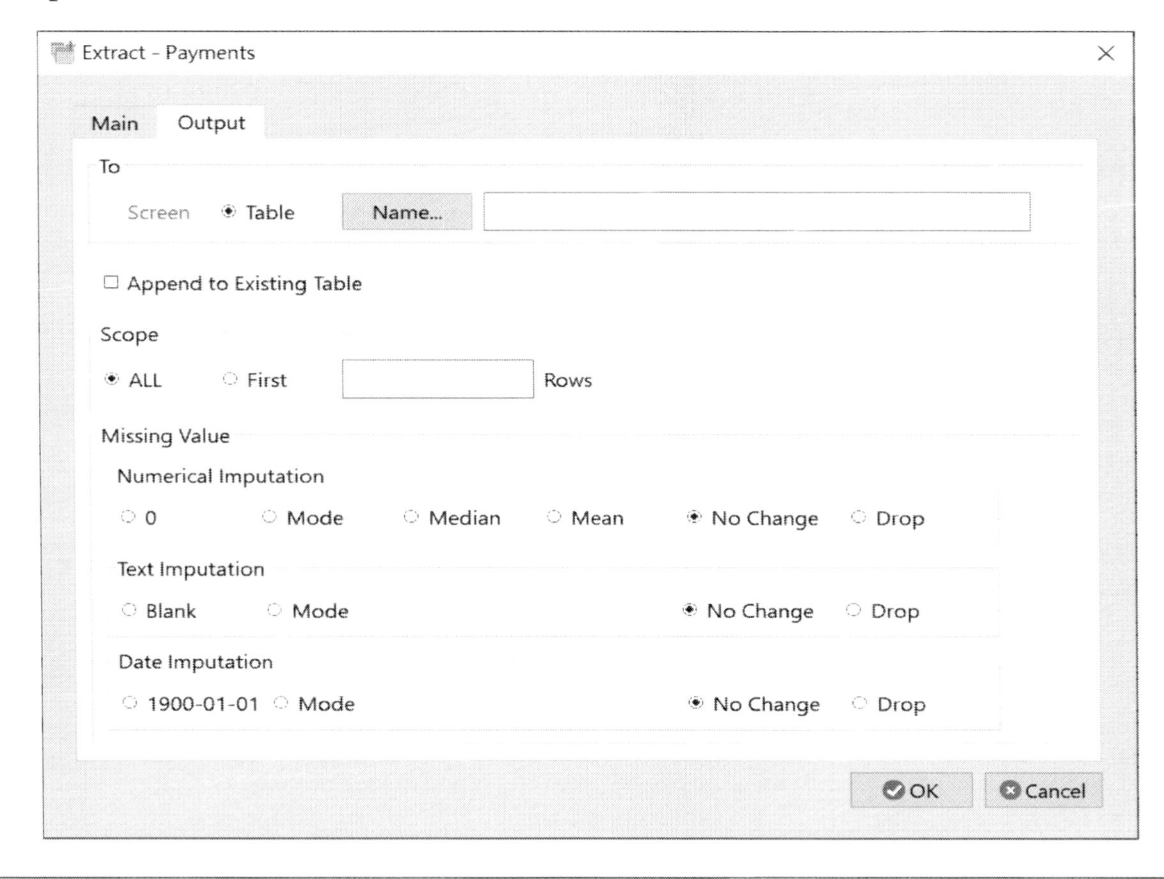

Determine the specific records and fields

- Specific records and fields
  - **Extract On...** to select the specific fields
  - **IF...** to filter the specific records.
  - **Scope** to focus on the specific records.
  - **Missing Value** to avoid non-clean data.
- Deal with type
  - **Values**: it means that you want to retrieve the actual data that is stored in the field, without any formulas or other calculations that may have been used to generate those values.
  - **Formulas**: it means that you want to retrieve the formula expressions that were used to calculate the values in the computed field, rather than the actual values themselves. Data fields are still treated as values.

Append to Existing Table

- The fields name, order, format and data type of the new table need all the same as the existing table.
- The width of the field data between new table and existing table can be different.

AP_Trans_January

Custno	Product	Inv_Date	Inv_Amount
01542	Printer	5-Jan-05	257.89
04723	Toner	18-Jan-05	39.99
29452	Printer	19-Jan-05	294.32
03914	Paper	25-Jan-05	15.86
46778	Scanner	30-Jan-05	125.99

Extract →

AP_Trans_February

Custno	Product	Inv_Date	Inv_Amount
04723	Paper	1-Feb-05	15.86
33397	Paper	15-Feb-05	84.33
46778	Toner	28-Feb-05	79.98

Extract / Append →

AP_Trans_March

Custno	Product	Inv_Date	Inv_Amount
01542	Flash Drive	2-Mar-05	99.99
88754	Toner	3-May-05	39.99
12679	Scanner	25-Mar-05	134.99
46778	Scanner	31-Mar-05	85.00

Extract / Append →

AP_Trans_Quarter_1

Custno	Product	Inv_Date	Inv_Amount
01542	Printer	5-Jan-05	257.89
04723	Toner	18-Jan-05	39.99
29452	Printer	19-Jan-05	294.32
03914	Paper	25-Jan-05	15.86
46778	Scanner	30-Jan-05	125.99
04723	Paper	1-Feb-05	15.86
33397	Paper	15-Feb-05	84.33
46778	Toner	28-Feb-05	79.98
01542	Flash Drive	2-Mar-05	99.99
88754	Toner	3-May-05	39.99
12679	Scanner	25-Mar-05	134.99
46778	Scanner	31-Mar-05	85.00

JCAATs - AI Audit Software

2. Merge Command

Merge

- Merge command allows you to combine two or more tables of data into a single table.
- It typically works by identifying a common field or set of fields in the tables being merged, and then creating a new table that includes all of the columns from the primary table.
- The rows in the new table are then created by matching the values in the common field(s) and combining the information from the corresponding rows in each original table.

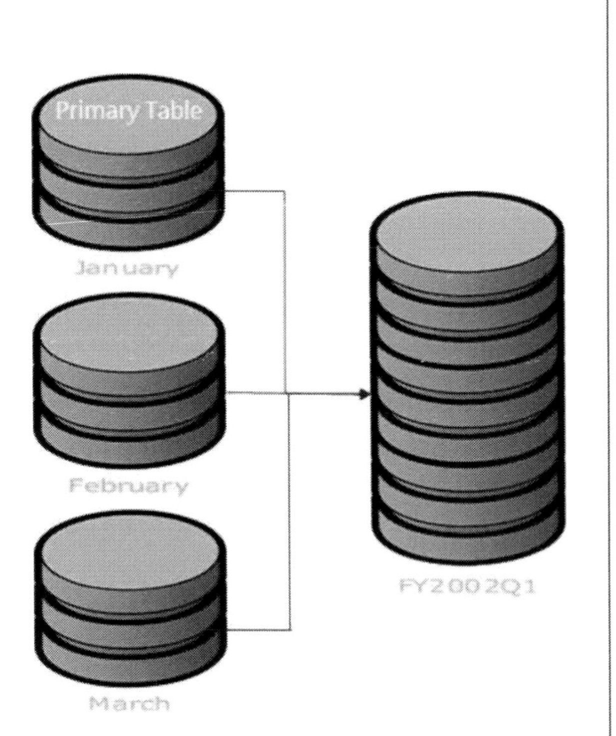

Main tab - Merge Command

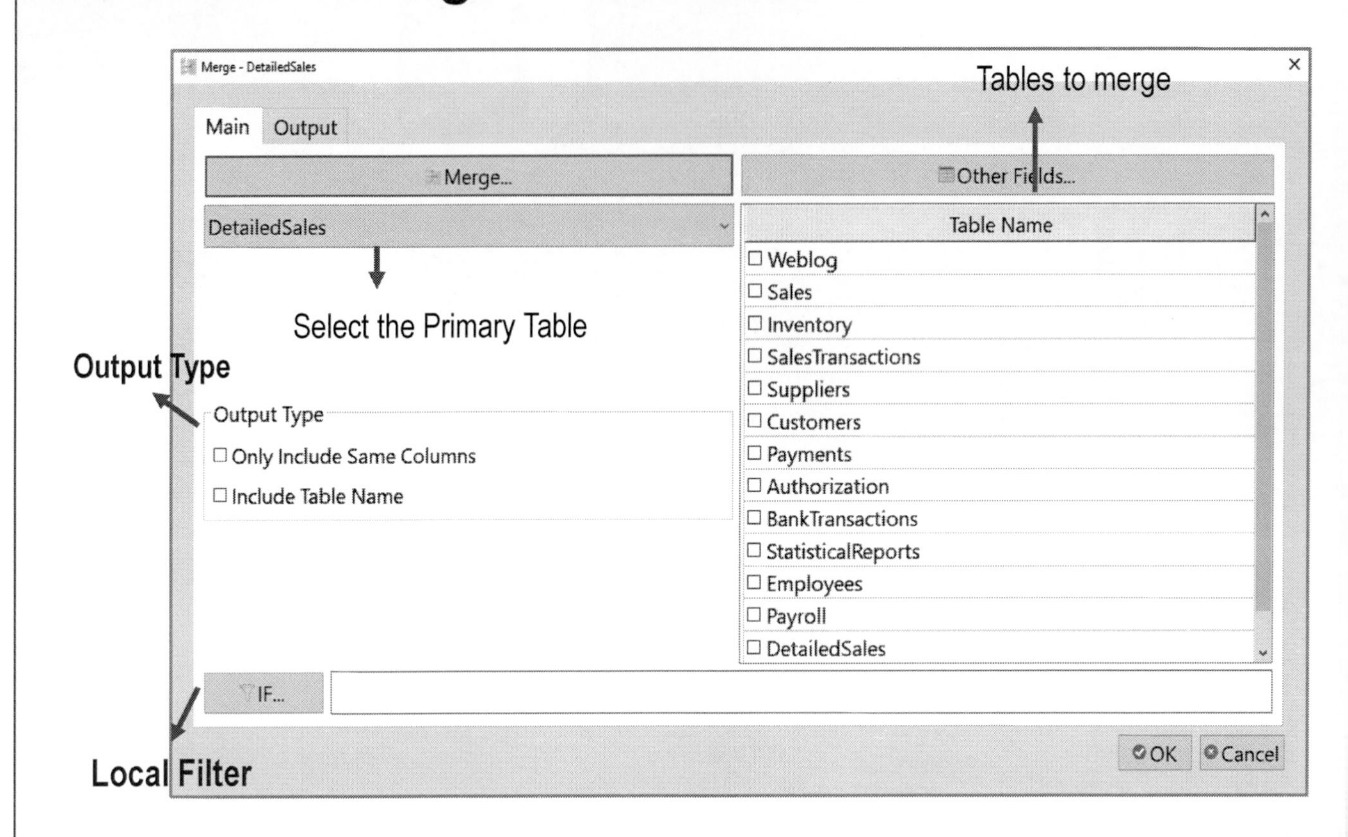

Output tab - Merge Command

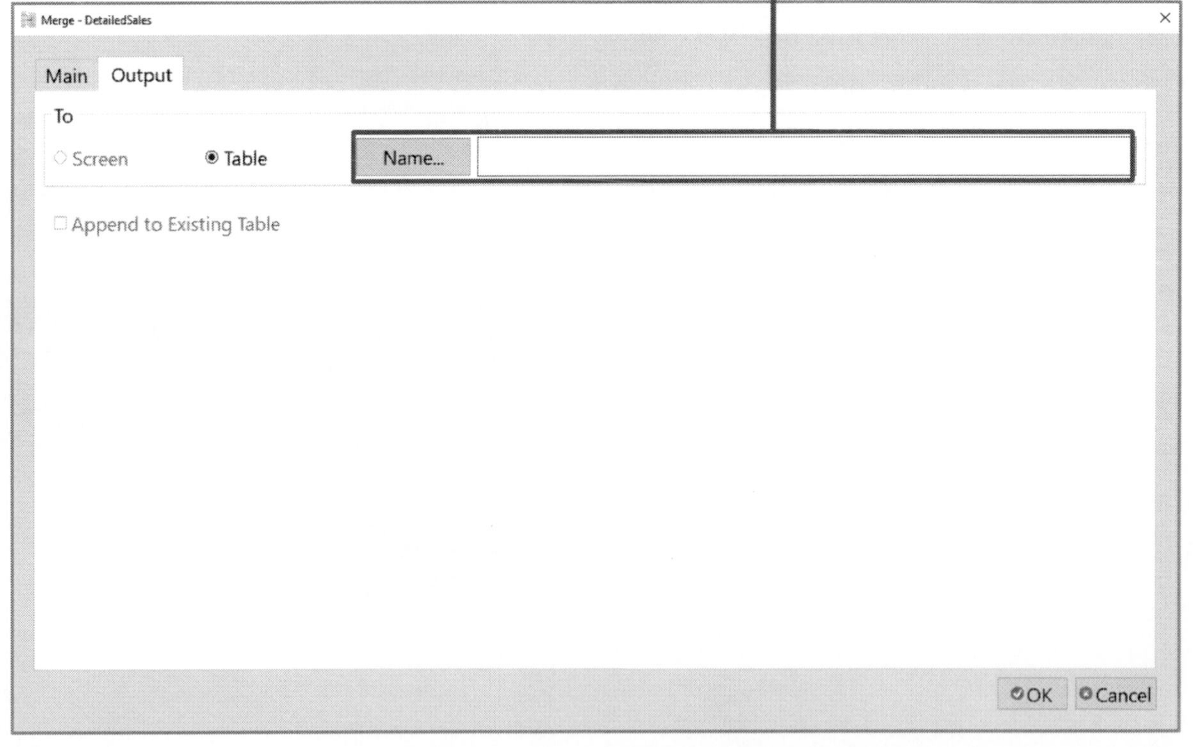

JCAATs - AI Audit Software

3. Export Command

Export Command

- "Export" command allows you to extract a table or dataset from JCAATs and save it in a different file format that can be used by other software applications.
- JCAATs support the following formats to Export data:
  - **CSV(*.csv):** CSV (Comma Separated Values) file
  - **Excel(*.xlsx):** Microsoft Excel file
  - **ODS(*.ods):** OpenDocument Spreadsheet Document format file
  - **JSON(*.json):** JavaScript Object Notation is an open standard file
  - **Text(*.txt):** A text document that contains plain text in the form of lines.
  - **XML(*.xml):** contains XML code and ends with the file extension ".xml"
- When exporting data, it's important to ensure that the data is properly formatted. Whether the exported data is a data field (Data) or a formula field (Computed), it is exported as a value

Main tab - Export

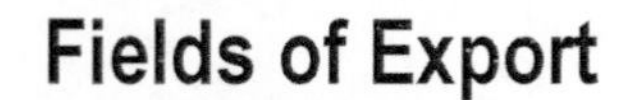

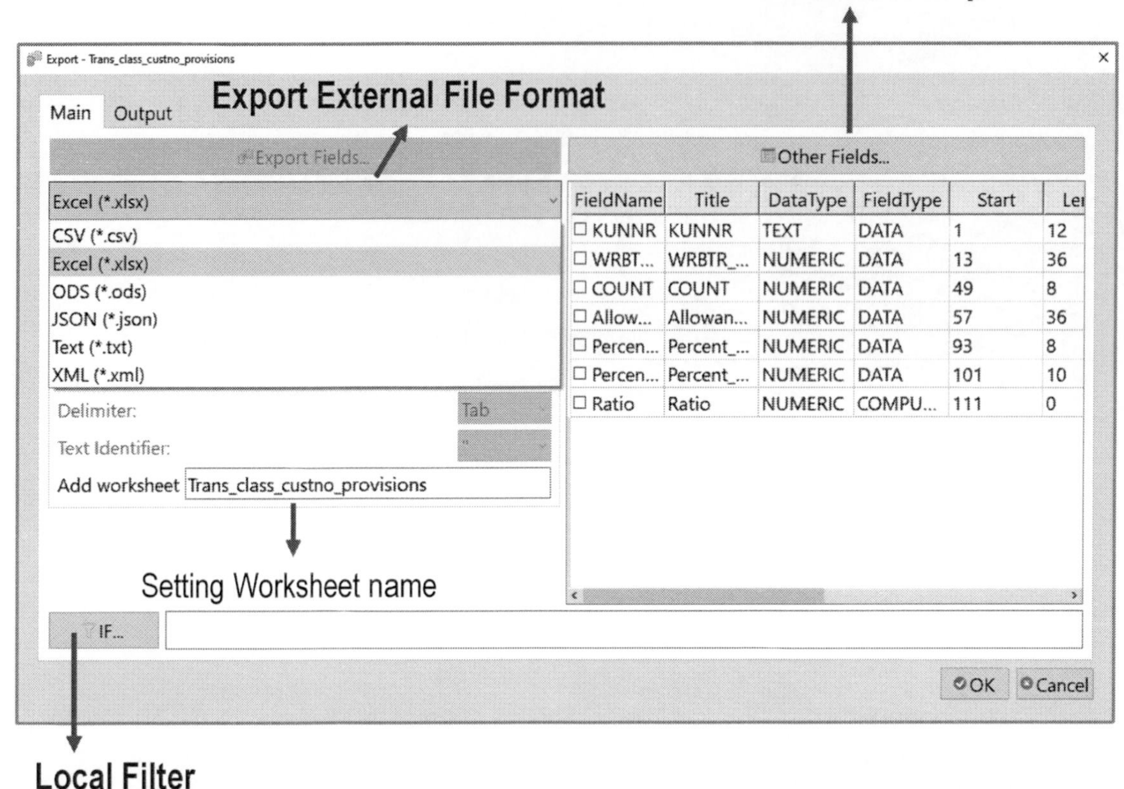

FieldName	Title	DataType	FieldType	Start	Le
☐ KUNNR	KUNNR	TEXT	DATA	1	12
☐ WRBT...	WRBTR_...	NUMERIC	DATA	13	36
☐ COUNT	COUNT	NUMERIC	DATA	49	8
☐ Allow...	Allowan...	NUMERIC	DATA	57	36
☐ Percen...	Percent_...	NUMERIC	DATA	93	8
☐ Percen...	Percent_...	NUMERIC	DATA	101	10
☐ Ratio	Ratio	NUMERIC	COMPU...	111	0

Output tab - Export Command

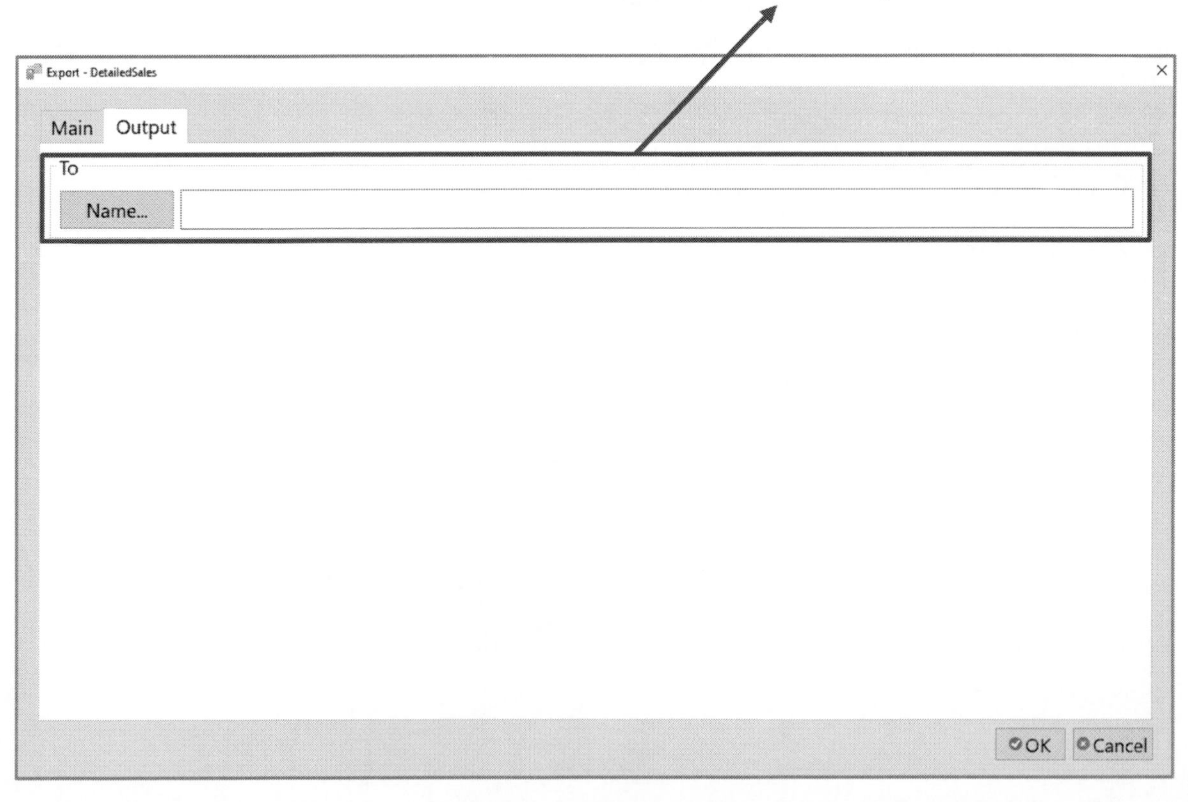

Result of Export Command

- Table Export to external file. Take Excel (.xlsx) as an example.

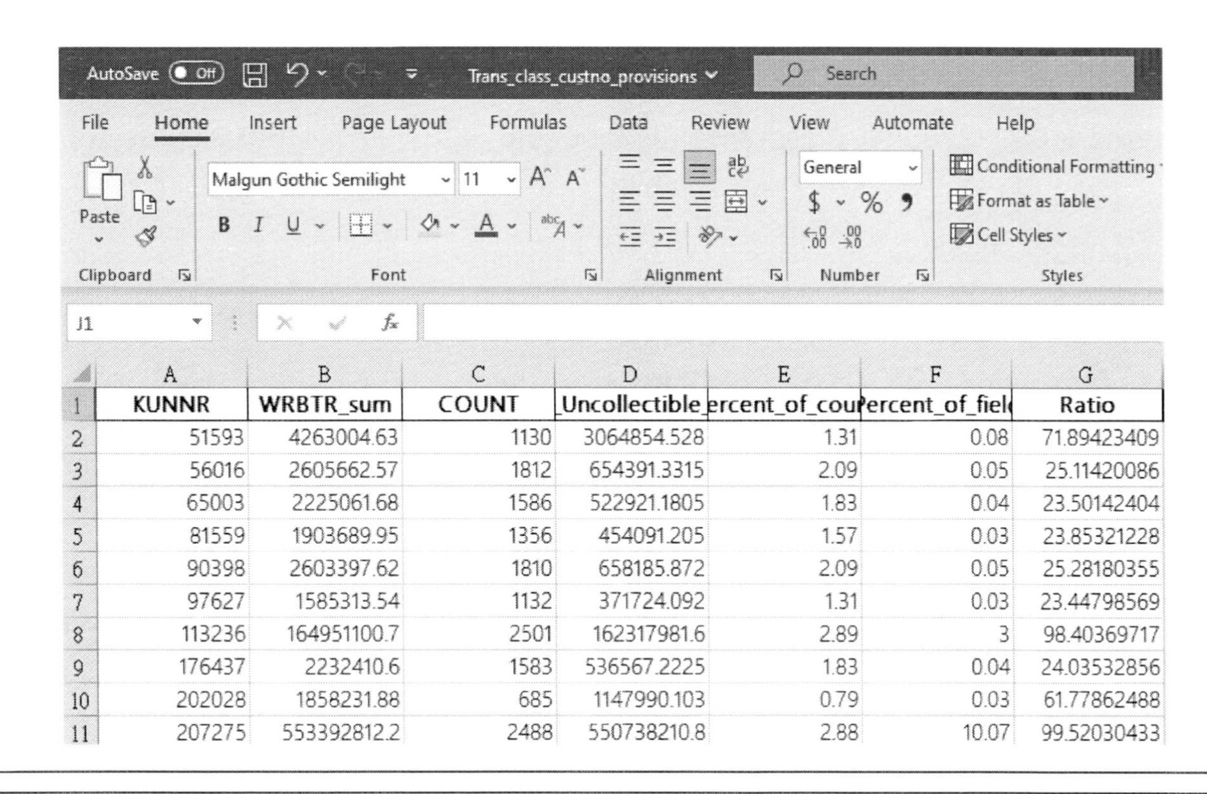

	A	B	C	D	E	F	G
1	KUNNR	WRBTR_sum	COUNT	Uncollectible	ercent_of_cou	ercent_of_fiel	Ratio
2	51593	4263004.63	1130	3064854.528	1.31	0.08	71.89423409
3	56016	2605662.57	1812	654391.3315	2.09	0.05	25.11420086
4	65003	2225061.68	1586	522921.1805	1.83	0.04	23.50142404
5	81559	1903689.95	1356	454091.205	1.57	0.03	23.85321228
6	90398	2603397.62	1810	658185.872	2.09	0.05	25.28180355
7	97627	1585313.54	1132	371724.092	1.31	0.03	23.44798569
8	113236	164951100.7	2501	162317981.6	2.89	3	98.40369717
9	176437	2232410.6	1583	536567.2225	1.83	0.04	24.03532856
10	202028	1858231.88	685	1147990.103	0.79	0.03	61.77862488
11	207275	553392812.2	2488	550738210.8	2.88	10.07	99.52030433

JCAATs - AI Audit Software

4. Chart Command

Chart Command

- Chart command is useful for data visualization because it allows you to quickly and easily understand large amounts of data by presenting it in a visual format that is easy to interpret.
- To create a chart, you typically need to select the data you want to display in the chart, and then choose the chart type that best represents the data. Common types of charts include:
  - **One Dimension**: **Pie Chart**, **Donut Chart**, **Tree Chart**
  - **Two Dimensions**: **Bar Chart**, **Dot Chart**, **Box Chart**
  - **Three Dimensions**: **Bubble Chart**
- Once you have created a chart, you can customize it by adjusting the colors, fonts, and other visual elements to make it more readable and attractive. You can also add labels, titles, and other annotations to help explain the data being presented.

Chart Interface

Selected Data

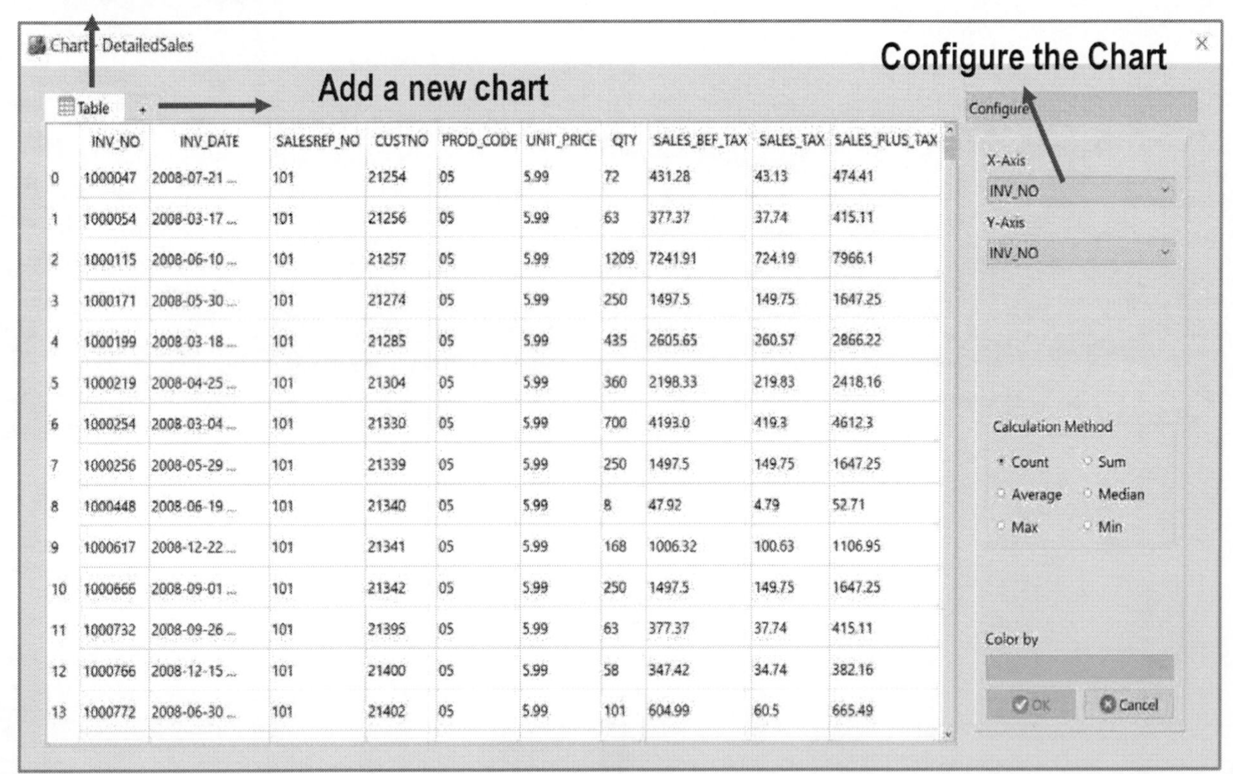

	INV_NO	INV_DATE	SALESREP_NO	CUSTNO	PROD_CODE	UNIT_PRICE	QTY	SALES_BEF_TAX	SALES_TAX	SALES_PLUS_TAX
0	1000047	2008-07-21 ...	101	21254	05	5.99	72	431.28	43.13	474.41
1	1000054	2008-03-17 ...	101	21256	05	5.99	63	377.37	37.74	415.11
2	1000115	2008-06-10 ...	101	21257	05	5.99	1209	7241.91	724.19	7966.1
3	1000171	2008-05-30 ...	101	21274	05	5.99	250	1497.5	149.75	1647.25
4	1000199	2008-03-18 ...	101	21285	05	5.99	435	2605.65	260.57	2866.22
5	1000219	2008-04-25 ...	101	21304	05	5.99	360	2198.33	219.83	2418.16
6	1000254	2008-03-04 ...	101	21330	05	5.99	700	4193.0	419.3	4612.3
7	1000256	2008-05-29 ...	101	21339	05	5.99	250	1497.5	149.75	1647.25
8	1000448	2008-06-19 ...	101	21340	05	5.99	8	47.92	4.79	52.71
9	1000617	2008-12-22 ...	101	21341	05	5.99	168	1006.32	100.63	1106.95
10	1000666	2008-09-01 ...	101	21342	05	5.99	250	1497.5	149.75	1647.25
11	1000732	2008-09-26 ...	101	21395	05	5.99	63	377.37	37.74	415.11
12	1000766	2008-12-15 ...	101	21400	05	5.99	58	347.42	34.74	382.16
13	1000772	2008-06-30 ...	101	21402	05	5.99	101	604.99	60.5	665.49

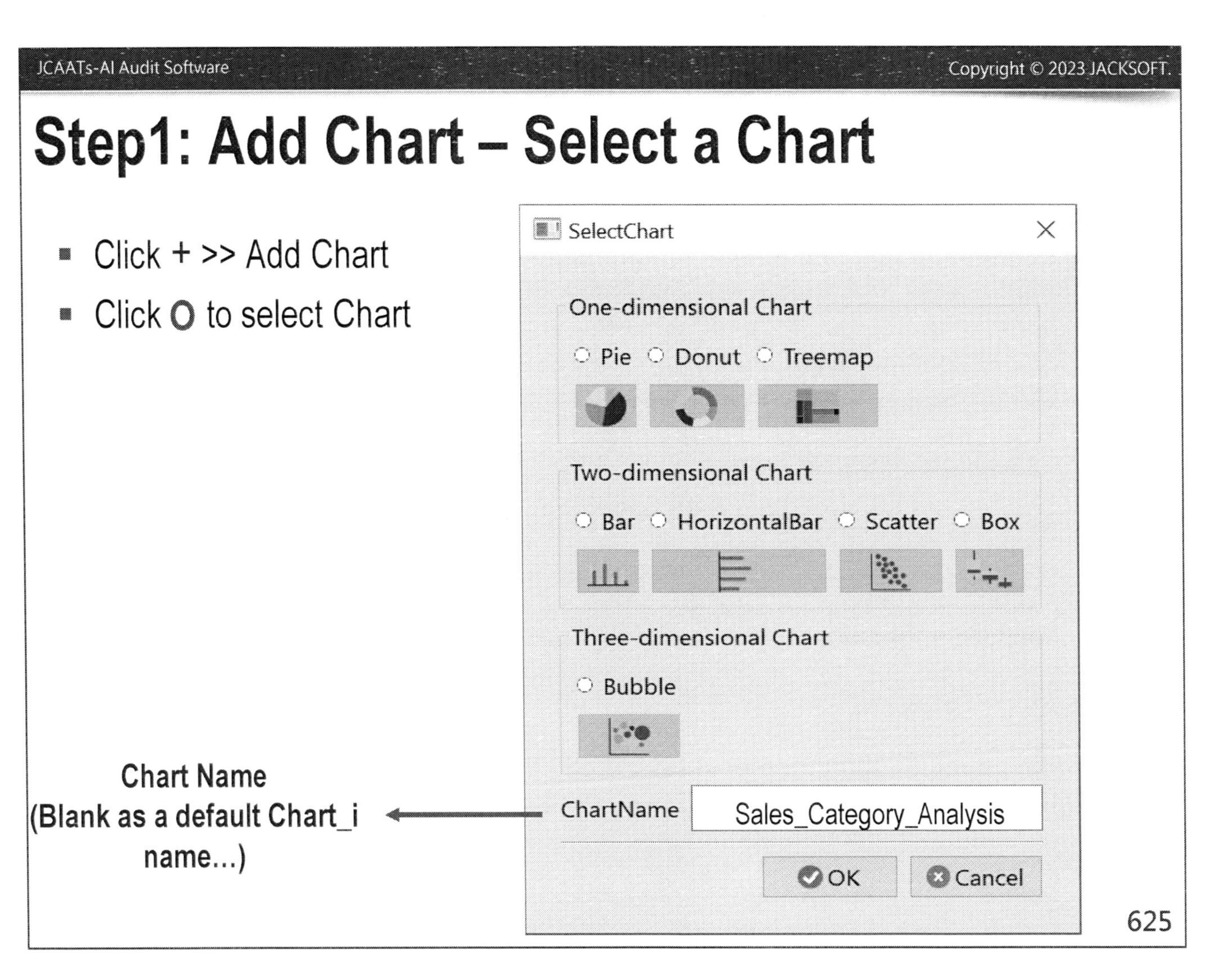
JCAATs-AI Audit Software
Copyright © 2023 JACKSOFT.
Step1: Add Chart – Select a Chart
Click + >> Add Chart
Click O to select Chart
SelectChart
One-dimensional Chart
Pie
Donut
Treemap
Two-dimensional Chart
Bar
HorizontalBar
Scatter
Box
Three-dimensional Chart
Bubble
Chart Name
(Blank as a default Chart_i name...)
ChartName
Sales_Category_Analysis
OK
Cancel
625

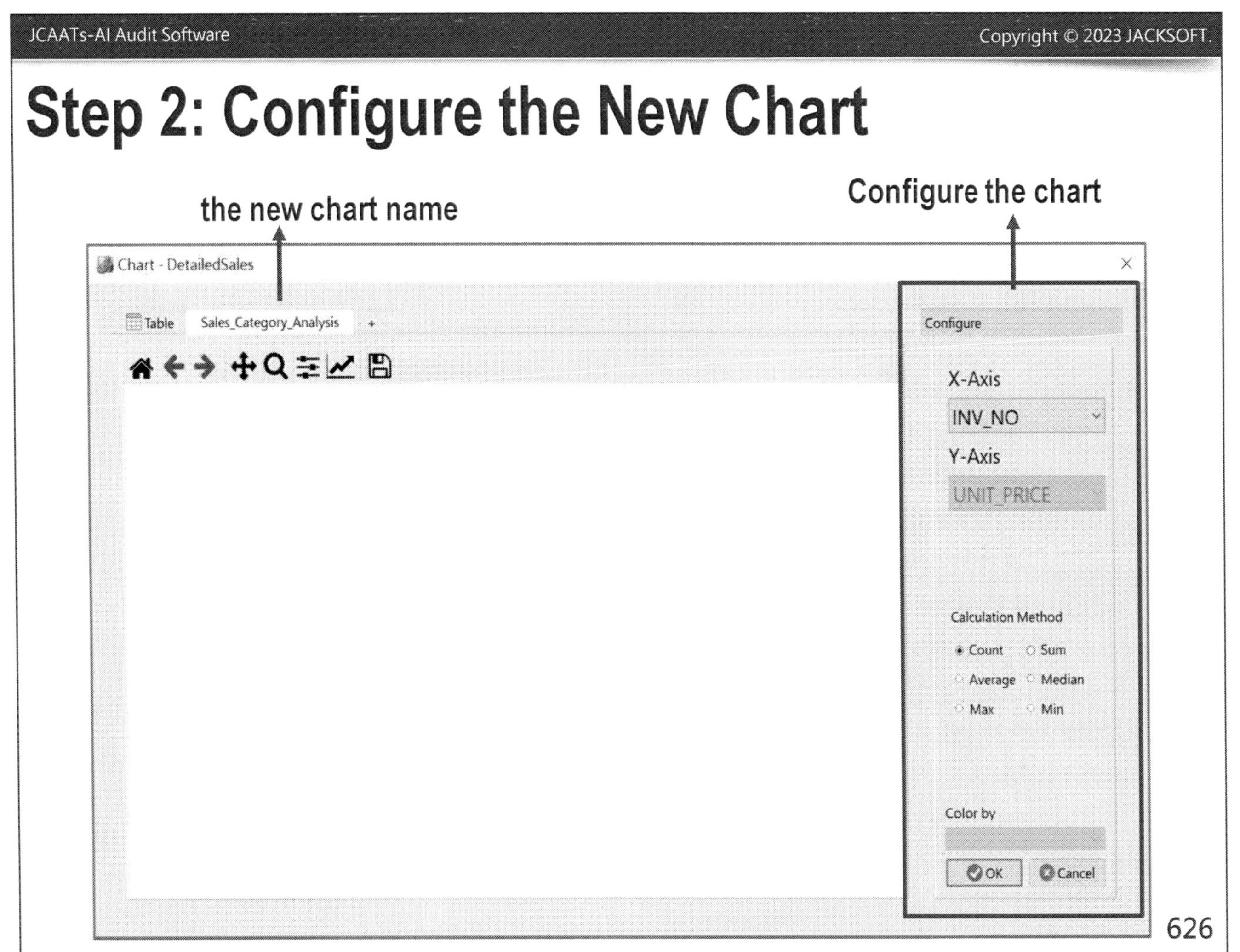
JCAATs-AI Audit Software
Copyright © 2023 JACKSOFT.
Step 2: Configure the New Chart
the new chart name
Configure the chart
Chart - DetailedSales
Table
Sales_Category_Analysis
Configure
X-Axis
INV_NO
Y-Axis
UNIT_PRICE
Calculation Method
Count
Sum
Average
Median
Max
Min
Color by
OK
Cancel
626

The process of Configure the Chart

- **X-axis** classify field: Select the data to be used as the X-axis
- **Y-axis** numeric field: Select the data to be used as the Y-axis
- **Calculation Method**
  - Value: The number of groups of field data
  - Sum: Grouped sum of Y-axis numeric fields
  - Average: Grouped average of Y-axis numeric fields
  - Median: Grouped median of Y-axis numeric fields
  - Max Value: Grouped max value of Y-axis numeric fields
  - Min Value: Grouped min value of Y-axis numeric fields

- Color by: Add a field to display by color. This will enable you to add a new dimension from the chart.

Step 3: Result Pie chart

- You can manage your chart by using the functions on the Char menu.

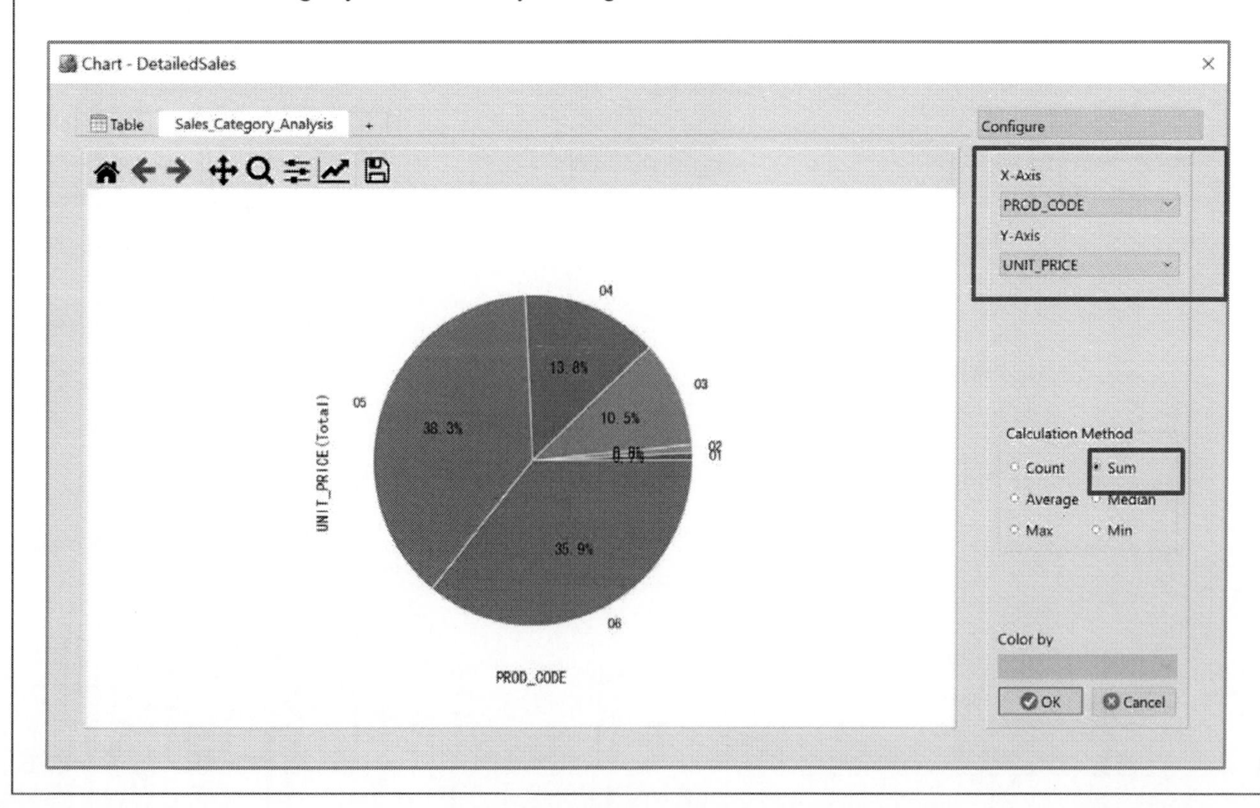

Functions on the Menu of Chart

- Reset original view
- Back to previous view
- Forward to next view
- Pan axes with left mouse and zoom with right
- Zoom to rectangle
- Configure subplot
- Edit axis, curve, and image parameters
- Save the figure

Example: Result of Bar chart

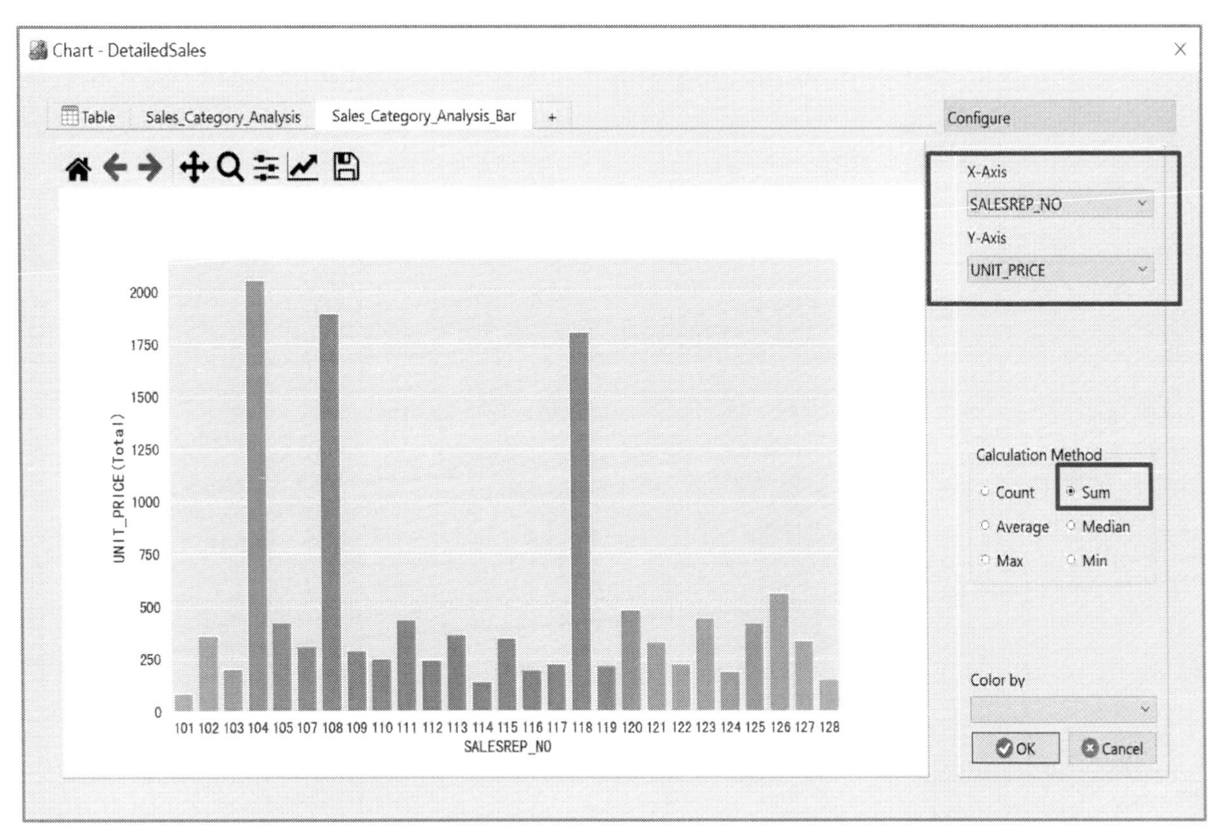

Ex: Result of Bar chart with Color by

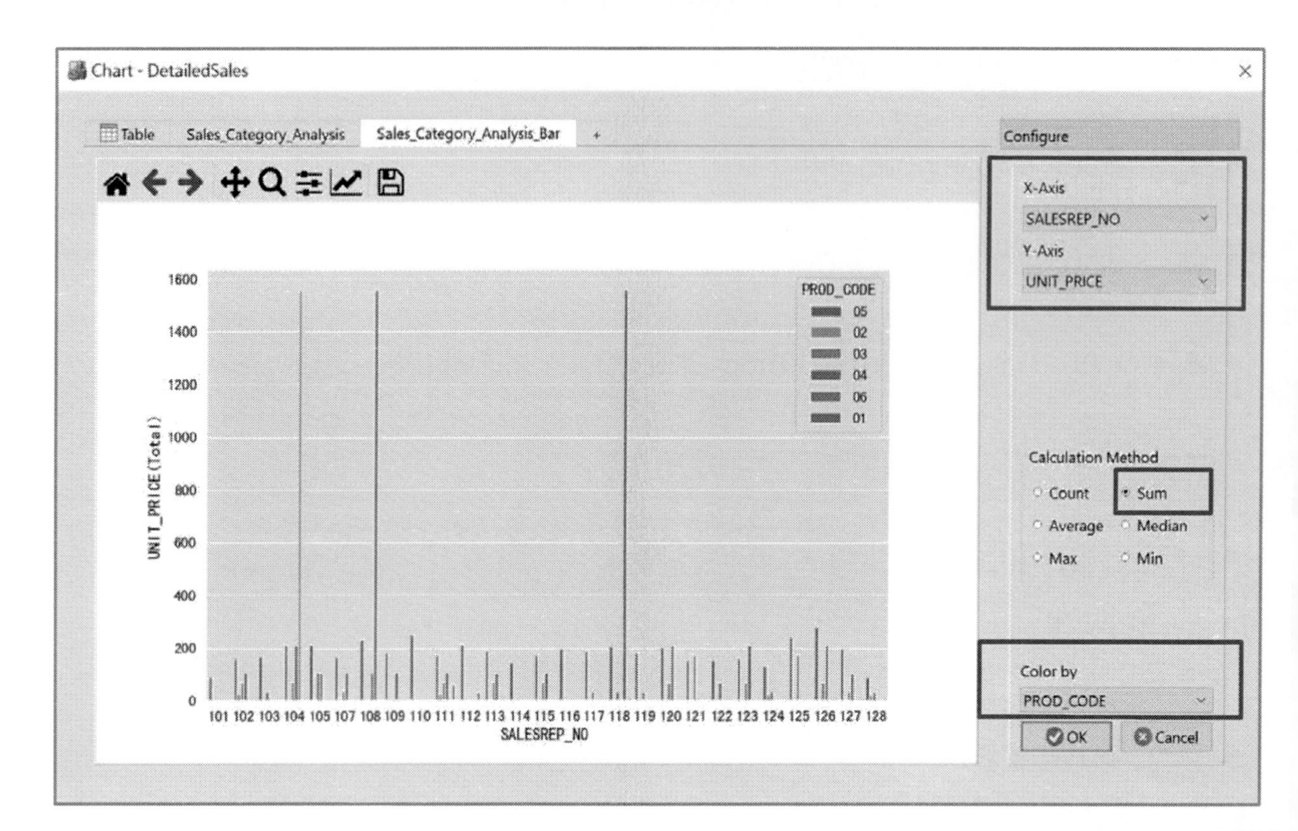

JCAATs Learning notes：

JCAATs - AI Audit Software

Python Based Computer-Assisted Audit Techniques (CAATs)

Data Analysis and Smart Audit

Chapter 12. Sampling

Outline:

1. Audit Sampling Overview
2. Random Sampling
3. Attribute Sampling
4. Monetary Sampling

JCAATs - AI Audit Software

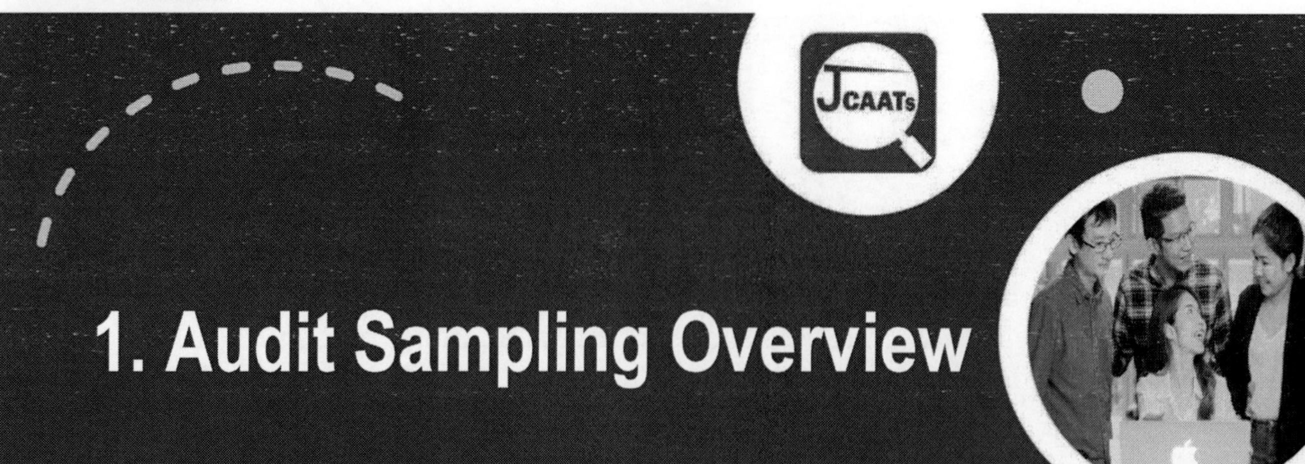

Audit Sampling Overview

- Since it is often impractical or impossible to review every single item in the population, auditors use sampling technique to select a smaller subset of data that can be analyzed and used to make conclusions about the entire population.

- Audit sampling typically involves selecting a random or systematic sample of data, using statistical methods to ensure that the sample is representative of the population.

- JCAATs provides three commonly used methods for auditing Sampling:
  - **Random Sampling**
  - **Attribute Sampling**
  - **Monetary Sampling**

JCAATs - AI Audit Software

2. Random Sampling

Random Sampling

- Random sampling is a statistical technique used to select a representative sample of data from a larger population. Each member of the population has an equal chance of being selected for the sample.
- The process of random sampling:
  - **Determine the sample size**: The step is to determine the sample size, which is the number of individuals or items that you will select from the population. **Randomly select items**: The step is to randomly select individuals or items from the population. You can a seed to perform a random number generator.
  - **Determine the Data Scope**: The step enable you to set the start and end range of your population data to make it randomly.
  - **Collect Data**: The step enable you to set how your sample has been collected from the items in the sample. This can be done through sort and no repeat.

Main tab - Random Sampling

- random sampling is a valuable tool in audit as it allows auditors to test a representative sample of items from a population.

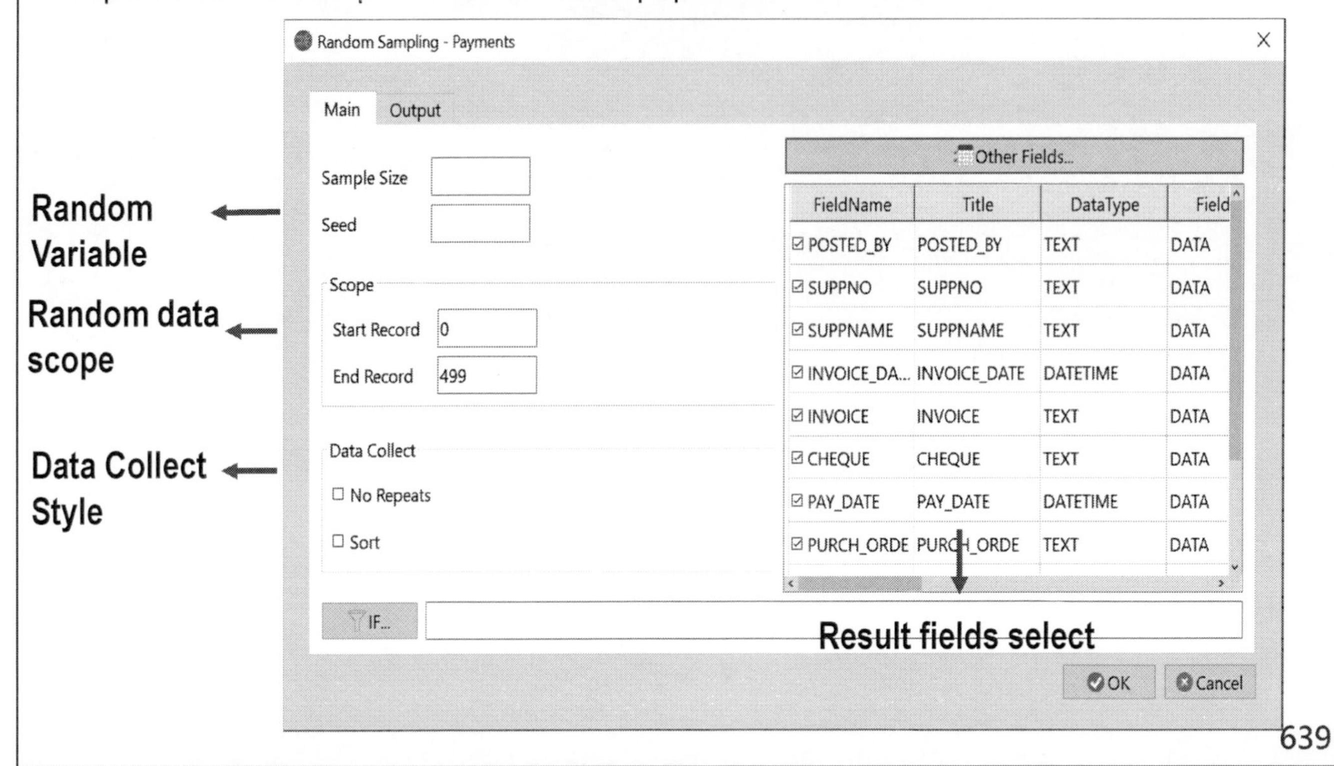

Output tab - Random Sampling

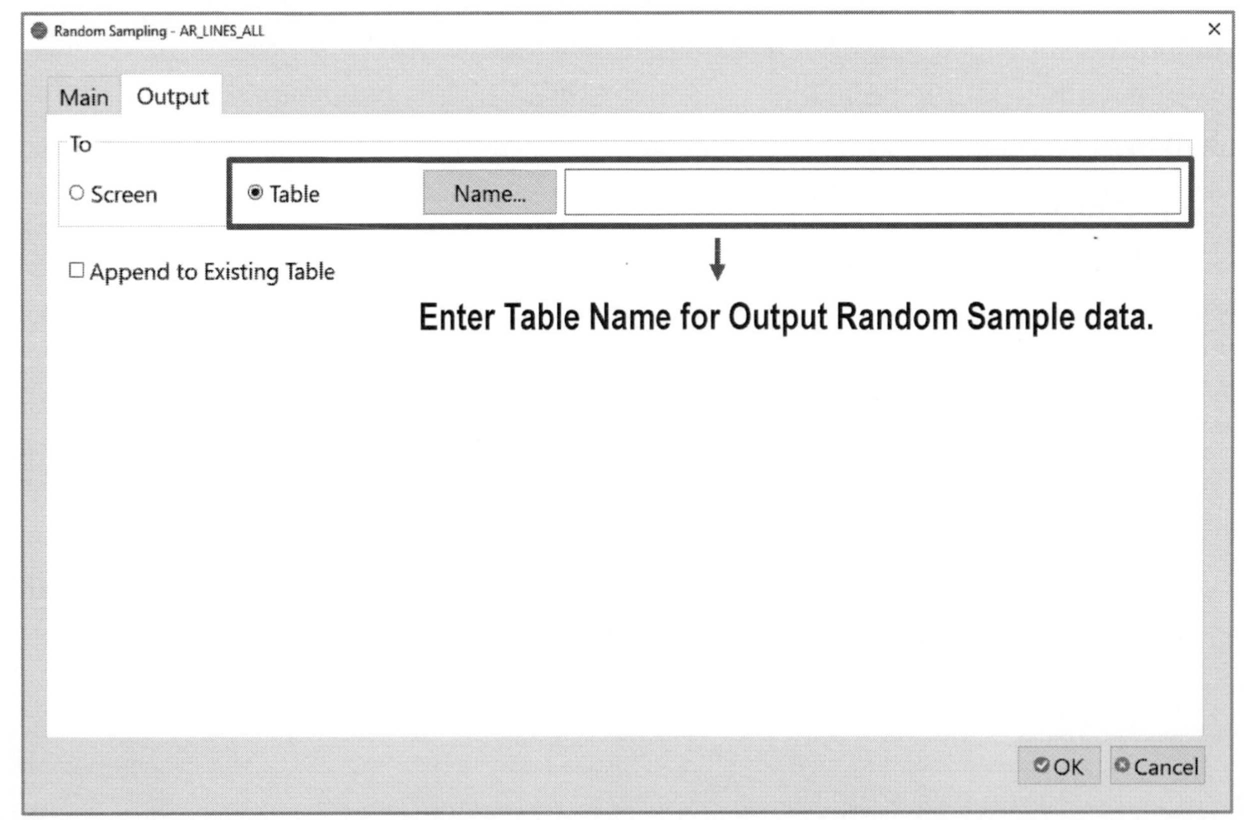

Result of Random Sampling Command

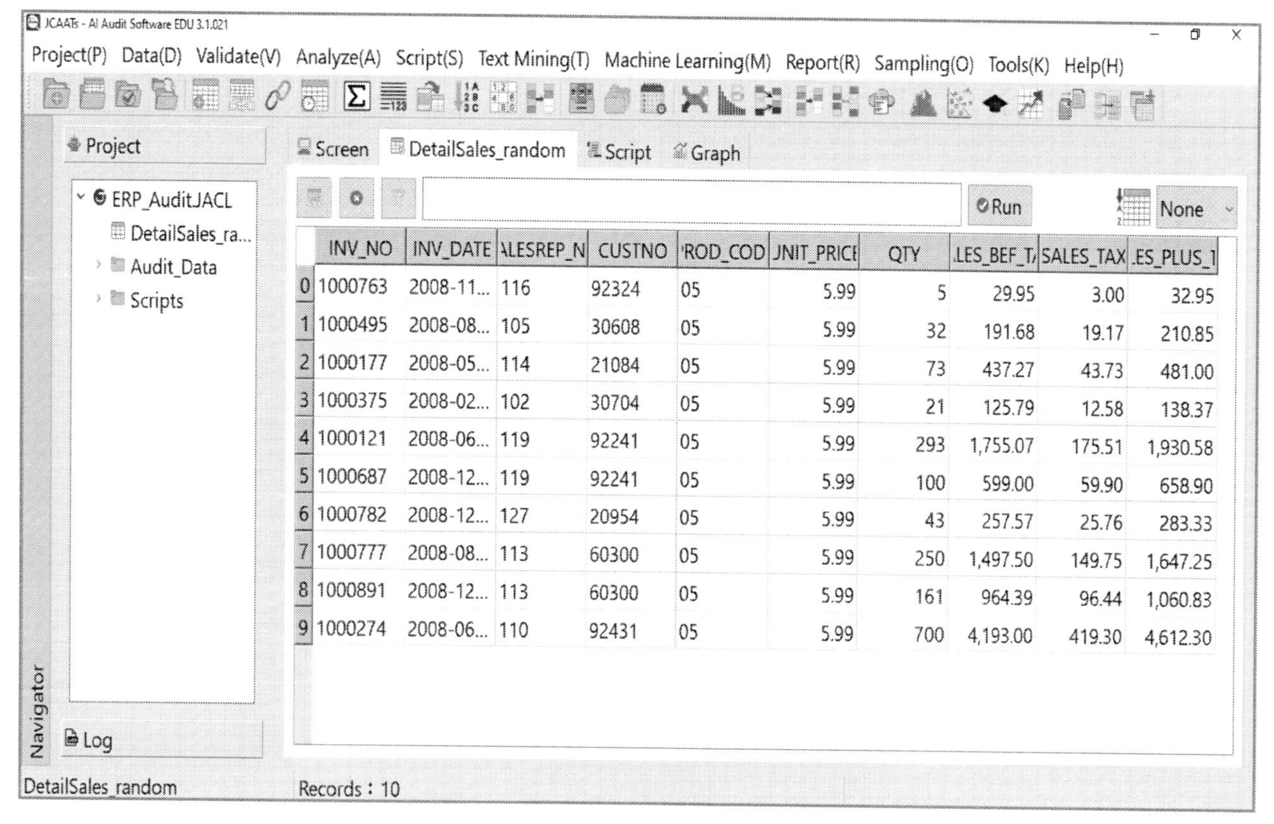

	INV_NO	INV_DATE	ALESREP_N	CUSTNO	ROD_COD	JNIT_PRICE	QTY	LES_BEF_T	SALES_TAX	ES_PLUS_T
0	1000763	2008-11...	116	92324	05	5.99	5	29.95	3.00	32.95
1	1000495	2008-08...	105	30608	05	5.99	32	191.68	19.17	210.85
2	1000177	2008-05...	114	21084	05	5.99	73	437.27	43.73	481.00
3	1000375	2008-02...	102	30704	05	5.99	21	125.79	12.58	138.37
4	1000121	2008-06...	119	92241	05	5.99	293	1,755.07	175.51	1,930.58
5	1000687	2008-12...	119	92241	05	5.99	100	599.00	59.90	658.90
6	1000782	2008-12...	127	20954	05	5.99	43	257.57	25.76	283.33
7	1000777	2008-08...	113	60300	05	5.99	250	1,497.50	149.75	1,647.25
8	1000891	2008-12...	113	60300	05	5.99	161	964.39	96.44	1,060.83
9	1000274	2008-06...	110	92431	05	5.99	700	4,193.00	419.30	4,612.30

JCAATs - AI Audit Software

3. Attribute Sampling

Attribute Sampling

- Attribute sampling is a statistical sampling method used in auditing to determine whether a population meets a certain characteristic or attribute. It is used to estimate the proportion of items in a population that have a specific attribute or characteristic, such as whether a financial transaction was properly authorized, whether an account is reconciled, or whether an inventory item is in the correct location.

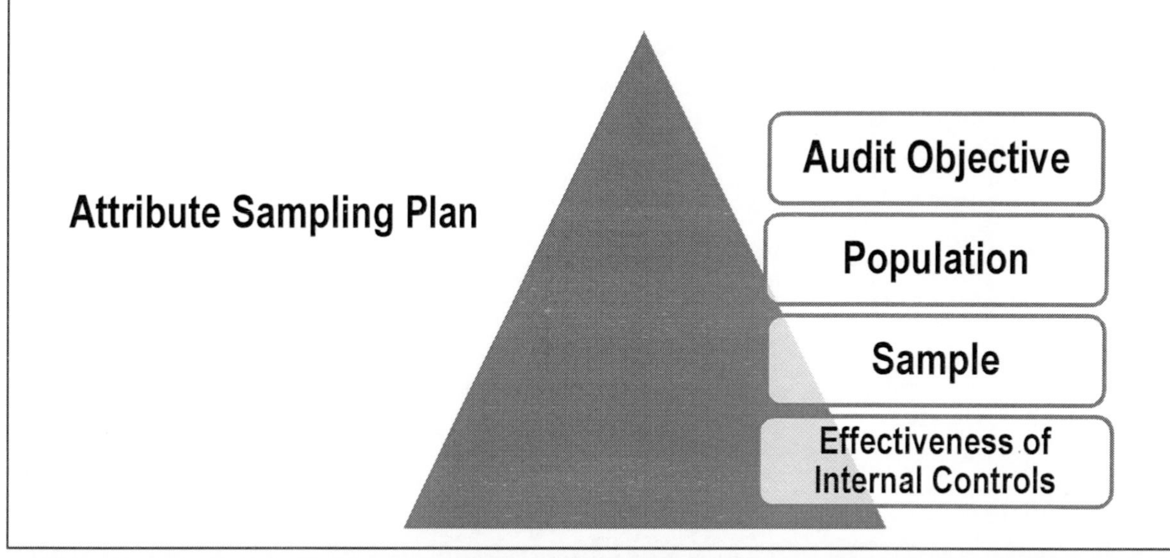

Attribute Sampling Parameters

- **Population size**: The total number of items in the population being sampled.
- **Sample size**: The number of items selected from the population for examination.
- **Confidence level**: The level of confidence the auditor wants to have in the sample results. It is typically expressed as a percentage, such as 95% or 99%.
- **Tolerable error (Materiality)**: The maximum error rate that the auditor is willing to accept in the sample without concluding that the population is materially misstated.
- **Expected error rate**: An estimate of the error rate in the population based on prior experience or other available information.

Select Samples from a Population

- **Interval Sampling:** The auditor selects items from the population at regular intervals. The auditor divides the population size by the desired sample size to determine the sampling interval. Then, the auditor selects items at fixed intervals throughout the population until the desired sample size is reached. Interval sampling is useful when the population is large and the items in the population are relatively homogenous.
- **Stratified Sampling**: The auditor divides the population into subpopulations or strata based on some relevant characteristic, such as size, location, or risk. Then, the auditor selects a sample from each stratum based on some predetermined criteria, such as a proportional allocation or a random selection within each stratum. Stratified sampling is useful when the population is diverse, and the auditor wants to ensure that the sample is representative of the entire population.
- **Random Sampling**: The auditor selects items from the population at random, with each item having an equal chance of being selected. The auditor uses a random number generator or some other random selection process to select the sample. Random sampling is useful when the population is large and diverse, and the auditor wants to ensure that the sample is representative of the entire population.

Main tab - Attribute Sampling

- Attribute sampling is only meaningful if used to audit internal controls that are correctly designed and efficiently executed.

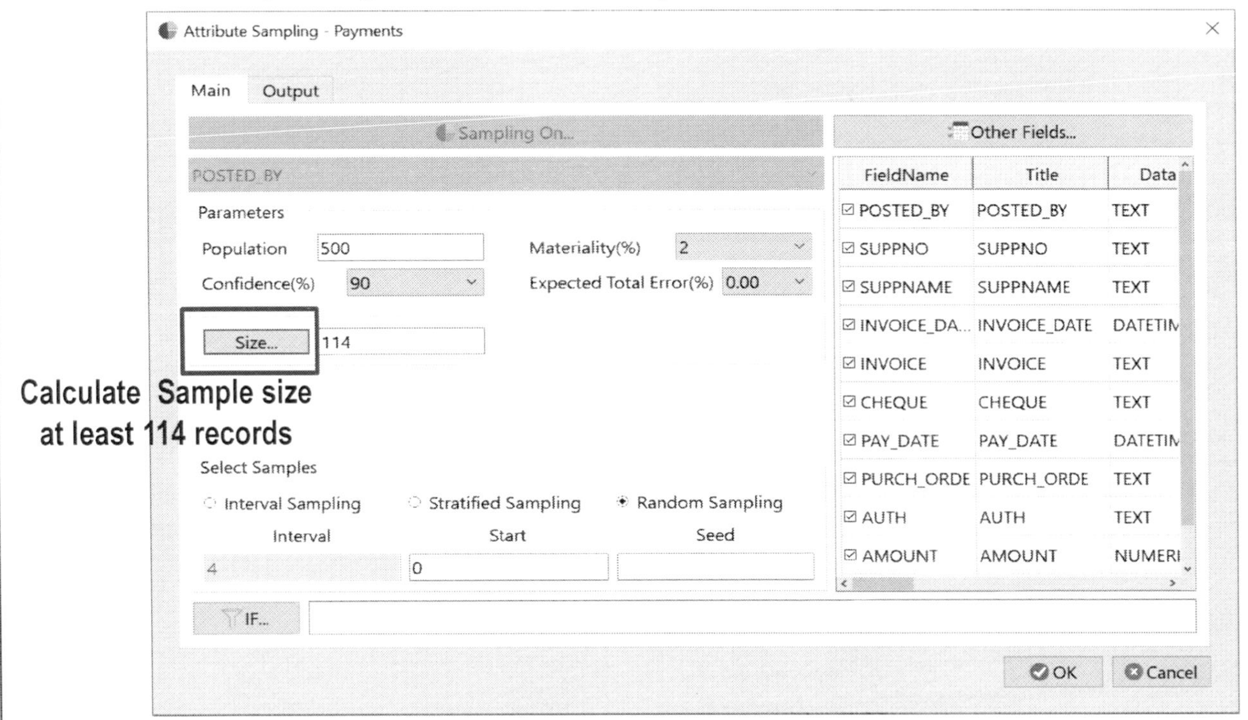

Output tab - Attribute Sampling

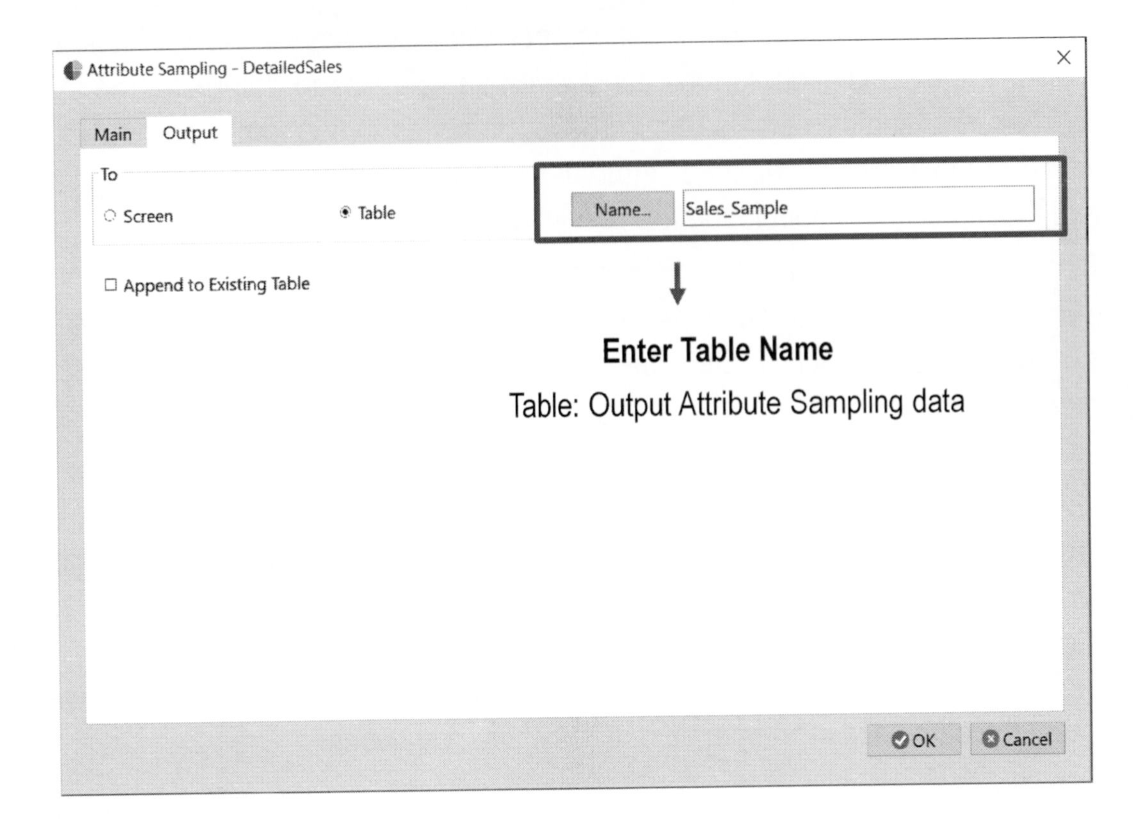

Result of Attribute Sampling Command

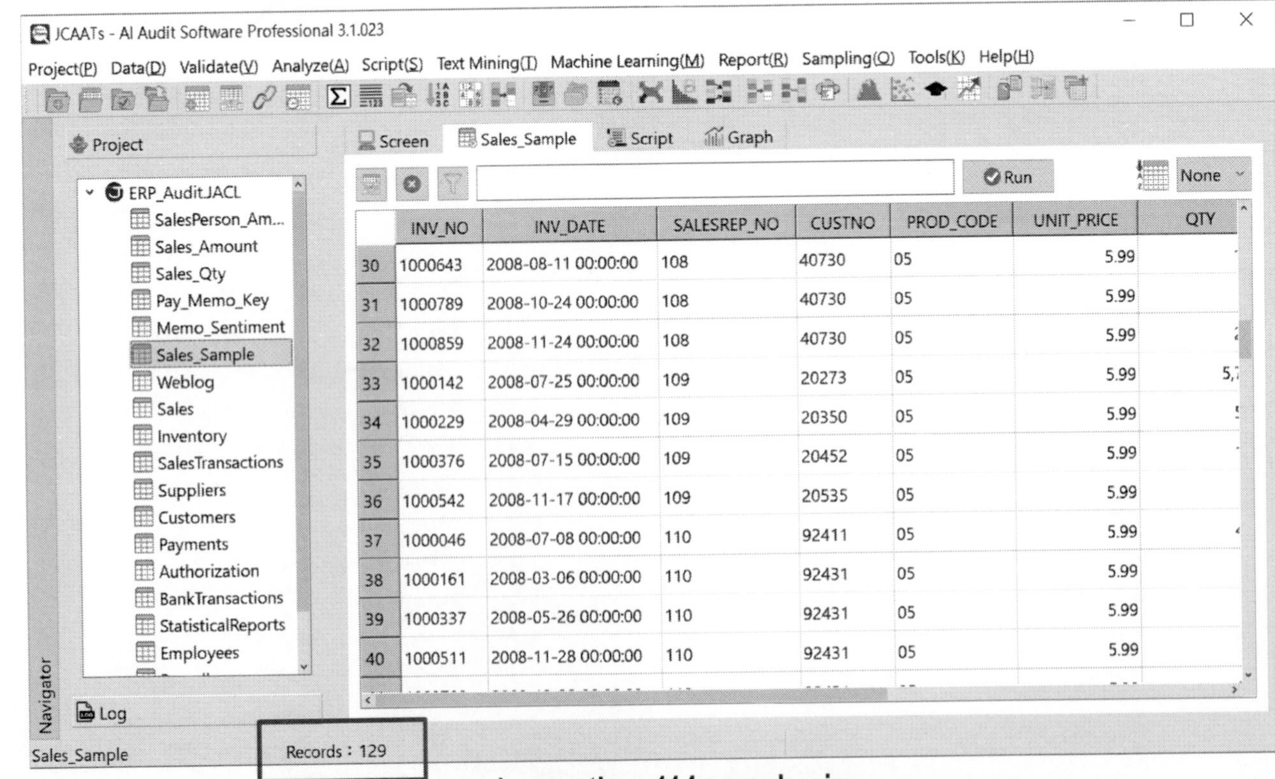

Larger than 114 sample size

JCAATs - AI Audit Software

3. Monetary Sampling

Monetary Sampling

- Monetary sampling is a statistical sampling technique used in auditing to select items from a population based on their dollar value. Monetary sampling is used when the auditor is interested in testing the monetary amount of errors or misstatements in the population.

- In monetary sampling, the auditor selects a sample of items based on their monetary value rather than their physical quantity. The sample is selected in such a way that the total monetary value of the sample is proportional to the total monetary value of the population. The auditor then examines the sample for errors or misstatements and extrapolates the results to the entire population based on the total monetary value of the sample and the population.

Main tab - Monetary Sampling

- For example, if the auditor suspects that there may be a few large-dollar misstatements in the population, monetary sampling would be an appropriate technique to use.

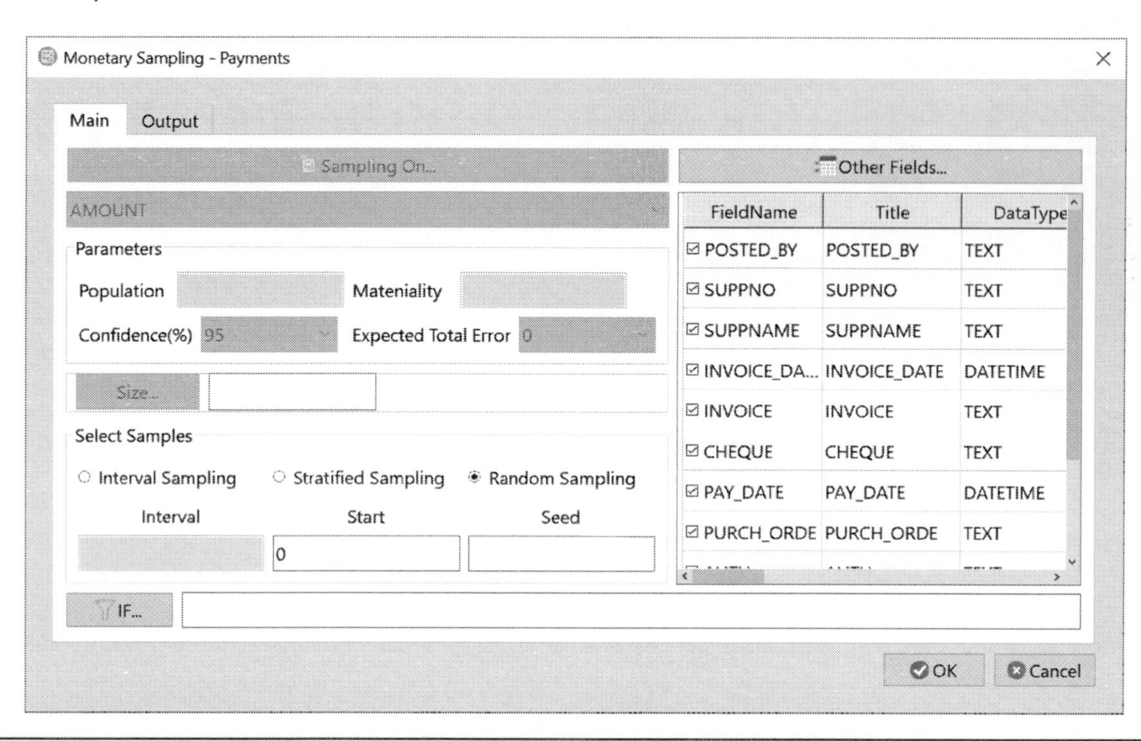

Result of Monetary Sampling

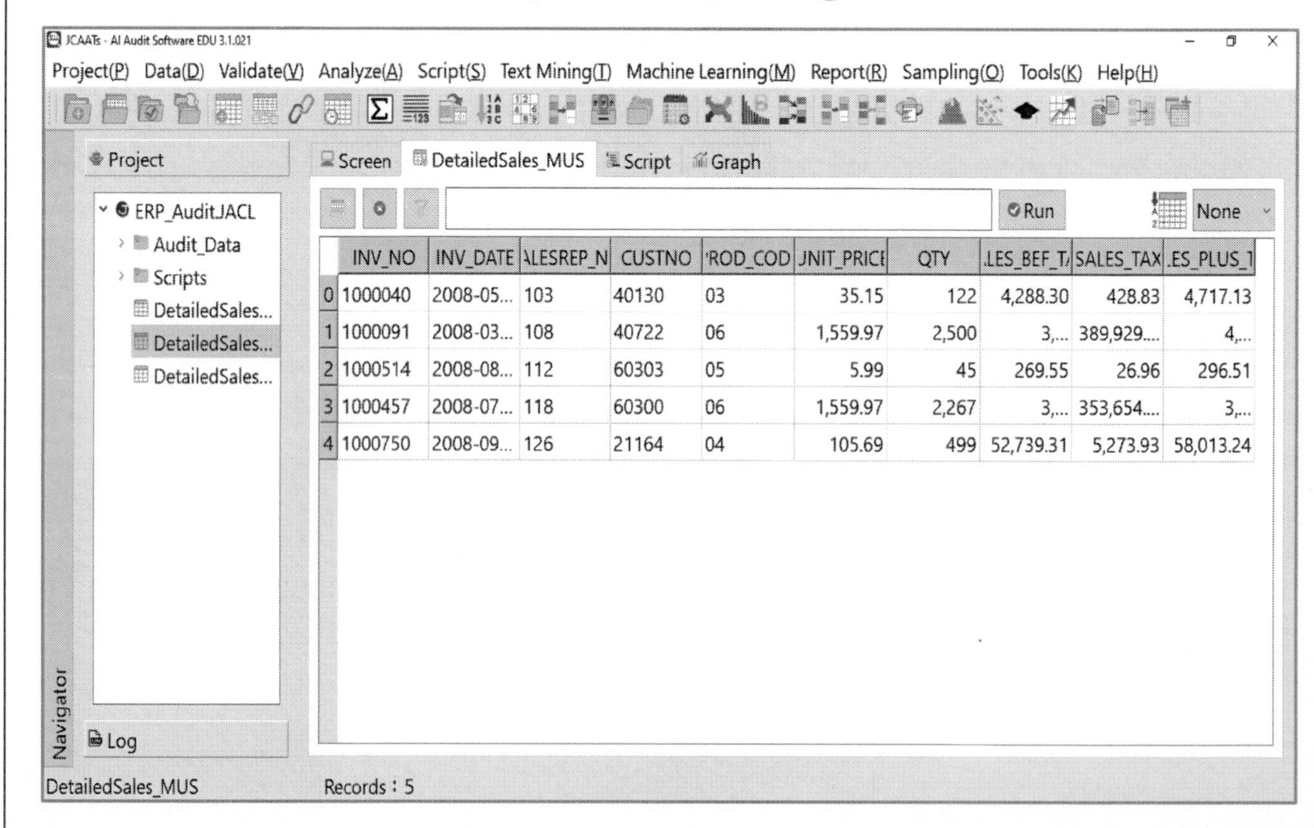

	INV_NO	INV_DATE	ALESREP_N	CUSTNO	PROD_COD	UNIT_PRICE	QTY	LES_BEF_T	SALES_TAX	ES_PLUS_T
0	1000040	2008-05...	103	40130	03	35.15	122	4,288.30	428.83	4,717.13
1	1000091	2008-03...	108	40722	06	1,559.97	2,500	3,...	389,929....	4,...
2	1000514	2008-08...	112	60303	05	5.99	45	269.55	26.96	296.51
3	1000457	2008-07...	118	60300	06	1,559.97	2,267	3,...	353,654....	3,...
4	1000750	2008-09...	126	21164	04	105.69	499	52,739.31	5,273.93	58,013.24

JCAATs - AI Audit Software

Python Based Computer-Assisted Audit Techniques (CAATs)

Data Analysis and Smart Audit

Chapter 13. Tools

Outline:

1. Dictionary Tool
2. Variable Tool
3. Index Tool

Tools

- "Tools" function is a list of tools to support smart project process in JCAATs. There are 3 tools as the following:
  - **Dictionary Tool:** Provide a custom dictionary function for uploading modifications to existing dictionary file for text mining features.
  - **Variable Tool:** Provide the ability to add, modify, and browse variables for the current JCAATs project. Variables are categorized into system (**SYSTEM**) variables, which are automatically generated by executing commands, and user-defined (**USER**) variables, which are defined by users.
  - **Index Tool:** Provide the ability to manage the indexes generated through **Index** commands for each table, including listing and deleting them. 。

JCAATs - AI Audit Software

Dictionary Tool Interface

- The dictionaries are for Text Mining process. You can custom your own text dictionary and upload to the JCAATs.
- Your dictionary file information will show on the dictionary information area.

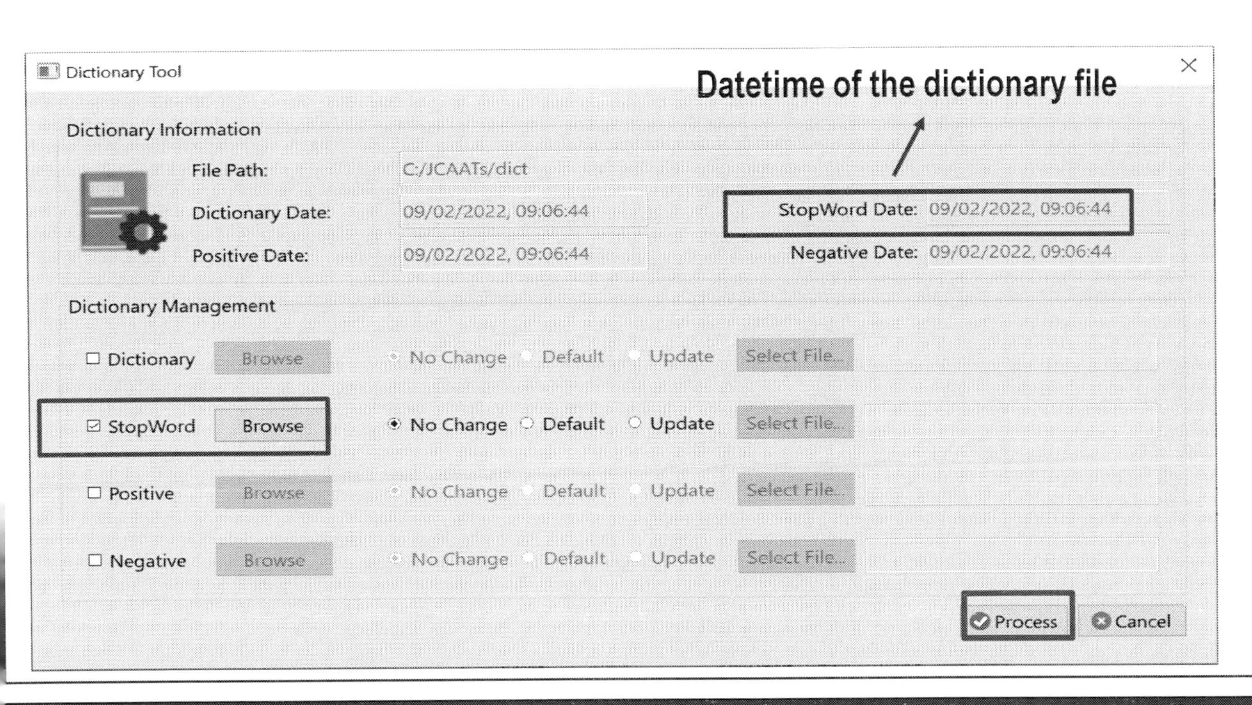

Browse - Stopword Dictionary

- You can browse the existing dictionary file and revise it to a new file for your audit purpose. A dictionary file is a text file with line for each phrase. .

```
nan
I
am
to
you
are
a
an
```

Upload a New Stopword Dictionary

- You can upload your dictionary file to make it change in your text mining process.

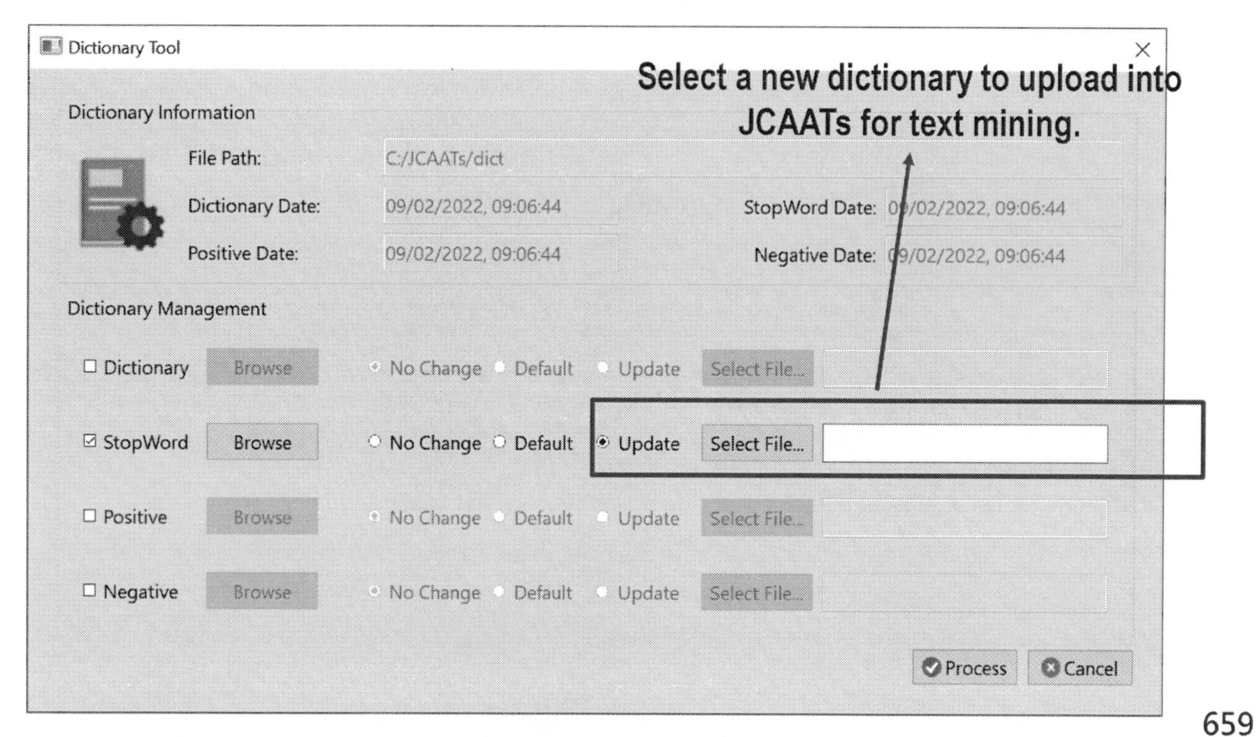

jacksoft
www.jacksoft.com.tw
JCAATs - AI Audit Software

2. Variable Tool

Variable Tool Interface

- JCAATs provide a flexible way to manage variables for a smart audit project.
- It can add, modify and delete USER variables. It only allow to modify the SYSTEM variables.

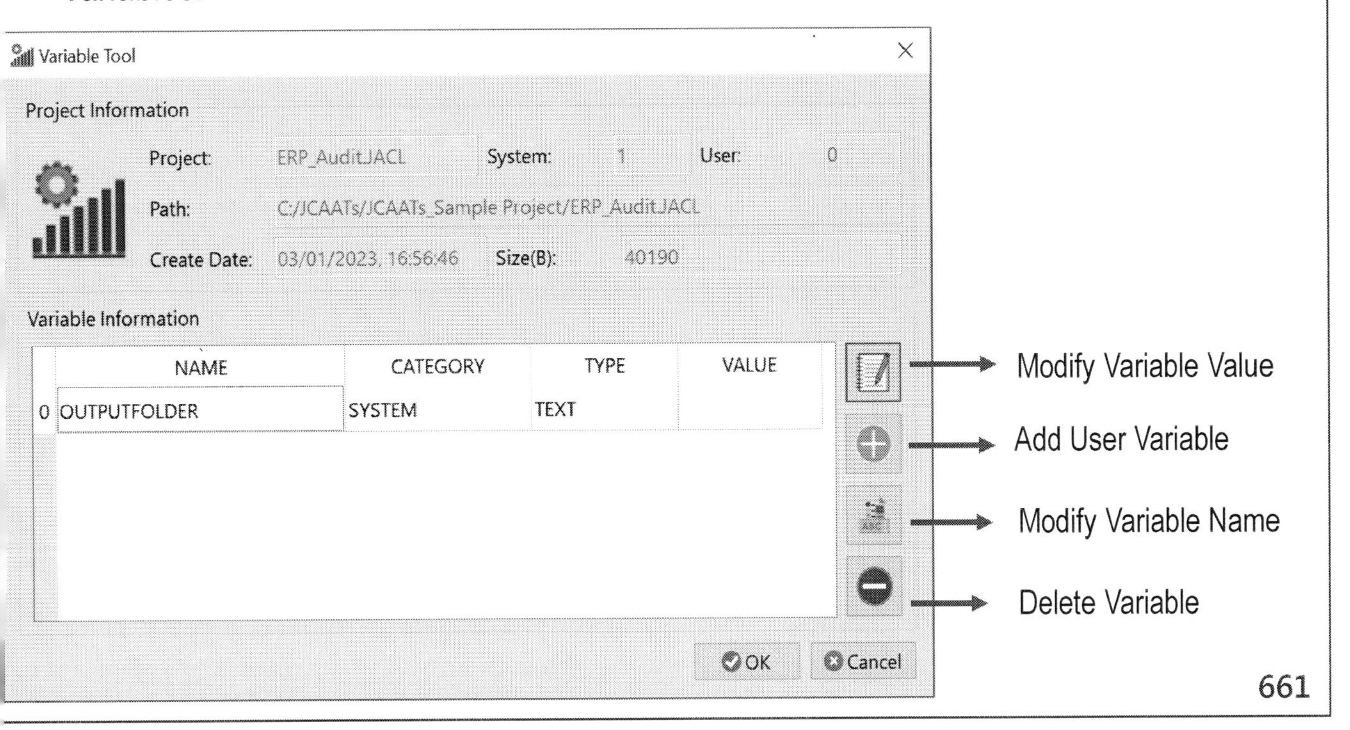

Add - Variable Tool

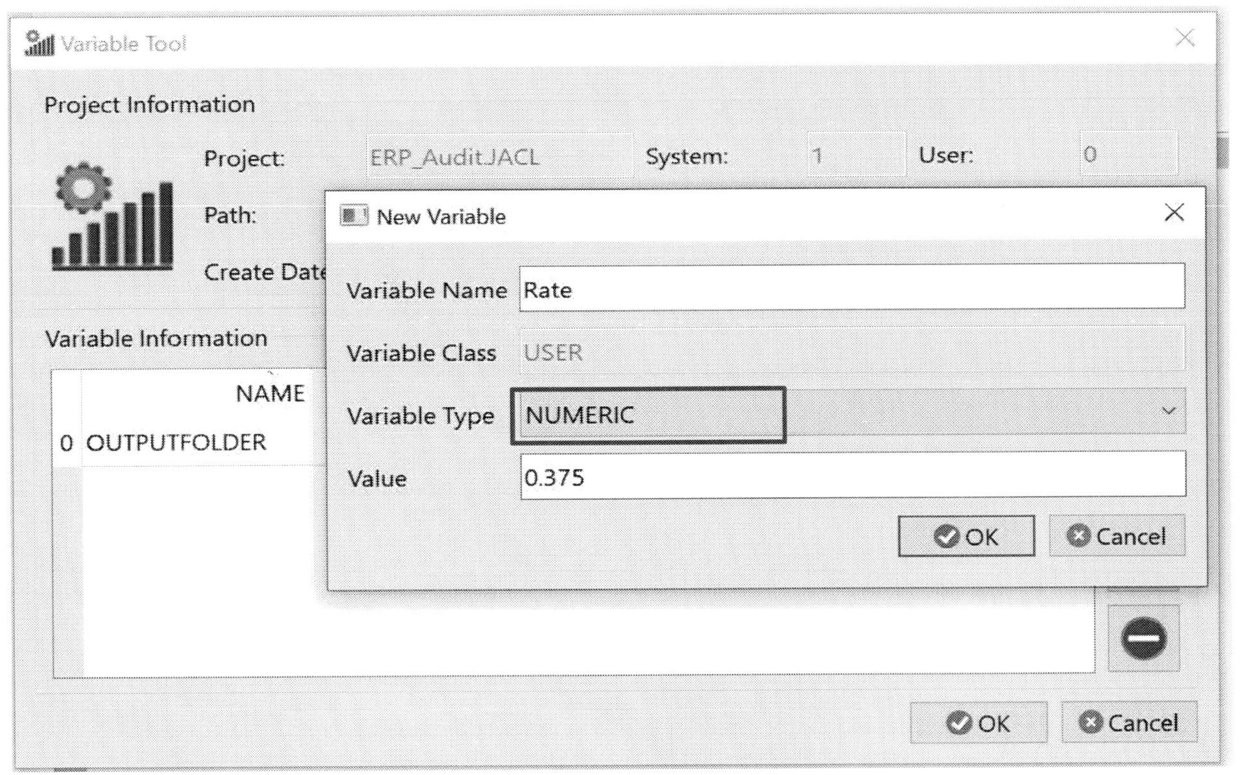

Result of Variable Tool

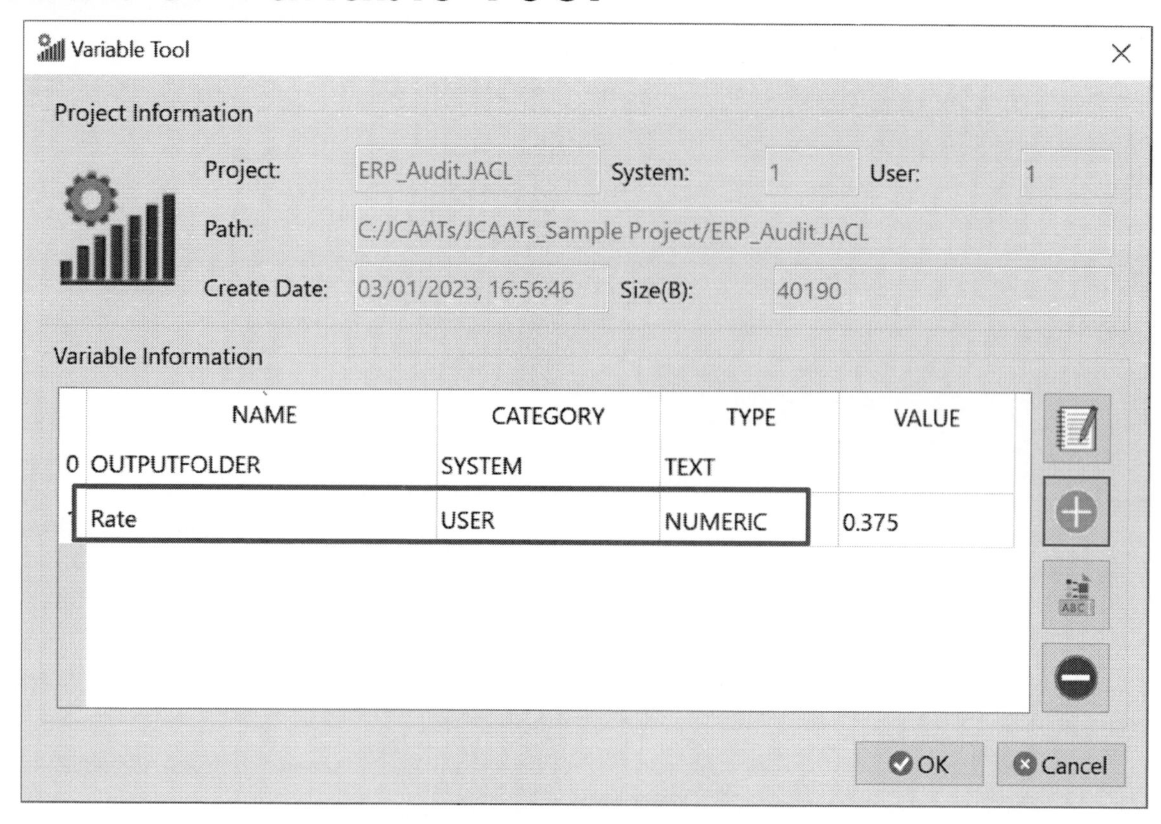

JCAATs - AI Audit Software

Index Tool Interface

- You can delete the selected index.

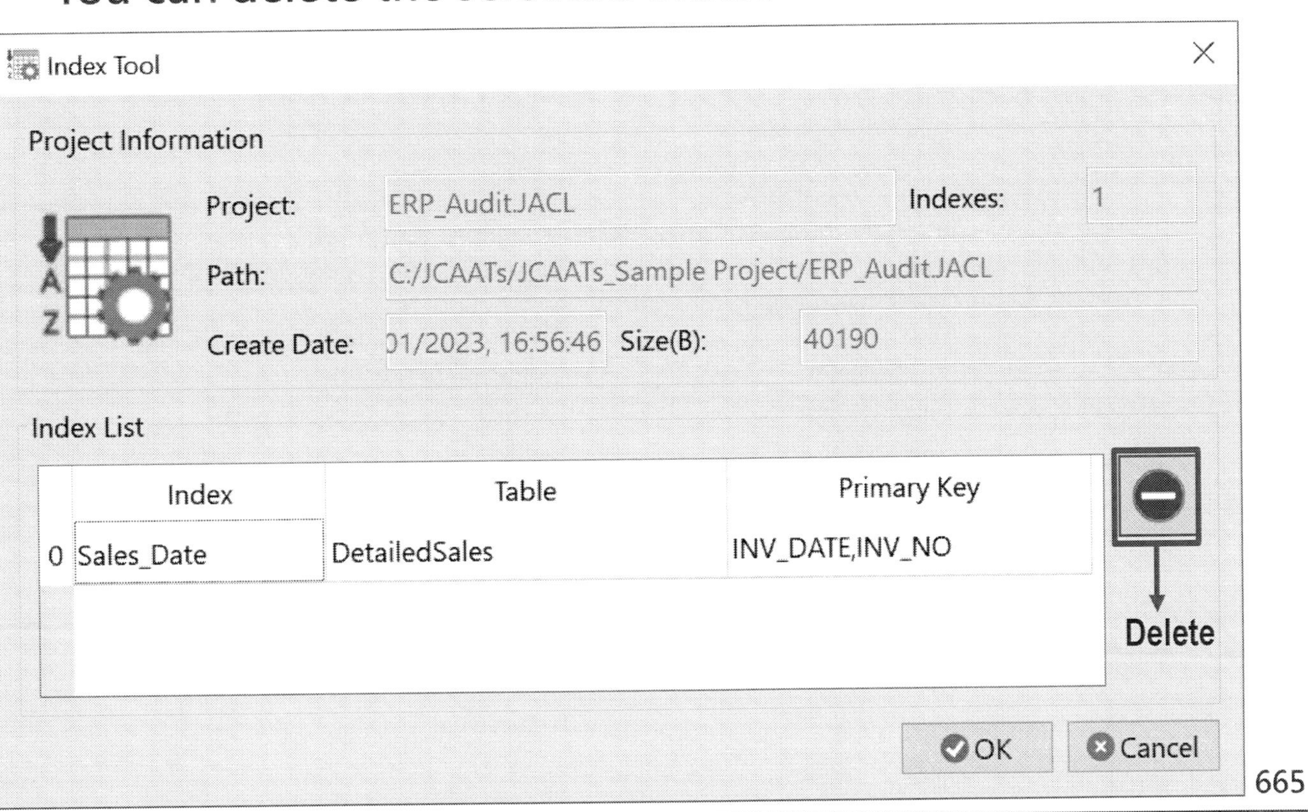

JCAATs Learning notes：

Appendix A References

1. AACSB (2014). AACSB International Accounting Accreditation Standard A7: Information Technology Skills and Knowledge for Accounting Graduates: An Interpretation, https://docplayer.net/16013752-Aacsb-international-accounting-accreditation-standard-a7-information-technology-skills-and-knowledge-for-accounting-graduates-an-interpretation.html

2. Adams, J. (2019, March 4). Continuous auditing monitoring architecture. Accounting Today. Retrieved from https://www.accountingtoday.com/magazine/

3. AICPA. (2019). Using Python for Data Analysis in Accounting and Auditing. Retrieved from https://us.aicpa.org/content/dam/aicpa/interestareas/frc/assuranceadvisoryservices/downloadabledocuments/ads-instructional-paper-python.pdf

4. AICPA. (2023). Audit Data Standards. Retrieved March 25, 2023, from https://us.aicpa.org/interestareas/frc/assuranceadvisoryservices/auditdatastandards

5. AICPA. (2021, March 5). AICPA CPA Exam changes. CPA Licensure One Step Closer to Change. Retrieved March 25, 2023, from https://www.aicpa.org/news/article/cpa-licensure-one-step-closer-to-change.html

6. Analytics Vidhya. (2022, August 8). Dealing with outliers using the Z-Score method. Retrieved from https://www.analyticsvidhya.com/blog/2022/08/dealing-with-outliers-using-the-z-score-method/

7. Anuganti Suresh (2020). What is a confusion matrix? https://medium.com/analytics-vidhya/what-is-a-confusion-matrix-d1c0f8feda5

8. APACCIOoutlook (2019), JACKSOFT: RegTech Bots in Action, https://www.apacciooutlook.com/jacksoft

9. Claire Reilly (2018), Robots don't want to take your miserable office job, https://www.cnet.com/science/robots-dont-want-to-take-your-miserable-office-job/

10. Cnyes News. (2020). If Auditors Cannot Go to Mainland China for Audit, FSC Allows Video Audit of Annual Reports. https://news.cnyes.com/news/id/4446011

11. David Denyer. (2017), Organizational Resilience, https://www.cranfield.ac.uk/-/media/images-for-new-website/som-media-room/images/organisational-report-david-denyer.ashx (Source: Cranfield University)

12. Delen, Dursun & Ram, Sudha. (2018). Research challenges and opportunities in business analytics. Journal of Business Analytics. 1. 2-12. 10.1080/2573234X.2018.1507324.

13. Financial News. (2020, February 26). Impact of the Epidemic on Financial Statement Announcements, FSC: Auditors can adopt alternative solutions. TechNews. https://finance.technews.tw/2020/02/26/accountants-can-use-alternatives-for-auditing-financial-statements/

14. Galvanize. (2021). Death of the Tick Mark. Retrieved from https://www.wegalvanize.com/assets/ebook-death-of-tickmark.pdf

15. H. J. Will, 1983, ACL: a language specific for auditors, Communications of the ACM, Volume 26 Issue 5 pp 356–361 https://doi.org/10.1145/69586.358138

16. Huang, S. F. (2011). JOIN Data Comparison Analysis - Analysis Report of Unauthorized False Transaction Audit Activity. Audit Automation, 013, ISSN:2075-0315.

17. Huang, S. M. (2022). ACL Data Analysis and Computer Audit Handbook (8th ed.). Chuan Hwa Book Co.. ISBN 9786263281691.

18. Huang, S. M., Yen, J. C., Ruan, J. S., et al. (2013). Computer Auditing: Theory and Practice (2nd ed.). Chuan Hwa Book Co.

19. Huang, S. M., Huang, S. F., & Chou, L. Y. (2013). Big Data Era: New Challenges for Audit Data Warehouse Construction and Application. Accounting Research Monthly, 337, 124-129.

20. Huang, S. M., Chou, L. Y., & Huang, S. F. (2013). Development Trends in Audit Automation. Accounting Research Monthly, 326.

21. Huang, S. & Huang, S. M. (2017). J-CAATs: a Cloud Data Analytic Platform for Auditors, 2017 International Conference on Computer Auditing, London, UK.

22. ICAEA (2023), ICAEA Code of Ethics and Professional Practice, https://www.icaea.net/English/Certification/Code_of_Ethics.php

23. ICAEA (2023). CAATs (Computer Assisted Audit Techniques) Training Courses. Retrieved from https://www.icaea.net/English/Training/CAATs_Courses_Free_JCAATs.php

24. ICAEA (2023), International Computer Auditing Education Association, https://www.icaea.net.

25. ICAEA (2019, October 2). Audit data analytic case contest [LinkedIn post]. Retrieved from https://www.linkedin.com/groups/10357617/

26. Liu, B. (2011, September 7). Nature language processing. Retrieved from https://www.cnblogs.com/yuxc/archive/2011/09/07/2170385.html

27. Nguyet, T. (2022, February 15). Learn and code confusion matrix with Python. Retrieved from https://www.nbshare.io/notebook/626706996/Learn-And-Code-Confusion-Matrix-With-Python/

28. Oxford Martin Programme on Technology and Employment. (2013). Future of employment. Oxford, UK: Author. Retrieved from https://www.oxfordmartin.ox.ac.uk/downloads/academic/future-of-employment.pdf

29. Phil Leifermann, Shagen Ganason. (2021). Internal Audit Department of Tomorrow, 2021 IIA International Conference, Singapore.

30. Python Software Foundation. (n.d.). Welcome to Python.org. Retrieved from https://www.python.org/

31. The Economist. (2014, January 18). The onrushing wave. Retrieved from https://www.economist.com/briefing/2014/01/18/the-onrushing-wave

32. U.S. Department of the Treasury. (2023). OPEN DATA - the SDN sanctions list from the Office of Foreign Assets Control (OFAC). Retrieved from https://home.treasury.gov/policy-issues/financial-sanctions/specially-designated-nationals-and-blocked-persons-list-sdn-human-readable-lists

33. Wang, T. and Huang, S. (2019), Computer Auditing: The Way Forward, International Journal of Computer Auditing, Vol.1, No.1, pp.1- 3. https://doi.org/10.53106/256299802019120101001

34. Wikipedia (2023, March 11). Tf-idf. In Wikipedia. Retrieved from https://en.wikipedia.org/wiki/Tf-idf

35. Wikipedia (2023, March 11). Benford's law. In Wikipedia. Retrieved from https://en.wikipedia.org/wiki/Benford%27s_law

36. Wikipedia (2023, March 11). Levenshtein distance, In Wikipedia. Retrieved from https://en.wikipedia.org/wiki/Levenshtein_distance

37. Yahoo! News. (2022, February 22). Exposure of False Accounts! Kangyou KY Caused Investors to Lose NT$4.7 Billion, Ernst & Young is ruled by the court for false seizure. https://tw.news.yahoo.com/news/%E5%85%A8%E6%96%87-%E5%81%87%E5%B8%B3%E6%9B%9D%E5%85%89-%E5%BA%B7%E5%8F%8Bky%E5%AE%B3%E6%8A%95%E8%B3%87%E4%BA%BA%E6%90%8D%E5%A4%B147%E5%84%84-%E5%8B%A4%E6%A5%AD%E7%9C%BE%E4%BF%A1%E9%81%AD%E6%B3%95%E9%99%A2%E8%A3%81%E5%AE%9A%E5%81%87%E6%89%A3%E6%8A%BC-215859185.html